Handbook of Flavoproteins

Hille, Miller, Palfey (Eds.)

Handbook of Flavoproteins

Oxidases, Dehydrogenases and Related Systems
Volume 1
Russ Hille, Susan M. Miller, Bruce Palfey (Eds.)
ISBN 978-3-11-026842-3
e-ISBN 978-3-11-026891-1
De Gruyter, Berlin 2013

Complex Flavoproteins, Dehydrogenases and Physical Methods
Volume 2
Russ Hille, Susan M. Miller, Bruce Palfey (Eds.)
ISBN 978-3-11- 029828-4
e-ISBN 978-3-11-029834-5
De Gruyter, Berlin 2013

Available as Set
ISBN 978-3-11-030089-5
e-ISBN 978-3-11-030090-1

Also of Interest

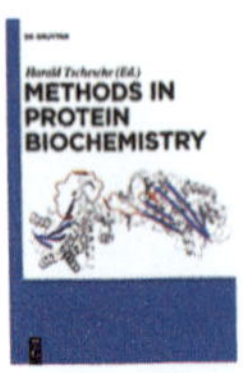

Methods in Protein Biochemistry
Tschesche (Ed.), 2011
ISBN 978-3-11-025233-0, e-ISBN 978-3-11-025236-1

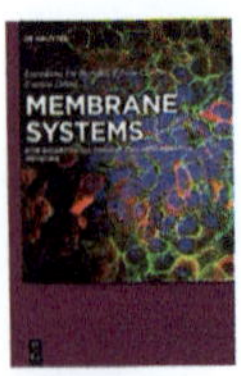

Membrane Systems
For Bioartificial Organs and Regenerative Medicine
De Bartolo, Curcio, Drioli, 2013
ISBN 978-3-11-026798-3, e-ISBN 978-3-11-026801-0

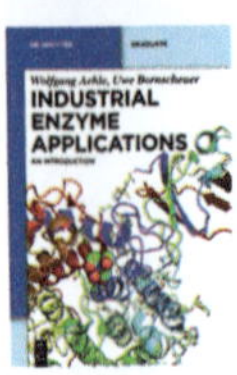

Industrial Enzyme Applications
An Introduction
Aehle, Bornscheuer, 2014
ISBN 978-3-11-026157-8, e-ISBN 978-3-11-026397-8

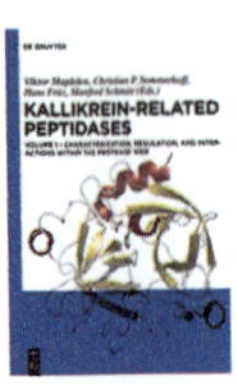

Kallikrein –Related Peptidases
Characterization, Regulation, and Interactions within the Protease Web
Magdolen, Sommerhoff, Fritz, Schmitt (Eds.), 2012
ISBN 978-3-11-026036-6, e-ISBN 978-3-11-026037-3

Handbook of Flavoproteins

Oxidases, Dehydrogenases and Related Systems

Volume 1

Edited by Russ Hille, Susan M. Miller, Bruce Palfey

DE GRUYTER

Editors

Prof. Russ Hille
University of California
Department of Biochemistry
1643 Boyce Hall
92521 Riverside, CA
USA
russ.hille@ucr.edu

Prof. Susan M. Miller
University of California
Department of Pharmaceutical Chemistry
600 16th Street
94158 San Francisco, CA
USA
smiller@cgl.ucsf.edu

Prof. Bruce Palfey
University of Michigan
Department of Biological Chemistry
1150 W. Medical Center Dr.
48109 Ann Arbor
USA
brupalf@umich.edu

ISBN 978-3-11-026842-3 • e-ISBN 978-3-11-026891-1

Library of Congress Cataloging-in-Publication Data
A CIP catalog record for this book has been applied for at the Library of Congress.

Bibliographic information published by the Deutsche Nationalbibliothek
The Deutsche Nationalbibliothek lists this publication in the Deutsche Nationalbibliografie; detailed bibliographic data are available in the Internet at http://dnb.dnb.de.

Typesetting: Compuscript, Shanon – Ireland
Printing and binding: Hubert & Co. GmbH & Co. KG, Göttingen
Cover image: The figure presents a crystal structure of a thymidylate synthase from *Thermotoga maritima* in the presence of the co-factor FAD and substrate 2′-Deoxyuridine-5′-Monophosphate. Kindly provided by Prof. Amnon Kohen

♾ Printed on acid-free paper
Printed in Germany
www.degruyter.com

Preface

Since the discovery of what was initially called the Yellow Enzyme by Otto Warburg in 1933 (and subsequently renamed Old Yellow Enzyme, a name the protein still retains) flavoproteins have been identified and characterized from virtually every organism known. More recent investigations have demonstrated that, in addition to their now well-established roles in catalyzing oxidation-reduction reactions of all sorts, flavoproteins also play important roles in a broad range of biological processes, including signal transduction, protein folding and photochemistry. In addition, biotechnological applications of flavoproteins have progressed significantly since the development of blood glucose tests based on the action of glucose oxidase. Finally, flavoproteins have provided the testing ground for some of the newest and most cutting edge experimental methodologies in biochemistry and biophysics, and have provided critical insight into, for example, the short-time molecular dynamics of flavoproteins and the manner in which flavoproteins activate triplet molecular oxygen for reaction with singlet molecules.

Given the significant progress in our understanding of both the diversity of processes in which flavoproteins are involved and the specific manner in which they carry out their roles within (or without) the cell, it seems both appropriate and timely to provide a survey of the field in this and the accompanying volume. We have attempted, within the constraints of space and availability of the material, to compile a comprehensive pair of volumes that together accurately reflect our present understanding of how a broad spectrum of flavoproteins work. It is our hope that these volumes will prove to be useful reference sources both for workers in the field and for instructors at both the undergraduate and graduate level not only in biochemistry and biophysics, but also pharmacology, medicine and bioengineering.

We wish to thank our contributors who have prepared such lucid accounts of their respective areas. We are also particularly indebted to three individuals at Walter de Gruyter: Dr. Stephanie Dawson for her enthusiasm and support for this project from the outset, Ms. Julia Lauterbach for her excellent editorial efforts that made publication of these volumes a reality, and to Ms. Sabina Dabrowski for coordinating technical production. Finally, we are of course grateful to our spouses (Kim Hille, Walter Moos and Kim Palfey) for their support and patience throughout our careers generally, and during the preparation of these volumes specifically. One of us (RH) would also like to thank the Alexander von Humboldt Foundation of Germany for its support during the editing of these volumes.

November 2012

Russ Hille
Susan M. Miller
Bruce Palfey

Contributing authors

Peter Macheroux
Institute of Biochemistry
Technical University Graz
Petersgasse 21/2
8010 Graz, Austria
e-mail: peter.macheroux@tugraz.at
chapter 1

Donald Becker
Department of Biochemistry
University of Nebraska
N200 Beadle Center, 1901 Vine St.
Lincoln, NE 68588, USA
e-mail: dbecker3@unlnotes.unl.edu
chapter 2

Hisashi Hemmi
Department of Applied Bioscience
Nagoya University
Furo-cho Chikusa-ku
464-8601 Nagoya, Japan
e-mail: hhemmi@agr.nagoya-u.csic.edu
chapter 3

Milagros Medina
Department of Biochemistry
University of Zaragoza
50009 Zaragoza, Spain
e-mail: mmedina@unizar.es
chapter 4

Shiao-Chun Tu
Department of Biology and Biochemistry
University of Houston
Houston, TX 77204, USA
e-mail: dtu@uh.edu
chapter 5

Paul Fitzpatrick
Department of Biochemistry
University of Texas Health Science Center
7703 Floyd Curl Dr., MC7760
San Antonio, TX 78229, USA
e-mail: fitzpatrick@biochem.uthscsa.edu
chapter 6

Dale Edmondson
Department of Biochemistry and Chemistry
Emory University
Atlanta, GA 30322, USA
e-mail: deedmon@emory.edu
chapter 7

Andrea Mattevi
Department of Genetics and Microbiology
University of Pavia
Via Ferrata 1
27100 Pavia, Italy
e-mail: andrea.mattevi@unipv.it
chapter 7

Giovanni Gadda
Department of Chemistry
Georgia State University
P.O. Box 4098
Atlanta, GA 30302, USA
e-mail: ggadda@gsu.edu
chapter 8

Pimchai Chaiyen
Department of Biochemistry
Mahidol University
272 Rama VI Road
10400 Bangkok, Thailand
e-mail: scpcy@mahidol.ac.th
chapter 9

Marilyn Jorns
Department of Biochemistry and Molecular Biology
Drexel University College of Medicine
245 N. 15th St MS 497
Philadelphia, PA 19102, USA
e-mail: Marilyn.Jorns@DrexelMed.edu
chapter 10

Jung-Ja Kim
Department of Biochemistry
Medical College of Wisconsin
8701 Watertown Plank Road
Milwaukee, WI 53226, USA
e-mail: jjkim@mcw.edu
chapter 11

Sandro Ghisla
Department of Biology
University of Konstanz
Universitätsstraße 10
78464 Konstanz, Germany
e-mail: sandro.ghisla@uni-konstanz.de
chapter 11

Colin Thorpe
Department of Chemistry and Biochemistry
University of Delaware
102 Brown Laboratory
Newark, DE 19716, USA
e-mail: cthorpe@udel.edu
chapter 12

Maria A. Vanoni
Department of Biochemistry
University of Milan
Via Celoria 26
20131 Milan, Italy
e-mail: maria.vanoni@unimi.it
chapter 13

Maria Cristina Nonato
Laboratory of Protein Crystallography
FCFRP-University of Sao Paulo
Sao Paulo, Brazil
e-mail: cristy@fcfrp.usp.br
chapter 14

Antonio J. da Costa Filho
Departamento de Fisica
FFCLRP/UPS
Av. Bandeirantes 3900
CEP 14040-901 Monte Alegre, Ribeirao Preto, SP, Brazil
e-mail: ajcosta@ffclrp.usp.br
chapter 14

Eduardo Ceccarelli
IBR - CONICET
National University of Rosario
Rosario, Argentina
e-mail: ceccarelli@ibr.gov.ar
chapter 15

Steven Rokita
Department of Chemistry
Johns Hopkins University
3400 N. Charles St.
Baltimore, MD 21218
e-mail: rokita@jhu.edu
chapter 16

Table of contents

1 Berberine bridge enzyme and the family of bicovalent flavoenzymes

Silvia Wallner, Corinna Dully, Bastian Daniel and Peter Macheroux

Abstract

Flavoproteins are a large and diverse group of proteins that employ either FMN or FAD for catalysis. In the majority of flavoproteins (~90%) the flavin coenzyme is tightly but non-covalently bound in the active site of the enzyme. In the 1950s the first example of a covalently attached flavocoenzyme was discovered in succinate dehydrogenase, followed by several more cases where either the 8α- or 6-position of the isoalloxazine forms a covalent bond to the amino acid side chain of a cysteine, histidine or tyrosine. Very recently, in 2005, the first representative of a bicovalently linked flavocoenzyme was discovered in glucooligosaccharide oxidase. Since then, the number of bicovalently linked flavoproteins has risen rapidly and appears to be more common than any of the monocovalent attachment modes. These enzymes have extraordinary properties, including unusually high reduction potentials, that enable catalysis of challenging chemical reactions in plants, fungi and bacteria. Moreover, bicovalent linkage appears to be crucial for structural integrity of the protein and correct positioning of the flavin in the active site. As a starting point, we review the wealth of information available for berberine bridge enzyme from *Eschscholzia californica* which can be considered a paradigm for the family of bicovalent flavoproteins. Moreover, we discuss the scope of reactions catalyzed by these enzymes in plants, fungi and bacteria. Finally, genomic information is explored to predict the number of bicovalent flavoenzymes present in nature. This analysis suggests that some genera are rich in bicovalent flavoproteins while others appear to have only a few members of this family of enzymes.

1.1 Introduction

The majority of flavoproteins (approximately 90%) contain non-covalently bound (dissociable) FMN or FAD as cofactor [1]. The first example of a covalently bound flavin was discovered in the 1950s by Singer and coworkers. Singer's group demonstrated that succinate dehydrogenase, a central enzyme of the tricarboxylic acid cycle and entry point of electrons into the mitochondrial electron transport chain ("complex II"), contains a covalently attached FAD [2]. Further studies demonstrated a linkage of the 8α-methyl group of the isoalloxazine ring to N-3 of a histidine residue of the protein [3]. In the following years several other covalent linkages in flavoenzymes have emerged, including tyrosinylation and cysteinylation at the 8α-methyl group and cysteinylation of C-6 [4,5]. At the end of the last century, Mewies *et al.* listed 24 flavoenzymes with a known linkage to either the 8α- or 6-position of the flavin [4,6]. A few years later, in 2005, the elucidation of the crystal structure of glucooligosaccharide oxidase (GOOX) provided the first example of a bicovalently linked FAD cofactor [7]. In this fungal enzyme the FAD is linked to a histidine and cysteine residue at the 8α-methyl and C-6 position of the flavin, respectively. In other words, in this flavoenzyme the already known monocovalent attachments are combined to give a bicovalently linked flavin. At the time, this finding may have been regarded as a peculiarity and unique to an enzyme that catalyzes the oxidation of a variety of mono- and oligosaccharides in a pathogenic fungus (for structures, see later). However, subsequent work has shown that bicovalent attachment occurs in many bacterial, fungal and plant enzymes and enzymes with bicovalent attachment may eventually outnumber those with monocovalent attachment. From the current list of 43 flavoproteins bearing a covalently attached flavin cofactor, eleven are reported or predicted to feature bicovalent flavin attachment (▶Tab. 1.1). Interestingly, on the basis of Blastp searches, bicovalent flavin attachment appears to be confined to bacteria, fungi and plants and is absent from archaea and the animal kingdom. It is also notable that all bicovalently attached flavins are found in proteins adopting a fold first observed in p-cresolmethylhydroxylase (PCMH) in the family (not clan!) FAD_binding_4 of the PCMH superfamily (or clan). With only a few exceptions, this structural environment appears to predispose covalent binding of the flavin cofactor in either a mono- (as in PCMH) or in a bicovalent fashion [1]. On the other hand, monocovalent attachment occurs in at least three different protein topologies: NADP Rossman and $GMC_{oxred}C$ for FAD-containing enzymes and TIM barrel for FMN-containing enzymes (▶Tab. 1.1).

Recent research offers several explanations as to why flavins might become covalently in flavoenzymes [8]. It increases the midpoint potential of the flavin cofactor to unprecedentedly high values and thus enables challenging substrate oxidations [9–11]. Furthermore, these enzymes typically have very bulky substrates and the binding cavity near the flavin cofactor is necessarily very large, potentially leading to high mobility of the isoalloxazine ring and/or lower affinity of the apoprotein for cofactor. Mobility is severely restricted by bicovalent linkage, freezing out translational modes that could compromise catalytic efficiency [8]. Indeed, these two factors play a role in chitooligosaccharide oxidase (ChitO) from *Fusarium graminearum* where bicovalent flavinylation modulates the reduction potential and is required for correct positioning of cofactor and substrate in the active site (and hence is a prerequisite for the formation of a catalytically competent Michaelis-Menten-complex) [10]. Finally, bicovalent linkage appears to be crucial for structural integrity, as evidenced by the low yields of protein variants, lacking one or both covalent sites [12]. Thus it appears that several factors are influenced by bicovalent linkage of the flavin, and their relative contribution may vary from one enzyme in the family to the next.

Tab. 1.1: Covalent attachment of FMN and FAD to flavoenzymes. Bicovalent attachment is marked by an asterisk

No.	E.C.	Enzyme	Cofactor	Structure Clan (Family)	Genus
1*	1.1.3.5	hexose oxidase	8α-(His, 6-Cys)-FAD	–	*Chondrus*
2	1.1.3.6	cholesterol oxidase	8α-(N1-His)-FAD	FAD_PCMH (FAD_binding_4)	*Brevibacterium*
3	1.1.3.8	L-gulonolactone oxidase	8α-(N1-His)-FAD	–	ubiquitous
4	1.1.3.10	pyranose 2-oxidase	8α-(N3-His)-FAD	GMC_oxred_C	*Trametes, Peniophora*
5	1.1.3.17	choline oxidase	8α-(N3-His)-FAD	NADP_Rossmann (GMC_oxred_N)	*Arthrobacter*
6	1.1.3.23	thiamine oxidase	8α-(N1-His)-FAD	–	soil bacteria
7	1.1.3.37	D-arabino-1,4-lactone oxidase	8α-(N1-His)-FAD	–	*Candida, Saccharomyces*
8	1.1.3.38	vanillyl-alcohol oxidase	8α-(N3-His)-FAD	FAD_PCMH (FAD_binding_4)	*Penicillium*
9	1.1.3.39	nucleoside oxidase (H_2O_2-forming)	FAD (covalent)	–	*Elizabethkingia*
10	1.1.3.41	alditol oxidase	8α-(N1-His)-FAD	FAD_PCMH (FAD_binding_4)	*Streptomyces*
11*	1.1.3.-	aclacinomycin oxidoreductase	8α-(N1-His, 6-Cys)-FAD	FAD_PCMH (FAD_binding_4)	*Streptomyces*
12*	1.1.3.-	chitooligosaccharide oxidase	8α-(N1-His, 6-Cys)-FAD	predicted[a]	*Fusarium*
13	1.1.3.-	eugenol oxidase	8α-(N3-His)-FAD	predicted[a]	*Rhodococcus*
14	1.1.3.-	D-gluconolactone oxidase	8α-(N3-His)-FAD	predicted[a]	*Penicillium*
15*	1.1.3.-	glucooligosaccharide oxidase	8α-(N1-His, 6-Cys)-FAD	FAD_PCMH (FAD_binding_4)	*Acremonium*
16*	1.1.3.-	glycopeptide hexose oxidase (Dbv29)	8α-(N1-His, 6-Cys)-FAD	FAD_PCMH (FAD_binding_4)	*Actinomadura*

(*Continued*)

(*Continued*)

No.	E.C.	Enzyme	Cofactor	Structure Clan (Family)	Genus
17*	1.1.3.-	10-hydroxy-dehydrogenase in tirandamycin biosynthesis (TrdL, TamL)	8α-(N1-His, 6-Cys)-FAD	FAD_PCMH (FAD_binding_4)	*Streptomyces*
18	1.1.3.-	nectarin V (putative sugar oxidase)	FAD (covalent)?	–	*Nicotiana*
19*	1.1.3.-	pregilvocarcin V dehydrogenase, GilR	8α-(N1-His, 6-Cys)-FAD	FAD_PCMH (FAD_binding_4)	*Streptomyces*
20	1.1.99.3	gluconate 2-dehydrogenase (acceptor)	8α-(N3-His)-FAD	–	bacteria
21	1.1.99.4	dehydrogluconate dehydrogenase	8α-(N3-His)-FAD	–	bacteria
22*	1.3.3.8	(*S*)-tetrahydroprotoberberine oxidase	8α-(N1-His, 6-Cys)-FAD	predicted from homology modelling[b]	*Argemone*
23	1.3.5.1	succinate dehydrogenase (ubiquinone)	8α-(N3-His)-FAD	NAPH_Rossmann (FAD_binding_2)	ubiquitous
24	1.3.99.1	succinate dehydrogenase / fumarate reductase	8α-(N3-His)-FAD	NADP_Rossmann (FAD_binding_2)	*Shewanella*, *Wolinella*
25	1.4.3.4	monoamine oxidase	8α-(Cys)-FAD	NADP_Rossmann (Amino_oxidase)	mammals
26	1.4.3.-	amino acid oxidase, NikD	8α-(Cys)-FAD	NADP_Rossmann (DAO)	*Streptomyces*

(*Continued*)

(*Continued*)

No.	E.C.	Enzyme	Cofactor	Structure Clan (Family)	Genus
27	1.5.3.1	sarcosine oxidase (monomeric) sarcosine oxidase (heterotetrameric)	8α-(Cys)-FAD 8α-(N3-His)-FMN	NADP_Rossmann (DAO) α and β interface	*Bacillus* *Corynebacterium*
28	1.5.3.6	(*R*)-6-hydroxynicotine oxidase	8α-(N1-His)-FAD	FAD_PCMH (FAD_binding_4)	*Arthrobacter*
29	1.5.3.7	L-pipecolate oxidase	8α-(Cys)-FAD ?	–	*Pseudomonas*
30	1.5.3.10	dimethylglycine oxidase	8α-(N3-His)-FAD	NADP_Rossmann (DAO)	*Arthrobacter*
31	1.5.3.-	γ-*N*-methylaminobutyrate oxidase	8α-(His)-FAD	–	*Arthrobacter*
32	1.5.8.1	dimethylamine dehydrogenase	6-(Cys)-FMN	–	*Hyphomicrobium*
33	1.5.8.2	trimethylamine dehydrogenase	6-(Cys)-FMN	TIM_barrel (Oxidored_FMN)	*Hyphomicrobium*
34	1.5.99.1	sarcosine dehydrogenase	8α-(N3-His)-FAD	–	ubiquitous
35	1.5.99.2	dimethylglycine dehydrogenase	8α-(N3-His)-FAD	–	ubiquitous
36	1.5.99.12	cytokinin dehydrogenase	8α-(N1-His)-FAD	FAD_PCMH (FAD_binding_4)	plants
37	1.14.14.-	halogenase in chloramphenicol biosynthesis (CmlS)	8α-(Asp)-FAD	NADP_Rossmann (Trp_halogenase)	*Streptomyces*
38	1.17.99.1	4-cresol dehydrogenase (hydroxylating)	8α-(*O*-Tyr)-FAD	FAD_PCMH (FAD_binding_4)	*Pseudomonas*

(*Continued*)

(*Continued*)

No.	E.C.	Enzyme	Cofactor	Structure Clan (Family)	Genus
39*	1.21.3.3	reticuline oxidase (berberine bridge enzyme)	8α-(N1-His, 6-Cys)-FAD	FAD_PCMH (FAD_binding_4)	*Eschscholzia*
40	1.21.99.1	β-cyclopiazonate dehydrogenase	8α-(N1-His)-FAD	–	*Aspergillus*
41*	1.-.-.-	Δ^1-tetrahydrocannabinolic acid synthase	8α-(N1-His, 6-Cys)-FAD	FAD_PCMH (FAD_binding_4)	*Cannabis*
42	--------	redox driven ion pumps, RnfG and RnfD	ribitylphosphate-(*O*-Thr)-FMN	–	*Vibrio*
43*	--------	pollen allergens (BBE-like proteins)	8α-(N1-His, 6-Cys)-FAD	FAD_PCMH (FAD_binding_4)	grasses

[a] Predicted (Leferink, N.G.H., Heuts, D.P.H.M., Fraaije, M.W., van Berkel, W.J.H.: The growing VAO flavoprotein family, Arch. Biochem. Biophys. 474:292–301, 2008)

[b] Predicted from homology modeling (Gesell, A., Diaz Chavez, M. L., Kramell, R., Piotrowski, M., Macheroux, P., Kutchan, T.: Heterologous expression of two FAD-dependent oxidases with (*S*)-tetrahydroprotoberberine oxidase activity from *A. mexicana* and *B. wilsoniae*, Planta, 233:1185–1197, 2011)

1.2 The paradigm of bicovalent flavoenzymes: Berberine bridge enzyme (BBE) from *Eschscholzia californica*

BBE is a branch point enzyme in benzylisoquinoline alkaloid biosynthesis [13]. It catalyzes the oxidative cyclization of *(S)*-reticuline to *(S)*-scoulerine, the first committed step in the biosynthesis of protoberberine, protobine and benzophenanthridine alkaloids (▶Fig. 1.1) [14,15].

The first isolation of the enzyme dates back to 1975, when Rink and Böhm reported the partial purification of BBE from *Macleaya microcarpa* cell cultures [16]. Subsequently Steffens *et al.* found BBE activity in 66 differentiated plants and cell suspension cultures, mainly from the families *Papaveraceae* and *Fumariaceae* [17]. Later the enzyme was successfully expressed in small amounts in insect cell culture and it was demonstrated that the FAD cofactor is covalently linked to a histidine residue via its 8α-position [18]. During our studies with BBE, expressed in the methylotrophic host *Pichia pastoris*, we noticed that photoreduction of the enzyme yielded 6-thio-FAD, a product known to form during photoreduction of 6-cysteinyl-flavins [19]. This observation prompted us to isolate the flavin-containing tryptic fragment and analyze its composition by amino acid sequencing and mass spectrometry [20]. This analysis demonstrated that the FAD bore two peptides and we proposed that His104 and Cys166 forms a linkage to position 8α and 6 of the flavin isoalloxazine ring system, respectively [20]. Later, the elucidation of the crystal structure confirmed this mode of bicovalent cofactor attachment [14].

Not surprisingly, the structural topology of BBE turned out to be very similar to GOOX, which is also a member of the PCMH clan (or superfamily) (FAD_binding_4). The structure is composed of a FAD and substrate binding domain, which is formed by a seven-stranded antiparallel β-sheet in an α/β-domain [14] (▶Fig. 1.2). It provides the ultimate proof for the bicovalent tethering of the flavin cofactor to His104 and Cys166 of the protein. The substrate *(S)*-reticuline is found in a relatively open substrate binding site sandwiched between the flavin cofactor and active site amino acid residues extending from the central β-sheet. The structure of BBE has enabled a comprehensive mutagenesis program geared towards gaining a better understanding of the mechanism of C-C bond formation. This work has involved replacement of active site residues Glu417, Tyr106, and His459 with Gln, Phe, and Ala, respectively, and determination of the kinetic parameters for all variant proteins (see ▶Figure 1.2 for an active site representation of BBE) [14]. Eventually, these studies have led to the formulation of a concerted reaction mechanism for the formation of the "berberine" bridge in the product *(S)*-scoulerine. In this mechanism, Glu417 acts as catalytic base by deprotonating the C3′ hydroxy group of *(S)*-reticuline and thereby increases the nucleophilicity of the C2′ atom, which further

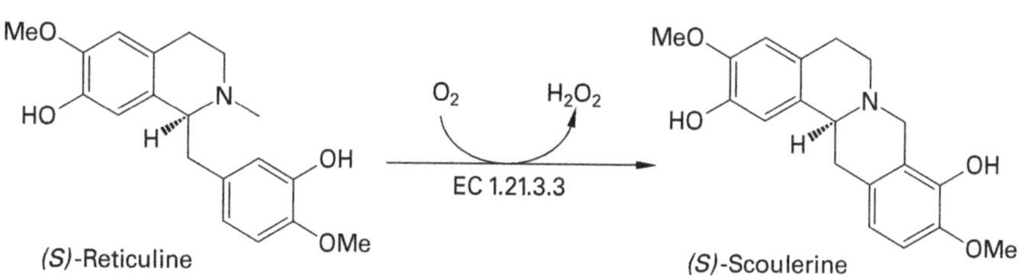

Fig. 1.1: The overall reaction catalyzed by BBE.

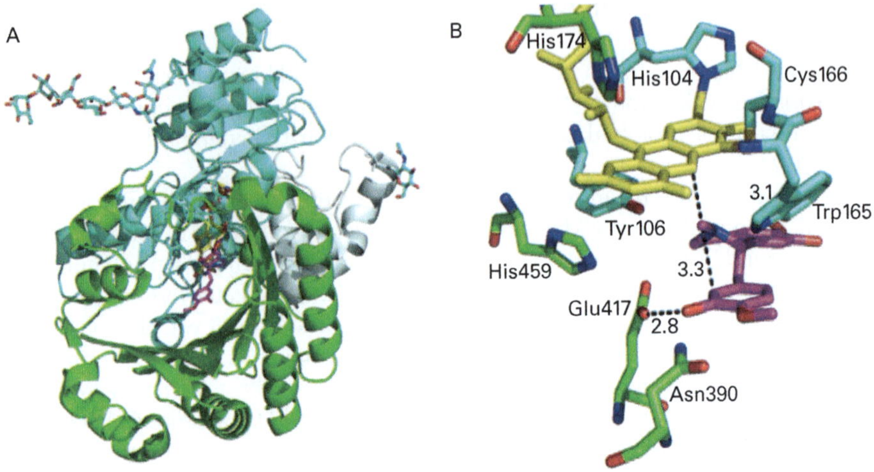

Fig. 1.2: Schematic representation of the structure of BBE. (A) Overall topology of BBE. The *N*-terminal FAD-binding subdomains are shown in cyan including the C-terminal α-helical stretch in pale blue. The central substrate binding domain is shown in green. N-linked sugar residues (blue), the FAD cofactor (yellow), and the substrate (*S*)-reticuline (magenta) are represented as stick models. (B) Active site of BBE in complex with (*S*)-reticuline. The flavin cofactor is shown in yellow with its bicovalent linkage to His104 and Cys166. Important active site amino acid residues are drawn in stick representation. The substrate (*S*)-reticuline is shown in magenta. Distances are indicated in Å.

performs a S_N2- nucleophilic attack on the *N*-methyl group of the substrate. Thus the methylene bridge of *(S)*-scoulerine is formed and a hydride is transferred to the flavin cofactor (►Fig. 1.3) [14].

Single variant proteins with either His104 or Cys166 replaced by alanine have been expressed and shown to result in a decrease in midpoint potential and concomitant reduction in k_{cat}-values compared to wild type enzyme [12]. Thus the bicovalent linkage in BBE increases the reduction potential of the flavin cofactor and is crucial for enzyme activity. Unfortunately, attempts to obtain a H104A/C166A double-variant have failed, presumably because covalent cofactor linkage is required for protein folding and/or stability [12].

Even before the first crystal structures of flavoproteins became available, the presence of a positive charge in vicinity to the N(1)-C(2)=O locus had been suggested to be a common feature of flavin-dependent oxidases [21–24]. Upon reduction the resulting negative charge at the N(1)-C(2)=O locus of the isoalloxazine ring system would thus be stabilized by a positive environment [8,25]. This also leads to the stabilization of anionic flavin semiquinones, N(5)-sulfite adducts as well as the deprotonated forms of artificial flavins bearing an ionizable group at positions 6 or 8 (*e.g.* 6- or 8-mercapto-flavin) [21–23,26–34,34]. So far, investigations on different flavoproteins suggest that this charge can be supplied by either a positively charged amino acid residue, such as a histidine [35–37], lysine [27,38–41], or arginine [42], or by the positive amino terminus of a helix dipole [8,43–45]. BBE exhibits the characteristic properties of a flavoprotein oxidase, including the stabilization of the anionic semiquinone and the ability to form

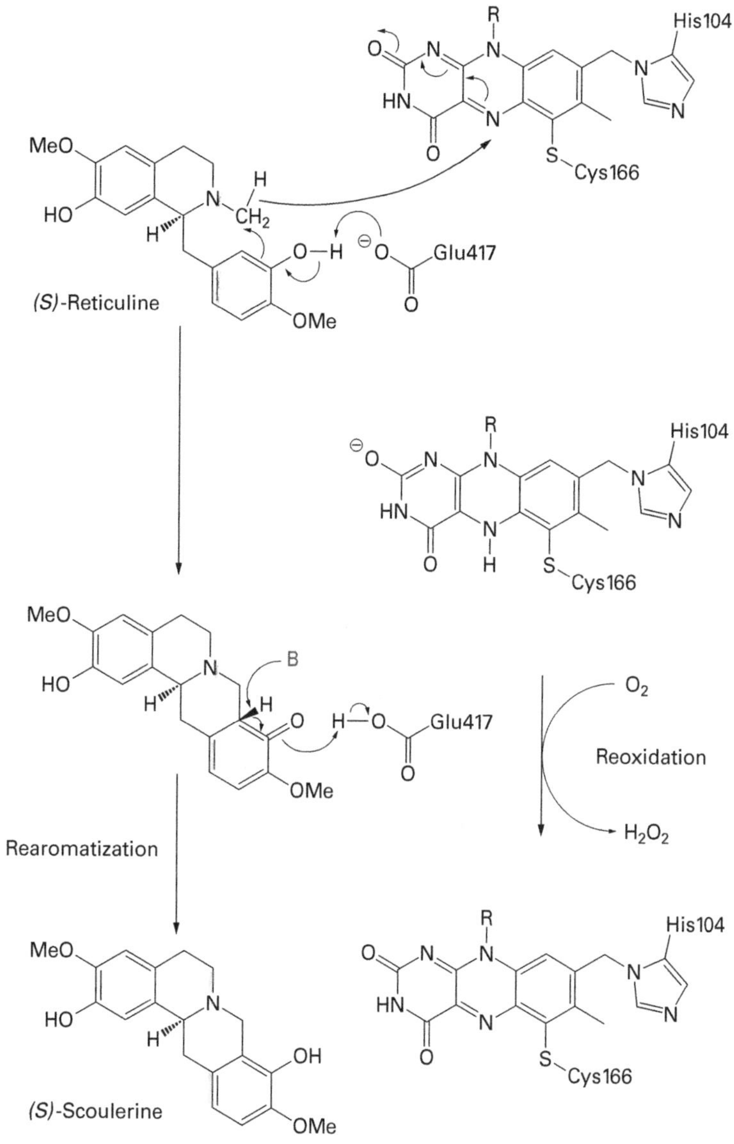

Fig. 1.3: A concerted mechanism for BBE.

N(5)-sulfite adducts (albeit with a dissociation constant that is notably higher than for other characterized flavin-dependent oxidases) [46]. His174 was identified to be the amino acid in BBE providing stabilization of negative charge. Interestingly, this residue does not directly interact with the N(1)-C(2)=O locus of the flavin, but instead forms a hydrogen bonding network via the C2′ hydroxyl group of the ribityl side chain of the flavin. Exchange of His174 to Ala had a pronounced effect on the catalytic activity by decreasing k_{cat} 120-fold (Wallner, S., Winkler, A., Riedl, S., Dully, C., Horvath, S., Gruber, K., Macheroux, P. Catalytic and structural role of a conserved active site histidine in berberine bridge enzyme, Biochemistry 2012;51:6134–6147).

Today, BBE is considered a paradigm of the entire family of bicovalent flavoenzymes as it is arguably one of the best characterized members. Our detailed understanding of the *E. californica* BBE has also led to the exploitation of the enzyme for the production of enantiomerically pure alkaloids with different pharmacological activities [47].

Recently Schrittwieser *et al.* applied BBE for the production of various (*S*)-berbines and (*R*)-benzylisoquinolines [48]. Berbines and benzylisoquinoline alkaloids are a related class of natural compounds [49] that share the isoquinoline heterocycle as a common structural feature (►Fig. 1.4).

These compounds exhibit a broad range of biological effects and have potential applications as pharmaceuticals with analgesic, sedative, hypnotic, antihypertensive, tranquilizing, or muscle relaxation activity [50–53]. Furthermore, some 1-benzyl-1,2,3,4,-tetrahydroisoquinolines show *in vitro* anti-HIV activity making these compounds potential lead structures for AIDS drugs [54]. Most berbine alkaloids show a relaxing effect on the central nervous system and are potent sedatives [55]. Berbine derivatives are also being investigated as drugs for the treatment of schizophrenia [56].

So far, there are several alkaloids with potential applications as pharmaceuticals. However, the isolation of these alkaloids from their natural source is very time consuming and the yields are very low. Moreover, total organic synthesis of those chiral and complex alkaloids is inefficient and cannot be accomplished with satisfactory enantiomeric specificity [57–59]. The biocatalytic approach of Schrittwieser *et al.* is thus an important step towards the development of efficient and cost-effective production methods for enantiomerically pure (*S*)-berbines and (*R*)-benzylisoquinolines [48]. A broad spectrum of substrates has been investigated and it has been shown that BBE accepts non-natural benzylisoquinolines in (*S*)-configuration (►Fig. 1.5).

The reaction has also been optimized regarding pH, temperature and solvent [47,60]. Due to the low solubility of benzylisoquinolines in aqueous buffer the use of cosolvents is required to allow substrate concentrations of at least 20 g L^{-1}. Interestingly, BBE is unexpectedly tolerant towards different organic solvents, with 70% vv^{-1} toluene being the best solvent concentration for efficient catalysis. High conversion rates can be achieved using pH values between 8 and 11 and a temperature from 30 to 50 °C. Using optimized reaction conditions the model reaction has been performed on a scale of 500 mg scale with 50% conversion in 24 h, giving enantiopure (*S*)-berbine (*ee* >97%) and (*R*)-benzylisoquinole (*ee* >97%) [60]. The conversion of different racemic benzylisoquinoles has also been studied (►Fig. 1.5). In all cases the (*S*)-enantiomer of the respective benzylisoquinole was accepted as substrate resulting in the formation of (*S*)-berbines and (*R*)-benzylisoquinolines in optically pure form (*ee* >97%) and in good to excellent yield (22–50%).

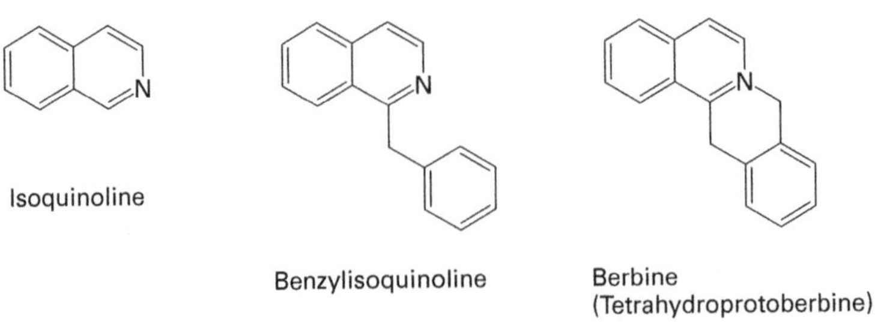

Fig. 1.4: The structure of benzylisoquinoline and berbine alkaloids.

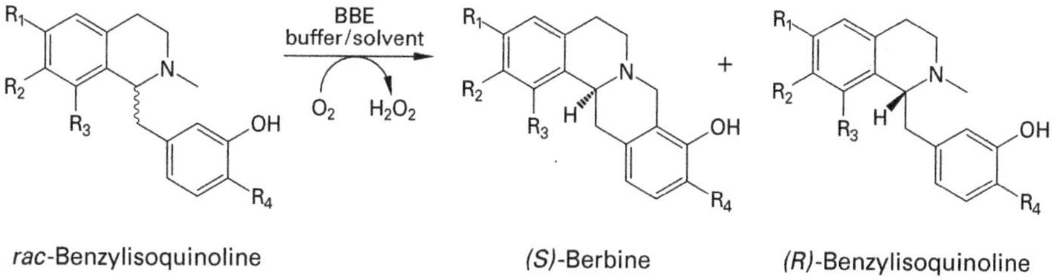

a: $R_1 = R_2 = OMe, R_3 = R_4 = H$
b: $R_1 = OMe, R_2 = R_3 = R_4 = H$
c: $R_1 = R_2 = R_3 = OMe, R_4 = H$
d: $R_1 = R_2 = OCH_2O, R_3 = R_4 = H$
e: $R_1 = OMe, R_2 = OH, R_3 = R_4 = H$
f: $R_1 = OMe, R_2 = OH, R_3 = H, R_4 = OMe$
g: $R_1 = R_2 = R_3 = R_4 = H$

Fig. 1.5: The kinetic resolution of racemic benzylisoquinolines catalysed by BBE.

It is noteworthy that BBE converts 1,2,3,4-tetrahydrobenzylisoqinolines mainly to the 9-hydroxy products and not to the 11-hydroxy regioisomers. In a further study Resch *et al.* has attempted to change the regioselectivity of BBE by medium engineering and by blocking the "normal" reaction centre with a flouro moiety. This has proven to be an efficient strategy for completely changing the regioselectivity, leading to the formation of new 11-hydroxy functionalized products (Resch, V., Lechner, H., Schrittwieser, H. J., Wallner, S., Gruber, K., Macheroux, P., Kroutil, W. Investigating the regioselectivity of the berberine bridge enzyme by employing customized fluorine-containing substrates, 2012;18:13173–13179).

In conclusion, studies with BBE as a potential industrial biocatalyst show that the enzyme can be successfully applied for the preparation of enantiomerically pure isoquinoline and berbine derivatives, and such applications hold considerable promise.

1.3 The family of BBE-like enzymes in the plant kingdom: how many and what for?

The group of BBE-like proteins is steadily growing and sequence alignments provide strong evidence that a variety of plant proteins possess conserved histidine and cysteine residues appropriate for bicovalent attachment of the cofactor. The availability of whole genomes for different plant species enabled a targeted search of BBE-homologues in 26 plant species as shown in ▶Tab. 1.2. Using the Blast tool provided on the phytozome website (http://www.phytozome.net/search.php) many BBE-homologues have thus been identified in the available plant genomes. The highest number of BBE homologues has been found in the poplar, *Populus trichocarpa*, whose genome apparently encodes 57 BBE-like proteins. To our knowledge poplar does not accumulate complex alkaloids, and these BBE-like proteins may have metabolic roles other than alkaloid biosynthesis. Furthermore, our analysis has revealed that a high number of BBE-homologues are also present in the orders of Fabales (*e.g.* soybean), Rutales (*e.g.* citrus fruits) and Brassicales (*e.g. Arabidopsis*), suggesting a high abundance of bicovalent flavoproteins in the plant kingdom. Poales, such as rice, sorghum or maize, possess comparatively few BBE-homologues, and the same applies for grape. Finally, in mosses and algae BBE-like proteins seem to be rare (▶Tab. 1.2).

Tab. 1.2: BBE-like proteins in various plant species and families

Plant	Family	Common name	BBE-like homologues
Populus trichocarpa	*Salicaceae*	Western balsam poplar	57
Glycine max	*Fabaceae*	soybean	46
Citrus clementina	*Rutaceae*	clementine	39
Prunus persica	*Rosaceae*	peach	30
Arabidopsis thaliana	*Brassicaceae*	mouse-ear cress	28
Arabidopsis lyrata	*Brassicaceae*		27
Eucalyptus grandis	*Myrtaceae*	eucalyptus	27
Mimulus guttatus	*Scrophulariaceae*	common monkey-flower	25
Citrus sinensis	*Rutaceae*	orange	24
Cucumis sativus	*Cucurbitaceae*	cucumber	23
Aquilegia coerulea	*Ranunculaceae*	Rocky mountain columbine	18
Medicago truncatula	*Fabaceae*	barrel medic	18
Manihot esculenta	*Euphorbiaceae*	cassava (manioc)	17
Zea mays	*Poaceae*	maize	16
Ricinus communis	*Euphorbiaceae*	ricinus	15
Setaria italica	*Poaceae*	foxtail millet	13
Sorghum bicolor	*Poaceae*	sorghum	13
Carica papaya	*Caricaceae*	papaya	12
Oryza sativa	*Poaceae*	rice	11
Brachypodium distachyon	*Poaceae*	purple false brome	10
Vitis vinifera	*Vitaceae*	grape	5
Physcomitrella patens	*Funariaceae*		1
Chlamydomonas reinhardtii	*Chlamydomonadaceae*		0
Selaginella moellendorffii	*Selaginellaceae*		0
Volvox carteri	*Volvocaceae*		0

In addition to this bioinformatical analysis, there is also experimental evidence that flavoproteins from different plant sources possess both a covalent histidinyl- and cysteinyl-linkage of the cofactor [20,61–65]. In plants bicovalent flavoproteins are involved in many different processes. They comprise oxidases in alkaloid biosynthesis

(BBE, STOX) [20,64], cannabinoid metabolism (THCA synthase) [66], or active plant defense (nectarin V, carbohydrate oxidases from *Helianthus annuus* and *Lactuca sativa*) [67,68]. In the seaweed *Chondrus crispus* a BBE-like enzyme has been found that catalyzes the oxidation of a variety of hexose sugars [69–72]. A further group of bicovalently linked plant proteins is known to act as pollen allergens and is found in various grasses, such as timothy (Phl p 4) [65] or Bermuda grass (Cyn d 4, earlier referred to as BG60) [73].

In *Papaveraceae* bicovalently linked flavoproteins are reported to catalyze crucial oxidations in alkaloid biosynthesis. As discussed before, BBE catalyzes the oxidative formation of the methylene bridge of *(S)*-scoulerine in different poppy species [17,74]. Moreover, *(S)*-tetrahydroprotoberberine oxidase (STOX) is described as a flavin containing oxidase accepting tetrahydroprotoberberines in *(S)*-configuration [75–77]. Due to the lack of a crystal structure of STOX there is no definitive evidence that the enzyme belongs to the family of bicovalent flavoproteins, although homology modelling and a phylogenetic analysis strongly suggests that the flavin cofactor in STOX is indeed bicovalently bound [64]. The substrate specificity of STOX from *Argemone mexicana* and *Berberis wilsoniae* have been investigated and both enzymes catalyze the oxidation of *(S)*-tetrahydropalmatine to palmatine (▶Fig. 1.6) [64]. Additionally, *(S)*-scoulerine and *(S)*-canadine are accepted as substrates for STOX from *B. wilsoniae*, and STOX from *A. mexicana* converts *(S)*-coreximine, respectively [64]. STOX thus catalyzes the oxidation of different tetrahydroprotoberberines in benzylisoquinoline producing plants.

Δ^1-Tetrahydrocannabinolic acid (THCA) synthase has been identified as an important enzyme in controlling the psychoactivity of *Cannabis sativa* [66]. It catalyzes the oxidative cyclization of the monoterpene moiety of cannabigerolic acid (CBGA) into THCA, which further decarboxylates in a non-enzymatic fashion yielding the psychoactive tetrahydrocannabinol (THC) (▶Fig. 1.7) [66,78]. Cannabinoids are exclusively found in *C. sativa* [79] and the pharmacological importance of these compounds [79–86] boosted investigations on THCA synthase [87]. The X-ray crystal structure of THCA synthase has been reported [61] and clearly demonstrates that the enzyme belongs to the PCMH superfamily of flavoproteins with a bicovalently tethered cofactor as seen in BBE [88]. THCA synthase catalyzes an oxidative ring closure reaction, which is similar to berberine bridge formation. However, Shoyama *et al.* have suggested a cationic mechanism for THCA synthase rather than the one-step concerted reaction proposed for BBE [61]. According to these authors, catalysis is initiated by a hydride transfer from the C-3 position of CBGA to the N-5 position of the isoalloxazine ring moiety of the FAD cofactor. The resulting carbocation then collapses to the product by nucleophilic attack of the oxyanion previously generated by proton abstraction. The

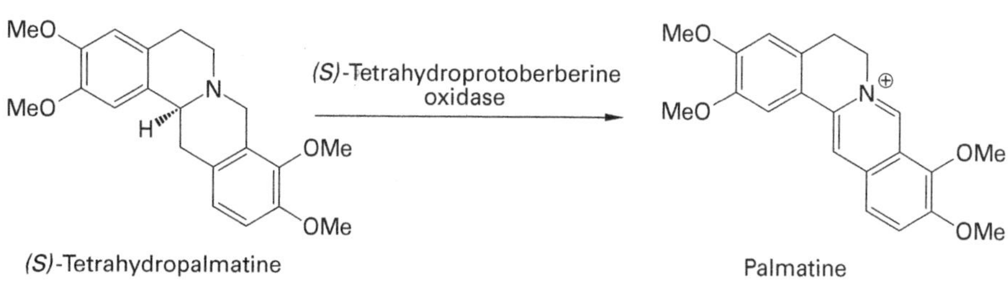

Fig. 1.6: The conversion of *(S)*-tetrahydropalmatine by STOX.

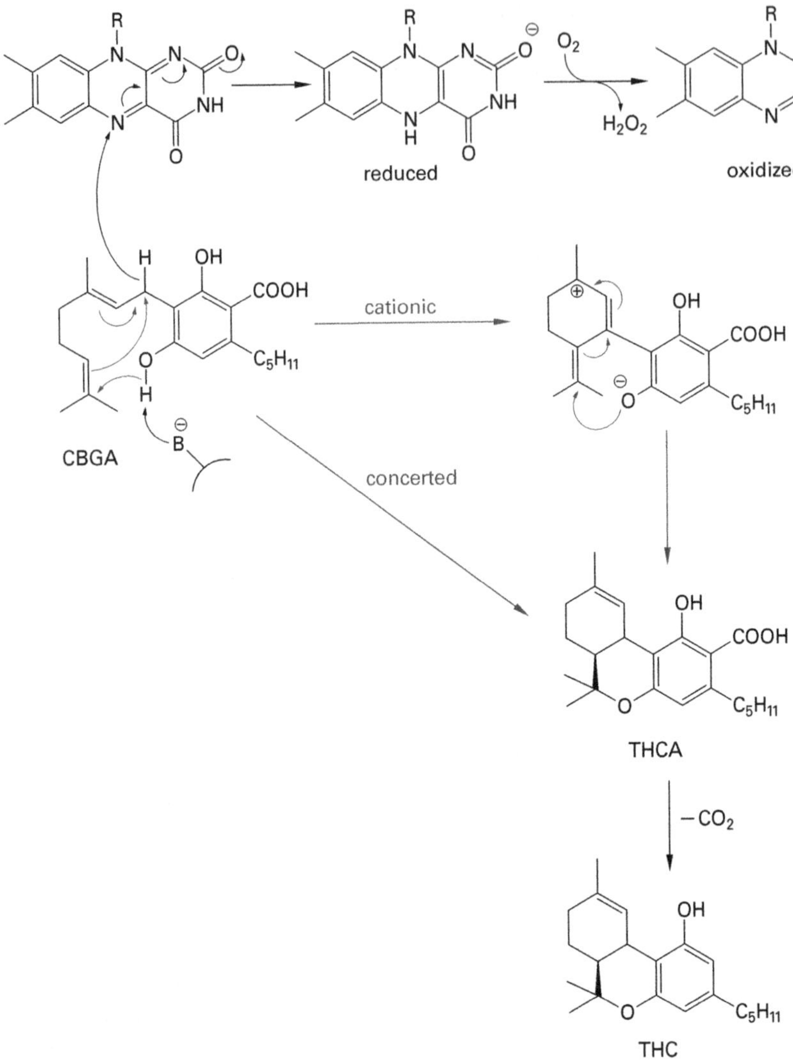

Fig. 1.7: The proposed mechanism for THCA synthase. Two possible routes for THCA formation, which both are initiated by proton abstraction from CBGA. In the mechanism suggested by Sirikantaramas *et al.* (red arrows) a cationic intermediate is formed, which further cyclizes yielding THCA, which further undergoes non-enzymatic decarboxylations. We rather suggest a concerted mechanism (blue arrow) as was already reported for BBE. In both cases the reduced flavin is regenerated with molecular oxygen [66].

reduced flavin is reoxidized by transferring a hydride to molecular oxygen resulting in the formation of hydrogen peroxide as a side product (▶Fig. 1.7) [66]. So far, there is no evidence for such a cation-based mechanism and we suggest as an alternative that the concerted mechanism of BBE might also be relevant for the C-C bond formation of Δ^1-tetrahydrocannabinolic acid (THCA) synthase. It is noteworthy that BBE and THCA synthase exhibit substantial sequence similarities and that the crucial catalytic glutamate residue is present in both enzymes, suggesting a similar path for catalysis in these two oxidases [14].

Interestingly, not all bicovalent flavoproteins are involved in secondary metabolite production in plants. Nectarin V (NEC5) is a BBE-like enzyme and is expressed in the nectary gland of ornamental tobacco [62,89]. The enzyme shows glucose oxidase activity by oxidizing glucose to gluconolactone and hydrogen peroxide [62]. Hydrogen peroxide in turn seems to play a vital role in defense against microbial attack by preventing microbe contamination of the floral reproductive tract [67,90–93]. There is also evidence that in contrast to the earlier discussed bicovalently flavinylated oxidases, NEC5 does not strictly depend on oxygen as electron acceptor for cofactor reoxidation. According to unpublished data of C. Carter and R.W. Thornburg mentioned in [67] dehydroascorbate is accepted instead of molecular oxygen resulting in flavin reoxidation and recycling of ascorbate from dehydroascorbate.

Enzymes involved in plant defense were also identified in other species, such as *H. annuus* (sunflower) or *L. sativa* (lettuce). Both plants express a carbohydrate oxidase with broad substrate specificity for the majority of mono- and disaccharides and with the production of H_2O_2 as a toxic reaction product [68].

Hexoses oxidase (HOX) from the seaweed *C. crispus* is another BBE-like enzyme accepting glucose as substrate [71]. However, this enzyme exhibits a broad substrate range compared to nectarin V, accepting various oligomeric sugars such as lactose, maltose, cellobiose and maltotriose as well as glucooligosaccharides with up to seven glucose units, in addition to glucose and galactose [71,72].

Another group of BBE-like proteins has yet to have a specific catalytic function assigned. These proteins are known to act as pollen allergens and they share the histidine and cysteine motifs for bicovalent flavinylation [65,73]. Pollen allergens are generally found in grasses of different sorts, as well as corn. BG60, also referred to as Cyn d4, is described as a covalently flavinylated pollen allergen in Bermuda grass [73,94]. Subsequently, antigens with homology to BG60 have been described in celery (Api g 5) and in oilseed rape pollen [95–97]. Moreover, Phl p 4 , discovered in two isoforms in timothy grass [98] has been identified as BBE-like protein [65]. The crystal structure of Phl p4 from *Phleum pratense* confirms the structural homology to BBE and bicovalent attachment of the FAD cofactor (pdb code 3tsh). Blast searches with the two experimentally determined full-length BBE-like pollen allergens BG60 and Phl p4 resulted in the identification of four additional BBE-like proteins, namely Sec c 4, Hor v 5, Tri a 4 and Lol p 4 from rye, barley, wheat and perennial ryegrass, respectively. Except for Lol p 4 these proteins show almost the same size and high sequence similarities.

These examples of structurally well-characterized BBE-like enzymes underscore that this group of bicovalently linked flavoproteins is involved in a variety of chemical reactions and biological functions. Thus it is obvious that specific enzyme functions cannot easily be predicted solely on the basis of sequence homology. Careful functional annotations are necessary to establish their physiological roles [99,100]. This has been confirmed by various examples, where marginal or even single amino acid exchanges are known to lead to a distinct substrate specificity [101,102]. For example, the substrate specificity of ChitO from *F. graminearum* is completely changed by a single amino acid replacement in the active site of the enzyme that leads to an increased acceptance of glucose and small oligosaccharides such as maltose or lactose [100]. Still, a comprehensive analysis of sequence homology and available biochemical data of already characterized bicovalent flavoproteins can result in reliable predictions of general enzyme function, which can further be established experimentally. One recent example for the practicability of a comprehensive theoretical analysis is the prediction of Gln268 in ChitO as an active site residue involved in substrate recognition. Homology modelling of ChitO and a detailed comparison with the elucidated crystal structure of GOOX led to the correct identification of this residue [100]. Here we present a detailed analysis of BBE homologues from *Arabidopsis thaliana,* which includes both comparisons of sequence homology and analysis of homology models in combination with structural and biochemical data available for BBE and GOOX.

To identify BBE-like enzymes in *A. thaliana* the sequence of BBE from *E. californica* was used for a Blastp search against the whole *Arabidopsis* genome. We find 28 BBE homologues in *A. thaliana,* encoded on chromosomes 1, 2, 4, and 5 (▶Tab. 1.3). As a consequence of this unexpectedly high number of BBE homologues, the question has to be addressed as to whether *A. thaliana* can produce complex alkaloids, or whether there are alternative functions for these enzymes *in planta* (for a detailed review on this issue see [99]). It is known that *Arabidopsis* species accumulate simple alkaloids such as camalexin [103]. However, complex alkaloids have not been detected in this plant so far (although this does not necessarily imply their absence in *Arabidopsis*). Toghe *et al.* describe the available chemical information regarding *Arabidopsis* metabolites as insufficient [104] and the recent discovery of volatile terpenoids in *A. thaliana* suggests the presence of a complex secondary metabolism, lending credence to the possible occurrence of yet undiscovered alkaloid pathways in the plant [99,105,106]. Moreover, the identification of a plant efflux carrier with the ability to transport berberine [107] can be taken as an indication for the presence of complex alkaloids in *Arabidopsis.*

We have analyzed the 28 homologous proteins of *A. thaliana* using the YASARA modelling tool. These BBE homologues all possess a conserved histidine residue at the site corresponding to His104 of BBE, indicating a covalent 8α-histidinyl linkage of the cofactor. Moreover, 24 of the 28 BBE homologues also possess the conserved cysteine residue for 6-*S*-cysteinylation, and it appears that *A. thaliana* possesses 24 bicovalent flavoenzymes! The remaining four BBE homologous (BBE-like 10, 25, 27, and 28) feature serine, tyrosine, or histidine residue at the position corresponding to Cys166 of BBE. So far, only 6-*S*-cysteinylation has been reported for bicovalently linked flavoproteins, although other modes of covalent modification cannot be ruled out. Whether these residues are capable of forming a covalent bond to C6 of the flavin cofactor in the respective protein remains to be seen.

Moreover, the homology models were scanned for BBE active site residues, such as Glu417, which is the catalytic base necessary for carbon-carbon bond formation in BBE [14]. Table 1.3 shows a compilation of predicted active site residues in the 28 BBE-like *Arabidopsis* proteins.

Interestingly, only three BBE-homologues, BBE-like proteins 10, 11 and 16, possess a glutamate residue at the respective site of Glu417 in BBE. Consequently, these proteins could possibly be involved in ring-closure reactions initiated by proton abstraction from the substrate by the catalytic glutamate residue. BBE-like 11 and BBE-like 16 show overall active site architectures which are highly similar to the structural environment in BBE suggesting that these enzymes could be involved in similar enzymatic reactions. BBE-like 10, however, possesses a serine residue in place of the cysteine required for cofactor attachment. Hence this serine residue could possibly influence the

Tab. 1.3: Predicted active site residues in putative BBE-like proteins of *A. thaliana* through homology modelling

Locus Name	Working Name	Accession Number	Comparison to Berberine Bridge Enzyme Active Site Residues							
			His104	Cys166	Glu417	Tyr106	His459	His174	Tyr456	Trp165
AT1G01980.1	BBE-like 1	Q9LPC3	His	Cys	Gln	Tyr	Tyr	His	Tyr	Val
AT1G11770.1	BBE-like 2	Q9SA99	His	Cys	Gln	Tyr	---[a]	His	Tyr	Val
AT1G26380.1	BBE-like 3	Q9FZC4	His	Cys	Gln	Asn	Tyr	His	Phe	Ile
AT1G26390.1	BBE-like 4	Q9FZC5	His	Cys	Gln	Tyr	---	His	---	---
AT1G26400.1	BBE-like 5	Q9FZC6	His	Cys	Gln	Tyr	Phe	His	Leu[b]	Val
AT1G26410.1	BBE-like 6	Q9FZC7	His	Cys	Gln	Asn	---	His	---	Val
AT1G26420.1	BBE-like 7	Q9FZC8	His	Cys	Gln	Leu	---	His	---	Val
AT1G30700.1	BBE-like 8	Q9SA85	His	Cys	Gln	Tyr	Tyr	His	Phe	Ile
AT1G30710.1	BBE-like 9	Q9SA86	His	Cys	Gln	Tyr	---	His	---	Ile
AT1G30720.1	BBE-like 10	Q9SA87	His	Ser	Glu	Phe	---	His	---	Leu
AT1G30730.1	BBE-like 11	Q9SA88	His	Cys	Glu	Phe	---	His	---	Leu
AT1G30740.1	BBE-like 12	Q9SA89	His	Cys	Gln	Tyr	---	His	Tyr	Val
AT1G30760.1	BBE-like 13	Q93ZA3	His	Cys	Gln	Tyr	---	His	Tyr	Leu
AT1G34575.1	BBE-like 14	Q9LNL9	His	Cys	Gln	Tyr	---	His	---	Ile
AT2G34790.1	BBE-like 15	O64743	His	Cys	Gln	Tyr	---	His	---	Leu
AT2G34810.1	BBE-like 16	O64745	His	Cys	Glu	Tyr	Tyr	His	Phe	Val
AT4G20800.1	BBE-like 17	Q9SVG7	His	Cys	Gln	Tyr	Tyr	His	Phe	Val
AT4G20820.1	BBE-like 18	Q9SVG5	His	Cys	Gln	Tyr	Leu	His	Tyr	Val

(*Continued*)

(*Continued*)

Locus Name	Working Name	Accession Number	Comparison to Berberine Bridge Enzyme Active Site Residues							
			His104	Cys166	Glu417	Tyr106	His459	His174	Tyr456	Trp165
AT4G20830.1[c]	BBE-like 19	Q9SVG4	His	Cys	Gln	Tyr	---	His	Tyr	Val
AT4G20830.2[c]	BBE-like 20	Q9SVG4	His	Cys	Gln	Tyr	---	His	Tyr	Val
AT4G20840.1	BBE-like 21	Q9SVG3	His	Cys	Gln	Tyr	---	His	Tyr	Val
AT4G20860.1	BBE-like 22	Q9SUC6	His	Cys	Gln	Tyr	---	His	---	Ile
AT5G44360.1	BBE-like 23	Q9FKV2	His	Cys	Gln	Tyr	Tyr	His	Tyr	Thr
AT5G44380.1	BBE-like 24	Q9FKV0	His	Cys	Leu	Phe	Tyr	Tyr	Tyr	Leu
AT5G44390.1	BBE-like 25	Q9FKU9	His	Tyr	Gln	Tyr	Tyr	His	Tyr	Val
AT5G44400.1	BBE-like 26	Q9FKU8	His	Cys	Gln	Tyr	Tyr	His	Tyr	Leu
AT5G44410.1	BBE-like 27	Q9FI25	His	Tyr	Gln	Tyr	Phe	Gln	Tyr	Leu
AT5G44440.1	BBE-like 28	Q9FI21	His	His	Gln	Tyr	Phe	Gln	Tyr	Ile

[a] Prediction of the respective active site residue not feasible due to different secondary structure elements in berberine bridge enzyme and the modelled homologue
[b] Prediction is not accurate since homology model of BBE-like 5 shows different secondary structure and is rather imprecise in the respective area
[c] AT4G20830.1 and AT4G20830.2 represent two isoforms of the respective BBE-like protein in *A. thaliana*

flavin cofactor in BBE-like 10 and modulate the redox potential for efficient catalysis by forming a hitherto uncharacterized novel covalent linkage or by influencing the electronic environment of the cofactor without covalent tethering.

Except for BBE-like 24, the remaining BBE-like proteins have a glutamine residue at the position of Glu417 in BBE (▶Tab. 1.3, ▶Fig. 1.8). Interestingly, multiple sequence alignments reveal that this active site glutamine residue is also present in the carbohydrate oxidases, such as GOOX (AAS79317), ChitO (XP_391174) and the carbohydrate oxidases from sunflower (AAL77103) and lettuce (AAL77102). The proposed reaction mechanism for GOOX suggests an active site tyrosine residue (Tyr429) as general base for catalysis, which initiates substrate conversion by proton abstraction from the

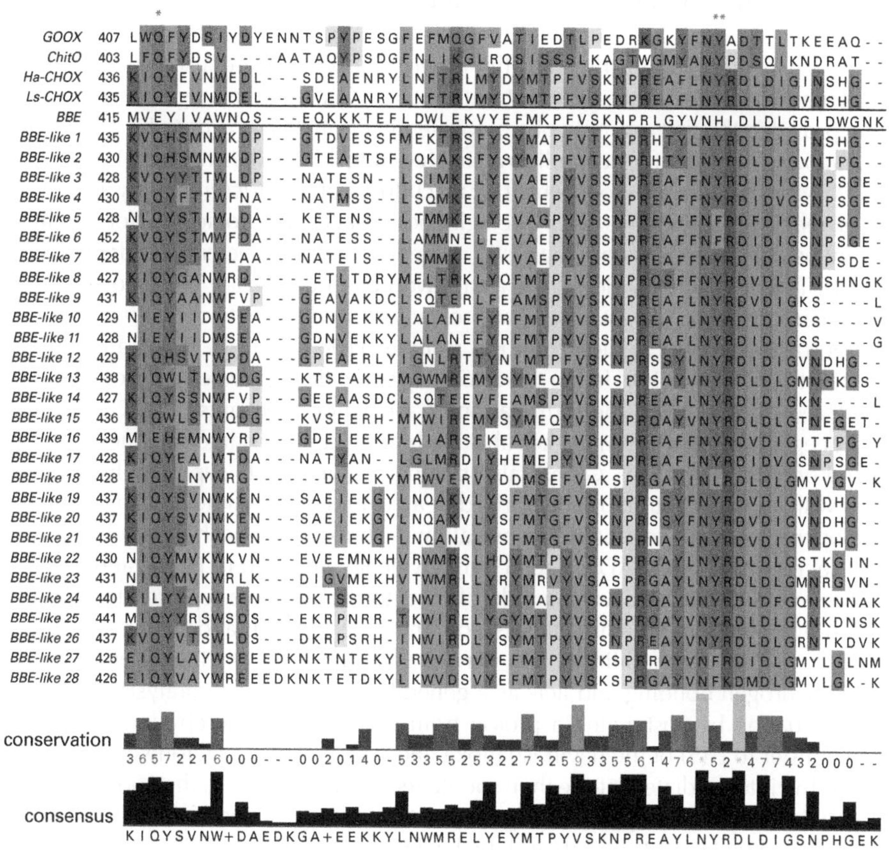

Fig. 1.8: Multiple sequence alignment of characterized bicovalently flavinylated oxidases and 28 predicted BBE-like proteins from *A. thaliana* using ClustalW. The sequence of BBE is displayed without colouring and the position of glutamate 417 is indicated (*). The conserved tyrosine residue in carbohydrate oxidases is indicated with **. Conserved amino acid residues are highlighted in coloured boxes. The amino acid sequences of following characterized oxidases were used: GOOX from *A. strictum* (AAS79317), ChitO from *F. graminearum* (XP_391174), carbohydrate oxidases from sunflower (AAL77103) and from lettuce (AAL77102) and BBE from *E. californica* (AC39358).

OH[1] group of the sugar moiety [7,108]. Multiple sequence alignments performed with ClustalW [109,110] and analyzed in Jalview 2.7 [111] have shown that this tyrosine residue is strictly conserved in carbohydrate oxidases and hence might be a reliable predictor for enzymatic function. 23 out of the 28 BBE homologues from *A. thaliana* possess this conserved tyrosine residue (▶Fig. 1.8), which supports the idea that some BBE homologues in *A. thaliana* are in fact carbohydrate oxidases. In plants, carbohydrate oxidases appear to play a vital role in active plant defense due to their ability to form hydrogen peroxide as a product of enzymatic turnover [67,68]. Thus it is conceivable that a portion of the BBE homologues in *Arabidopsis* acts as carbohydrate oxidases involved, for example, in defense against pathogen attack (see above).

1.4 The occurrence of BBE-like enzymes in fungi

BBE-like enzymes are not restricted to the plant kingdom and are also found in fungi and bacteria. Elucidation of the crystal structure of GOOX from the fungus *Acremonium strictum* provided the first example of a bicovalently flavinylated oxidase [7] and ChitO from *F. graminearum* is a further well-investigated fungal BBE-like enzyme [10].

GOOX was first isolated from wheat bran cultures of *A. strictum* T1 by Lin *et al.* during screening experiments for novel glucooligosaccharide oxidases for application as industrial biocatalysts or diagnostic reagents for alternative carbohydrate assays [112,113]. The enzyme was described as a flavoprotein catalyzing the oxidation of a variety of carbohydrates like glucose, maltose, lactose, and a variety of oligosaccharides composed of 1,4-linked D-glucopyranosyl residues [108,112,114]. The availability of the crystal structure of GOOX led to the discovery of the first bicovalent cofactor attachment. Moreover, it enabled the formulation of a reaction mechanism for GOOX and led to a deeper insight in the mode of substrate binding and substrate specificity [7]. As discussed previously for BBE, GOOX also belongs to the PCMH superfamily of flavoproteins with an overall structure comprised of a flavin- and substrate-binding domains with an open carbohydrate binding groove allowing bulky oligosaccharide substrates to access the active site. The FAD cofactor is bicovalently attached to His70 and Cys130 of the enzyme and for GOOX this mode of cofactor linkage was reported to tune the reduction potential for efficient catalysis, to contribute to flavin and substrate binding and to increase protein stability [7,11]. Tyr429 is activated by Asp355 through water-mediated hydrogen bonding and acts as a general base crucial for catalysis [7,108]. It initiates oxidation by abstracting a proton from substrate and thereby initiates direct hydride transfer to N(5) of the flavin isoalloxazine ring, yielding a lactone and reduced flavin [7]. The intermediate lactone then spontaneously hydrolyses to the corresponding acid. Stabilization of the negative charge at the N(1)-C(2)=O locus of the isoalloxazine ring is achieved by His138, Tyr426 and the backbone NH of Tyr144 [7]. In the remainder of the catalytic sequence, the reduced flavin cofactor is reoxidized in a reaction with molecular oxygen to form hydrogen peroxide (▶Fig. 1.9).

Recently, Foumani *et al.* reported the successful expression of a GOOX variant (GOOX-VN) from a different strain of *A. strictum* with altered substrate specificity [115] in which fifteen amino acid substitutions relative to the sequence of GOOX from *A. strictum* T1 were made. Additionally to glucose, maltose, lactose, and oligosaccharides composed of 1,4-linked D-glucopyranosyl residues, GOOX-VN accepts galactose

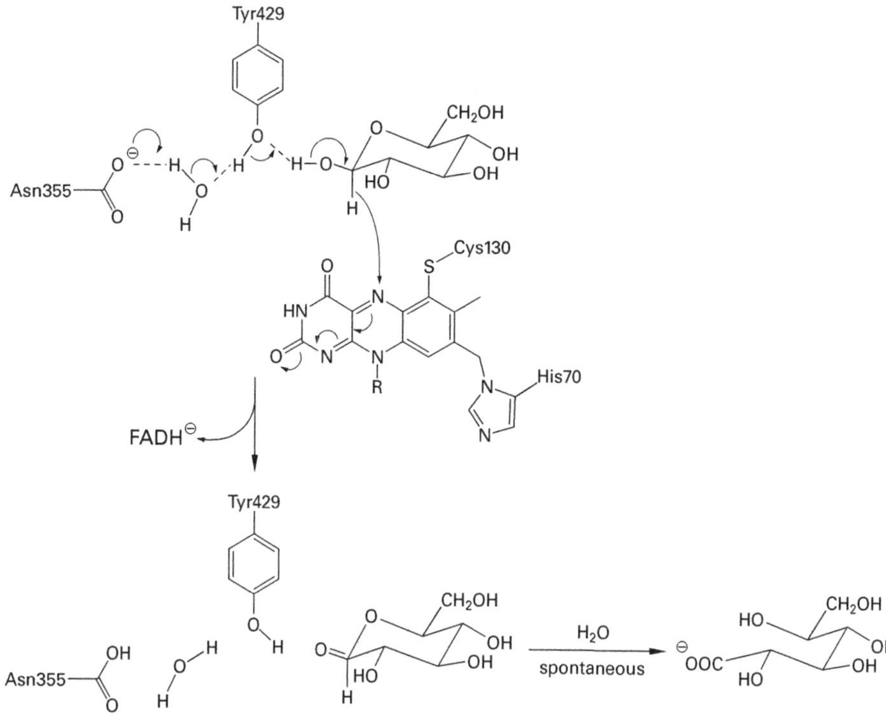

Fig. 1.9: The proposed mechanism for GOOX. Tyr429, which is activated by Asn355 via water mediated hydrogen bonding, subtracts a proton from the OH[1] group of the substrate thereby causing a direct transfer of a hydride to the N(5) atom of the flavin isoalloxazine ring. The resulting glucono-1,5-lactone spontaneously hydrolyzes to gluconic acid. The reduced flavin cofactor gets reoxidized by molecular oxygen. Figure 1.9 was modified according to [7].

and xylo-oligosaccharides with similar catalytic efficiency as for cello-oligosaccharides. This indicates that directed evolution on GOOX can lead to improved biocatalysts for oxidative modification of cello- and xylo-oligosaccharides [115].

GOOX is of special interest since many putative bicovalently linked flavoproteins show a tyrosine residue at the position corresponding to the catalytic Tyr429 in GOOX. Hence it is likely that these proteins are carbohydrate oxidases with various functions in the respective organism (see discussion in previous section).

Blast searches against the non redundant protein sequence database on NCBI with GOOX or BBE as query identifies a great number of putative bicovalently linked flavoprotein oxidases in fungi. Molds and pathogenic fungi which affect plants or insects in particular are found to possess proteins homologous to BBE and GOOX. Blastp searches clearly reveal the high abundance of putative bicovalently flavinylated proteins among Ascomycota and Basidiomycota. *Aspergillus niger, Aspergillus oryzae,* and the basidiomycete *Melampsora laricis* have 9, 7, and 10 BBE-homologs, respectively, the highest number of putative bicovalently linked flavoproteins.

Heuts *et al.* have identified six new glucooligosaccharide oxidase homologues with altered substrate preference [100,108], and a putative oxidase (XP_391174) from *F. graminearum* has been selected for detailed investigations [10,100]. Surprisingly, this new oxidase displays a distinct substrate preference as compared to known oligosaccharide oxidases, catalyzing the regioselective oxidation of *N*-acetylated oligosaccharides; it has been designated chitooligosaccharide oxidase (ChitO). ChitO is a bicovalently linked flavoprotein belonging to the same structural class as GOOX or BBE. For ChitO, a glutamine (Gln268) has been identified as responsible for recognition of the *N*-acetyl moiety. Indeed, mutation of Gln268 to arginine has a drastic effect on the K_M value for all substrates, leading to an increased acceptance of cello-oligosaccharides and a reduced conversion rate for ChitO, again suggesting a possible application of site-directed mutagenesis for the production of improved oligosaccharide oxidases with tailored substrate specificity.

1.5 BBE-like enzymes in bacteria: oxidative power for the biosynthesis of antibiotics

In recent years the number of bacterial flavoenzymes identified as possessing a bicovalently attached flavin cofactor has steadily risen [70,116–121]. In contrast to plants, none of these bacterial enzymes appear to be involved in cyclization reactions. Instead, the enzymes typically catalyze the oxidation of primary or secondary alcohol groups in complex and bulky molecules, and are part of biosynthetic pathways leading to antibiotics such as aclacinomycin from *Streptomyces galileus* [116], glycopeptide A40926 (a teicoplanin homologue) from *Nonomuraea* sp. [118], tirandamycin [116,118] and gilvocarcin V from *Streptomyces* sp. (►Fig. 1.10) [120,122]. The first two reactions comprise a four electron oxidation, *i.e.* the FAD cofactor runs twice through the catalytic cycle of two electron reduction and reoxidation by molecular oxygen in the course of the reaction. Aclacinomycin oxidoreductase (AknOx) first oxidizes the alcohol group at C-4 of the terminal pyranose ring and in the second round of oxidation a C-C double bond is introduced between C-2 and C-3 thus yielding a α,β-unsaturated ketone (►Fig. 1.10A). On the other hand, the reaction catalyzed by the hexose oxidase Dbv29 is a four-electron oxidation of a primary alcohol group to a carboxylic acid at C-6 of the sugar moiety (►Fig. 1.10B). In both cases, the enzyme acts on the carbohydrate moiety of the molecule rather than its aglycone. This is in contrast to the reactions in tirandamycin and gilvocarcin V biosynthesis, where the oxidation concerns a secondary alcohol group. In the latter case, the oxidation yields a lactone in a complex polyketide molecule whereas the former reaction leads to a ketone group in the bicyclic ketal moiety of the antibiotic (►Fig. 1.10C and D). It is noteworthy that in all of the reactions considered here, the flavoenzyme catalyzes one of the terminal reactions in the biosynthetic pathways, *i.e.* occurs at a stage where the skeleton of the target molecule is already established. This contrasts to the reactions catalyzed by BBE from *E. californica* and THCA synthase from *C. sativa* where the oxidation reaction is central to the generation of the carbon skeleton of the natural compound produced by the plant (compare Figure 1.1 and Figure 1.7).

Similarity searches against the non-redundant protein database on NCBI has led to the discovery of many putative BBE homologues in the bacterial kingdom. BBE homologues are found in various *Streptomyces* species, for example, which according to sequence alignments possess the conserved cysteine and histidine residue for bicovalent flavin attachment. These enzymes are suggested to be oxidoreductases, berberine

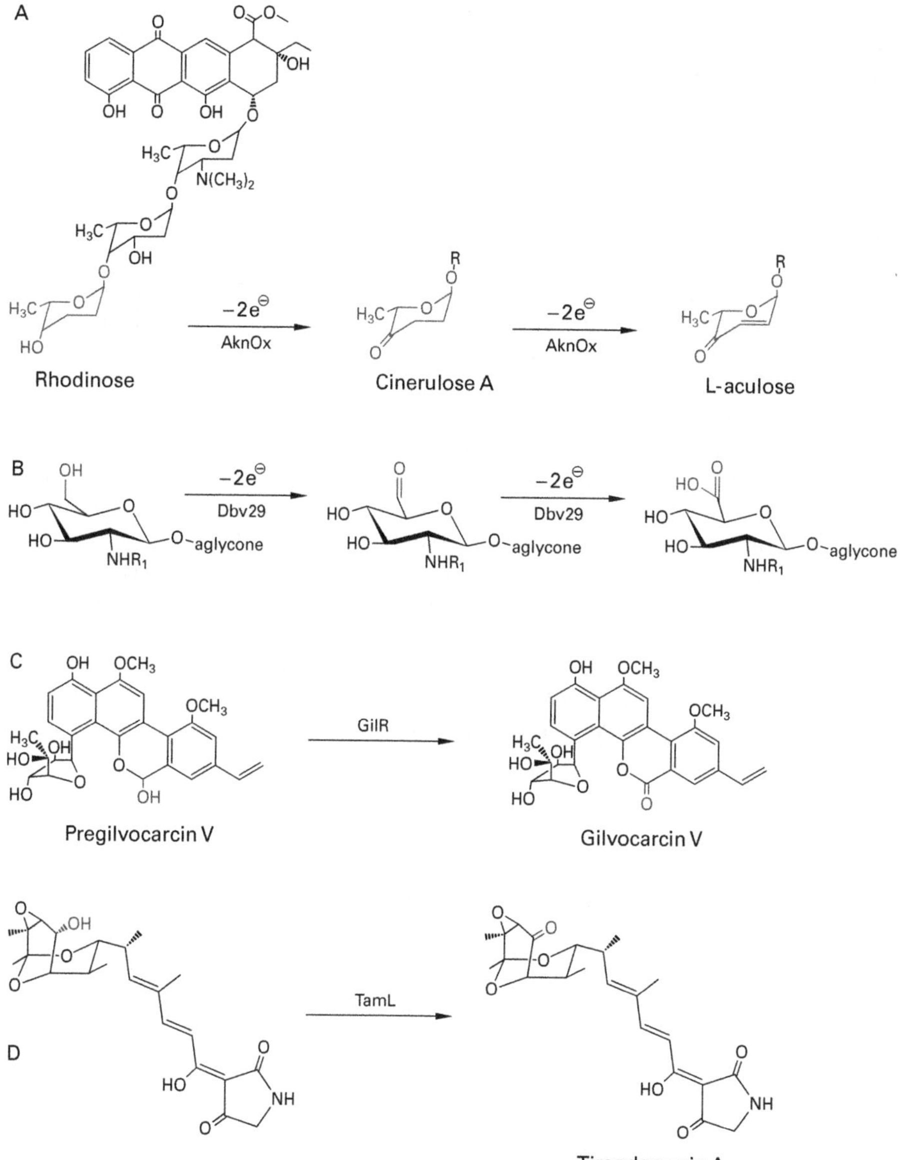

Fig. 1.10: Reactions catalyzed by bicovalent flavoenzymes in the bacterial kingdom. Reactions catalyzed by (A) AknOx, aclacinomycin reductase from *Streptomyces galileus,* (B) Dbv29 from *Nonomuraea* sp., (C) GilR, pregilvocarcin V dehydrogenase from *Streptomyces,* and (D) TamL, 10-hydroxy-dehydrogenase in tirandamycin biosynthesis from *Streptomyces.*

bridge enzymes and FAD binding proteins; many possess the twin-arginine translocation signal for transport into the periplasmic space. Interestingly, a relatively large number of the BBE homologues identified in *Streptomyces* appear to be lipoproteins. A sequence alignment of these putative lipoproteins shows high sequence identity of these enzymes in different *Streptomyces* species. A similar search in the genomes of *Pseudomonas* does not detect BBE homologues, and it thus appears that BBE-like enzymes are not uniformly distributed in eubacteria but appear to be present only in distinct genera. Apart from *Streptomyces*, BBE-like proteins were also found in *Bacillus, Mycobacterium, Ruegeria* and *Frankia* species.

1.6 Conclusions

In the last five years, bicovalent flavinylation has emerged as a novel mode of cofactor linkage in a surprisingly large number of proteins. Currently, ~25% of covalently linked flavoenzymes feature bicovalent flavinylation, with a large number of bacterial, fungal and plant genes predicted to belong to this family of "BBE-like proteins". So far, bicovalent flavin attachment is confined to bacteria, fungi, and plants, where some genera seem to be particularly rich in bicovalent flavoproteins while others appear to have only a few members of this family. Bicovalent flavoproteins appear to be totally absent from archea and the animal kingdom. In nature, these enzymes adopt different roles in the metabolism of plants, fungi, and bacteria. In plants, bicovalent flavoenzymes catalyze complex cyclization reactions in natural product biosynthesis, such as alkaloids and terpenes. Moreover, these enzymes act as sugar and alcohol oxidases and thus fulfil different roles in the respective organism. As sugar oxidases bicovalent flavoproteins are involved in plant or fungal defense against pathogens and as alcohol oxidases they catalyze the final step(s) in antibiotic biosynthesis in bacteria. Hence these enzymes catalyze reactions with considerable potential to produce novel natural products and antibiotics in a biocatalytic setting.

1.7 Acknowledgments

We like to thank the Austrian Science Fonds (FWF) for financial support through the PhD programme "Molecular Enzymology" (W9) and Prof. Karl Gruber, Dr. Gustav Oberdorfer and Dr. Georg Steinkellner for their assistance using the modelling tool YASARA.

1.8 References

[1] Macheroux P, Kappes B, Ealick SE. Flavogenomics - a genomic and structural view of flavin-dependent proteins. FEBS J 2011;278:2625–34.
[2] Singer TP, Kearney EB. Biochem Biophys Acta 1956;15:151.
[3] Salach J, Walker WH, Singer TP, et al. Studies on succinate dehydrogenase. Site of attachment of the covalently-bound flavin to the peptide chain. Eur J Biochem 1972;26:267–78.
[4] Kearney EB, Salach JI, Walker WH, et al. The covalently-bound flavin of hepatic monoamine oxidase. 1. Isolation and sequence of a flavin peptide and evidence for binding at the 8alpha position. Eur J Biochem 1971;24:321–7.
[5] McIntire W, Edmondson DE, Hopper DJ, Singer TP. 8α-(O-tyrosyl)flavin adenine dinucleotide, the prosthetic group of bacterial *p*-cresol methylhydroxylase. Biochemistry 1981; 20:3068–75.

[6] Mewies M, McIntire WS, Scrutton NS. Covalent attachment of flavin adenine dinucleotide (FAD) and flavin mononucleotide (FMN) to enzymes: The current state of affairs. Protein Science 1998;7:7–20.

[7] Huang CH, Lai WL, Lee MH, et al. Crystal structure of glucooligosaccharide oxidase from *Acremonium strictum*: A novel flavinylation of 6-S-cysteinyl, 8α-N1-histidyl FAD. J Biol Chem 2005;280:38831–8.

[8] Heuts DPHM, Scrutton NS, McIntire WS, Fraaije MW. What's in a covalent bond?: On the role and formation of covalently bound flavin cofactors. FEBS J 2009;276:3405–27.

[9] Winkler A, Kutchan TM, Macheroux P. 6-S-cysteinylation of bi-covalently attached FAD in berberine bridge enzyme tunes the redox potential for optimal activity. J Biol Chem 2007;282:24437–43.

[10] Heuts DPHM, Winter RT, Damsma GE, Janssen DB, Fraaije MW. The role of double covalent flavin binding in chito-oligosaccharide oxidase from *Fusarium graminearum*. Biochem J 2008;413:175–83.

[11] Huang CH, Winkler A, Chen CL, et al. Functional roles of the 6-S-cysteinyl, 8α-N1-histidyl FAD in glucooligosaccharide oxidase from *Acremonium strictum*. J Biol Chem 2008;283:30990–6.

[12] Winkler A, Motz K, Riedl S, Puhl M, Macheroux P, Gruber K. Structural and mechanistic studies reveal the functional role of bicovalent flavinylation in berberine bridge enzyme. J Biol Chem 2009;284:19993–20001.

[13] Kutchan TM. Molecular genetics of plant alkaloid biosynthesis. In: Cordell GA, ed. The Alkaloids: Chemistry and Biology. San Diego, CA, USA: Academic Press, 1998;50:257–316.

[14] Winkler A, Łyskowski A, Riedl S, et al. A concerted mechanism for berberine bridge enzyme. Nat Chem Biol 2008;4:739–41.

[15] Liscombe DK, Facchini PJ. Evolutionary and cellular webs in benzylisoquinoline alkaloid biosynthesis. Curr Opin Biotechnol 2008;19:173–80.

[16] Rink E, Bohm H. Conversion of reticuline into scoulerine by a cell free preparation from *Macleaya microcarpa* cell suspension cultures. FEBS Lett 1975;49:396–9.

[17] Steffens P, Nagakura N, Zenk MH. Purification and characterization of the berberine bridge enzyme from *Berberis beaniana* cell cultures. Phytochem 1985;24:2577–83.

[18] Kutchan TM, Dirtrich H. Characterization and mechanism of the berberine bridge enzyme, a covalently flavinylated oxidase of benzophenanthridine alkaloid biosynthesis in plants. J Biol Chem 1995;270:24475–81.

[19] Ghisla S, Kenney WC, Knappe WR, McIntire W, Singer TP. Chemical synthesis and some properties of 6-substituted flavins. Biochemistry 1980;19:2537–44.

[20] Winkler A, Hartner F, Kutchan TM, Glieder A, Macheroux P. Biochemical evidence that berberine bridge enzyme belongs to a novel family of flavoproteins containing a bi-covalently attached FAD cofactor. J Biol Chem 2006;281:21276–85.

[21] Massey V, Ghisla S, Moore EG. 8-Mercaptoflavins as active site probes of flavoenzymes. J Biol Chem 1979;254:9640–50.

[22] Fitzpatrick PF, Massey V. The reaction of 8-mercaptoflavins and flavoproteins with sulfite. Evidence for the role of an active site arginine in D-amino acid oxidase. J Biol Chem 1983;258:9700–5.

[23] Massey V, Muller F, Feldberg R, et al. The reactivity of flavoproteins with sulfite. Possible relevance to the problem of oxygen reactivity. J Biol Chem 1969;244:3999–4006.

[24] Massey V, Hemmerich P. Active-site probes of flavoproteins. Biochem Soc Trans 1980; 8:246–57.

[25] Fraaije MW, Mattevi A. Flavoenzymes: diverse catalysts with recurrent features. Trends Biochem Sci 2000;25:126–32.

[26] Macheroux P, Kieweg V, Massey V, Soderlind E, Stenberg K, Lindqvist Y. Role of tyrosine 129 in the active site of spinach glycolate oxidase. Eur J Biochem 1993;213:1047–54.

[27] Wagner MA, Trickey P, Che ZW, Mathews FS, Jorns MS. Monomeric sarcosine oxidase: 1. Flavin reactivity and active site binding determinants. Biochemistry 2000;39:8813–24.
[28] Ohta-Fukuyama M, Miyake Y, Emi S, Yamano T. Identification and properties of the prosthetic group of choline oxidase from *Alcaligenes* sp. J Biochem 1980;88:197–203.
[29] Gomez-Moreno C, Choy M, Edmondson DE. Purification and properties of the bacterial flavoprotein: thiamin dehydrogenase. J Biol Chem 1979;254:7630–5.
[30] Bruhmuller M, Mohler H, Decker K. Covalently bound flavin in D-6-hydroxynicotine oxidase from *Arthrobacter oxidans*. Z Naturforsch 1972;27:1073–4.
[31] Gadda G, Wels G, Pollegioni L, et al. Characterization of cholesterol oxidase from *Streptomyces hygroscopicus* and *Brevibacterium sterolicum*. Eur J Biochem 1997;250:369–76.
[32] Muller F, Massey V. Flavin-sulfite complexes and their structures. J Biol Chem 1969; 244:4007–16.
[33] Gadda G, Fitzpatrick PF. Biochemical and physical characterization of the active FAD-containing form of nitroalkane oxidase from *Fusarium oxysporum*. Biochemistry 1998; 37:6154–64.
[34] Ghisla S, Massey V, Yagi K. Preparation and some properties of 6-substituted flavins as active site probes for flavin enzymes. Biochemistry 1986;25:3282–9.
[35] Ghanem M, Gadda G. Effects of reversing the protein positive charge in the proximity of the flavin N(1) locus of choline oxidase. Biochemistry 2006;45:3437–47.
[36] Hecht HJ, Kalisz HM, Hendle J, Schmid RD, Schomburg D. Crystal structure of glucose oxidase from *Aspergillus niger* refined at 2.3 A resolution. J Mol Biol 1993;229:153–72.
[37] Wohlfahrt G, Witt S, Hendle J, Schomburg D, Kalisz HM, Hecht HJ. 1.8 and 1.9 A resolution structures of the *Penicillium amagasakiense* and *Aspergillus niger* glucose oxidases as a basis for modelling substrate complexes. Acta Crystallogr D Biol Crystallogr 1999;55:969–77.
[38] Lindqvist Y, Branden CI. The active site of spinach glycolate oxidase. J Biol Chem 1989;264:3624–8.
[39] Trickey P, Wagner MA, Jorns MS, Mathews FS. Monomeric sarcosine oxidase: Structure of a covalently flavinylated amine oxidizing enzyme. Structure 1999;7:331–45.
[40] Muh U, Massey V, Williams Jr. CH. Lactate monooxygenase. I. Expression of the mycobacterial gene in *Escherichia coli* and site-directed mutagenesis of lysine 266. J Biol Chem 1994;269:7982–8.
[41] Xia ZX, Mathews FS. Molecular structure of flavocytochrome b2 at 2.4 A resolution. J Mol Biol 1990;212:837–63.
[42] Efimov I, Cronin CN, Bergmann DJ, Kuusk V, McIntire WS. Insight into covalent flavinylation and catalysis from redox, spectral, and kinetic analyses of the R474K mutant of the flavoprotein subunit of *p*-cresol methylhydroxylase. Biochemistry 2004;43:6138–48.
[43] Vrielink A, Lloyd LF, Blow DM. Crystal structure of cholesterol oxidase from *Brevibacterium sterolicum* refined at 1.8 A resolution. J Mol Biol 1991;219:533–54.
[44] Hallberg BM, Henriksson G, Pettersson G, Divne C. Crystal structure of the flavoprotein domain of the extracellular flavocytochrome cellobiose dehydrogenase. J Mol Biol 2002; 315:421–34.
[45] Mattevi A, Vanoni MA, Todone F, et al. Crystal structure of D-amino acid oxidase: a case of active site mirror-image convergent evolution with flavocytochrome b2. Proc Natl Acad Sci U S A 1996;93:7496–501.
[46] Macheroux P, Massey V, Thiele DJ, Volokita M. Expression of spinach glycolate oxidase in *Saccharomyces cerevisiae*: purification and characterization. Biochemistry 1991;30: 4612–9.
[47] Schrittwieser JH, Resch V, Sattler JH, et al. Biocatalytic enantioselective oxidative C-C coupling by aerobic C-H activation. Angew Chem Int Ed 2011;50:1068–71.
[48] Schrittwieser JH, Resch V, Wallner S, et al. Biocatalytic organic synthesis of optically pure (*S*)-scoulerine and berbine and benzylisoquinoline alkaloids. J Org Chem 2011;76: 6703–14.

[49] Bentley KW. The isoquinoline alkaloids. Amsterdam, NL: Harwood Academic Publishers, 1998.

[50] Eisenreich WJ, Hofner G, Bracher F. Alkaloids from *Croton flavens* L. and their affinities to GABA-receptors. Nat Prod Res 2003;17:437–40.

[51] Gao JM, Liu WT, Li ML, Liu HW, Zhang XC, Li ZX. Preparation and structural elucidation of (–)-tetrahydroberberine-(+)-2,3-di(*p*-toluyl) tartaric acid complex. J Mol Struct 2008;892:466–9.

[52] Martin ML, Diaz MT, Montero MJ, Prieto P, San Roman L, Cortes D. Antispasmodic activity of benzylisoquinoline alkaloids analogous to papaverine. Planta Med 1993;59:63–7.

[53] Chulia S, Ivorra MD, Lugnier C, Vila E, Noguera MA, D'Ocon P. Mechanism of the cardiovascular activity of laudanosine: comparison with papaverine and other benzylisoquinolines. Br J Pharmacol 1994;113:1377–85.

[54] Kashiwada Y, Aoshima A, Ikeshiro Y, et al. Anti-HIV benzylisoquinoline alkaloids and flavonoids from the leaves of *Nelumbo nucifera*, and structure-activity correlations with related alkaloids. Bioorg Med Chem 2005;13:443–8.

[55] Yamahara J, Konoshima T, Sakakibara Y, Ishiguro M, Sawada T. Central depressant action of tetrahydroberberine and its derivatives. Chem Pharm Bull 1976;24:1909–12.

[56] Li J, Jin G, Shen J, Ji R. l-Chloroscoulerine mesylate. Drugs Future 2006;31:379–84.

[57] Barton DH, Kirby GW, Steglich W, et al. Investigations on the biosynthesis of morphine Alkaloids. J Chem Soc 1965;65:2423–38.

[58] Chrzanowska M, Rozwadowska MD. Asymmetric synthesis of isoquinoline alkaloids. Chem Rev 2004;104:3341–70.

[59] Matulenko MA, Meyers AI. Total synthesis of (-)-tetrahydropalmatine via chiral formamidine carbanions: unexpected behavior with certain ortho-substituted electrophiles. J Org Chem 1996;61:573–80.

[60] Resch V, Schrittwieser JH, Wallner S, Macheroux P, Kroutil W. Biocatalytic oxidative C-C bond formation catalysed by the berberine bridge enzyme: Optimal reaction conditions. Adv Synth Catal 2011;353:2377–83.

[61] Shoyama Y, Takeuchi A, Taura F, et al. Crystallization of Δ1-tetrahydrocannabinolic acid (THCA) synthase from *Cannabis sativa.* Acta Crystallogr Sect F Struct Biol Cryst Commun 2005;61:799–801.

[62] Carter CJ, Thornburg RW. Tobacco nectarin V is a flavin-containing berberine bridge enzyme-like protein with glucose oxidase activity. Plant Physiol 2004;134:460–9.

[63] Nandy A, Petersen A, Wald M, et al. Primary structure, recombinant expression, and molecular characterization of Phl p 4, a major allergen of timothy grass (*Phleum pratense*). Biochem Biophys Res Commun 2005;337:563–70.

[64] Gesell A, Chávez MLD, Kramell R, Piotrowski M, Macheroux P, Kutchan TM. Heterologous expression of two FAD-dependent oxidases with (*S*)-tetrahydroprotoberberine oxidase activity from *Argemone mexicana* and *Berberis wilsoniae* in insect cells. Planta 2011;233:1185–97.

[65] Dewitt AM, Andersson K, Peltre G, Lidholm J. Cloning, expression and immunological characterization of full-length timothy grass pollen allergen Phl p 4, a berberine bridge enzyme-like protein with homology to celery allergen Api g 5. Clin Exp Allergy 2006;36:77–86.

[66] Sirikantaramas S, Morimoto S, Shoyama Y, et al. The gene controlling marijuana psychoactivity. Molecular cloning and heterologous expression of Δ1-tetrahydrocannabinolic acid synthase from *Cannabis sativa* L. J Biol Chem 2004;279:39767–74.

[67] Carter C, Thornburg RW. Is the nectar redox cycle a floral defense against microbial attack? Trends Plant Sci 2004;9:320–4.

[68] Custers JH, Harrison SJ, Sela-Buurlage MB, et al. Isolation and characterisation of a class of carbohydrate oxidases from higher plants, with a role in active defence. Plant J 2004; 39:147–60.

[69] Hansen OC, Stougaard P. Hexose oxidase from the red alga *Chondrus crispus.* Purification, molecular cloning, and expression in *Pichia pastoris.* J Biol Chem 1997;272:11581–7.

[70] Rand T, Qvist KB, Walter CP, Poulsen CH. Characterization of the flavin association in hexose oxidase from *Chondrus crispus*. FEBS J 2006;273:2693–703.
[71] Poulsen C, Høstrup PB. Purification and characterization of a hexose oxidase with excellent strengthening effects in bread. Cereal Chem 1998;75:51–7.
[72] Van der Lugt JP. Evaluation of methods for chemical and biological carbohydrate oxidation. Delft, NL: Delft University of Technology 1998:1–148.
[73] Liaw SH, Lee DY, Chow LP, Lau GX, Su SN. Structural characterization of the 60-kDa Bermuda grass pollen isoallergens, a covalent flavoprotein. Biochem Biophys Res Commun 2001;280:738–43.
[74] Steffens P, Nagakura N, Zenk MH. The berberine bridge forming enzyme in tetrahydroprotoberberine biosynthesis. Tetrahedron Lett 1984;25:951–2.
[75] Amann M, Nagakura N, Zenk MH. (*S*)-tetrahydroprotoberberine oxidase the final enzyme in protoberberine biosynthesis. Tetrahedron Lett 1984;25:953–4.
[76] Amann M, Nagakura N, Zenk MH. Purification and properties of (*S*)-tetrahydroprotoberberine oxidase from suspension-cultured cells of *Berberis wilsoniae*. Eur J Biochem 1988; 175:17–25.
[77] Chou WM, Kutchan TM. Enzymatic oxidations in the biosynthesis of complex alkaloids. Plant J 1998;15:289–300.
[78] Taura F, Morimoto S, Shoyama Y, Mechoulam R. First direct evidence for the mechanism of Δ1-tetrahydrocannabinolic acid biosynthesis. J Am Chem Soc 1995;117 (38):9766–7.
[79] Turner CE, Elsohly MA, Boeren EG. Constituents of *Cannabis sativa* L. XVII. A review of the natural constituents. J Nat Prod 1980;43:169–234.
[80] Pertwee RG. Cannabinoid pharmacology: the first 66 years. Br J Pharmacol 2006;147 Suppl 1:S163–71.
[81] Pertwee RG. Pharmacological actions of cannabinoids. Handb Exp Pharmacol 2005; 168:1–51.
[82] Pertwee RG. The pharmacology of cannabinoid receptors and their ligands: an overview. Int J Obes 2006;30 Suppl 1:13–8.
[83] Pertwee RG. Cannabinoids and multiple sclerosis. Mol Neurobiol 2007;36:45–59.
[84] Grotenhermen F. Cannabinoids. Curr Drug Targets CNS Neurol Disord 2005;4:507–30.
[85] Guzman M. Cannabinoids: potential anticancer agents. Nat Rev Cancer 2003;3:745–55.
[86] Baker D, Pryce G, Giovannoni G, Thompson AJ. The therapeutic potential of cannabis. Lancet Neurol 2003;2:291–8.
[87] Taura F, Dono E, Sirikantaramas S, Yoshimura K, Shoyama Y, Morimoto S. Production of Δ1-tetrahydrocannabinolic acid by the biosynthetic enzyme secreted from transgenic *Pichia pastoris*. Biochem Biophys Res Commun 2007;361:675–80.
[88] Shoyama Y, Tamada T, Kurihara K, et al. Structure and function of Δ1-tetrahydrocannabinolic acid (THCA) synthase, the enzyme controlling the psychoactivity of *Cannabis sativa*. J Mol Biol 2012;423:96–105.
[89] Carter C, Graham RA, Thornburg RW. Nectarin I is a novel, soluble germin-like protein expressed in the nectar of *Nicotiana* sp. Plant Mol Biol 1999;41:207–16.
[90] Chamnongpol S, Willekens H, Moeder W, et al. Defense activation and enhanced pathogen tolerance induced by H_2O_2 in transgenic tobacco. Proc Natl Acad Sci U S A 1998; 95:5818–23.
[91] Bolwell GP, Wojtaszek P. Mechanisms for the generation of reactive oxygen species in plant defence - A broad perspective. Physiol Mol Plant Pathol 1997;51 (6):347–66.
[92] Wu G, Shortt BJ, Lawrence EB, Levine EB, Fitzsimmons KC, Shah DM. Disease resistance conferred by expression of a gene encoding H_2O_2-generating glucose oxidase in transgenic potato plants. Plant Cell 1995;7:1357–68.
[93] Wojtaszek P. Oxidative burst: an early plant response to pathogen infection. Biochem J 1997;322:681–92.
[94] Liaw SH, Lee DY, Yang SY, Su SN. Crystallization and preliminary diffraction data of 60-kDa glycosylated pollen isoallergens from Bermuda grass. J Struct Biol 1999;127:83–7.

[95] Ganglberger E, Radauer C, Grimm R, et al. N-terminal sequences of high molecular weight allergens from celery tuber. Clin Exp Allergy 2000;30:566–70.
[96] Bublin M, Radauer C, Wilson IB, et al. Cross-reactive N-glycans of Api g 5, a high molecular weight glycoprotein allergen from celery, are required for immunoglobulin E binding and activation of effector cells from allergic patients. FASEB J 2003;17:1697–9.
[97] Chardin H, Mayer C, Senechal H, Tepfer M, Desvaux FX, Peltre G. Characterization of high-molecular-mass allergens in oilseed rape pollen. Int Arch Allergy Immunol 2001; 125:128–34.
[98] Fischer S, Grote M, Fahlbusch B, Müller WD, Kraft D, Valenta R. Characterization of Phl p 4, a major timothy grass (*Phleum pratense*) pollen allergen. J Allergy Clin Immunol 1996;98:189–98.
[99] Facchini PJ, Bird DA, St-Pierre B. Can *Arabidopsis* make complex alkaloids? Trends Plant Sci 2004;9:116–22.
[100] Heuts DPHM, Janssen DB, Fraaije MW. Changing the substrate specificity of a chitooligosaccharide oxidase from *Fusarium graminearum* by model-inspired site-directed mutagenesis. FEBS Lett 2007;581:4905–9.
[101] Frick S, Kutchan TM. Molecular cloning and functional expression of *O*-methyltransferases common to isoquinoline alkaloid and phenylpropanoid biosynthesis. Plant J 1999; 17:329–39.
[102] Frey M, Chomet P, Glawischnig E, et al. Analysis of a chemical plant defense mechanism in grasses. Science 1997;277:696–9.
[103] Tsuji J, Jackson EP, Gage DA, Hammerschmidt R, Somerville SC. Phytoalexin accumulation in *Arabidopsis thaliana* during the hypersensitive reaction to *Pseudomonas syringae* pv *syringae*. Plant Physiol 1992;98:1304–9.
[104] Tohge T, Yonekura-Sakakibara K, Niida R, Watanabe-Takahashi A, Saito K. Phytochemical genomics in *Arabidopsis thaliana*: A case study for functional identification of flavonoid biosynthesis genes. Pure Appl Chem 2007;79:811–23.
[105] Chen F, Tholl D, D'Auria JC, Farooq A, Pichersky E, Gershenzon J. Biosynthesis and emission of terpenoid volatiles from *Arabidopsis flowers*. Plant Cell 2003;15:481–94.
[106] Aharoni A, Giri AP, Deuerlein S, et al. Terpenoid metabolism in wild-type and transgenic *Arabidopsis* plants. Plant Cell 2003;15:2866–84.
[107] Li L, He Z, Pandey GK, Tsuchiya T, Luan S. Functional cloning and characterization of a plant efflux carrier for multidrug and heavy metal detoxification. J Biol Chem 2002;277:5360–8.
[108] Lee MH, Lai WL, Lin SF, Hsu CS, Liaw SH, Tsai YC. Structural characterization of glucooligosaccharide oxidase from *Acremonium strictum*. Appl Environ Microbiol 2005;71:8881–7.
[109] Larkin MA, Blackshields G, Brown NP, et al. Clustal W and Clustal X version 2.0. Bioinformatics 2007;23:2947–8.
[110] Goujon M, McWilliam H, Li W, et al. A new bioinformatics analysis tools framework at EMBL-EBI. Nucleic Acids Res 2010;38:695–9.
[111] Waterhouse AM, Procter JB, Martin DM, Clamp M, Barton GJ. Jalview Version 2 - a multiple sequence alignment editor and analysis workbench. Bioinformatics 2009;25:1189–91.
[112] Lin SF, Yang TY, Inukai T, Yamasaki M, Tsai YC. Purification and characterization of a novel glucooligosaccharide oxidase from *Acremonium strictum* T1. Biochim Biophys Acta 1991;1118:41–7.
[113] Lin SF, Hu HM, Inukal T, Tsai YC. Production of novel oligosaccharide oxidase by wheat bran solid-state fermentation. Biotechnol Adv 1993;11:417–27.
[114] Fan Z, Oguntimein GB, Reilly PJ. Characterization of kinetics and thermostability of *Acremonium strictum* glucooligosaccharide oxidase. Biotechnol Bioeng 2000;68:231–7.
[115] Foumani M, Vuong TV, Master ER. Altered substrate specificity of the gluco-oligosaccharide oxidase from *Acremonium strictum*. Biotechnol Bioeng 2011;108:2261–9.
[116] Alexeev I, Sultana A, Mäntsälä P, Niemi J, Schneider G. Aclacinomycin oxidoreductase (AknOx) from the biosynthetic pathway of the antibiotic aclacinomycin is an unusual flavoenzyme with a dual active site. Proc Natl Acad Sci U S A 2007;104:6170–5.

[117] Carlson JC, Li S, Gunatilleke SS, et al. Tirandamycin biosynthesis is mediated by co-dependent oxidative enzymes. Nature Chem 2011;3:628–33.
[118] Li YS, Ho JY, Huang CC, et al. A unique flavin mononucleotide-linked primary alcohol oxidase for glycopeptide A40926 maturation. J Am Chem Soc 2007;129:13384–5.
[119] Mo X, Huang H, Ma J, et al. Characterization of TrdL as a 10-hydroxy dehydrogenase and generation of new analogues from a tirandamycin biosynthetic pathway. Org Lett 2011;13:2212–5.
[120] Noinaj N, Bosserman MA, Schickli MA, et al. The crystal structure and mechanism of an unusual oxidoreductase, GilR, involved in gilvocarcin V biosynthesis. J Biol Chem 2011;286:23533–43.
[121] Abad S, Nahalka J, Bergler G, et al. Stepwise engineering of a *Pichia pastoris* D-amino acid oxidase whole cell catalyst. Microb Cell Fact 2010;9:24.
[122] Kharel MK, Pahari P, Lian H, Rohr J. GilR, an unusual lactone-forming enzyme involved in gilvocarcin biosynthesis. ChemBioChem 2009;10:1305–8.

2 PutA and proline metabolism

John J. Tanner and Donald F. Becker

Abstract

Proline metabolism has a central role in carbon and nitrogen flux as well as a profound influence on cellular redox balance, apoptosis, and survival. Proline is catabolized to glutamate in two sequential steps by the enzyme actions of proline dehydrogenase (PRODH) and Δ^1-pyrroline-5-carboxylate (P5C) dehydrogenase (P5CDH). PRODH utilizes a flavin cofactor to perform the initial step of proline oxidation to generate P5C. P5C then undergoes a non-enzymatic hydrolysis to form γ-glutamate semialdehyde (GSA), which is oxidized to glutamate by NAD^+-dependent P5CDH. In Gram-negative bacteria, PRODH and P5CDH are combined into a single protein known as proline utilization A (PutA), which facilitates channeling of the P5C/GSA intermediate between PRODH and P5CDH domains. A sub-family of PutA proteins are also endowed with DNA binding capacity and regulate *put* gene transcription necessitating functional switching between DNA-binding and membrane-associated enzyme activity. Structural and biochemical studies have provided novel insights into the molecular details of PutA enzyme activity, substrate channeling, PutA-DNA interactions, and redox-linked conformational changes. Important questions, however, still remain such as identifying the binding sites on PutA for ubiquinone and the membrane, examining the coordination of active sites during channeling, and mapping interaction networks that link flavin reduction to functional switching in PutA.

2.1 Importance of proline metabolism

L-proline has multifaceted roles in primary carbon and nitrogen metabolism, osmotic and oxidative stress protection, protein chaperoning, cell signaling, programmed cell death, and nutrient adaptation and survival [1–4]. As a carbon and nitrogen source, L-proline supports the growth of bacteria such as *Escherichia coli, Pseudomonas putida, Bradyrhizobium japonicum,* and *Helicobacter pylori* [5–9]. A unique function of proline is its cytoprotective role observed in a broad range of organisms [3,10–12]. Proline is a well-known osmoprotectant and has been shown to act as a thermoprotectant by diminishing protein aggregation during heat stress [13–15]. Proline also combats oxidative stress in various organisms [3,11,16]. In the fungal pathogen, *Colletotrichum trifolii,* proline has been reported to increase catalase expression leading to enhanced resistance against oxidative stress [16]. In mammalian cells, proline minimizes apoptosis induced by different oxidative stress agents [17]. The mechanisms by which proline protects various organisms against stress are not fully known, but protein chaperoning, direct scavenging of reactive oxygen species (ROS), enhanced activity of antioxidant enzymes, and maintenance of key redox-active molecules such as glutathione and NADPH/NADP+ have been postulated to be involved [3,13,16–18].

Proline metabolism involves the interconversion of proline and glutamate by four enzymes (►Fig. 2.1). In certain organisms, pairs of these enzyme activities are combined into a bifunctional enzyme. For example, as indicated above proline dehydrogenase (PRODH) and Δ^1-pyrroline-5-carboxylate (P5C) dehydrogenase (P5CDH) are combined in some bacteria into the single protein known as proline utilization A (PutA) [19]. On the proline biosynthesis side, P5C synthetase (P5CS) is a bifunctional enzyme in plants and animals with γ-glutamyl kinase (GK) and γ-glutamyl phosphate reductase (GPR) activities, whereas in lower eukaryotes such as yeast and in prokaryotes, GK and GPR are separate enzymes [19].

PRODH is the lone flavin enzyme in the proline metabolic pathway and catalyzes the first step of proline oxidation. PRODH is a peripherally membrane-associated enzyme, and in eukaryotes is localized in the inner mitochondrial membrane [20,21]. The oxidation of proline is directly coupled to the electron transport chain via reduction of membrane-bound ubiquinone [22,23]. Glutamate generated by proline oxidation enters the tricarboxylic acid cycle after enzymatic conversion to α-ketoglutarate. Thus, proline catabolism provides important energy for the cell, particularly under nutrient depletion conditions. Some pathogens, including *Helicobacter pylori* [24] and procyclic trypanosomatids [25], have evolved to occupy proline-rich environmental niches and use proline as a major source of energy.

In addition to contributing to energy needs, proline metabolism in eukaryotes has been implicated in ROS formation, which has profound effects on oxidation-reduction homeostasis and cell survival. In humans, increased expression of PRODH, also known as proline oxidase (POX), causes significant ROS formation, which activates intrinsic

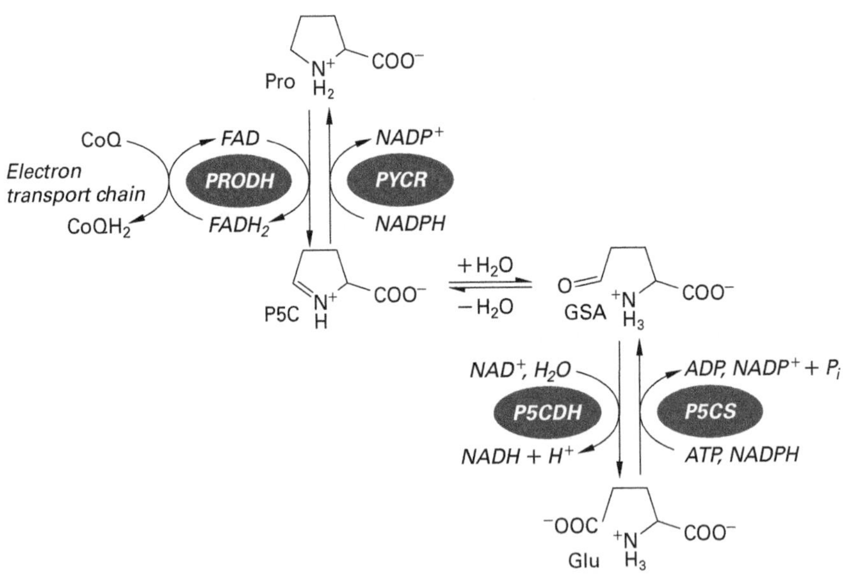

Fig. 2.1: The reactions of proline metabolism.

and extrinsic apoptotic pathways [26,27]. The gene encoding PRODH (*PRODH1*, NCBI RefSeq number NM_016335) is inducible by p53 with PRODH activity helping to prevent metastasis and tumor cell growth [28,29]. Paradoxically, PRODH is also capable of producing transient ROS that induces survival pathways [30]. In *Caenorhabditis elegans*, PRODH has been reported to generate transient ROS signals and induce the worm homologues of p38 MAP kinase and Nrf2, leading to increased expression of antioxidant enzymes and lifespan [30]. Thus, PRODH appears to play a pivotal role in cell signaling processes that impact cell survival and cell death.

2.2 Proline utilization A (PutA) proteins

As mentioned above, PRODH and P5CDH are combined into the single polypeptide known as proline utilization A (PutA), which refers to its key role in using environmental proline as a source of energy.

The available information on amino acid sequences and three-dimensional structures of PutAs and PutA domains suggest that there are five different subfamilies of PutA. Multiple sequence alignments reveal a phylogenetic tree having two main PutA branches (►Fig. 2.1, Branch 1 PutAs and Branch 2 PutAs) in addition to branches corresponding to organisms in which PRODH and P5CDH are separate enzymes (►Fig. 2.1, blue area). Branch 1 includes PutAs from α-, β-, and γ-proteobacteria. Branch 2 consists of PutAs from Gram-negative cyanobacteria, δ- and ε-proteobacteria, and corynebacterium. Thus, PutAs are primarily restricted to Gram-negative bacteria. Within Branch 1, the pairwise amino acid sequence identities exceed 38%, which indicates a closely related group of proteins. In contrast, the pairwise sequence identity within Branch 2 can be as low as 23%, suggesting that Branch 2 is more diverse than Branch 1. Between Branches 1 and 2, the pairwise sequence identities are typically less than 30%, which indicates that Branches 1 and 2 PutAs exhibit significant differences. Within each PutA branch, a further sub-classification based on the functional domains present in the polypeptide chain is evident, which leads to further division of PutAs into subfamilies denoted 1A, 1B, 1C, 2A, and 2B. Branch 1 PutAs can be subdivided into short bifunctional (or minimalist) PutAs (sub-family 1A), long bifunctional PutAs (1B), and trifunctional PutAs (1C). Within Branch 2, we see only short (2A) and long bifunctional PutAs (2B). Some of the defining characteristics of these PutA sub-families are listed in ►Tab. 2.1.

Minimalist PutAs (sub-families 1A and 2A), as the name implies, are the simplest type of PutA. They contain just two functional domains, the PRODH and P5CDH catalytic domains, and at 980–1100 residues, they are the smallest of all PutAs (►Fig. 2.3). The best characterized minimalist PutA is *B. japonicum* PutA (BjPutA), which belongs to Branch 1. The crystal structure of BjPutA has been determined and will be described below. Other members of the Branch 1A sub-family that are currently under investigation include the PutAs from *Rhodopseudomonas palustris* (1004 residues, NCBI RefSeq number NP_946926.1) and *Legionella pneumophila* (LpPutA, 1054 residues, YP_095723.1). Minimalist PutAs are also found in Branch 2 (subfamily 2A). Representative Branch 2 minimalist PutAs include the enzymes from *Bdellovibrio bacteriovorus* (982 residues, NP_968157.1), *Geobacter sulfurreducens* PCA (GsPutA, 1004 residues, NP_954435.1), and *Desulfovibrio vulgaris* (1006 residues, YP_012527.1).

Tab. 2.1: Defining characteristics of PutA sub-families

Sub-family	Phylogenetic branch	Functional designation	Functional domains	Representative members	Oligomeric state
1A	1	Bifunctional	PRODH, P5CDH	BjPutA, LpPutA	Tetramer[1], dimer[2]
1B	1	Bifunctional	PRODH, P5CDH, CTD	RcPutA	Monomer[3]
1C	1	Trifunctional	RHH, PRODH, P5CDH, CTD	EcPutA	Dimer[4]
2A	2	Bifunctional	PRODH, P5CDH	GsPutA	Dimer[5]
2B	2	Bifunctional	PRODH, P5CDH, CTD	HpPutA	Dimer[3]

[1]Determined from SAXS [38]
[2]Determined from SAXS (T. Pemberton and J. J. Tanner, unpublished results)
[3]Determined from SAXS (M. Luo, R. Singh, and J. J. Tanner, unpublished results)
[4]Determined from SAXS [39]
[5]Determined from X-ray crystallography (H. Singh and J. J. Tanner, unpublished results)

Long bifunctional PutAs (sub-families 1B and 2B) are distinguished from minimalist PutAs by the presence of a ~175-residue domain of unknown function in the C-terminal third of the polypeptide chain (CTD, see ▶Fig. 2.3). The typical polypeptide chain length for long bifunctional PutAs is in the range of 1100–1200 residues. *Helicobacter* PutAs (sub-family 2B) are the best studied of the long bifunctional PutAs [31,32]. Representative examples of Branch 1 long bifunctional PutAs include the PutAs from *Rhodobacter capsulatus* (1127 residues, YP_003578784.1) and *Azoarcus* sp. BH72 PutA (1221 residues, GenBank CAL96369.1).

Trifunctional PutAs (Branch 1C) are distinguished from all other PutAs by the presence of a DNA-binding domain (▶Fig. 2.3). Note that trifunctional PutAs also have the CTD (▶Fig. 2.3). The DNA-binding domain exhibits the ribbon-helix-helix (RHH) fold and is located in the first 50 residues of the polypeptide chain. Trifunctional PutAs are limited to a small section of branch 1 of the PutA/PRODH/P5CDH tree (▶Fig. 2.2) and have polypeptide chain lengths of ~1300–1400. The best characterized representative of this sub-family is PutA from *E. coli* (EcPutA).

The RHH domain endows PutA with a third function that is not associated with bifunctional PutAs, namely, autogenous transcription repressor. Although we have described trifunctional PutAs last in our classification scheme here, in fact, they were the first PutAs to be studied and remain the best characterized of all PutAs. Roth's group discovered trifunctional PutAs in the late 1970s during their studies of proline utilization in *Salmonella typhimurium* [33]. Shortly thereafter, Wood's group described an analogous proline utilization system in *Escherichia coli* [9,34]. In these and related bacteria, the proline utilization (*put*) regulon consists of the genes encoding PutA and the proline transporter PutP, along with a regulatory region situated between the divergently transcribed *putA* and *putP* genes [35].

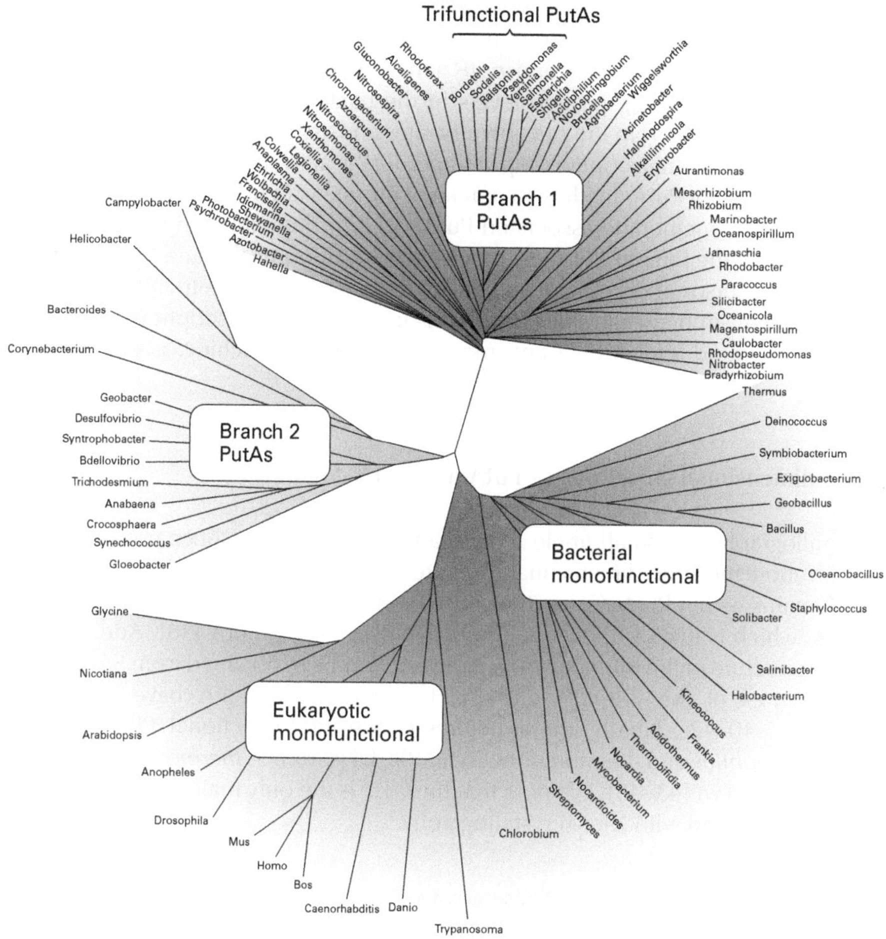

Fig. 2.2: Phylogenetic tree of PutA/PRODH/P5CDH proteins. Figure adapted from [46].

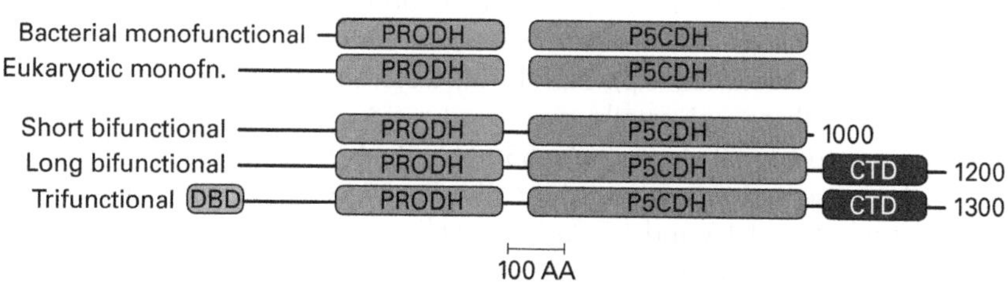

Fig. 2.3: Cartoon representations of the polypeptide chains of monofunctional proline catabolic enzymes and PutA proteins. Abbreviations used: DBD, DNA-binding domain; PRODH, proline dehydrogenase; P5CDH, 1-pyrroline-5-carboxylate dehydrogenase; CTD, C-terminal domain of unknown function.

Trifunctional PutAs are remarkable in that they link transcription and metabolism in response to an environmental cue (proline level). When proline levels are low, PutA represses transcription of the *putA* and *putP* genes by binding to five sites in the *put* intergenic region [35]. Rising proline levels in the bacterium's external environment induce PutA to dissociate from the *put* operator sites and associate with the inner bacterial membrane [36,37]. Transcription of *putA* and *putP* is activated by this change in the intracellular location of PutA, leading to uptake of proline via PutP and subsequent oxidation of proline by membrane-associated PutA [36,37]. The oxidation of proline supplies electrons, via the FAD in PutA, to the electron transport chain, and the glutamate produced is a precursor of α-ketoglutarate, a key intermediate in the tricarboxylic acid cycle. The mechanism by which PutA transforms from being a transcriptional repressor to a membrane-bound bifunctional enzyme, aka functional switching, is discussed in section 2.6.

2.3 Three-dimensional structures of PutA and PutA domains

X-ray crystallography and small-angle X-ray scattering (SAXS) have been used to obtain information about the three-dimensional structures of PutA domains and the arrangement of those domains within PutA. Currently, only one crystal structure of a full-length PutA is available, which is that of BjPutA, a minimalist (Branch 1A) PutA [38]. Additionally, the solution structure of the trifunctional *E. coli* PutA has been obtained from SAXS [39]. Crystal structures of the PRODH and RHH domains of EcPutA have also been determined [35,40–42]. Finally, the structures of the monofunctional PRODH from *Thermus thermophilus* and the monofunctional P5CDHs from *T. thermophilus*, human, and mouse are known [43–45]. We note that the CTD is the only major domain of PutA yet to be characterized with X-ray crystallography.

2.3.1 Structures of the catalytic domains of PutA

The PRODH domain is the flavoenzyme component of PutA. The PRODH active site is situated in a $(\alpha\beta)_8$ barrel (▶Fig. 2.4A). The non-covalently bound FAD is located at the C-termini of the strands of the barrel, with the *re* face of the isoalloxazine packed tightly against strands 4–6 of the barrel and the *si* face open to the substrate binding site.

The PRODH barrel is unusual in that the last helix of the fold, α8, sits above the barrel and perpendicular to the barrel axis rather than alongside β8 and parallel to the axis as in the classic triosphosphate isomerase barrel. For this reason, the PRODH fold is sometimes referred to as a *distorted* $(\alpha\beta)_8$ barrel. This distortion is essential for catalysis because α8 contributes three universally conserved residues that bind the substrate proline, corresponding to Tyr552, Arg555, Arg556 of EcPutA (▶Fig. 2.4B). The distorted barrel fold is likely conserved throughout the entire PutA/PRODH superfamily, based on its presence in both Branch1A and 1C PutAs as well as in monofunctional PRODH.

Insight into the basis of P5CDH activity has been obtained from crystal structures of BjPutA [38] as well as the monofunctional P5CDHs from human, mouse, and *T. thermophilus* [44,45]. These structures exhibit the classic aldehyde dehydrogenase (ALDH) fold, which consists of an N-terminal Rossmann NAD^+-binding domain, a

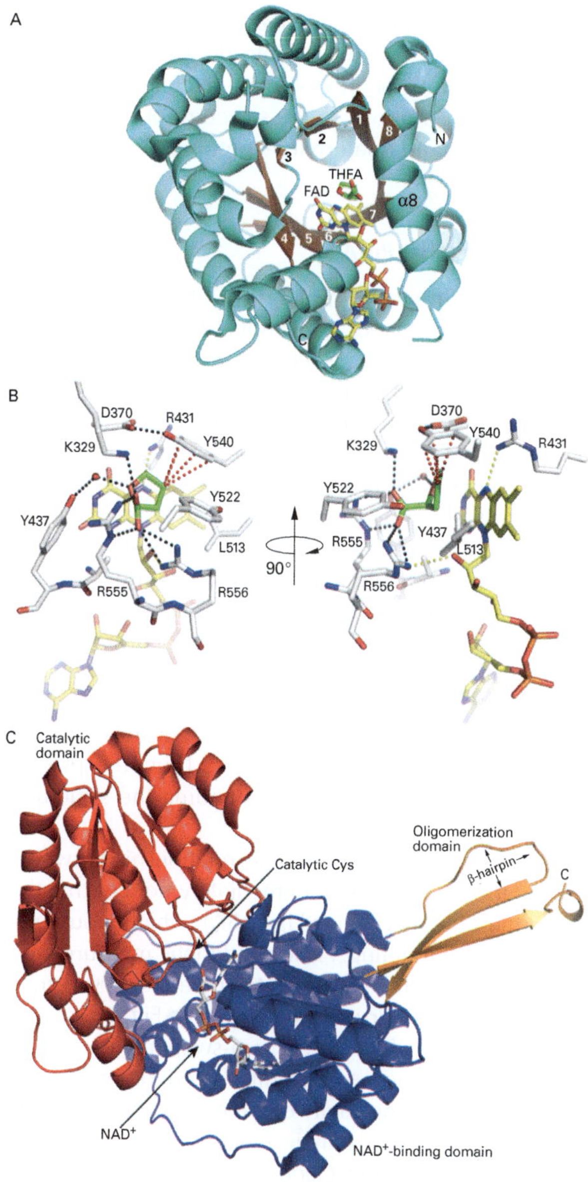

Fig. 2.4: The structures of the catalytic domains of PutA. (A) The PRODH barrel of EcPutA complexed with the proline analog THFA ([41], PDB code 1TIW). FAD and THFA are colored yellow and green, respectively. The β-strands of the barrel are labeled 1–8. The C-terminal α-helix of the fold is labeled α8. (B) Close-up view of the proline binding site of EcPutA. FAD and the proline analog THFA are colored yellow and green, respectively. Black dots denote enzyme-THFA electrostatic interactions. Red dots denote nonpolar contacts between the enzyme and THFA. Yellow dots indicate FAD-enzyme electrostatic interactions that have been implicated in functional switching. (C) The P5CDH part of BjPutA ([38], PDB code 3HAZ). The NAD^+-binding, catalytic, and oligomerization domains are colored blue, red, and orange, respectively. NAD^+ is shown in gray.

C-terminal α/β catalytic domain, and an oligomerization domain (►Fig. 2.4C). The oligomerization domain consists of a β-hairpin protruding from the NAD^+-binding domain and a β-strand at the C-terminus of the polypeptide chain (►Fig. 2.4C). As will be described below, PutAs exhibit variations in the structure of the β-hairpin, which results in different oligomeric states and quaternary structures.

2.3.2 Crystal structure of a minimalist PutA

The crystal structure of BjPutA, a Branch 1 minimalist PutA (sub-family 1A), shows how the two catalytic domains are oriented with respect to each other, which is relevant for understanding substrate channeling (section 2.4.2). The structure shows that PutAs are not simply a fusion of PRODH and P5CDH. In fact, the BjPutA protomer comprises seven distinct domains: arm, α-domain, PRODH barrel, linker, NAD^+-binding domain, P5CDH catalytic domain, and oligomerization domain (►Fig. 2.5A). The two active sites face each other and are separated by 41 Å. Furthermore, they are connected by a large, irregularly shaped internal cavity (silver surface in ►Fig. 2.5) that likely functions in substrate channeling. Although the presence of non-catalytic components in PutA was evident from sequence alignments, the function of these extra structural components was unknown prior to determination of the BjPutA structure. We now see that these ancillary domains – arm, α-domain, and linker – are structural components that help orient the two active sites toward each other and create the substrate-channeling cavity. These domains are conserved among all PutAs, which suggests that the spatial arrangement of the two active sites and the structure of the intervening cavity are also common to all PutAs. Thus, the protomer of the minimalist BjPutA is a good starting point for understanding the structures of more complex PutAs.

The BjPutA structure beautifully illustrates the intimate relationship between oligomerization and function. Viewing the protomer, one immediately notices that the substrate channeling cavity is open to the bulk medium (►Fig. 2.5A), which would result in loss of the intermediate and inefficient channeling. However, kinetic measurements indicate a significant degree of sequestration of the intermediate and efficient substrate channeling [38]. Dimerization solves this conundrum. As shown in ►Fig. 2.5B, the oligomerization domain of one BjPutA protomer forms a domain-swapped association with the other protomer to form a U-shaped dimer. As a result, the oligomerization domain of each protomer covers the substrate binding cavity of the other protomer (►Fig. 2.5B,C). Thus, dimerization is essential for proper catalytic function. The recently determined crystal structure of the Branch 2A PutA from *Geobacter sulfurreducens* PCA (GsPutA, PDB code 4F9I, H. Singh and J.J. Tanner, unpublished results) also exhibits this domain-swapped oligomeric structure, indicating that this mode of dimerization is shared by minimalist PutAs.

Curiously, two BjPutA dimers combine to form a ring-shaped tetramer, based on crystallographic and SAXS data [38]. The functional relevance of the tetramer is not obvious. Interestingly, preliminary data suggest that GsPutA and the short bifunctional PutA from *Legionella pneumophila* (LpPutA) both form the U-shaped minimalist dimer but do not assemble further into tetramers, indicating that the ring-shaped tetramer is not conserved by minimalist PutAs (►Tab. 2.1).

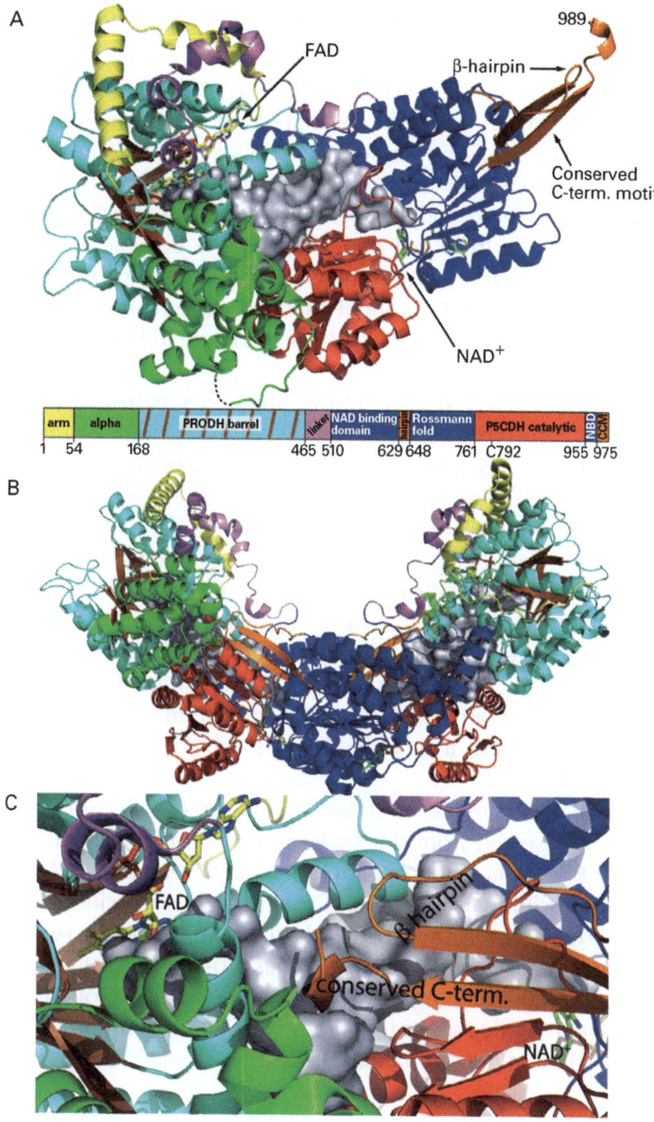

Fig. 2.5: Structure of the minimalist PutA, BjPutA (PDB code 3HAZ). (A) Structure of the protomer with the domains colored according to the domain diagram. FAD and NAD^+ are drawn as sticks in yellow and green, respectively. Abbreviations used in the domain diagram: NBD, NAD^+-binding domain; CCM, conserved C-terminal motif. (B) Structure of the domain-swapped dimer. (C) Close-up view of the dimer interface highlighting how oligomerization domain (orange) of one protomer seals the substrate-channeling cavity of the other protomer. This figure was adapted from a figure in [38] © 2010 by National Academy of Sciences and was originally published in the Journal of Biological Chemistry. R.K. Singh, J.D. Larson, W. Zhu, R.P. Rambo, G.L. Hura, D.F. Becker, and J.J. Tanner, Small-angle X-ray Scattering Studies of the Oligomeric State and Quaternary Structure of the Trifunctional Proline Utilization A (PutA) Flavoprotein from Escherichia coli. *J. Biol. Chem.* 2011;286:43144–43153 [39]. © the American Society for Biochemistry and Molecular Biology.

2.3.3 Solution structure of a trifunctional PutA and the role of the CTD

Although a crystal structure for trifunctional PutAs is not available, a low resolution solution structure for EcPutA, the archetypal Branch 1C PutA, has been determined by SAXS [39]. EcPutA forms a symmetric V-shaped dimer having dimensions of 205 Å × 85 Å × 55 Å (►Fig. 2.6). The particle consists of two large lobes connected by a 30-Å diameter cylinder. Domain deletion analysis showed that the RHH domain alone mediates dimerization. Rigid body modeling suggests that the catalytic domains reside in the large lobes, while the RHH dimeric assembly occupies the connecting cylinder. The SAXS model does not include the CTD because the structure of this domain is not available. Nevertheless, the model shows good agreement with the SAXS envelope (►Fig. 2.6) and SAXS profile [39]. DNA is predicted to bind in the groove of the V (►Fig. 2.6C).

The mode of dimerization observed in EcPutA is quite different from that of the short bifunctional BjPutA. This result reflects a peculiarity of the oligomerization domain in PutAs that have the CTD, *i.e.*, long bifunctional and trifunctional PutAs. Recall that the oligomerization domain of minimalist PutAs, and indeed all other ALDHs, consists of a β-hairpin protruding from the NAD^+-binding domain and a β-strand at the C-terminus of the polypeptide chain (►Fig. 2.4C, ►Fig. 2.5A,C). However, bioinformatics analyses suggest that the β-hairpin is truncated in long bifunctional and trifunctional PutAs [39,46]. Interestingly, it appears that the truncated β-hairpin always appears in combination with the CTD. Because the β-hairpin mediates dimerization in short bifunctional PutAs (►Fig. 2.5B,C), it is unlikely that CTD-containing PutAs will form the classic ALDH domain-swapped dimer that is seen in the minimalist PutAs. Indeed, this prediction has been confirmed for trifunctional PutAs, which dimerize via the DNA-binding domain (►Fig. 2.6).

The unexpected dimerization mode of EcPutA raised the question of how the substrate channeling cavity is sealed from the bulk medium in trifunctional PutAs [39]. As indicated in ►Fig. 2.6B, the cavity is open to the bulk solvent in the SAXS model of EcPutA. One intriguing possibility is that the CTD, which was not included in the SAXS model, functions as a lid that covers the cavity. This idea is supported by remote homology detection calculations indicating that the CTD has a β-hairpin that is analogous to that of ALDHs. This raises the intriguing possibility that the β-hairpin of the CTD forms an *intramolecular* lid that is analogous to the *intermolecular* lid of BjPutA. Validation of this hypothesis awaits structure determination of PutAs that contain the CTD.

2.4 Reaction kinetics of PutA

The reaction catalyzed by PRODH involves two half-reactions as shown in ►Fig. 2.1. In the reductive half-reaction, electrons are transferred to the flavin to generate reduced flavin and P5C. The oxidative half-reaction involves electron transfer from reduced flavin to ubiquinone. Reactivity with ubiquinone has been shown for PRODH from *Saccharomyces cerevisiae* (Put1p) and PutA [22,23]. Presumably, human PRODH also utilizes ubiquinone as the oxidizing substrate.

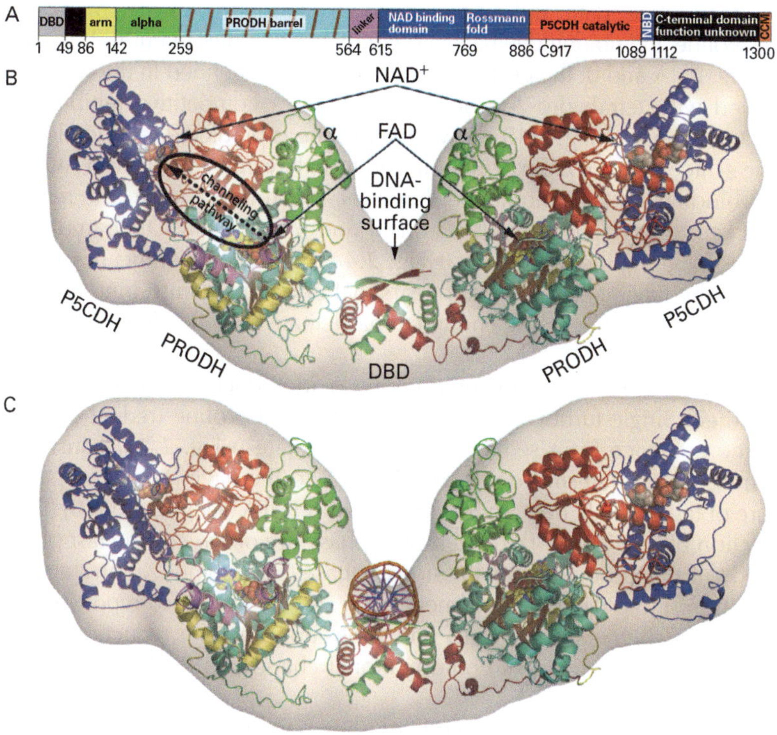

Fig. 2.6: Solution structure of EcPutA, a trifunctional PutA [39]. (A) Schematic diagram of the domain organization of EcPutA. Abbreviations used: DBD, DNA-binding domain; PRODH, proline dehydrogenase; NBD, NAD^+-binding domain; CCM, conserved C-terminal motif. The coloring of domains is that same as that of ►Fig. 2.5. (B) Rigid body model of EcPutA. The locations of the PRODH, α, and P5CDH domains are noted. The dashed oval shows the location of the putative substrate-channeling cavity. As modeled, the cavity is open to bulk solvent. It is hypothesized that the CTD, which has not been included in this model, covers the cavity to allow efficient substrate channeling. (C) Model of DNA bound to EcPutA. These figures were originally published in the Journal of Biological Chemistry. R.K. Singh, J.D. Larson, W. Zhu, R.P. Rambo, G.L. Hura, D.F. Becker, and J.J. Tanner, Small-angle X-ray Scattering Studies of the Oligomeric State and Quaternary Structure of the Trifunctional Proline Utilization A (PutA) Flavoprotein from Escherichia coli. *J. Biol. Chem.* 2011;286:43144–43153 [39]. © the American Society for Biochemistry and Molecular Biology.

2.4.1 Proline:ubiquinone oxidoreductase activity

The mechanism of proline oxidation has been studied in greatest detail with the PRODH domain of PutA from *E. coli* (EcPutA). A ping-pong mechanism was first described for PutA from *Salmonella typhimurium* and later confirmed in EcPutA by dead-end and product inhibition studies [23,47]. Using CoQ_1 as an electron acceptor, the kinetic parameters for PutA are $k_{cat} = 5.2\ s^{-1}$, $K_{m(pro)} = 42$ mM, and $K_{m(CoQ)} = 112\ \mu M$ [23]. The k_{cat} parameter is only slightly influenced by increased solvent viscosity, indicating that substrate binding and product release do not significantly limit the overall rate of the reaction [23].

Altogether, the steady-state data indicate a rapid equilibrium two-site ping-pong mechanism with proline and ubiquinone having separate binding sites in PutA [23].

As shown in ▶Fig. 2.4A and B, the proline binding site is well defined in the structure of PutA complexed with tetrahydro-2-furoic acid (THFA), a competitive inhibitor of PutA with respect to proline. ▶Fig. 2.4B shows that proline is predicted to bind closely to the FAD on the *si* face of the isoalloxazine ring and form key interactions with Arg555 and Arg556. The binding site for ubiquinone, however, remains unknown. The *re* face of the FAD is packed tightly against the β-strands 5 and 6 of the PRODH barrel with no apparent space for ubiquinone binding (▶Fig. 2.4A). Barring a significant conformational change in the flavin active site, it appears that ubiquinone must bind at a site more distant from the FAD but yet close enough to participate in facile electron transfer steps during catalytic turnover.

Rapid reaction kinetics of the proline:ubiquinone oxidoreductase activity of PutA have been performed. Single-turnover reactions of PutA with proline (reductive half-reaction) shows no evidence for a semiquinone intermediate (▶Fig. 2.7A) consistent with a hydride transfer mechanism from the proline C5 to the flavin N5 as proposed from the structure of the PRODH domain in complex with THFA (▶Fig. 2.4B) [40]. During proline reduction of the flavin a fast and a slow phase are observed. The fast phase has been assigned to flavin reduction and the slower phase to an isomerization step that occurs after reduction of the flavin. The isomerization step is thought to report on the reversible conformational changes that PutA undergoes upon proline reduction of the flavin [48]. Details of redox-dependent conformational changes in PutA are provided in section 2.6.2.

Interestingly, the reductive half-reaction has been shown to be reversible. In stopped-flow experiments, rapid mixing of reduced PutA with the product P5C leads to increased absorbance at 451 nm indicative of oxidized flavin [48]. Similar to the forward reaction, no semiquinone species are detected in the reverse reaction with P5C.

The oxidative half-reaction has been explored by reacting different concentrations of CoQ_1 (▶Fig. 2.7B), with reduced PutA prepared by titrating with proline. Similar to the reductive half-reaction, the data are best fit with two observed rate constants, and flavin

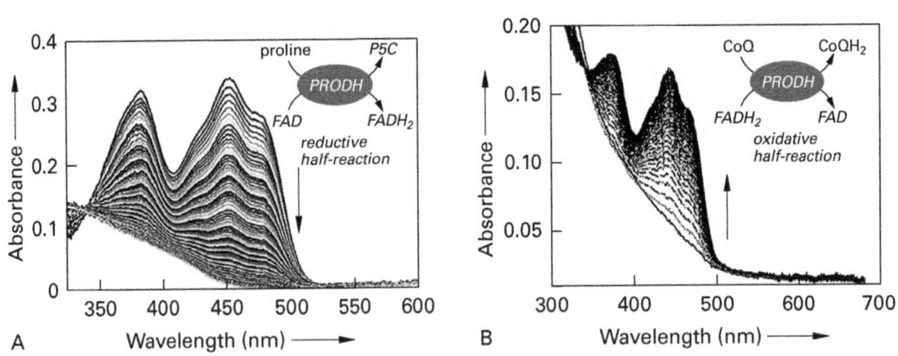

Fig. 2.7: PutA reductive and oxidative half-reactions monitored by stopped-flow kinetics. (A) Reaction of oxidized PutA (27 μM after mixing) with 100 mM proline (after mixing). Spectral traces were recorded at 0.025–10 sec after mixing. (B) Reaction of proline reduced PutA (14 μM after mixing) with 60 μM CoQ_1 (after mixing). Spectra shown were recorded at 0.025–10 sec after mixing. Figure adapted from [48]. © the American Chemical Society.

semiquinone is again not detected in the absorbance traces of the reaction between reduced PutA and CoQ_1. An apparent rate constant of 7.5 s^{-1} has been determined for the oxidation of reduced flavin with CoQ_1 in the faster phase of the reaction, and one of 4 s^{-1} was assigned to an isomerization step for the slower phase [48].

Data from the reverse and forward reactions of the reductive and oxidative half-reactions can be fit globally to the mechanism shown in ▶Fig. 2.8 using the KinTek Global Explorer software package [49]. This analysis has enabled estimation of the microscopic rate constants for the individual steps in the mechanism (▶Fig. 2.8). The microscopic rates constants for PutA proline:ubiquinone oxidoreductase activity with CoQ_1 show that the oxidative step is rate-limiting overall with k_{cat} (5.4 s^{-1}) during catalytic turnover with CoQ_1. The proposed isomerization step does not limit the steady-state turnover rate in vitro with soluble CoQ_1 as a conformational change is not required for enzymatic cycling. In the cell, however, the isomerization step may become rate-limiting as membrane binding is required for PutA to access ubiquinone. A k_{cat} of 0.6 s^{-1} is observed using *E. coli* membrane inverted vesicles as the electron acceptor, significantly lower than the k_{cat} value with soluble CoQ_1 [23]. The lower k_{cat} could be due to the conformational change required for PutA to bind the membrane which would be similar to the *in vivo* situation. More discussion of the conformational change follows in section 2.6.1.

2.4.2 Substrate channeling

Substrate channeling was first described in PutA from *S. typhimurium* [50]. Substrate channeling, however, has been best characterized in PutA from *B. japonicum* (BjPutA). An X-ray crystal structure of the full-length enzyme provides a remarkable view of a cavity linking the PRODH and P5CDH active sites (▶Fig. 2.9) [38]. The substrate-channeling cavity spans 41 Å and has a volume of 1400 Å^3, which is equivalent to about 170 water molecules [38]. The cavity has an irregular shape and does not resemble the cylindrical tunnels of bifunctional enzymes such as tryptophan synthase and carbamoyl phosphate synthetase [51–53].

The cavity has three sections, corresponding to the two active sites and a large, intervening chamber (▶Fig. 2.9). This chamber is 24 Å by 14 Å in its two largest dimensions,

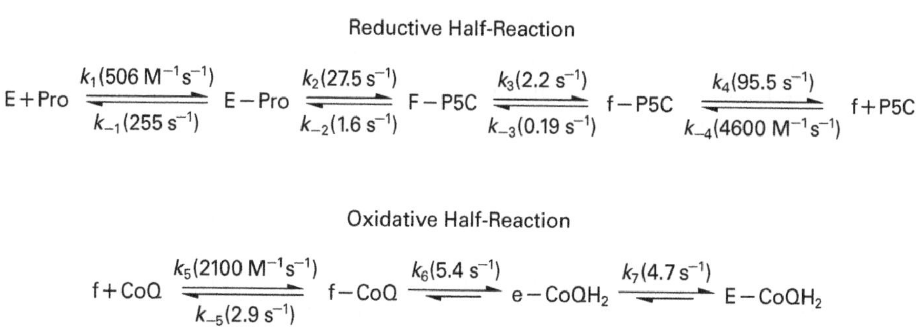

Fig. 2.8: Mechanisms of the reductive and oxidative half-reactions of PutA used for fitting the rapid reaction kinetic data. E, oxidized PutA conformer 1; F, reduced PutA conformer 1; f, reduced PutA conformer 2; e, oxidized PutA conformer 2. Figure adapted from [48]. © the American Chemical Society.

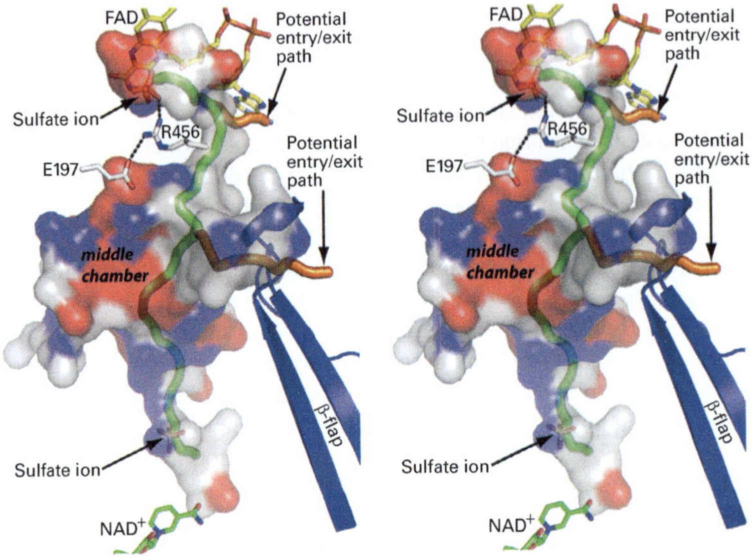

Fig. 2.9: Stereographic view of the substrate-channeling cavity of BjPutA. The cavity is represented as a semitransparent surface and is colored to indicate the locations of positively (blue) and negatively (red) charged residues lining the cavity. The tubes represent pathways identified by the program MOLE. The green tube guides the eye from the PRODH active site (Top) to the P5CDH active site (Bottom). The orange tubes represent possible pathways leading to the bulk medium. This figure was originally published in Srivastava *et al.* [38]. © 2010 by National Academy of Sciences.

and 3–7 Å in the third dimension. It has a volume of about 1325 Å^3, which represents over 90% of the total volume of the entire cavity system. Thus, the middle chamber is large enough to accommodate P5C (molecular volume 102 Å^3) and glutamate semialdehyde (120 Å^3). The walls of the cavity are lined with twelve arginine, three lysine, ten glutamate, and seven aspartate residues; thus, the cavity itself is very hydrophilic [38], consistent with the fact that the cavity must serve as a reaction vessel for the hydrolysis of P5C.

The connection between the PRODH active site and the middle chamber is particularly interesting because of its circuitous path (▶Fig. 2.9). It runs down the ribityl chain of the FAD and passes over a potential opening to the bulk medium and under the side chain of Arg456 before connecting to the middle chamber. A more direct path into the middle chamber appears to be blocked by the ion-pairing residues Arg456 and Glu197 (▶Fig. 2.9). Both residues are strictly conserved in PutAs, and Arg456 plays an essential role in substrate binding by forming a critical ionic interaction with the carboxylate of the substrate. This ion pair has been proposed to function as a gate that closes and opens in response to the binding of proline and release of P5C, respectively.

To support the structural evidence for channeling in BjPutA, a kinetic test for substrate channeling using mixed variants has been devised [38]. ▶Fig. 2.10A shows the strategy for testing substrate channeling in BjPutA. Monofunctional variants of BjPutA were engineered that lacked PRODH or P5CDH activity to generate a non-channeling PutA

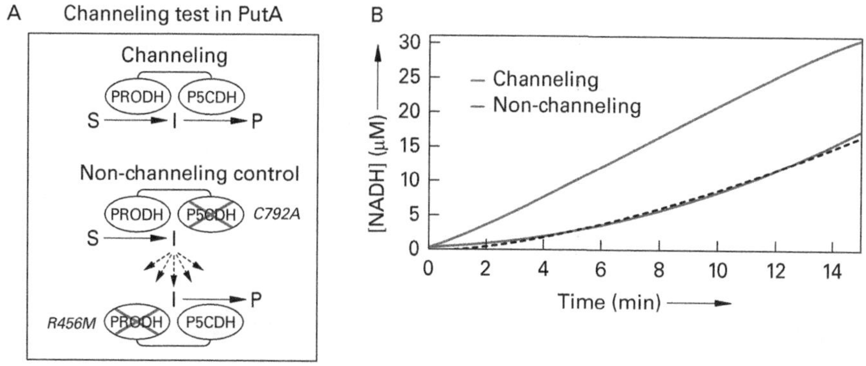

Fig. 2.10: Substrate channeling kinetics of BjPutA. (A) Strategy for analyzing channeling in PutA. In a channeling mechanism, the P5C intermediate (I) is transferred directly between the PRODH and P5CDH active sites of PutA. For a non-channeling control reaction, monofunctional variants are used that can only catalyze the coupled reaction via a diffusion mechanism. In BjPutA, monofunctional variants were generated by mutating active site residues in the PRODH (R456M) and P5CDH (C792A) domains. (B) NADH formation using proline as a substrate by native BjPutA (solid black curve) and an equimolar mixture of monofunctional variants R456M and C792A (solid grey curve). Assays were performed at pH 7.5. The dashed line overlaying the grey curve of the non-channeling control reaction was simulated using the kinetic parameters of PRODH and P5CDH as described previously [38] and using the equation: $[NADH] = v_1 t + (v_1/v_2)K_m(e^{-v_2 t/K_m} - 1)$ [53]. Figure adapted from [19].

system as a control. The Arg456Met mutation in the PRODH active site disables proline binding whereas the Cys792Ala mutation eliminates the essential catalytic cysteine in the P5CDH active site. The overall conversion of proline to glutamate is then compared for wild-type BjPutA and an equimolar mixture of the monofunctional variants. ▸Fig. 2.10B shows the reaction progress curves, which monitor NADH formation at 340 nm. Immediately apparent for wild-type BjPutA is the steady-state formation of NADH. In the assay with the mixed variants, however, a significant lag phase (~8 min) is observed before achieving steady-state NADH formation. The non-channeling control assay fits well to an equation describing the activity of two consecutive enzyme steps without channeling [53]. Once steady-state is achieved, the turnover number for the overall channeling reaction (~0.5 s^{-1}) is similar for wild-type PutA and the PutA mixed variants. Thus, kinetic evidence supports that the cavity identified in the BjPutA structure transports P5C between the PRODH and P5CDH active sites.

2.5 DNA and membrane binding of trifunctional PutA

2.5.1 DNA binding

As indicated above, trifunctional PutAs are transcriptional repressors of the *put* regulon. This aspect of PutA function has been best studied for EcPutA, so we will focus on EcPutA in this section. However, the themes described here are thought to apply generally to other trifunctional PutAs.

The *putA and putP* genes in *E. coli* are transcribed in opposite directions from different promoters in a 419-bp intergenic control region (▶Fig. 2.11A) [35]. Trifunctional PutA binds specifically to a GTTGCA consensus sequence that is found at five different operator sites (O1–O5) in the *put* control DNA (▶Fig. 2.11B) [35]. PutA binding to sites O1 and O2 represses *putP* expression, and PutA binding to sites O3, O4, and O5 leads to repression of the *putA* gene [35]. Gel-mobility shift assays with the isolated

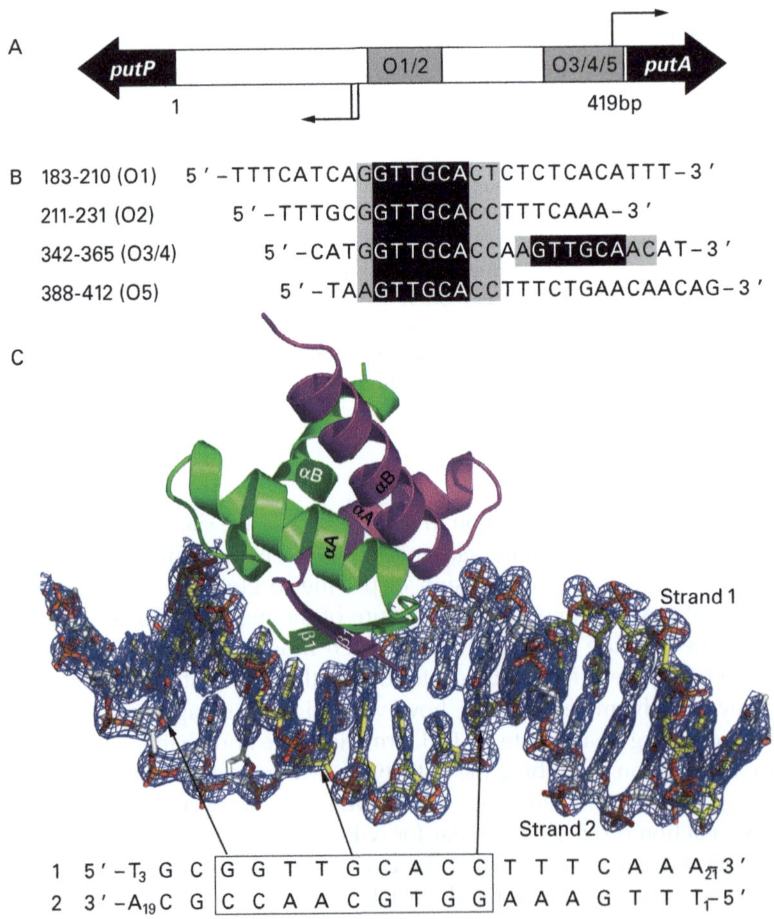

Fig. 2.11: Structural basis of the transcriptional repressor function of PutA. (A) Schematic representation of the intergenic *put* control region in *E. coli* and transcriptional directions of *putP* and *putA*. The black arrows show the transcription start sites. The five PutA binding sites are labeled as O1, O2, O3, O4 and O5. (B) Sequences of the five PutA binding sites in the *put* regulon. The consensus sequence GTTGCA is highlighted in black. Flanking nucleotides predicted to contact PutA based on the crystal structure are shaded in gray. (C) Overall structure of PutA52 bound to O2. PutA52 chains A and B are colored green and magenta, respectively. DNA is represented as sticks, with strand 1 colored yellow and strand 2 colored white. The cage represents an electron density map contoured at 1 α. The box encloses the sequence of the 9-bp fragment contacted by PutA52. These illustrations were originally published in Zhou *et al.* [35].

DNA binding domain of PutA demonstrate that all five operator sites can be occupied simultaneously [35]. Whether full occupation of the operator sites by PutA is necessary for repression of the *putP* and *putA* genes *in vivo* is not known. Analysis of bacterial genomes that share the same genetic organization as *E. coli* has demonstrated that the GTTGCA consensus sequence is conserved in other *put* control DNA regions indicating that it is a fundamental element in the transcriptional regulation of the *put* genes.

Insights into DNA recognition by PutA have been revealed by an X-ray crystal structure of the DNA binding domain of PutA (PutA52) in complex with O2 (▶Fig. 2.11C) [35]. PutA52 adopts the RHH fold, which consists of a β-strand (β1) followed by two α-helices (αA, αB). The two protein chains assemble into a dimer featuring an intermolecular two-stranded antiparallel β-sheet, and the β-sheet inserts into the DNA major groove (▶Fig. 2.11C). Residues of the sheet contact specific DNA bases, imparting sequence specificity, while residues near the N-terminus of αB interact with the DNA backbone. This general mode of binding is typical for RHH proteins [54].

The structure shows that PutA contacts with the DNA involve a 9-bp sequence of GGTTGCACC (▶Fig. 2.11C) [35]. The base pairs flanking the 6-bp consensus sequence form a palindromic cap and are involved in critical hydrogen bond interactions with Lys9, a strictly conserved residue in trifunctional PutAs. Substitution of Lys9 with a Met residue abolishes PutA-DNA binding. The 9-bp sequence is not conserved in the other operator sites. Because of the major hydrogen bond interactions observed between Lys9 and the pair of guanine bases at the 5′-end of both strands, it has been proposed that the PutA binding affinity at the different operator sites may vary depending on the presence of the GG/CC palindromic cap. Isothermal titration calorimetry experiments have shown that this is indeed the case: PutA binding is 15-fold stronger to O2 (K_D = 210 nM) than to O4 (K_D = 3.1 μM), an operator site that lacks the GG/CC cap (AGTTGCAAC) [35].

The overall binding affinity of PutA for the entire *put* intergenic DNA (K_D ~45 nM) is about five-fold tighter than with O2 alone [56]. Initially it was thought that reduction of the FAD cofactor might have a profound effect on PutA-DNA interactions. The DNA binding affinity, however, only decreases twofold (K_D ~100 nM) upon reduction of the FAD [55]. These results demonstrate that FAD reduction alone does not disrupt PutA-DNA binding and indicate that other factors are involved in the mechanism by which proline relieves PutA repression of the *put* genes. As described next, reduction of the FAD induces tight association of PutA to the membrane, which indirectly relieves transcriptional repression of the *put* regulon.

2.5.2 Membrane association

Membrane binding interactions have been studied in detail for EcPutA and BjPutA. EcPutA membrane binding is highly regulated by proline and reduction of the flavin cofactor, whereas BjPutA membrane interactions are not tightly regulated [56,57]. ▶Fig. 2.12 illustrates a surface plasmon resonance (SPR) experiment using oxidized and proline-reduced EcPutA [56]. Oxidized EcPutA does not bind to the lipid bilayer surface prepared from *E. coli* polar lipid extracts. Proline, however, induces strong interactions between EcPutA and the membrane surface with a K_D of <0.01 nM estimated for the EcPutA-lipid complex [56]. Reduction of the FAD cofactor with sodium dithionite also induces strong EcPutA-membrane interactions indicating that flavin reduction alone is sufficient for

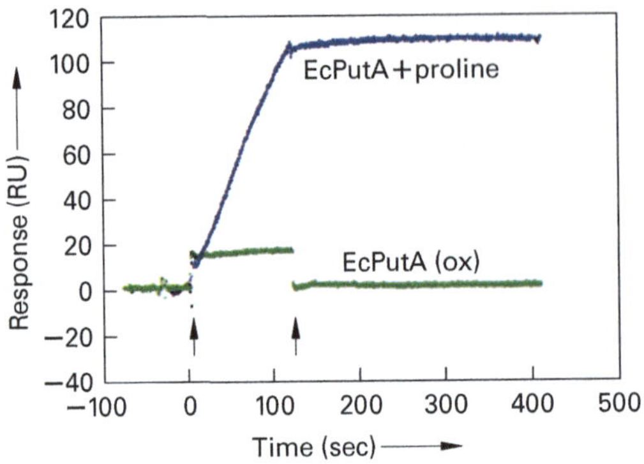

Fig. 2.12: SPR sensorgrams of PutA binding to lipids. Sensorgrams of PutA (20 nM) in the absence (green) and presence of 5 mM proline (blue) injected onto an L1 chip coated with *E. coli* polar lipid vesicles (HEPES N buffer, pH 7.4). The arrows indicate the beginning and end of the PutA sample injection. PutA binding to the lipid bilayer is observed in the presence of proline as indicated by the significant increase in response units (RU) after injection of PutA. Figure adapted from [56]. © the American Chemical Society.

activating EcPutA-membrane binding [56]. EcPutA membrane binding induced by proline is similar with differently charged lipids suggesting that electrostatic effects do not contribute significantly to the stabilization of EcPutA-membrane interactions. Also, EcPutA binding to the membrane also does not appear to require other protein components [58].

As noted above, BjPutA is not highly regulated by proline. BjPutA binds to lipid vesicles in the absence and presence of proline with dissociation constants of 4.3 and 1.1 nM, respectively, as determined by SPR [57]. Because BjPutA lacks a transcriptional repressor function, regulation of BjPutA membrane interactions does not appear to be physiologically required. Isothermal titration calorimetry experiments reveal an endothermic BjPutA membrane binding reaction, suggesting penetration into the lipid bilayer and a substantial hydrophobic component to the protein-membrane interface [57]. Thus, although the membrane binding of EcPutA and BjPutA are quite different as regards proline regulation, both proteins appear to form strong hydrophobic interactions that stabilize the complex with the membrane.

Identifying the specific amino acid residues involved in the PutA-membrane interactions has been challenging. Analysis of PutA sequences and structures does not readily reveal a hydrophobic patch or region suitable for membrane binding. Thus, in addition to not knowing the ubiquinone binding site, the residues responsible for membrane binding have not yet been identified. There is some evidence, however, that implicates residues in the C-terminal domain in membrane binding. A highly conserved 17-residue motif is present in the C-terminal end of Branch 1 PutAs. This motif, denoted CCM, has the consensus sequence of Exxxxv[N or D]t[T or A]AaGGnaxL [46]. The CCM of EcPutA corresponds to residues 1300–1316 and has the sequence of ERSLSVNTAAAGGNASL.

The impact of removing residues in the CCM on EcPutA binding to membranes has been examined by functional membrane association assays. Deleting residues 1295–1320 (EcPutA1–1294) eliminated EcPutA-membrane binding [58]. However, despite having full DNA-binding and PRODH activities, EcPutA1–1294 lacked P5CDH activity, making it difficult to draw any conclusions. Future studies are needed to show whether residues in the CCM are directly involved in membrane binding.

2.6 PutA functional switching

As describe above, PutA switches from a DNA-binding protein to a membrane bound PRODH in response to increased intracellular proline levels. Muro-Pastor *et al.* have shown that the DNA and membrane binding of *S. typhimurium* PutA are mutually exclusive, significant changes in macromolecular interactions must therefore occur for PutA to fulfill its distinct enzymatic and transcriptional regulatory roles [59]. Wood originally proposed an oxidation-reduction mechanism by which PutA switched between enzymatic and regulatory functions [36]. As discussed above, later work has confirmed that chemical reduction of the FAD cofactor is adequate for inducing tight PutA-membrane interactions [56]. Insights into how changes in flavin oxidation state induce dramatic changes in PutA-membrane binding affinity and regulate PutA function are now emerging from biochemical and X-ray crystallography studies.

2.6.1 Redox-linked global conformational changes

Janet Wood's group has identified a global conformational change in EcPutA that occurs upon treatment with proline [60]. Later it has been shown that this global conformational change is due solely to reduction of the FAD. A reduction potential of –58 mV (pH 7.5) has been determined for the global conformational change from limited proteolysis performed under different controlled reduction potentials [61]. The reduction potential associated with the conformational change is near that for the bound flavin cofactor (–76 mV, pH 7.5), indicating that reduction of the flavin alone induces a global conformational change in PutA [55,61]. Mapping of the polypeptides generated from limited proteolysis has revealed a new cleavage site in the reduced PutA enzyme in the α-domain (see ▶Fig. 2.6), indicating that this part of EcPutA is flexible [61].

The dynamics of PutA conformational changes have been examined in intrinsic Trp fluorescence studies of a truncated form of EcPutA that contains residues 86–601 (EcPutA86–601), including the arm, α-domain, PRODH domain, and part of the linker domain (▶Fig. 2.6) [62]. Addition of proline to EcPutA86–601 decreases Trp fluorescence by 36%, consistent with a significant conformational change upon reduction of the flavin by proline [62]. The proline-induced change in Trp fluorescence occurs with an apparent rate constant of 0.59 s^{-1} as determined by stopped-flow fluorescence [62]. Site-directed mutagenesis studies have identified Trp194 and Trp211 of the α-domain as the sources of the fluorescence signal [62]. This study has further confirmed the that redox-linked conformational changes occur in the linker region near the PRODH domain.

As mentioned above, stopped-flow kinetic analysis of the proline:ubiquinone oxidoreductase reaction reveal that an isomerization step occurs after flavin reduction, with a microscopic rate constant of 2.2 s^{-1} [48]. From the different stopped-flow experiments, it appears that PutA undergoes a redox-dependent conformational change with a rate constant of 0.6–2.2 s^{-1}. Thus, PutA switching from a transcriptional repressor to a membrane-bound enzyme most likely occurs on the timescale of seconds, which is consistent with movements of large domains and the folding and unfolding of sections of the polypeptide chain.

For proline to regulate PutA function, conformational changes must occur that lead to enhanced PutA-membrane binding. It is not clear, however, how proline-dependent changes in the linker region near PRODH contribute to increased PutA-membrane interactions. Most likely exposure of a membrane binding peptide occurs as a result of the conformational change. Proline has been reported to increase the overall hydrophobicity of PutA, consistent with an increase in hydrophobic surface area for membrane binding [63]. It is possible that the CCM in PutA, as noted above, has a membrane binding function that is activated upon flavin reduction.

2.6.2 Local structural changes near the flavin

Structural insights into the conformation changes that accompany flavin reduction have been obtained from crystal structures of EcPutA PRODH domain constructs reduced with dithionite [64] or inactivated with *N*-propargylglycine (PPG) [65]. PPG covalently modifies the FAD in PutA by forming a three-carbon link between active site residue K329 and the FAD N(5) atom [65]. Full-length EcPutA inactivated with PPG exhibits lipid binding similar to proline-reduced EcPutA, and PPG-inactivated EcPutA adopts a conformation that is similar to that of proline-reduced EcPutA as determined by limited proteolysis [65]. Thus, EcPutA inactivated with PPG mimics the proline-reduced, membrane bound form of EcPutA.

Conformational changes associated with FAD reduction have been deduced by comparing the structure of the oxidized PRODH construct EcPutA86–669 complexed with the competitive inhibitor L-THFA to the structures of dithionite-reduced EcPutA86–669 and PPG-inactivated EcPutA86–630. In the oxidized, inhibited enzyme, the carboxylate group of THFA forms ion pairs with Arg555 and Arg556 suggesting these residues are critical for proline binding (►Fig. 2.4B). Other interactions that turn out to be relevant for functional switching include hydrogen bonds between Arg431 and the FAD N(5), and between Arg556 and the FAD 2′-OH (►Fig. 2.13).

Reduction of the FAD induces major conformational changes in the FAD itself (►Fig. 2.13). Reduction by dithionite or inactivation by PPG causes a significant "butterfly" bend (22–35°) of the isoalloxazine ring and crankshaft rotation of the FAD ribityl chain [64,65]. These changes break interactions of the FAD with Arg556 and Arg431 and result in formation of new hydrogen bonds that link the FAD 2′-OH ribityl group with the FAD N(1) and Gly435.

The structure of PPG-inactivated EcPutA86–630 has also revealed new orientations of active site residues relative to the THFA-bound enzyme. A comparison of the oxidized THFA bound and PPG-inactivated enzyme structures shows that Arg431 and Asp370 have different orientations (►Fig. 2.13). For Arg431, the distance to the FAD N(5) is increased from 3.1 Å in the THFA-bound structure to 4.2 Å in the PPG structure [65]. Arg431 is thus no longer within hydrogen bonding distance to the FAD N(5), suggesting that reduction of the flavin disrupts interactions with Arg431. Changes in hydrogen bonding are also observed for Asp370, which in the THFA structure hydrogen bonds to Arg431, but in the PPG structure is rotated 180° away from Arg431. The change in orientation of Asp370 enables it to hydrogen bond with Glu372 in the PPG structure. Glu372 also appears to interact with Glu373 via a water mediated hydrogen bond.

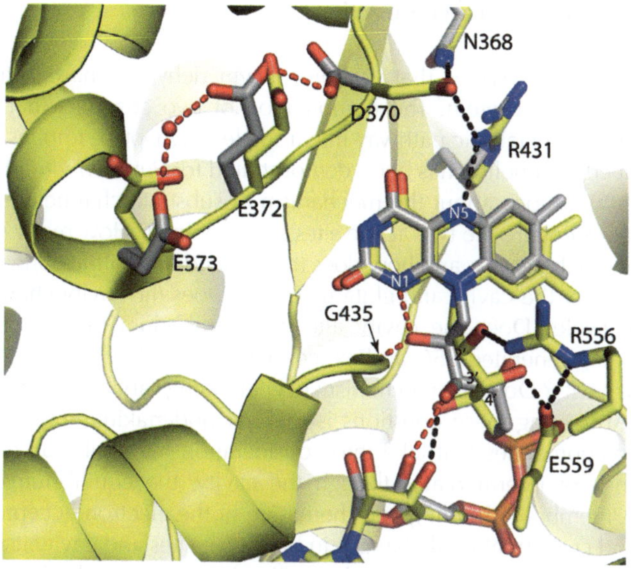

Fig. 2.13: Structural changes induced by reduction of the FAD in PutA. The structure of the oxidized EcPutA PRODH domain complexed with THFA (yellow, PDB code 1TIW) is compared with that of the PPG-inactivated (reduced) PRODH domain (gray, PDB 3ITG). Black dashes denote electrostatic interactions in the oxidized enzyme, while red dashes indicate those formed in the inactivated, reduced enzyme. Note that none of these interactions are found in both states of the enzyme.

Of note, Asp370, Glu372, and Glu373 are part of the β_3-α_3 loop on the surface of the PRODH domain. From the analysis of the PPG structure, Arg431 and Asp370 have been postulated to have a role in transmitting redox-linked signals from the FAD N(5) to a distal membrane binding domain on PutA [65].

2.6.3 Residues important for functional switching

Two key interactions identified in the dithionite and PPG structures have been tested to determine their impact on redox-linked functional switching of PutA. The roles of the ribityl 2′-OH group and Arg431 were tested by substituting normal FAD in PutA with 2′-deoxyFAD and by site-directed mutagenesis [64]. PutA reconstituted with 2′-deoxyFAD is found to be severely defective in membrane binding [64], and it has been concluded that this functional group is essential for controlling binding of PutA to membranes by acting as molecular switch between Arg556 and the FAD N(1) atom in the oxidized and reduced forms, respectively, of the flavin. Similarly, replacing Arg431 with Met in the PutA mutant R431M abolished redox regulation of PutA-membrane binding [64]. Reduction of the PutA mutant R431M failed to activate PutA-membrane binding, consistent with redox-linked signals being transmitted from the FAD N(5) atom to the distal membrane-association domain via Arg431 [64]. Mutagenesis of Asp370 to Ala also leads to loss of redox regulation of PutA membrane binding (unpublished data). Thus, it appears that Arg431 and Asp370 are necessary for transmitting redox signals from the FAD to residues involved in membrane binding.

2.7 Conclusions and future research directions

PutAs are remarkable flavoenzymes that combine flavin dehydrogenase activity and ALDH activity into a single protein. The combination of two enzymes that catalyze consecutive reactions of a metabolic pathway into a single polypeptide implies substrate channeling, and indeed channeling has been demonstrated for PutAs *in vitro*. However, there are many open questions about the mechanism of substrate channeling in PutA. For example, to what extent are the two active sites synchronized? How many GSA/P5C molecules are present in the cavity, and how does this depend on proline concentration? Where does proline enter the cavity and glutamate exit? Does the cavity change shape during the catalytic cycle? Does the cavity alter the equilibrium between P5C and GSA? Do endogenously generated P5C and exogenously added P5C traverse the same pathway through the cavity? Does the mechanism of channeling vary among the different PutA sub-families? Answering these questions is a major undertaking and will require sophisticated kinetic approaches that take into consideration the inhibition of P5CDH by proline, the transit of intermediates though the cavity, the equilibrium between P5C and GSA, PutA membrane binding and the nature of the electron acceptor for the PRODH reaction, and conformational changes in the active sites and cavity itself.

Furthermore, the importance of substrate channeling in proline catabolism *in vivo* is largely unknown. Curiously, studies with BjPutA suggest that channeling does not provide faster throughput of proline to glutamate. This observation suggests that sequestration of the intermediate may be the primary advantage of substrate channeling. For example, the PutA cavity may provide a physical separation between proline catabolism and proline biosynthesis, which both utilize P5C/GSA as the intermediate.

The observation of substrate channeling in PutA suggests the hypothesis that monofunctional PRODH and P5CDH physically interact and engage in some form of substrate channeling. Interactions between these enzymes in humans may be particularly important because P5C is a signaling molecule in eukaryotes. A reasonable starting point for this new phase of proline catabolism research is to test substrate channeling using bacterial monofunctional enzymes, such as those from *T. thermophilus*, whose crystal structures have been determined.

Because of the transcriptional repressor function, trifunctional PutAs have extra layers of complexity beyond substrate channeling. Functional switching is the most intriguing and challenging aspect of trifunctional PutAs. Some insights into the mechanism of functional switching have emerged, but this information mostly concerns redox-linked events that occur proximal to the FAD. How flavin redox signals are transmitted to regulatory and functional domains in PutA remains a significant outstanding question. A major challenge of future research will be to obtain a more global view of how redox-based conformational dynamics in PutA elicit membrane binding and ultimately functional switching of PutA. A central part of this challenge will be to determine the three-dimensional structural basis for membrane association.

The diversity of PutAs could not have been envisioned when PutAs were discovered over two decades ago. Bioinformatics studies have revealed a complex lineage of PutAs (►Fig. 2.2 and ►Tab. 2.1), which provides a roadmap for future functional and structural studies of PutA. Crystallography will play an important role moving forward as it provides hypotheses that can be tested using biochemical and biophysical methods. At present, only the minimalist 1A and 2A PutAs have yielded to crystal structure determination. Thus, a major goal in the PutA field will be to determine crystal structures of longer PutAs that contain the CTD, which is the only major PutA domain that has not been resolved with crystallography. Lastly, crystal structure determination of a trifunctional PutA remains the holy grail of proline catabolism structural biology.

2.8 Acknowledgements

This work was supported, in whole or in part, by National Institutes of Health Grants GM065546 and GM061068.

2.9 References

[1] Blake RL, Hall JG, Russell ES. Mitochondrial proline dehydrogenase deficiency in hyperprolinemic PRO/Re mice: genetic and enzymatic analyses. Biochem Genet 1976;14:739–57.

[2] Phang JM, Pandhare J, Liu Y. The metabolism of proline as microenvironmental stress substrate. J Nutr 2008;138:2008S–2015S.

[3] Szabados L, Savoure A. Proline: a multifunctional amino acid. Trends Plant Sci 2010; 15:89–97.

[4] Wu G, Bazer FW, Burghardt RC, Johnson GA, Kim SW, Knabe DA, et al. Proline and hydroxyproline metabolism: implications for animal and human nutrition. Amino Acids 2011; 40:1053–63.

[5] Nagata K, Nagata Y, Sato T, Fujino MA, Nakajima K, Tamura T. L-Serine, D- and L-proline and alanine as respiratory substrates of *Helicobacter pylori*: correlation between in vitro and in vivo amino acid levels. Microbiology 2003;149:2023–2030.

[6] Curtis J, Shearer G, Kohl DH. Bacteriod proline catabolism affects N_2 fixation rate of drought-stressed soybeans. Plant Physiol 2004;136:3313–3318.

[7] Kohl DH, Schubert KR, Carter MB, Hagedorn CH, Shearer G. Proline metabolism in N_2-fixing root nodules: Energy transfer and regulation of purine synthesis. Proc Natl Acad Sci USA 1988;85:2036–2040.

[8] Vílchez S, Manzanera M, Ramos J. Control of expression of divergent *Pseudomonas putida put* promoters for proline catabolism. Appl Environ Microbiol 2000;66:5221–5225.

[9] Wood JM. Genetics of L-proline utilization in *Eschericia coli.* J Bacteriol 1981;146:895–901.

[10] Sharma S, Villamor JG, Verslues PE. Essential role of tissue-specific proline synthesis and catabolism in growth and redox balance at low water potential. Plant Physiol 2011;157: 292–304.

[11] Cecchini NM, Monteoliva MI, Alvarez ME. Proline dehydrogenase contributes to pathogen defense in *Arabidopsis.* Plant Physiol 2011;155:1947–1959.

[12] Morita Y, Nakamori S, Takagi H. L-proline accumulation and freeze tolerance of *Saccharomyces cerevisiae* are caused by a mutation in the PRO1 gene encoding gamma-glutamyl kinase. Appl Environ Microbiol 2003;69:212–219.

[13] Chattopadhyay MK, Kern R, Mistou MY, Dandekar AM, Uratsu SL, Richarme G. The chemical chaperone proline relieves the thermosensitivity of a *dnaK* deletion mutant at 42 degrees C. J Bacteriol 2004;186:8149–8152.

[14] Csonka LN. Proline over-production results in enhanced osmotolerance in *Salmonella typhimurium.* Mol Gen Genet 1981;182:82–86.

[15] Wood JM. Proline porters effect the utilization of proline as a nutrient of osmoprotectant for bacteria. J Membr Biol 1988;106:183–202.

[16] Chen C, Dickman MB. Proline suppresses apoptosis in the fungal pathogen of *Colletotrichum trifolii*. Proc Natl Acad Sci USA 2005;102:3459–3464.

[17] Krishnan N, Dickman MB, Becker DF. Proline modulates the intracellular redox environment and protects mammalian cells against oxidative stress. Free Radic Biol Med 2008;44: 671–81.

[18] Wondrak GT, Jacobson MK, Jacobson EL. Identification of quenchers of photoexcited states as novel agents for skin photoprotection. J Pharmacol Exp Therapeut 2005;312:482–491.

[19] Arentson BW, Sanyal N, Becker DF. Substrate channeling in proline metabolism. Front Biosci 2012;17:375–88.

[20] Wang SS, Brandriss MC. Proline utilization in Saccharomyces cerevisiae: sequence, regulation, and mitochondrial localization of the PUT1 gene product. Mol Cell Biol 1987;7:4431–40.

[21] Reinders J, Zahedi RP, Pfanner N, Meisinger C, Sickmann A. Toward the complete yeast mitochondrial proteome: multidimensional separation techniques for mitochondrial proteomics. J Proteome Res 2006;5:1543–54.

[22] Wanduragala S, Sanyal N, Liang X, Becker DF. Purification and characterization of Put1p from Saccharomyces cerevisiae. Arch Biochem Biophys 2010;498:136–42.

[23] Moxley MA, Tanner JJ, Becker DF. Steady-state kinetic mechanism of the proline:ubiquinone oxidoreductase activity of proline utilization A (PutA) from Escherichia coli. Arch Biochem Biophys 2011;516:113–20.

[24] Nakajima K, Natsu S, Mizote T, Nagata Y, Aoyania K, Fukuda Y, et al. Possible involvement of putA gene in *Helicobacter pylori* colonization in the stomach and motility. Biomedical Research-Tokyo 2008;29:9–18.

[25] Lamour N, Riviere L, Coustou V, Coombs GH, Barrett MP, Bringaud F. Proline metabolism in procyclic *Trypanosoma brucei* is down-regulated in the presence of glucose. J Biol Chem 2005;280:11902–10.

[26] Donald SP, Sun XY, Hu CA, Yu J, Mei JM, Valle D, et al. Proline oxidase, encoded by p53-induced gene-6, catalyzes the generation of proline-dependent reactive oxygen species. Cancer Res 2001;61:1810–5.

[27] Liu Y, Borchert GL, Surazynski A, Hu CA, Phang JM. Proline oxidase activates both intrinsic and extrinsic pathways for apoptosis: the role of ROS/superoxides, NFAT and MEK/ERK signaling. Oncogene 2006;25:5640–7.

[28] Polyak K, Xia Y, Zweier JL, Kinzler KW, Vogelstein B. A model for p53-induced apoptosis. Nature 1997;389:300–5.

[29] Liu Y, Borchert GL, Donald SP, Diwan BA, Anver M, Phang JM. Proline oxidase functions as a mitochondrial tumor suppressor in human cancers. Cancer Res. 2009;69:6414–22.

[30] Zarse K, Schmeisser S, Groth M, Priebe S, Beuster G, Kuhlow D, et al. Impaired insulin/IGF1 signaling extends life span by promoting mitochondrial L-proline catabolism to induce a transient ROS signal. Cell Metab 2012;15:451–65.

[31] Krishnan N, Becker DF. Oxygen reactivity of PutA from *Helicobacter* species and proline-linked oxidative stress. J Bacteriol 2006;188:1227–1235.

[32] Krishnan N, Doster AR, Duhamel GE, Becker DF. Characterization of a *Helicobacter hepaticus putA* mutant strain in host colonization and oxidative stress. Infect Immun 2008;76:3037–3044.

[33] Ratzkin B, Roth J. Cluster of genes controlling proline degradation in Salmonella typhimurium. J Bacteriol 1978;133:744–54.

[34] Wood JM, Zadworny D. Amplification of the put genes and identification of the put gene products in Escherichia coli K12. Can J Biochem 1980;58:787–96.

[35] Zhou Y, Larson JD, Bottoms CA, Arturo EC, Henzl MT, Jenkins JL, et al. Structural basis of the transcriptional regulation of the proline utilization regulon by multifunctional PutA. J Mol Biol 2008;381:174–188.

[36] Wood J. Membrane association of proline dehydrogenase in *Escherichia coli* is redox dependent. Proc Natl Acad Sci USA 1987;84:373–377.

[37] Ostrovsky De Spicer P, Maloy S. PutA protein, a membrane-associated flavin dehydrogenase, acts as a redox-dependent transcriptional regulator. Proc Natl Acad Sci USA 1993;90:4295–4298.

[38] Srivastava D, Schuermann JP, White TA, Krishnan N, Sanyal N, Hura GL, et al. Crystal structure of the bifunctional proline utilization A flavoenzyme from *Bradyrhizobium japonicum.* Proc Natl Acad Sci USA 2010;107:2878–83.

[39] Singh RK, Larson JD, Zhu W, Rambo RP, Hura GL, Becker DF, et al. Small-angle X-ray scattering studies of the oligomeric state and quaternary structure of the trifunctional proline utilization A (PutA) flavoprotein from *Escherichia coli.* J Biol Chem 2011;286:43144–53.

[40] Lee YH, Nadaraia S, Gu D, Becker DF, Tanner JJ. Structure of the proline dehydrogenase domain of the multifunctional PutA flavoprotein. Nat Struct Biol 2003;10:109–14.

[41] Zhang M, White TA, Schuermann JP, Baban BA, Becker DF, Tanner JJ. Structures of the *Escherichia coli* PutA proline dehydrogenase domain in complex with competitive inhibitors. Biochemistry 2004;43:12539–48.

[42] Larson JD, Jenkins JL, Schuermann JP, Zhou Y, Becker DF, Tanner JJ. Crystal structures of the DNA-binding domain of *Escherichia coli* proline utilization A flavoprotein and analysis of the role of Lys9 in DNA recognition. Protein Sci 2006;15:2630–41.

[43] White TA, Krishnan N, Becker DF, Tanner JJ. Structure and kinetics of monofunctional proline dehydrogenase from *Thermus thermophilus.* J Biol Chem 2007;282:14316–27.

[44] Inagaki E, Ohshima N, Takahashi H, Kuroishi C, Yokoyama S, Tahirov TH. Crystal structure of Thermus thermophilus Delta(1)-pyrroline-5-carboxylate dehydrogenase. J Mol Biol 2006;362:490–501.

[45] Srivastava D, Singh RK, Moxley MA, Henzl MT, Becker DF, Tanner JJ. The three-dimensional structural basis of Type II hyperprolinemia. J Mol Biol 2012;420:176–89.

[46] Singh RK, Tanner JJ. Unique structural features and sequence motifs of proline utilization A (PutA). Front Biosci 2012;17:556–68.

[47] Menzel R, Roth J. Enzymatic properties of the purified putA protein from Salmonella typhimurium. J Biol Chem 1981;256:9762–6.

[48] Moxley MA, Becker DF. Rapid reaction kinetics of proline dehydrogenase in the multifunctional proline utilization A protein. Biochemistry 2012;51:511–20.

[49] Johnson KA, Simpson ZB, Blom T. Global kinetic explorer: a new computer program for dynamic simulation and fitting of kinetic data. Anal Biochem 2009;387:20–9.

[50] Surber MW, Maloy S. The PutA protein of Salmonella typhimurium catalyzes the two steps of proline degradation via a leaky channel. Arch Biochem Biophys 1998;354:281–7.

[51] Miles EW, Rhee S, Davies DR. The molecular basis of substrate channeling. J Biol Chem 1999;274:12193–6.

[52] Huang X, Holden HM, Raushel FM. Channeling of substrates and intermediates in enzyme-catalyzed reactions. Annu Rev Biochem 2001;70:149–80.

[53] Meek TD, Garvey EP, Santi DV. Purification and characterization of the bifunctional thymidylate synthetase-dihydrofolate reductase from methotrexate-resistant Leishmania tropica. Biochemistry 1985;24:678–86.

[54] Schreiter ER, Drennan CL. Ribbon-helix-helix transcription factors: variations on a theme. Nat Rev Microbiol 2007;5:710–20.

[55] Becker DF, Thomas EA. Redox properties of the PutA protein from Escherichia coli and the influence of the flavin redox state on PutA-DNA interactions. Biochemistry 2001;40: 4714–21.

[56] Zhang W, Zhou Y, Becker DF. Regulation of PutA-membrane associations by flavin adenine dinucleotide reduction. Biochemistry 2004;43:13165–74.

[57] Zhang W, Krishnan N, Becker DF. Kinetic and thermodynamic analysis of Bradyrhizobium japonicum PutA-membrane associations. Arch Biochem Biophys 2006;445:174–83.

[58] Zhou Y, Zhu W, Bellur PS, Rewinkel D, Becker DF. Direct linking of metabolism and gene expression in the proline utilization A protein from *Escherichia coli*. Amino Acids 2008;35:711–8.
[59] Muro-Pastor AM, Ostrovsky P, Maloy S. Regulation of gene expression by repressor localization: biochemical evidence that membrane and DNA binding by the PutA protein are mutually exclusive. J Bacteriol 1997;179:2788–91.
[60] Brown ED, Wood JM. Conformational change and membrane association of the PutA protein are coincident with reduction of its FAD cofactor by proline. J Biol Chem 1993;268:8972–9.
[61] Zhu W, Becker DF. Flavin redox state triggers conformational changes in the PutA protein from Escherichia coli. Biochemistry 2003;42:5469–77.
[62] Zhu W, Becker DF. Exploring the proline-dependent conformational change in the multifunctional PutA flavoprotein by tryptophan fluorescence spectroscopy. Biochemistry 2005;44:12297–306.
[63] Despicer PO, Maloy S. Puta Protein, a Membrane-Associated Flavin Dehydrogenase, Acts as a Redox-Dependent Transcriptional Regulator. Proc Natl Acad Sci U S A 1993;90:4295–98.
[64] Zhang W, Zhang M, Zhu W, Zhou Y, Wanduragala S, Rewinkel D, et al. Redox-induced changes in flavin structure and roles of flavin N(5) and the ribityl 2′-OH group in regulating PutA--membrane binding. Biochemistry 2007;46:483–91.
[65] Srivastava D, Zhu W, Johnson WH, Jr., Whitman CP, Becker DF, Tanner JJ. The structure of the proline utilization a proline dehydrogenase domain inactivated by N-propargylglycine provides insight into conformational changes induced by substrate binding and flavin reduction. Biochemistry 2010;49:560–9.

3 Flavoenzymes involved in non-redox reactions

Hisashi Hemmi

Abstract

Flavoenzymes are generally recognized as catalyzing oxidation-reduction reactions, but some are known to be involved in non-redox reactions. In this chapter, the biological origin, physiological importance, biochemical properties, structural information, and the catalytic mechanism of several of such unique flavoenzymes are outlined. In most cases, the catalytic roles of the flavin are likely based on its oxidation-reduction properties (*i.e.*, transitions between the oxidized, semiquinone, or reduced states), and resemble those in more typical flavoenzymes. In two exceptional examples, however, the flavin cofactors has been suggested to play a completely non-redox role: reduced FMN as a general acid/base catalyst in type 2 isopentenyl diphosphate isomerase, and reduced FAD as a nucleophilic catalyst in UDP-galactopyranose mutase. These findings have expanded our knowledge about the flavin chemistry. The roles of the flavin cofactors remain uncertain in some flavoenzymes that catalyze non-redox reactions, and these are also summarized here.

3.1 Introduction

Flavin cofactors, *i.e.*, FMN and FAD, are recognized as playing redox roles in enzyme reactions. Indeed, most flavoenzymes catalyze oxidation-reduction reactions (substrate reduction, oxidation/dehydrogenation, hydroxylation and so on). However, a small number of flavoenzymes are known to be involved in non-redox reactions. Studies of such enzymes have given us a new perspective concerning the chemistry of flavoenzymes. In particular, some of the recently discovered enzymes have expanded the body of knowledge concerning flavin chemistry beyond the context of well-known oxidation-reduction. In this chapter, several examples of such flavoenzymes, *i.e.*, those catalyzing no net oxidation-reduction, are introduced. They are, for convenience, classified by the catalytic roles of their flavin cofactors. This chapter, however, does not include the flavoenzymes for which flavin cofactors are considered from structural data to play non-catalytic, only structural roles, such as acetohydroxyacid synthase [1] and hydroxynitrile lyase [2]. The biological origin, physiological importance, properties, structure (if available), and – the most important aspect – the catalytic mechanism of the flavoenzymes will be described. It should be noted that for several of these enzymes the reaction mechanisms remain controversial and only those mechanisms for which the experimental evidence is persuasive are illustrated here because of page limitations.

3.2 Flavoenzymes for which flavin cofactors likely play redox-based catalytic roles

The reaction mechanisms of most flavoenzymes involved in non-redox reactions can be understood in the context of the well-studied mechanisms of more typical redox-catalyzing flavoenzymes. In these cases, the flavin cofactor catalyzes the transfer of the same number of electrons (sometimes in the form of a hydrogen radical or a hydride) in first one and then the reverse directions, with the overall reaction being (apparently) non-redox. The mechanisms of these enzymes can be divided into half-reactions, *i.e.*, oxidative and reductive, similar to the reactions catalyzed by the usual redox-catalyzing flavoenzymes. Here I offer several examples of such non-redox catalyzing flavoenzymes, for which flavin cofactors are considered to play general oxidation-reduction roles.

3.2.1 Chorismate synthase

Chorismate synthase (CS) is involved in the biosynthesis of various aromatic compounds by catalyzing the last step of the shikimate pathway [3]. The enzyme is found in bacteria, archaea, fungi, plants, and apicomplexan protozoa parasites. Because of its absence in humans, CS is a potential target for antibiotic development. CS catalyzes the transformation of 5-enolpyruvylshikimate 3-phosphate (EPSP) into chorismate and inorganic phosphate. The 1,4-elimination of the 3-phosphate group and the *proR* hydrogen at position 6 of EPSP occurs in an anti-manner. Reduced FMN is strictly required for the activity of CS [4]. Monofunctional CSs such as those from bacteria and plants are activated by a trans-acting flavin reductase, while bifunctional CSs (such as those from fungi and protozoa) are able to catalyze the reduction of FMN by NADPH [5].

A radical-based mechanism, involving a single electron transfer (SET) from reduced FMN to form a radical intermediate and FMNH·, has been suggested by the inactivity of CS reconstituted with a flavin analogue, 5-deaza-FMN [6,7], which is incapable of SET. Experiments using the substrate analogues, (6*S*)-6-fluoro-EPSP and (6*R*)-6-fluoro-EPSP, have contributed significantly to our understanding of the catalytic mode of CS. The former analogue is converted to 6-fluorochorismate by *Escherichia coli* CS, at a rate reduced by two-orders of magnitude slower than EPSP itself [8]. This suggests that the reaction proceeds via a stepwise 1,4-elimination involving either an allylic radical or carbocation intermediate. The later substrate analogue turns reduced FMN in *E. coli* CS into stable FMN semiquinone in the presence of dithonite, which supports the radical mechanism [9]. The analogue is converted to chorismate stoichiometrically in the active site of *E. coli* CS, probably through the reduction of a radical intermediate by dithionite, and strongly inhibits the enzyme by forming a stable complex [7].

The X-ray crystal structure of the enzyme from *Streptococcus pneumoniae* (PDB ID: 1qxo) in complex with oxidized FMN and EPSP has provided important information about the catalytic mechanism of CS [10]. EPSP binds on the *si*-face of the isoalloxazine ring of FMN, extending its C6-*proR* hydrogen toward N5 of FMN. A site-directed mutagenesis study has suggested that an aspartate residue acts as a base to help N5 of FMN withdraw the *proR* proton [11]. Considering the structural and mutagenic

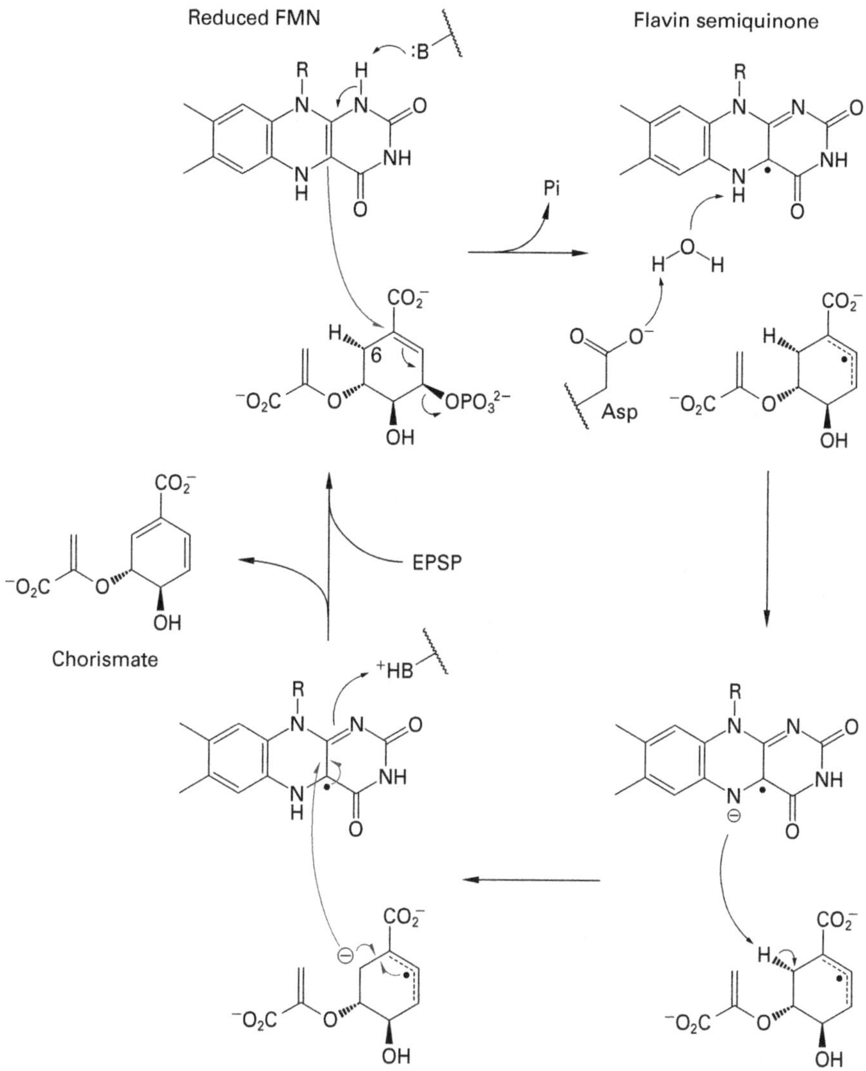

Fig. 3.1: Probable catalytic mechanism of CS. Half arrows in blue show the movement of a single electron.

information, the radical mechanism of CS is likely to proceed as depicted in ▶Fig. 3.1. The SET from reduced FMN to EPSP induces elimination of the phosphate group, which forms an allylic radical intermediate and a flavin semiquinone. Via an intervening water molecule, the aspartate residue deprotonates N5 of the flavin semiquinone, which then withdraws the C6-*proR* proton from the intermediate. SET from the resultant intermediate back to the flavin semiquinone yields chorismate and regenerates reduced FMN.

3.2.2 4-Hydroxybutyryl-CoA dehydratase

4-Hydroxybutyryl-CoA dehydratase (4-BUDH) is involved in the catabolic degradation of γ-aminobutyrate (GABA) by *Clostridium aminobutyricum* [12]. GABA, which is usually produced by other microorganisms such as lactic acid bacteria [13], is converted into 4-hydroxybutyryl-CoA via succinate semialdehyde and 4-hydroxybutyrate. The enzyme then catalyzes the dehydration of 4-hydroxybutyryl-CoA to yield crotonyl-CoA, which is later metabolized to butyrate and acetate. *Clostridium kluyveri* also uses the enzyme for succinate reduction [14]. 4-BUDH from *C. aminobutyricum* is a homotetrameric enzyme containing FAD [15] and [4Fe-4S] clusters [16,17]. Although FAD is easily reduced to the semiquinone form, full reduction of FAD inactivates the enzyme. The enzyme irreversibly loses activity if exposed to air, due to degradation of the [4Fe-4S] cluster. There is no oxidation-reduction equilibrium between the FAD and the [4Fe-4S] cluster. EPR analysis of the enzyme incubated with 4-hydroxybutyryl-CoA yields signals due to flavin semiquinone and what appears to be a substrate-derived radical interacting with the [4Fe-4S] cluster [18].

A chemically reasonable radical mechanism has been proposed for the enzyme (►Fig. 3.2) [19]. The reaction starts with abstraction of the activated 2-*proR* proton of 4-hydroxybutyryl-CoA to form an enolate intermediate. SET from the enolate to oxidized FAD yields an enoxy radical intermediate and FAD semiquinone. The C3 proton is then activated in the enoxy radical, with a pK_a of ~14 if in the free form. Abstraction of the 3-*proS* proton gives a ketyl radical anion [20], which enables the [4Fe-4S] cluster to eliminate the 4-hydroxyl group. The resultant dienoxy radical accepts an electron from FAD semiquinone, regenerating oxidized FAD and producing crotonyl-CoA via a dienolate intermediate.

Such a mechanism is supported by the X-ray crystal structure of *C. aminobutyricum* 4-BUDH (PDB ID: 1u8v) [21]. In this structure, one FAD molecule and one [4Fe-4S] cluster (coordinated by three cysteines and one histidine) exist in each catalytic site. The distance from the isoalloxazine ring of FAD to the [4Fe-4S] cluster is approximately 7 Å, and several water molecules are sandwiched between them. Based on the structural similarity with medium-chain acyl-CoA dehydrogenase, a model structure of 4-BUDH binding 4-hydroxybutyryl-CoA, which substituted for the water molecules, was constructed. In the model, the 2-*proR* hydrogen of the substrate is within hydrogen-bonding distance to a histidine residue, which probably acts as a base for deprotonation. The 3-*proS* hydrogen interacts with N5 of FAD, and the C4-hydroxyl group coordinates with the histidine-coordinated iron in the [4Fe-4S] cluster.

A similar radical reaction is catalyzed by 2-hydroxyacyl-CoA dehydratases [22]. Some of the enzymes reportedly contain flavins such as FMN and riboflavin. However, the contents of the cofactors varied among them, and the 2-hydroxyacyl-CoA dehydratase from *Clostridium difficile* requires no flavin cofactor at all for activity [23]. It thus appears unlikely that the flavin cofactors play important catalytic roles in these enzymes. On the basis of the X-ray crystal structure of the *C. difficile* enzyme, it appears that a [4Fe-4S] cluster probably catalyzes both the one-electron oxidation of the substrate, which yields a ketyl radical anion, and subsequent elimination of the 2-hydroxy group [24].

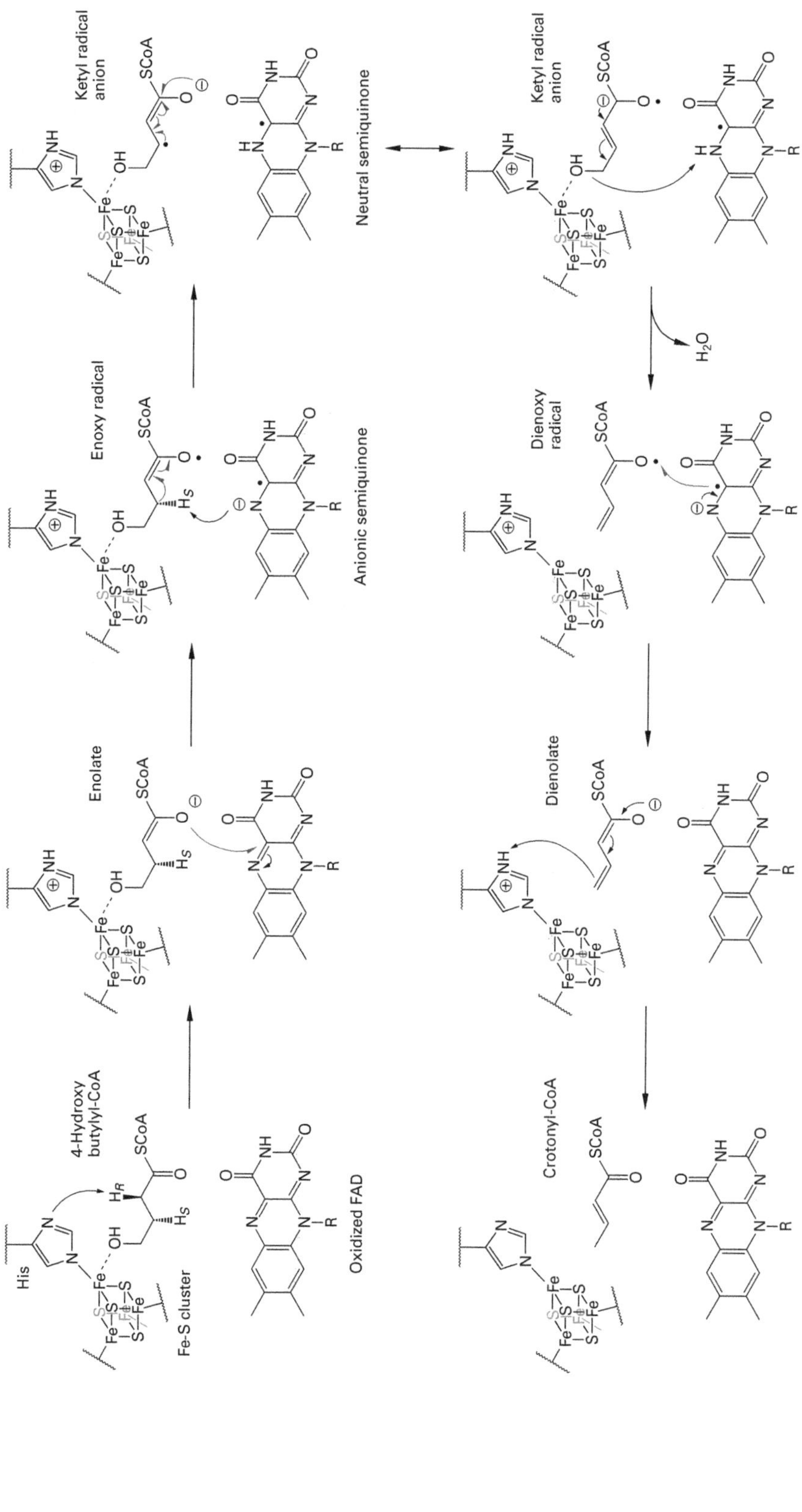

Fig. 3.2: Probable catalytic mechanism of 4-BUDH. Half arrows in blue show the movement of a single electron.

3.2.3 Polyunsaturated fatty acid isomerase

Polyunsaturated fatty acid (PUFA) isomerase is involved in a metabolic pathway called "biohydrogenation", which eventually saturates intrinsic or extrinsic PUFAs for further metabolism [25]. The pathway is found in rumen bacteria and protozoa, and also in PUFA-producing marine algae, but varies according to the organism. PUFA isomerase activity is usually membrane-associated, except for the enzymes from an anaerobic Gram-positive bacterium, *Propionibacterium acnes*, and a red marine alga, *Ptilota filicina* [26]. Thus, detailed enzymatic studies have for the most part been performed with these soluble flavoenzymes, which have subsequently been shown to share little homology with the membranous enzymes (with the exception of a conserved dinucleotide-binding motif). *P. acnes* PUFA isomerase accepts either C18 or C16 fatty acids, and typically catalyzes the transformation of linoleic acid [LA, (9*Z*,12*Z*)-octadecadienoic acid] to its isomer, (10*E*,12*Z*)-conjugated LA [(10*E*,12*Z*)-CLA] [27]. The enzyme from *P. filicina* prefers longer PUFAs such as docosahexaenoate [DHA, (4*Z*,7*Z*,10*Z*,13*Z*,16*Z*,19*Z*)-docosahexaenoic acid] and eicosapentaenoic acid to shorter ones, and typically catalyzes conversion of DHA into the (4*Z*,7*Z*,9*E*,11*E*,16*Z*,19*Z*)-isomer [28,29]. In the reaction of *P. acnes* PUFA isomerase, the deuterium label at C11 of LA migrates to C9, which means that the reaction is an intramolecular hydrogen transfer [30]. The absence of incorporation of deuterium during the reaction when carried out in D_2O further supports the mechanism.

The X-ray crystal structure of *P. acnes* PUFA isomerase has been elucidated as the complex with (10*E*,12*Z*)-CLA (PDB ID: 2bab) [31]. The product binds in the active site in a U-configuration, with C9 to C13 sandwiched between the isoalloxazine ring of FAD and a phenylalanine residue, placing C11 at a distance of 3.2 Å from N5 of FAD. Significantly, there is no amino acid residue that can catalyze proton transfer around C11. The structural information suggests a catalytic mechanism in which FAD first removes and subsequently returns the same hydrogen (as a hydride anion or hydrogen radical) in the course of the reaction. A hydride transfer mechanism, which involves the formation of a bis-allylic carbocation intermediate, as shown in ▸Fig. 3.3, is supported by the facts that the characteristic absorption features of neither the neutral nor anionic semiquinone have been observed in a stopped-flow experiment with a time scale from 2 ms to 100 min, and that no paramagnetic species has been detected in the EPR analysis of frozen reaction mixtures [30]. Nevertheless, the inactivity of the *P. acnes* enzyme reconstituted with 5-deaza-FAD suggests that the reaction does indeed proceed via a radical transfer mechanism.

3.2.4 4′-Phosphopantothenoylcysteine decarboxylase

4′-Phosphopantothenoylcysteine decarboxylase (PPC-DC) is a FMN-dependent flavoenzyme that exists in almost all organisms and catalyzes the final decarboxylation step in the biosynthetic pathway of 4′-phosphopantetheine, which is a functional prosthetic group of coenzyme A and acyl-carrier proteins [32]. PPC-DC from *E. coli* is a bifunctional enzyme fused with 4′-phosphopantothenoylcysteine (PPC) synthase, which catalyzes the CTP-dependent peptidyl-bond formation between 4′-phophopantothenate and cysteine [33], while PPC-DCs from both *Arabidopsis thaliana* and humans are monofunctional [34,35]. The reaction of PPC-DC can be separated into two steps. In an

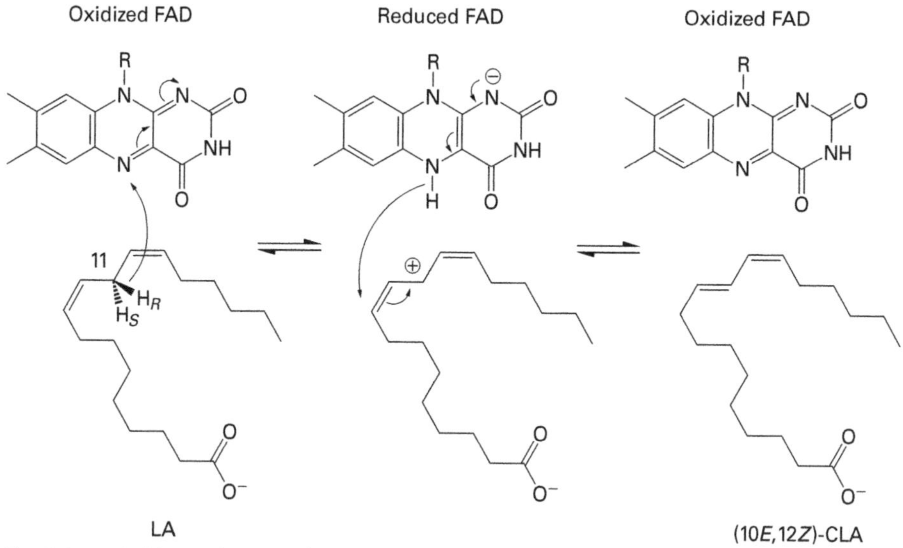

Fig. 3.3: Probable catalytic mechanism of PUFA isomerase.

initial oxidative half-reaction, PPC-DC decarboxylates PPC and forms an intermediate 4′-phosphopantothenoylaminoethenethiolate. The subsequent reductive half-reaction involves reduction of a double bond of the enethiolate intermediate. Though the overall reaction is redox-neutral, oxidized FMN is temporarily reduced and then re-oxidized. Such a two-step mechanism has been supported by the existence of oxidative peptidyl-cysteine decarboxylases such as FAD-dependent MsrD from a *Bacillus* strain and FMN-dependent EpiD from *Staphylococcus epidermidis*, which are involved in the formation of lantibiotic peptides [36]. These enzymes are homologous to PPC-DC, but only catalyze reactions analogous to the oxidative half-reaction of PPC-DC. The C175S mutant of *A. thaliana* PPC-DC, in which an active residue is mutated, also catalyzes only the oxidative half-reaction [35].

On the basis of the observed kinetic isotopic effects using deuterated substrates as well as the observed the substrate/product-induced formation of charge transfer complexes, a catalytic mechanism for PPC-DC has been proposed that involves the formation of adducts between FMN and intermediates [37] (▶Fig. 3.4, route A, shown in black). In the oxidative half-reaction, the thiolate substrate reacts with oxidized FMN and forms a C4a-adduct. Decomposition of the adduct yields reduced FMN and an activated thioaldehyde intermediate, which is easily decarboxylated to form 4′-phosphopantothenoylaminoethenethiolate. The formation of the enethiolate intermediate is supported by the observation of an absorption spectrum attributable to the charge transfer complex between the intermediate and FMN, which has been confirmed with a mutant of human PPC-DC that lacks the active cysteine residue and a substrate analogue containing a cyclopropyl group for trapping an enethiolate intermediate [38]. In the reductive half-reaction, 4′-phosphopantothenoylaminoethen ethiolate reacts with reduced FMN to form another C4a-adduct. Decomposition of the adduct intermediate, which is likely to be dependent on functional cysteine, yields

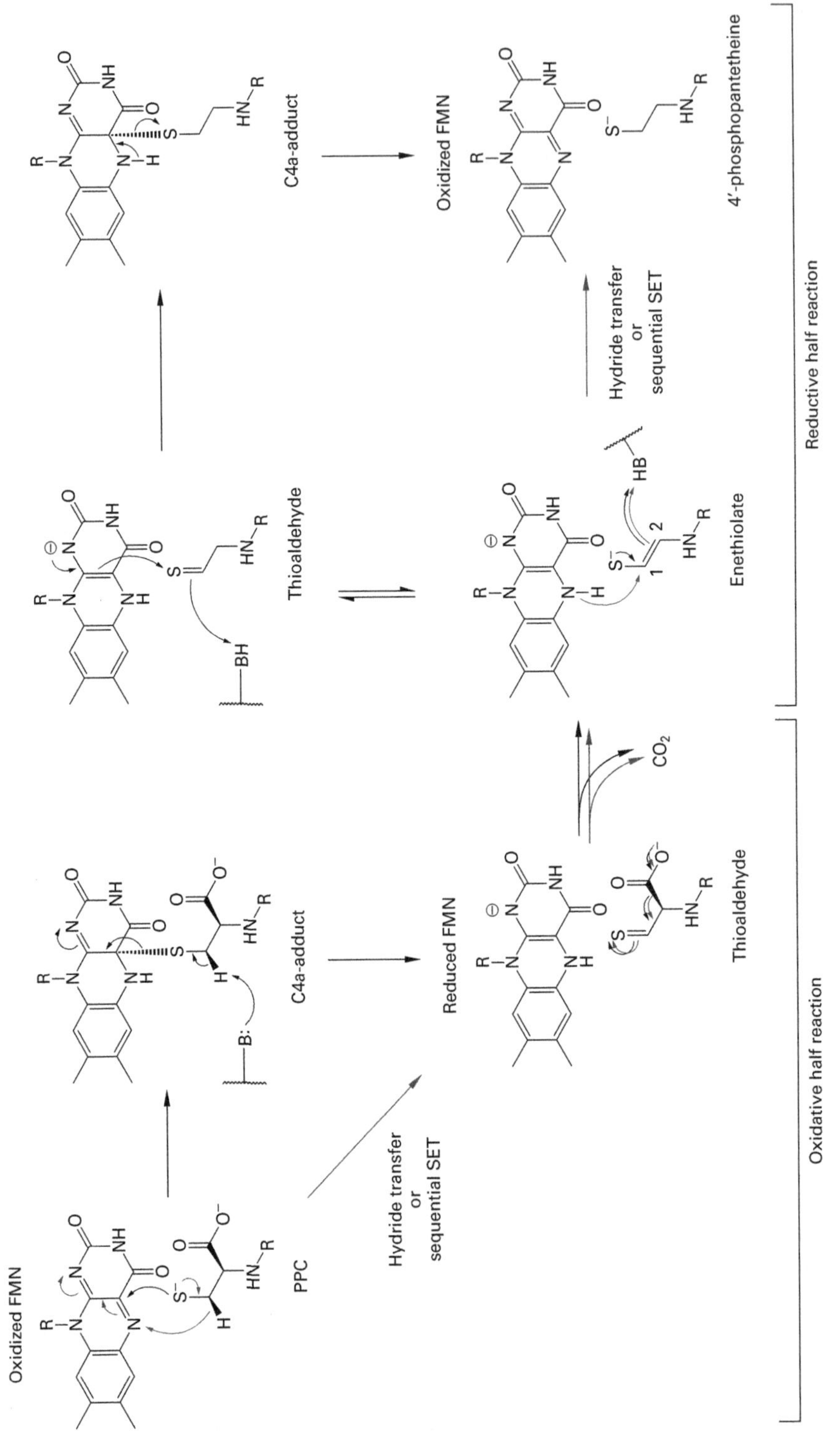

Fig. 3.4: Probable catalytic mechanisms of PPC-DC. Arrows in black and red indicate route A and B, respectively.

the product 4′-phosphopantetheine and oxidized FMN. Still, formation of the adduct has not been confirmed.

The crystal structure of the C175S mutant of *A. thaliana* PPC-DC in complex with reduced FMN and pantothenoylaminoethenethiolate (PDB ID: 1mvn) suggests a different mechanism [39]. The thiolate sulfur of the ligand points to C4a of FMN from the *re*-face of the isoalloxazine ring, as expected in the adduct-formation mechanism. However, the hydroxyl group of Ser175, which corresponds with the Cys residue that probably acts as the general acid for protonating the enethiolate ligand in the reductive half-reaction of the wild type PPC-DC, is 4.2 Å from C2 of the enethiolate moiety of the intermediate (corresponding with Cα of the cysteine moiety). On the other hand, C1 of the enethiolate (corresponding with Cβ of the cysteine moiety) is at 4 Å from N5 of FMN. These results imply the direct reduction of the C1=C2 double bond by hydride transfer or by sequential SET rather than via adduct formation (▸Fig. 3.4, route B, shown in red). Similarly, formation of the thioaldehyde intermediate in the oxidative half-reaction can be explained by direct hydride transfer or SET mechanism.

3.2.5 Other examples

In addition to the examples listed above, there are other flavoenzymes that catalyze non-redox reactions, but with redox-based functions for the flavin cofactors. Due to page limitations, only the catalytic mechanisms proposed for them, which partly resemble those for the flavoenzymes described above, will be concisely introduced.

DNA photolyases are widely distributed and dependent on reduced FAD [40,41]. SET from reduced FAD initiates repair of the DNA lesion (*i.e.*, cyclobutane pyrimidine dimer or (6-4) photoproduct, ▸Fig. 3.5), by inducing a rearrangement reaction that cleaves the dimer. After the repair reaction, reverse SET from a radical intermediate regenerates fully reduced FAD. These radical reactions appear similar to that of CS, but are in fact distinct because the initial SET reaction proceeds via the photoexcited state of the reduced flavin.

Alkyl-dihydroxyacetone phosphate (DHAP) synthase, which requires oxidized FAD for activity, catalyzes exchange of the acyl group of the substrate, acyl-DHAP, with an alkyl group in the biosynthesis of eukaryotic ether phospholipids [42,43]. The reaction is considered to proceed in a ping-pong manner via the formation of a covalent adduct

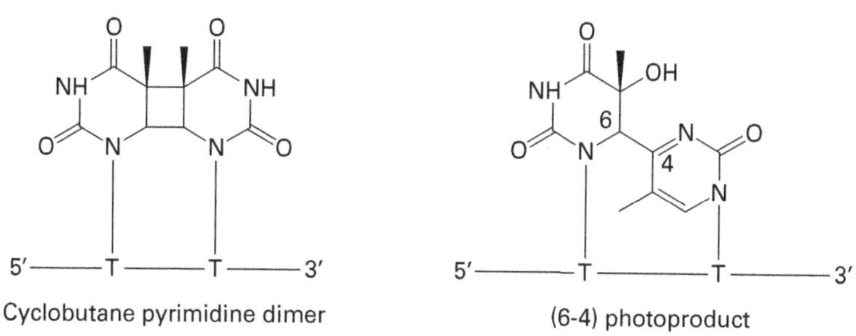

Fig. 3.5: Typical structures of DNA lesions repaired by photolyases.

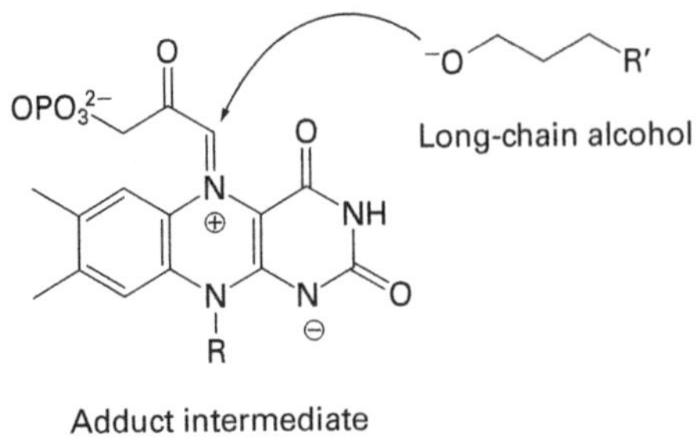

Fig. 3.6: Hypothetical structure of the adduct intermediate between FAD and DHAP moiety.

intermediate between FAD and the DHAP moiety (▶Fig. 3.6). This requires that FAD be temporarily reduced in the course of the reaction. However, the proposed position of FAD for the adduct formation in alkyl-DHAP synthase is N5, not C4a as proposed for the reaction of PPC-DC.

3.3 Flavoenzymes for which flavin cofactors likely play non-redox catalytic roles

As described above, the reaction mechanisms of most flavoenzymes that catalyze reactions involving no net oxidation-reduction have been considered on the basis of redox-catalyzing flavoenzymes. However, recent studies have revealed a few specific examples of flavoenzymes in which flavin appears not to have any formal redox role, but instead serves as a general acid/base catalyst and as a nucleophilic catalyst. There is no doubt that these studies have greatly expanded our knowledge of flavin chemistry.

3.3.1 Type 2 isopentenyl diphosphate isomerase

Isopentenyl diphosphate isomerase catalyzes the non-redox conversion between isopentenyl diphosphate (IPP) and dimethylallyl diphosphate (DMAPP), both of which serve as 5-carbon active units for isoprenoid biosynthesis [44]. The enzyme is required for the growth of organisms that produce isoprenoid via the mevalonate pathway. Flavin-dependent type 2 isopentenyl diphosphate isomerase (IDI-2) is found in almost all archaea and some bacteria, whereas the flavin-independent isozyme, type 1 isopentenyl diphosphate isomerase (IDI-1), is found in eukaryotes and other bacteria.

The cofactor requirement of IDI-2, which catalyzes no net redox reaction, has attracted interest. Although the enzyme requires FMN, NAD(P)H and Mg^{2+} [45], NAD(P)H is not stoichiometrically consumed in the course of the isomerization reaction [46] and can be replaced by sodium dithionite [47]. These results show that reduced FMN and Mg^{2+} are required for activity. The facts that apo-IDI-2 is inactive when reconstituted with 5-deaza-FMN [47,48] and that neutral flavin semiquinone is stabilized by IDI-2 in the presence of the substrate [49,50] have suggested a radical-based mechanism for the enzyme [47,49]. However, such a mechanism has been ruled out by a consideration of the properties of IDI-2: 1) EPR analysis of IDI-2 in the reaction does not reveal any substrate-derived radical species [48]; and, 2) a radical clock-type substrate analogue having a cyclopropyl group undergoes isomerization by IDI-2, without cleavage of the cyclopropyl ring via radical rearrangement [51,52]. Instead, several results support the notion that IDI-2 utilizes a protonation-deprotonation mechanism similar to that of IDI-1, in which conserved cysteine and glutamate residues act as general acid/base catalysts [53,54]. This includes measurement of deuterium kinetic isotopic effects in D_2O or with (*R*)-[2-^{2}H]-IPP [48] and studies using alkyne/allene diphosphate substrate analogues [55].

X-ray crystal structures of IDI-2 in complex with reduced FMN, Mg^{2+} and the substrates have been solved for the enzyme from *Sulfolobus shibatae* (PDB ID: 3b05 and 3b06 for IPP- and DMAPP-complexes, respectively) [56,57]. In these structures, the substrate binds in proximity of the isoalloxazine ring of reduced FMN, on its *si*-face. Their juxtaposition had been expected on the basis of the known adduct formation between reduced FMN and the substrate analogues containing a reactive epoxy-, vinyl-, or fluoride-group [58,59]. The crystal structures as well as the results from a series of alanine-replacement mutagenesis experiments have shown that there is no amino acid residue in proximity to the substrate that can catalyze the protonation-deprotonation reaction, and that only the reduced FMN can act as a general acid/base catalyst [56]. An acid/base role of the flavin cofactor is supported by the results of a linear free energy relationship (LFER) study using 7- and 8-substituted FMN analogues [60]. Adduct formation between reduced FMN and the substrate analogues has been observed in the crystals of IDI-2, and the structures demonstrate that the new covalent bond is formed at C4a of the flavin cofactor [57].

Considering these results, a mechanism has been proposed in which reduced FMN acts as both the acid/base catalyst and as a stabilizer for a cationic intermediate [57] (►Fig. 3.7). In a first step, the zwitterionic 5,5-dihydroflavin tautomer of neutral reduced FMN is proposed to protonate C4 of IPP to form the carbocation intermediate. N5 of the zwitterionic flavin has been suggested to have a pKa of ≥4 [61]. The cationic intermediate is proposed to be stabilized by the negative charge delocalized on the anionic reduced FMN, possibly including a charge transfer interaction between the two. Finally, N5 of the reduced FMN accepts a proton from C2 of the intermediate to yield DMAPP and to regenerate the zwitterionic flavin species.

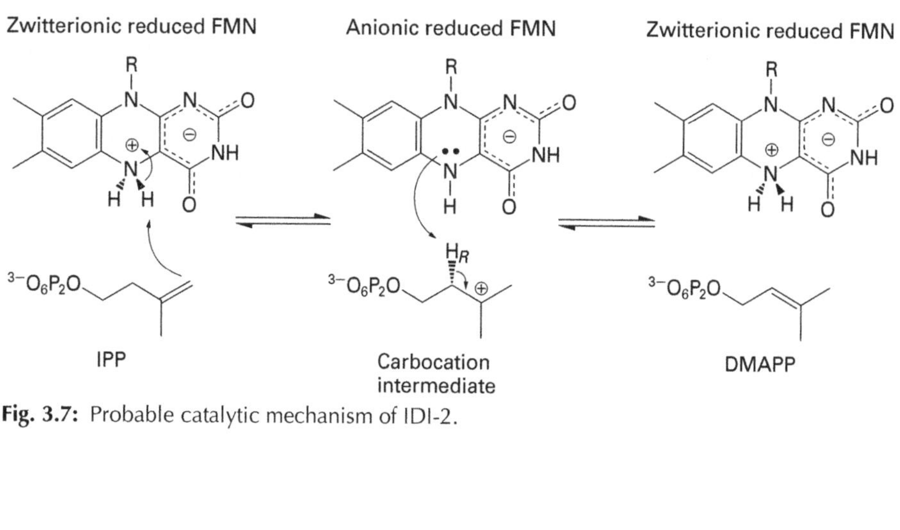

Fig. 3.7: Probable catalytic mechanism of IDI-2.

3.3.2 UDP-galactopyranose mutase

UDP-galactopyranose mutase (UGM) catalyzes the interconversion between UDP-galactopyranose (Gal*p*) and UDP-galactofuranose (Gal*f*). UDP-Gal*f*, which is less-favored at equilibrium, is a precursor of specific polysaccharide components of microbial cell walls, such as galactan from *Klebsiella pneunomiae*, arabinogalactan from *Mycobacterium tuberculosis*, and galactomannan from *Aspergillus fumigatus* [62]. Because some of these polysaccharides are important for pathogen virulence, and are absent in mammals, the enzymes involved in their biosynthesis are considered good targets for drug development.

UGM binds FAD and requires a reductant such as NAD(P)H to be activated [63,64]. The fact that 2- or 3-fluorinated (deoxyfluoro-) substrate analogues react with UGM rules out redox-based mechanisms involving 2- or 3-keto intermediates, which would initially require oxidized FAD [65,66]. However, fluorination decreased reactivity. The effect is more significant in the 2-substituted analogues than in the 3-substituted versions. In agreement with these results, cleavage of the anomeric C1-O1 bond during catalysis, which generates an oxocarbenium-type intermediate/transition state that would be destabilized by fluorination, has been confirmed by the observation of positional isotope exchange at O1 [67]. Various reaction mechanisms for UGM have been proposed, but most have been disproved. For example, the formation of a bicyclic 1,4-anhydrogalactose as an intermediate, formed by direct nucleophilic attack of the 4-hydroxy group on C1 and the accompanying elimination of the UDP group, has been ruled out because reaction of the putative intermediate with UGM does not yield the expected products [68]. Mechanisms involving the formation of an adduct intermediate by the attack of a nucleophilic catalyst on C1 of the substrate had been considered most likely, while the roles of reduced FAD in the mechanisms remained unclear.

A clue as to the catalytic role of the flavin cofactor was provided by a trial experiment to trap the adduct intermediate using sodium cyanoborohydride. Reaction of the enzyme-substrate complex with this reagent yield a reduced FAD that is covalently bonded at its N5 position with galactose-derived sugar [69]. The structure of the isolated compound suggests an adduct-formation mechanism in which N5 of reduced FAD acts as a nucleophilic catalyst (►Fig. 3.8). This hypothesis is supported by the X-ray crystal structure of reduced *K. pneumoniae* UGM binding UDP-Gal*p* (PDB ID: 3int) [70]. In this structure, N5 of FAD is 3.6 Å from C1 of the galactopyranose moiety, appropriate for nucleophilic attack. The UDP moiety extends from C1 in the direction opposite the isoalloxazine ring of FAD in a configuration appropriate for a leaving group.

Still, the mechanism to generate the flavin-substrate adduct remains controversial. The elimination of the UDP group with the cleavage of the anomeric C1-O1 bond, and the nucleophilic attack by N5 of reduced FAD can occur in concert (S_N2), or in an ordered manner via an oxocarbenium ion intermediate (S_N1). Further, adduct formation can also be explained by collapse of a radical pair that arises from SET between reduced FAD and the oxocarbenium intermediate. All of these variations are consistent with the inactivity of UGM that has been reconstituted with 5-deaza-FAD, which is incompatible with mechanisms based on either a nucleophilic role of N5 or SET [71]. The result of the LFER study on UGM using 7- and 8-substituted FAD analogues, as performed for IDI-2, is in good agreement with the S_N2 type reaction mechanism, as depicted in ►Fig. 3.8 [72].

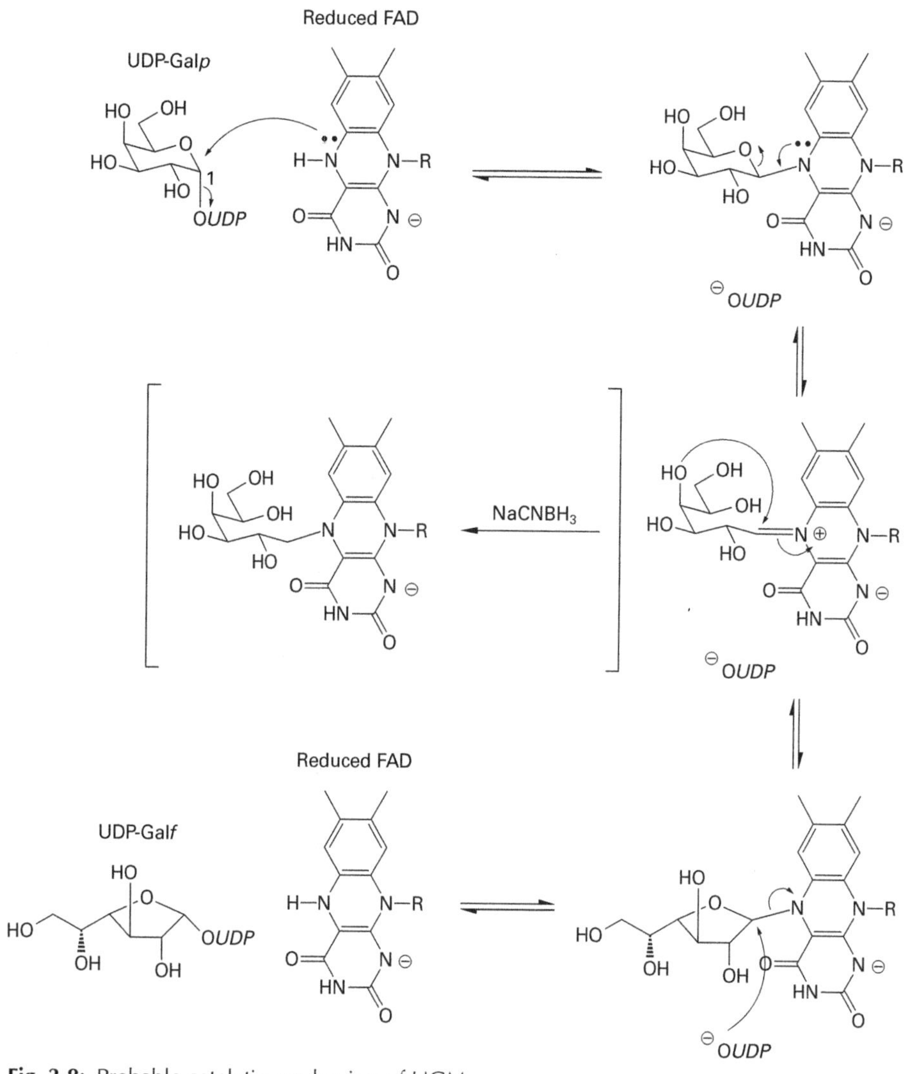

Fig. 3.8: Probable catalytic mechanism of UGM.

3.4 Flavoenzymes for which flavin cofactors play uncertain, but probably catalytic roles

The catalytic mechanisms of several, recently discovered flavoenzymes involved in non-redox reactions are yet to be characterized. Although hypothetical mechanisms have been proposed for them, the proofs now seem insufficient. Thus, the flavoenzymes are temporally categorized as those for which flavin cofactors play uncertain roles.

3.4.1 Lycopene cyclase

Lycopene cyclases catalyze the cyclization of one or both ends of the linear carotenoid compound lycopene to form cyclic carotenoids, *i.e.*, α-, β-, γ-, δ-, or ε-carotene (▸Fig. 3.9A) [73]. The enzymes are classified into 4 distinct groups, and 3 of them, *i.e.*, CrtY-, CrtL-, and CruA-type lycopene cyclases, possess the conserved N-terminal flavin-binding motif [74]. Exceptionally, heterodimer-type cyclases, which make up the remaining group, do not have the flavin-binding motif. All known lycopene cyclases are membranous and difficult to characterize biochemically; however, the binding of and dependence on flavins has been confirmed with lycopene β-cyclase CrtY from *Pantoea ananatis* [75] and CrtL-type capsanthin-capsorubin synthase (CCS) from pepper [76]. Although pepper CCS is not a genuine lycopene cyclase in catalyzing isomerization of the epoxy-containing rings of violaxanthin to form κ-cyclic carotenoids, it also has lycopene β-cyclase activity. Both *P. ananatis* CrtY and pepper CCS contain one molecule of FAD per monomer and are activated by a reductant such as NAD(P)H or Ti(III) citrate, suggesting reduced FAD is required [75–77]. Apo-CrtY is inactive, while the apoenzyme reconstituted with 5-deaza-FAD shows comparable, even higher, activity than the FAD-binding enzyme [75].

In addition, the activity of the enzyme is inversely related with the reduction potential of 8-substituted (and 5-deaza) FAD analogues. These results show that reduced FAD does not act as a general acid/base by the mechanism proposed for IDI-2. Rather, mutagenic studies suggest that a glutamate residue likely acts as the general acid/base catalyst [75,76]. Catalytic mechanisms for *P. ananatis* CrtY have been proposed in which anionic reduced FAD stabilizes either a cationic intermediate or transition state by charge transfer interaction [75]. However, detailed structural studies, particularly as concerns the distance between FAD and the substrate, are needed to elucidate the role of the flavin cofactor in lycopene cyclases.

3.4.2 Carotene *cis-trans* isomerase

Carotene *cis-trans* isomerase (CRTISO) is involved in carotenoid biosynthesis in plants [78,79] and in cyanobacteria [80], in which all-*trans* lycopene is produced via its *cis*-isomers. The enzyme catalyzes the *cis*-to-*trans* isomerization of all *cis* double bonds in prolycopene (7,9,9′,7′-tetra-*cis*-lycopene) to form all-*trans* lycopene (▸Fig. 3.9B). Interestingly, there is a sequence similarity between CRTISO and CrtI-type phytoene desaturase from archaea, bacteria, and fungi, which catalyzes desaturation (and, if 15-*cis*-double bond exists, isomerization) of phytoene to yield all-*trans* lycopene. The biochemical characterization of CRTISO from tomato [81], recombinantly expressed in *E. coli* and then solubilized and purified, shows that the enzyme contains non-covalently bound FAD. A reductant such as NADH or sodium dithionite is required for activity, as seen with the lycopene cyclase that catalyzes the immediately downstream reaction in the pathway. However, apo-CRTISO reconstituted with 5-deaza-FAD is inactive. CRTISO resembles IDI-2 in these properties, suggesting an acid/base role for the reduced FAD in CRTISO. Spectroscopic and structural analyses are required to better understand the catalytic role of the flavin cofactor, difficulties arising from it being a membrane protein notwithstanding.

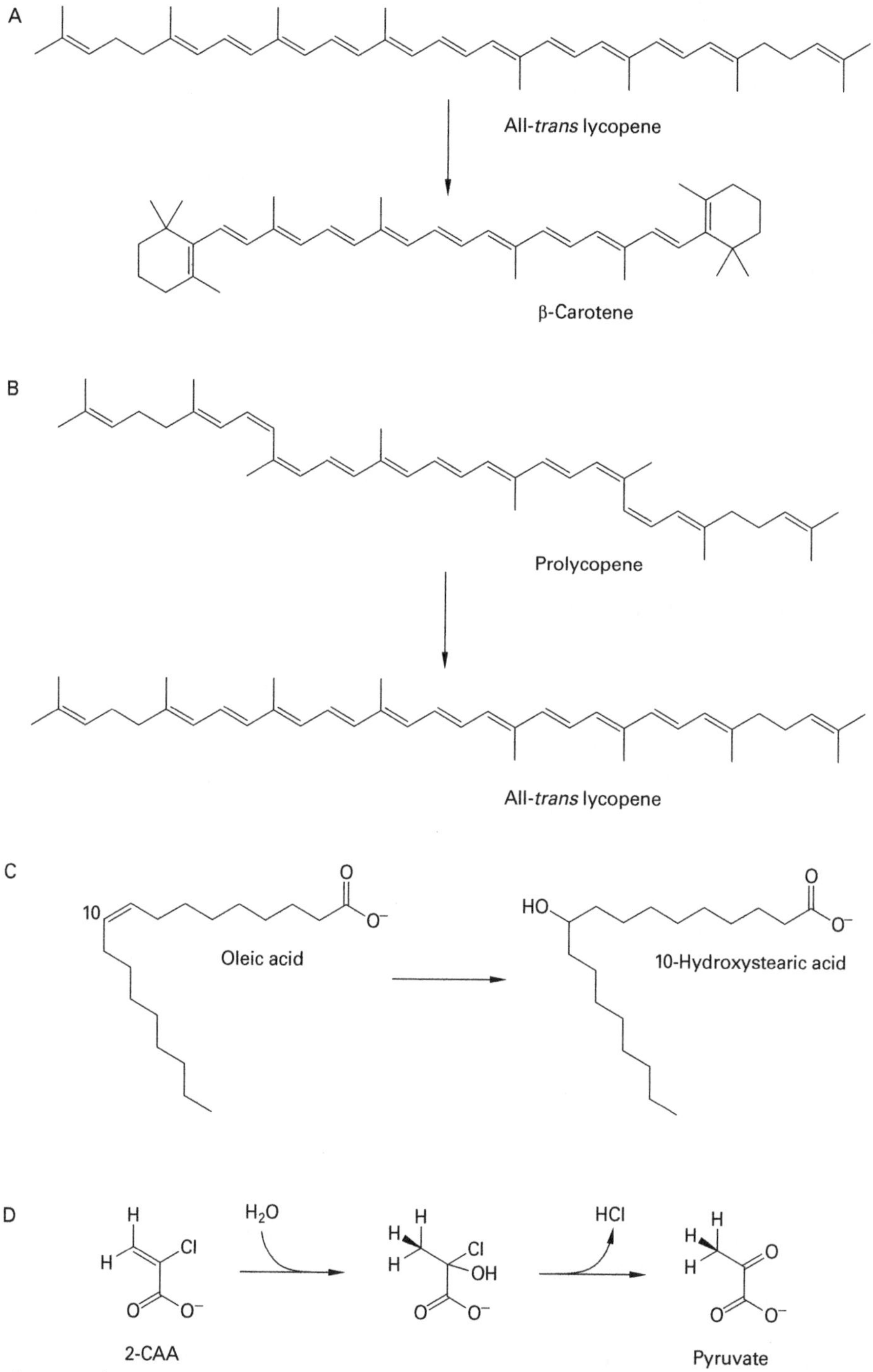

Fig. 3.9: The non-redox reactions catalyzed by flavoenzymes for which flavin cofactors play uncertain roles. (A) lycopene β-cyclase; (B) CRTISO; (C) fatty acid hydratase; (D) 2-haloacrylate hydratase.

3.4.3 Fatty acid hydratase

From the PUFA biohydrogenation pathways, another flavoenzyme that catalyzes a non-redox reaction has been identified. The enzyme, fatty acid hydratase, catalyzes the hydration of double bonds in fatty acids, including PUFAs. The enzyme from *Pseudomonas* species strain NRRL B-3266 has been cloned and characterized. It acts on the (9*Z*)-double bond of oleic acid and forms 10-hydroxystearic acid (▸Fig. 3.9C) [82]. The homologues of the enzyme from *Streptococcus pyrogenes* [83] and *Bifidobacterium breve* [84], which have been regarded as myosin cross-reactive antigen proteins, also have similar activity. The enzyme from *S. pyrogenes* acts on the (9*Z*)- and (12*Z*)-double bonds of C16 or C18 fatty acids, and typically produces 10-hydroxy and 10,13-dihydroxy fatty acids. The reaction in $H_2{}^{18}O$ has confirmed that the enzyme catalyzes hydration. *S. pyrogenes* fatty acid hydratase has been shown to contain oxidized FAD, and stopped-flow experiments on a time scale longer than 10 ms have shown that the UV-visible spectra of the enzyme do not change significantly in the course of the reaction, suggesting a non-redox (and possibly only structural) role for FAD. It must be recognized, however, that information is still limited and other possibilities are worth examining.

3.4.4 2-Haloacrylate hydratase

2-Haloacrylate hydratase from *Pseudomonas* sp. YL was identified as a protein induced by 2-chloroacrylate (2-CAA) [85]. The enzyme recombinantly expressed in *E. coli* contained FAD. To catalyze conversion of 2-CAA into pyruvate (▸Fig. 3.9D), the enzyme requires a reductant such as NAD(P)H or sodium dithionite, even though the reductant is not consumed stoichiometrically in the course of the reaction. The enzyme also accepts 2-bromoacrylate, but 2-fluoroacrylate, methacrylate, acrylate, 2-chloroacrylonitrile, 2-cloro-1-propene, fumarate, and phosphoenolpyruvate are inert as substrates. The product pyruvate contains ^{18}O when the reaction is performed in the presence of $H_2{}^{18}O$, indicating that the enzyme is a hydratase. Although the catalytic role of reduced FAD in the enzyme remains unclear, catalytic mechanisms have been proposed in which reduced FAD acts as a general acid/base catalyst to protonate at C3 or to help water attack at C2, or a SET from reduced FAD stimulates hydration of the substrate. These mechanisms yield a common intermediate 2-chloro-2-hydroxypropionate, which would decompose to the observed product pyruvate.

3.5 Conclusions

Study of the flavoenzymes involved in non-redox reactions, in particular those for which flavin cofactors play redox-independent roles, has led us to a better understanding of the depth of the (bio)chemical versatility of flavins. Their roles as general acid/base catalysts or nucleophilic catalysts are astonishing and far removed from their well-studied redox-based functions. New examples of such flavoenzymes continue to be discovered, and await future investigation. New and as-yet unidentified roles of flavin cofactors might be discovered in these enzymes.

3.6 References

[1] Pang SS, Duggleby RG, Guddat LW. Crystal structure of yeast acetohydroxyacid synthase: a target for herbicidal inhibitors. J Mol Biol 2002;317:249–62.

[2] Dreveny I, Andryushkova AS, Glieder A, Gruber K, Kratky C. Substrate binding in the FAD-dependent hydroxynitrile lyase from almond provides insight into the mechanism of cyanohydrin formation and explains the absence of dehydrogenation activity. Biochemistry 2009;48:3370–7.

[3] Macheroux P, Schmid J, Amrhein N, Schaller A. A unique reaction in a common pathway: mechanism and function of chorismate synthase in the shikimate pathway. Planta 1999;207:325–34.

[4] Welch GR, Cole KW, Gaertner FH. Chorismate synthase of *Neurospora crassa*: a flavoprotein. Arch Biochem Biophys 1974;165:505–18.

[5] Ehammer H, Rauch G, Prem A, Kappes B, Macheroux P. Conservation of NADPH utilization by chorismate synthase and its implications for the evolution of the shikimate pathway. Mol Microbiol 2007;65:1249–57.

[6] Lauhon CT, Bartlett PA. Substrate analogs as mechanistic probes for the bifunctional chorismate synthase from *Neurospora crassa*. Biochemistry 1994;33:14100–8.

[7] Osborne A, Thorneley RN, Abell C, Bornemann S. Studies with substrate and cofactor analogues provide evidence for a radical mechanism in the chorismate synthase reaction. J Biol Chem 2000;275:35825–30.

[8] Bornemann S, Ramjee MK, Balasubramanian S, et al. *Escherichia coli* chorismate synthase catalyzes the conversion of (6*S*)-6-fluoro-5-enolpyruvylshikimate-3-phosphate to 6-fluorochorismate. Implications for the enzyme mechanism and the antimicrobial action of (6*S*)-6-fluoroshikimate. J Biol Chem 1995;270:22811–5.

[9] Ramjee MN, Balasubramanian S, Abell C, et al. Reaction of (6*R*)-6-F-EPSP with recombinant *Escherichia coli* chorismate synthase generates a stable flavin mononucleotide semiquinone radical. J Am Chem Soc 1992;114:3151–3.

[10] Maclean J, Ali S. The structure of chorismate synthase reveals a novel flavin binding site fundamental to a unique chemical reaction. Structure 2003;11:1499–511.

[11] Rauch G, Ehammer H, Bornemann S, Macheroux P. Mutagenic analysis of an invariant aspartate residue in chorismate synthase supports its role as an active site base. Biochemistry 2007;46:3768–74.

[12] Willadsen P, Buckel W. Assay of 4-hydroxybutyryl-CoA dehydratase from *Clostridium aminobutyricum*. FEMS Microbiol Lett 1990;70:187–92.

[13] Li H, Cao Y. Lactic acid bacterial cell factories for γ-aminobutyric acid. Amino Acids 2010;39:1107–16.

[14] Scherf U, Söhling B, Gottschalk G, Linder D, Buckel W. Succinate-ethanol fermentation in *Clostridium kluyveri*: Purification and characterization of 4-hydroxybutyryl-CoA dehydratase/vinylacetyl-CoA Δ^3- Δ^2-isomerase. Arch Microbiol 1994;161:239–45.

[15] Scherf U, Buckel W. Purification and properties of an iron-sulfur and FAD-containing 4-hydroxybutyryl-CoA dehydratase vinylacetyl-CoA Δ^3- Δ^2-isomerase from *Clostridium aminobutyricum*. Eur J Biochem 1993;215:421–9.

[16] Müh U, Çinkaya I, Albracht SPJ, Buckel W. 4-Hydroxybutyryl-CoA dehydratase from *Clostridium aminobutyricum*: Characterization of FAD and iron-sulfur clusters involved in an overall non-redox reaction. Biochemistry 1996;35:11710–8.

[17] Müh U, Buckel W, Bill E. Mössbauer study of 4-hydroxybutyryl-CoA dehydratase - Probing the role of an iron-sulfur cluster in an overall non-redox reaction. Eur J Biochem 1997;248:380–4.

[18] Näser U, Pierik AJ, Scott R, Çinkaya I, Buckel W, Golding BT. Synthesis of ^{13}C-labeled γ-hydroxybutyrates for EPR studies with 4-hydroxybutyryl-CoA dehydratase. Bioorg Chem 2005;33:53–66.

[19] Buckel W, Keese R. One-electron redox reactions of CoASH esters in anaerobic-bacteria - a mechanistic proposal. Angew Chem Int Edit 1995;34:1502–6.
[20] Scott R, Näser U, Friedrich P, Selmer T, Buckel W, Golding BT. Stereochemistry of hydrogen removal from the 'unactivated' C-3 position of 4-hydroxybutyryl-CoA catalysed by 4-hydroxybutyryl-CoA dehydratase. Chem Commun 2004:1210–1.
[21] Martins BM, Dobbek H, Çinkaya I, Buckel W, Messerschmidt A. Crystal structure of 4-hydroxybutyryl-CoA dehydratase: radical catalysis involving a [4Fe-4S] cluster and flavin. Proc Natl Acad Sci USA 2004;101:15645–9.
[22] Kim J, Hetzel M, Boiangiu CD, Buckel W. Dehydration of (*R*)-2-hydroxyacyl-CoA to enoyl-CoA in the fermentation of α-amino acids by anaerobic bacteria. FEMS Microbiol Rev 2004;28:455–68.
[23] Kim J, Darley D, Buckel W. 2-Hydroxyisocaproyl-CoA dehydratase and its activator from *Clostridium difficile*. FEBS J 2005;272:550–61.
[24] Knauer SH, Buckel W, Dobbek H. Structural basis for reductive radical formation and electron recycling in (*R*)-2-hydroxyisocaproyl-CoA dehydratase. J Am Chem Soc 2011;133:4342–7.
[25] Or-Rashid MM, Wright TC, McBride BW. Microbial fatty acid conversion within the rumen and the subsequent utilization of these fatty acids to improve the healthfulness of ruminant food products. Appl Microbiol Biotechnol 2009;84:1033–43.
[26] Liavonchanka A, Feussner I. Biochemistry of PUFA double bond isomerases producing conjugated linoleic acid. ChemBioChem 2008;9:1867–72.
[27] Deng M-D, Grund AD, Schneider KJ, et al. Linoleic acid isomerase from *Propionibacterium acnes*: Purification, characterization, molecular cloning, and heterologous expression. Appl Biochem Biotechnol 2007;143:199–211.
[28] Wise ML, Hamberg M, Gerwick WH. Biosynthesis of conjugated triene-containing fatty acids by a novel isomerase from the red marine alga *Ptilota filicina*. Biochemistry 1994;33:15223–32.
[29] Wise ML, Rossi J, Gerwick WH. Characterization of the substrate binding site of polyenoic fatty acid isomerase, a novel enzyme from the marine alga *Ptilota filicina*. Biochemistry 1997;36:2985–92.
[30] Liavonchanka A, Rudolph MG, Tittmann K, Hamberg M, Feussner I. On the mechanism of a polyunsaturated fatty acid double bond isomerase from *Propionibacterium acnes*. J Biol Chem 2009;284:8005–12.
[31] Liavonchanka A, Hornung E, Feussner I, Rudolph MG. Structure and mechanism of the *Propionibacterium acnes* polyunsaturated fatty acid isomerase. Proc Natl Acad Sci USA 2006;103:2576–81.
[32] Begley TP, Kinsland C, Strauss E. The biosynthesis of coenzyme A in bacteria. Vitam Horm 2001;61:157–71.
[33] Strauss E, Kinsland C, Ge Y, McLafferty FW, Begley TP. Phosphopantothenoylcysteine synthetase from *Escherichia coli*. Identification and characterization of the last unidentified coenzyme A biosynthetic enzyme in bacteria. J Biol Chem 2001;276:13513–6.
[34] Daugherty M, Polanuyer B, Farrell M, et al. Complete reconstitution of the human coenzyme A biosynthetic pathway via comparative genomics. J Biol Chem 2002;277:21431–9.
[35] Hernández-Acosta P, Schmid DG, Jung G, Culiáñez-Macià FA, Kupke T. Molecular characterization of the *Arabidopsis thaliana* flavoprotein AtHAL3a reveals the general reaction mechanism of 4′-phosphopantothenoylcysteine decarboxylases. J Biol Chem 2002;277:20490–8.
[36] Majer F, Schmid DG, Altena K, Bierbaum G, Kupke T. The flavoprotein MrsD catalyzes the oxidative decarboxylation reaction involved in formation of the peptidoglycan biosynthesis inhibitor mersacidin. J Bacteriol 2002;184:1234–43.
[37] Strauss E, Begley TP. Mechanistic studies on phosphopantothenoylcysteine decarboxylase. J Am Chem Soc 2001;123:6449–50.

[38] Strauss E, Zhai H, Brand LA, McLafferty FW, Begley TP. Mechanistic studies on phosphopantothenoylcysteine decarboxylase: trapping of an enethiolate intermediate with a mechanism-based inactivating agent. Biochemistry 2004;43:15520–33.
[39] Steinbacher S. Crystal structure of the plant PPC decarboxylase AtHAL3a complexed with an ene-thiol reaction intermediate. J Mol Biol 2003;327:193–202.
[40] Mees A, Klar T, Gnau P, et al. Crystal structure of a photolyase bound to a CPD-like DNA lesion after in situ repair. Science 2004;306:1789–93.
[41] Li J, Liu Z, Tan C, et al. Dynamics and mechanism of repair of ultraviolet-induced (6-4) photoproduct by photolyase. Nature 2010;466:887–90.
[42] de Vet EC, Hilkes YH, Fraaije MW, van den Bosch H. Alkyl-dihydroxyacetonephosphate synthase. Presence and role of flavin adenine dinucleotide. J Biol Chem 2000;275:6276–83.
[43] Razeto A, Mattiroli F, Carpanelli E, et al. The crucial step in ether phospholipid biosynthesis: structural basis of a noncanonical reaction associated with a peroxisomal disorder. Structure 2007;15:683–92.
[44] Kuzuyama T, Hemmi H, Takahashi S. Mevalonate Pathway in Bacteria and Archaea. In: Mander L, Liu H-W, eds. Comprehensive Natural Products II Chemistry and Biology. Oxford, UK: Elsevier, 2010:493–516.
[45] Kaneda K, Kuzuyama T, Takagi M, Hayakawa Y, Seto H. An unusual isopentenyl diphosphate isomerase found in the mevalonate pathway gene cluster from *Streptomyces* sp. strain CL190. Proc Natl Acad Sci USA 2001;98:932–7.
[46] Yamashita S, Hemmi H, Ikeda Y, Nakayama T, Nishino T. Type 2 isopentenyl diphosphate isomerase from a thermoacidophilic archaeon *Sulfolobus shibatae*. Eur J Biochem 2004;271:1087–93.
[47] Hemmi H, Ikeda Y, Yamashita S, Nakayama T, Nishino T. Catalytic mechanism of type 2 isopentenyl diphosphate:dimethylallyl diphosphate isomerase: verification of a redox role of the flavin cofactor in a reaction with no net redox change. Biochem Biophys Res Commun 2004;322:905–10.
[48] Thibodeaux CJ, Mansoorabadi SO, Kittleman W, Chang WC, Liu HW. Evidence for the involvement of acid/base chemistry in the reaction catalyzed by the type II isopentenyl diphosphate/dimethylallyl diphosphate isomerase from *Staphylococcus aureus*. Biochemistry 2008;47:2547–58.
[49] Kittleman W, Thibodeaux CJ, Liu YN, Zhang H, Liu HW. Characterization and mechanistic studies of type II isopentenyl diphosphate:dimethylallyl diphosphate isomerase from *Staphylococcus aureus*. Biochemistry 2007;46:8401–13.
[50] Rothman SC, Helm TR, Poulter CD. Kinetic and spectroscopic characterization of type II isopentenyl diphosphate isomerase from *Thermus thermophilus*: evidence for formation of substrate-induced flavin species. Biochemistry 2007;46:5437–45.
[51] Johnston JB, Walker JR, Rothman SC, Poulter CD. Type-2 isopentenyl diphosphate isomerase. Mechanistic studies with cyclopropyl and epoxy analogues. J Am Chem Soc 2007;129:7740–1.
[52] Walker JR, Rothman SC, Poulter CD. Synthesis and evaluation of substrate analogues as mechanism-based inhibitors of type II isopentenyl diphosphate isomerase. J Org Chem 2008;73:726–9.
[53] Wouters J, Oudjama Y, Barkley SJ, et al. Catalytic mechanism of *Escherichia coli* isopentenyl diphosphate isomerase involves Cys-67, Glu-116, and Tyr-104 as suggested by crystal structures of complexes with transition state analogues and irreversible inhibitors. J Biol Chem 2003;278:11903–8.
[54] Wouters J, Oudjama Y, Stalon V, Droogmans L, Poulter CD. Crystal structure of the C67A mutant of isopentenyl diphosphate isomerase complexed with a mechanism-based irreversible inhibitor. Proteins 2004;54:216–21.
[55] Sharma NK, Pan JJ, Poulter CD. Type II isopentenyl diphosphate isomerase: Probing the mechanism with alkyne/allene diphosphate substrate analogues. Biochemistry 2010;49:6228–33.

[56] Unno H, Yamashita S, Ikeda Y, et al. New role of flavin as a general acid-base catalyst with no redox function in type 2 isopentenyl-diphosphate isomerase. J Biol Chem 2009;284:9160–7.
[57] Nagai T, Unno H, Janczak MW, Yoshimura T, Poulter CD, Hemmi H. Covalent modification of reduced flavin mononucleotide in type-2 isopentenyl diphosphate isomerase by active-site-directed inhibitors. Proc Natl Acad Sci USA 2011;108:20461–6.
[58] Hoshino T, Tamegai H, Kakinuma K, Eguchi T. Inhibition of type 2 isopentenyl diphosphate isomerase from *Methanocaldococcus jannaschii* by a mechanism-based inhibitor of type 1 isopentenyl diphosphate isomerase. Bioorg Med Chem 2006;14:6555–9.
[59] Rothman SC, Johnston JB, Lee S, Walker JR, Poulter CD. Type II isopentenyl diphosphate isomerase: irreversible inactivation by covalent modification of flavin. J Am Chem Soc 2008;130:4906–13.
[60] Thibodeaux CJ, Chang WC, Liu HW. Linear Free Energy Relationships Demonstrate a Catalytic Role for the Flavin Mononucleotide Coenzyme of the Type II Isopentenyl Diphosphate:Dimethylallyl Diphosphate Isomerase. J Am Chem Soc 2010;132:9994–6.
[61] Macheroux P, Ghisla S, Sanner C, Ruterjans H, Muller F. Reduced flavin: NMR investigation of N5-H exchange mechanism, estimation of ionisation constants and assessment of properties as biological catalyst. BMC Biochem 2005;6:26.
[62] Richards MR, Lowary TL. Chemistry and biology of galactofuranose-containing polysaccharides. ChemBioChem 2009;10:1920–38.
[63] Köplin R, Brisson JR, Whitfield C. UDP-galactofuranose precursor required for formation of the lipopolysaccharide O antigen of *Klebsiella pneumoniae* serotype O1 is synthesized by the product of the $rfbD_{KPO1}$ gene. J Biol Chem 1997;272:4121–8.
[64] Sanders DAR, Staines AG, McMahon SA, McNeil MR, Whitfield C, Naismith JH. UDP-galactopyranose mutase has a novel structure and mechanism. Nat Struct Biol 2001;8:858–63.
[65] Barlow JN, Blanchard JS. Enzymatic synthesis of UDP-(3-deoxy-3-fluoro)-D-galactose and UDP-(2-deoxy-2-fluoro)-D-galactose and substrate activity with UDP-galactopyranose mutase. Carbohyd Res 2000;328:473–80.
[66] Zhang Q, Liu HW. Mechanistic investigation of UDP-galactopyranose mutase from *Escherichia coli* using 2- and 3-fluorinated UDP-galactofuranose as probes. J Am Chem Soc 2001;123:6756–66.
[67] Barlow JN, Girvin ME, Blanchard JS. Positional isotope exchange catalyzed by UDP-galactopyranose mutase. J Am Chem Soc 1999;121:6968–9.
[68] Caravano A, Sinaÿ P, Vincent SP. 1,4-Anhydrogalactopyranose is not an intermediate of the mutase catalyzed UDP-galactopyranose/furanose interconversion. Bioorg Med Chem Lett 2006;16:1123–5.
[69] Soltero-Higgin M, Carlson EE, Gruber TD, Kiessling LL. A unique catalytic mechanism for UDP-galactopyranose mutase. Nat Struct Mol Biol 2004;11:539–43.
[70] Gruber TD, Westler WM, Kiessling LL, Forest KT. X-ray crystallography reveals a reduced substrate complex of UDP-galactopyranose mutase poised for covalent catalysis by flavin. Biochemistry 2009;48:9171–3.
[71] Huang Z, Zhang Q, Liu HW. Reconstitution of UDP-galactopyranose mutase with 1-deaza-FAD and 5-deaza-FAD: analysis and mechanistic implications. Bioorg Chem 2003;31:494–502.
[72] Sun HG, Ruszczycky MW, Chang WC, Thibodeaux CJ, Liu HW. Nucleophilic participation of reduced flavin coenzyme in mechanism of UDP-galactopyranose mutase. J Biol Chem 2012;287:4602–8.
[73] Cunningham FX, Jr., Gantt E. One ring or two? Determination of ring number in carotenoids by lycopene ε-cyclases. Proc Natl Acad Sci USA 2001;98:2905–10.
[74] Maresca JA, Graham JE, Wu M, Eisen JA, Bryant DA. Identification of a fourth family of lycopene cyclases in photosynthetic bacteria. Proc Natl Acad Sci USA 2007;104:11784–9.

[75] Yu Q, Schaub P, Ghisla S, Al-Babili S, Krieger-Liszkay A, Beyer P. The lycopene cyclase CrtY from *Pantoea ananatis* (formerly *Erwinia uredovora*) catalyzes an FAD_{red}-dependent non-redox reaction. J Biol Chem 2010;285:12109–20.

[76] Mialoundama AS, Heintz D, Jadid N, et al. Characterization of plant carotenoid cyclases as members of the flavoprotein family functioning with no net redox change. Plant Physiol 2010;153:970–9.

[77] Schnurr G, Misawa N, Sandmann G. Expression, purification and properties of lycopene cyclase from *Erwinia uredovora*. Biochem J 1996;315:869–74.

[78] Isaacson T. Cloning of *tangerine* from tomato reveals a carotenoid isomerase essential for the production of β-carotene and xanthophylls in plants. Plant Cell 2002;14:333–42.

[79] Park H. Identification of the carotenoid isomerase provides insight into carotenoid biosynthesis, prolamellar body formation, and photomorphogenesis. Plant Cell 2002;14:321–32.

[80] Masamoto K, Wada H, Kaneko T, Takaichi S. Identification of a gene required for *cis*-to-*trans* carotene isomerization in carotenogenesis of the cyanobacterium *Synechocystis* sp. PCC 6803. Plant Cell Physiol 2001;42:1398–402.

[81] Yu Q, Ghisla S, Hirschberg J, Mann V, Beyer P. Plant carotene *cis-trans* isomerase CRTISO: a new member of the FAD_{red}-dependent flavoproteins catalyzing non-redox reactions. J Biol Chem 2011;286:8666–76.

[82] Bevers LE, Pinkse MW, Verhaert PD, Hagen WR. Oleate hydratase catalyzes the hydration of a nonactivated carbon-carbon bond. J Bacteriol 2009;191:5010–2.

[83] Volkov A, Liavonchanka A, Kamneva O, et al. Myosin cross-reactive antigen of *Streptococcus pyogenes* M49 encodes a fatty acid double bond hydratase that plays a role in oleic acid detoxification and bacterial virulence. J Biol Chem 2010;285:10353–61.

[84] Rosberg-Cody E, Liavonchanka A, Göbel C, et al. Myosin-cross-reactive antigen (MCRA) protein from *Bifidobacterium breve* is a FAD-dependent fatty acid hydratase which has a function in stress protection. BMC Biochem 2011;12:9.

[85] Mowafy AM, Kurihara T, Kurata A, Uemura T, Esaki N. 2-Haloacrylate hydratase, a new class of flavoenzyme that catalyzes the addition of water to the substrate for dehalogenation. Appl Environ Microbiol 2010;76:6032–7.

4 Enzymes of FMN and FAD Metabolism

Milagros Medina

Abstract

Organisms from all kingdoms require the conversin of endogenous or exogenous RF into FMN and FAD to serve as cofactors for their flavoproteins and flavoenzymes. Therefore, knowledge of the genetics, biochemistry, regulation and structure-function relationships of the enzymes involved in the FMN and FAD metabolic pathways, as well as the differences in these processes between species, may improve the use of these proteins in biotechnological processes and therapeutic applications. In this chapter an overview is provided of the current knowledge of the proteins involved in flavin metabolism in different organisms. For those cases where biochemical and structural data are available, structure-function relationships are described. Finally, insights into the subcellular localization of the enzymes involved in flavin metabolism, as well as systems for flavin transport between organelles are discussed.

4.1 Introduction

Riboflavin (Vitamin B_2, RF) is the universal precursor of flavin mononucleotide (FMN) and flavin adenine dinucleotide (FAD) cofactors, which are essential for the action of flavoproteins and flavoenzymes. RF is produced exclusively in microorganisms, yeast and plants, but must be taken up by mammals through their diet (►Fig. 4.1) [1–4]. In addition to its natural production, a large amount of this vitamin is biotechnologically manufactured (more than 3000 metric tons per year) to satisfy the need for RF in human and animal nutrition, and as a food colorant [5]. These aspects were an important driving force for understanding the RF biosynthesis pathway. The overall steps of this process have been reviewed on several occasions [5–9], and this knowledge has allowed RF production by fermentation to essentially replace its chemical synthesis.

Flavoproteins and flavoenzymes in all organisms participate in numerous key metabolic pathways involving one- and two-electron oxido-reduction processes that include mitochondrial electron transport, photosynthesis, fatty-acid oxidation, denitrification, sulfur respiration, and the metabolism of vitamin B_6, vitamin B_{12} and folates. This is due to the unique properties and the chemical versatility that the FMN and FAD cofactors confer on these proteins [10–13]. In addition to the redox enzymes, 10% of FMN- and FAD-dependent enzymes are involved in important non-redox reactions including light sensing, bioluminescence, circadian time-keeping, and DNA repair [14]. Thus, organisms from all kingdoms require conversion of endogenous or exogenous RF into FMN and FAD to serve as cofactors for their flavoproteins and flavoenzymes. In addition to their catalytic roles, flavin cofactors can also play regulatory roles, affecting protein

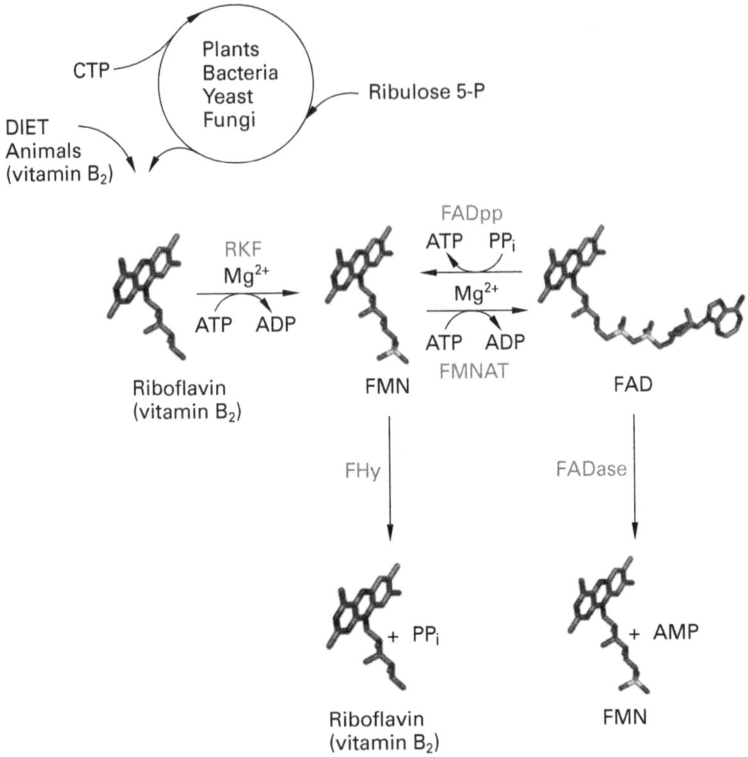

Fig. 4.1: Scheme for the RF production/uptake and the reactions in which FMN and FAD can be involved during their metabolism in different organisms. Names for each reaction are in grey, while those for substrates and products are in black.

expression and stability [15–20]. In eukaryotes, most flavoenzymes are compartmentalized in cellular organelles, where they ensure the functionality of particular processes. However, despite the crucial roles of flavin cofactors, interest in understanding the behavior of the enzymes responsible for the production of FMN and FAD, their hydrolysis, their subcellular localization, and their transport between organelles, has only been fuelled in the last 10–15 years. This chapter will predominantly review the current knowledge about FMN and FAD metabolism in different organisms, particularly the biosynthesis of FMN and FAD from RF.

4.2 Enzymes involved in the production of FMN and FAD in different organisms

All organisms are able to transform RF, first into FMN, and then into FAD, by the sequential action of two enzymatic activities, an ATP:riboflavin kinase (RFK, EC 2.7.1.26) and an ATP:FMN adenylyltransferase (FMNAT, also known as FAD synthetase, EC 2.7.7.2) (▶Fig. 4.1) [21]. Despite the large number of flavoproteins involved in the cellular metabolism of all types of organisms, as well as the fact that deficiency of RFK or FMNAT activities prevents the assembly of essential holoproteins, inducing metabolic alterations linked to physiological or pathological disorders, relatively few studies on these activities were reported in previous decades [1,3,22-24]. Fortunately this situation has changed considerably with more recent recognition of the key roles of these activities. A significant advance in studies of these enzyme activities came with the discovery that most prokaryotic organisms depend on a single bifunctional enzyme (named bifunctional FAD synthetase (FADS)) that catalyzes both activities (▶Figs. 4.1 and 4.2) [25–31], while eukaryotes generally use two different enzymes for FMN and FAD production. Recently, the increased availability of genome sequences and of methods to analyze them has considerably contributed to clarification of the distribution of the RFK and FMNAT activities in different organisms, as well as to identification of other enzymes involved in flavin metabolism (▶Fig. 4.2).

Yeasts, humans, rats and, in general, non-photosynthetic eukaryotes have independent monofunctional enzymes with either RFK or FMNAT activity. The first eukaryotic gene with putative FMNAT activity was identified in the *Saccharomyces cerevisiae* genome. It was designated *FAD1* and corresponded to a 35 kDa soluble cytosolic protein [32]. The first eukaryotic gene for an RFK activity (*FMN1*) coding for a membrane-bound protein putatively located in microsomes and in the inner mitochondrial membrane was also identified in *S. cerevisiae* [33]. Later, orthologous sequences were found in other eukaryotes, including humans, *Schizosaccharomyces pombe, Drosophila melanogaster* and *Caenorhabditis elegans*, and sequences showing relatively high identity

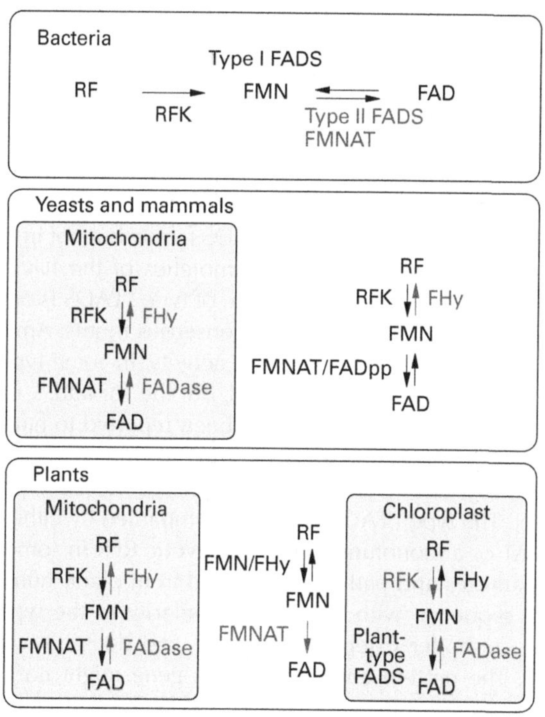

Fig. 4.2: Distribution of the activities present in subcellular localizations of different organisms. Black arrows and names indicate enzymes that have been either cloned and the recombinant proteins biochemically characterized, or whose existence is supported by biochemical evidences. Grey arrows and names indicate enzymes whose existence is supported by bioinformatics analysis.

and either lacking or including predicted mitochondrial localization sequences were also identified [26,33]. Therefore, subcellular localization of RFK and FMNAT activities and their dependence on different isoforms have been a matter of debate in recent years [23,26,33–35]. Eukaryotic FMNAT is classified as a member of the 3′-phosphoadenosine 5′-phosphosulfate reductase-like family (PAPS) belonging to the "adenosine nucleotide α-hydrolase-like" superfamily. Bioinformatics analyses detected two isoforms for the FMNAT activity in humans that differed by an extra-sequence of 97 residues at the N-terminus of isoform 1 (hFADS1) that is not present in isoform 2 (hFADS2). The first 17 residues of hFADS1 constitute a cleavable mitochondrial targeting sequence, suggesting that hFADS1 represents a mitochondrial isoenzyme [35]. In both isoforms, the hFADS sequence is organized into two regions: a phosphoadenosine phosphosulfate reductase at the C-terminus, and an N-terminus of unknown function resembling a molybdo-pterin-binding domain [35]. At least two more isoforms might be predicted in humans by gene sequences, but they remain uncharacterized and enigmatic [36].

Sequence analyses indicate that most prokaryotes contain a single bifunctional FADS herein labeled as type I FADS and represented by that found in *Corynebacterium ammoniagenes* (herein *Ca*FADS) [37,38]. These analyses also pointed out that each of the prokaryotic type I FADS activities was mainly associated with one of the protein regions: the RFK at the C-terminal region (residues 187–338, *Ca*FADS numbering) and the FMNAT at the N-terminal region (residues 1–186) [38,39]. The FMNAT N-terminal region of FADS shares little or no sequence similarity to eukaryotic FMNATs, as these enzymes belong to two different protein superfamilies that require different active-site residues [35,38,40]. Thus, while the eukaryotic FMNAT sequence is classified as a member of PAPS family, the FMNAT-region of type I FADS showed remote similarity to the nucleotidyltransferase (NT) superfamily. The FMNAT region of type I FADS contains the typical motifs involved in binding nucleotide and phosphate groups in NTs (28-(H/T)xGH-31, 123-Gx(D/N)-125 and 162-xSS(T/S)xxR-168), as well as additional residues that are conserved, or conservatively substituted, in the FADS family, but not in NTs [37,38]. This suggested that residues conserved in the FADS family but not in NTs might contribute to binding of the ribityl and isoalloxazine moieties of the flavin ligands, which are not recognized by other NTs. The RFK module of type I FADS is orthologous to monofunctional eukaryotic RFKs, including several consensus motifs. Among them, the 207-PTAN-210 motif is proposed to be key for RFK activity. In some type I FADSs and RFKs, sequence variants in this region can be found, but the Thr and Asn positions are always conserved [37,38]. Only four bacteria have been reported to have the RFK and FMNAT activities separated in monofunctional proteins, with their sequences corresponding to the RFK C-terminal and FMNAT N-terminal regions of the prokaryotic bifunctional type I FADS [37]. The type I FADS is also accompanied by either a monofunctional prokaryotic FMNAT or a monofunctional prokaryotic RFK in some bacteria. A small group of bacterial parasites and pathogens isolated from plant, human or soil material have an additional sequence with significant similarity to the type I FADS. However, this sequence has a shorter C-terminal module and does not contain the conserved PTAN motif [37]. The protein encoded by this gene might not have RFK activity and might constitute another divergent type of FADS. These sequences have been named type II FADS. Only four micro-organisms appear to contain exclusively type II FADS to carry out the FMNAT activity [37]. Finally, sequences of type I FADS have also been found in some primitive multi-cellular animals [37].

Green plants have sequences homologous with the monofunctional enzymes with RFK and FMNAT activities found in mammals and yeast, but they also have sequences with high similarity to prokaryotic FADS [41]. Different studies suggest that the RFK and FMNAT activities are localized in the cytosol as well as in chloroplasts and mitochondria [28,41,42]. The lack of organellar localization sequences (analogous to the mammal and yeast ones) in the monofunctional enzymes suggested cytosolic localization for these enzymes in plants, while the presence of localization sequences in the bifunctional sequences suggests they localize in organelles. Bioinformatics approaches also suggested that in several plant species, the sequences homologous with RFK from yeasts and mammals contain an additional N-terminal region that belongs to the haloacid dehalogenase superfamily (HAD) related with a FMN hydrolase (FHy) activity [41]. Finally, other bifunctional plant sequences display an N-terminal region with high similarity to the FMNAT module of the prokaryotic type I FADS, but have a shorter C-terminal region that lacks the PTAN motif and contains motifs that are not observed in prokaryotic FADS. Therefore, a different enzymatic activity from RFK is suggested for the C-terminal module in these enzymes that have been labeled as plant-type FADS [37].

4.3 FMN and FAD metabolism in yeasts and mammals

In yeasts and mammals FMN and FAD are mainly required in mitochondria, where they act as redox cofactors in a number of important reductases, dehydrogenases and oxidases that have crucial functions in bioenergetics and regulation. Therefore, the mechanisms involved in the uptake, biogenesis and/or processing of these cofactors by mitochondria are of significant interest. Initially, synthesis of FMN and FAD from RF was identified to occur in the cytosol [43,44], but later it was evidenced that mitochondria isolated from both rat liver and *S. cerevisiae* were also able to synthesize FAD from FMN or RF [26,45,46]. These results indicated that mitochondria also have enzymes designed for the synthesis of both flavin cofactors (▶Fig. 4.2).

Enzymes with RFK activity have been characterized from several yeast species, including *Pichia guilliermondii* and *S. cerevisiae* (herein *Pg*RFK, and *Sc*RFK, respectively) [33,46,47], as well as from fungi, *e.g., Neurospora crassa* (*Ns*RFK) [48], and mammals, including rats and humans (rRFK and hRFK, respectively) [49,50]. Different isoforms varying in their physico-chemical properties have been reported for organisms such as *Pg*RFK [51]. Moreover, RFK activity has been found at the cytosolic, microsomal and mitochondrial level in different eukaryotes [46]. Purified *Pg*RFK and *Sc*RFK (encoded by *FMN1*) are monofunctional enzymes that efficiently catalyze the formation of FMN from RF and ATP, without accompanying FMNAT or phosphatase activities. The active forms of these enzymes were described as monomers ranging from ~25 to 28 kDa that required the presence of divalent cations, particularly Mg^{2+}, to exhibit maximal activity. These enzymes also catalyze the phosphorylation of RF analogues with substitutions of the methyl groups at positions 7 and 8 of the flavin ring, and in some cases can use UTP, GTP, ADP and CTP, besides ATP, as phosphate donors. AMP generally inhibited the enzyme activity, which appears maximal at pH ~8.7 and with temperature optima in the 45–52 °C range. *Sc*RFK has been localized on the cytoplasmic face of the endoplasmic reticulum membranes, as well as in the inner membrane of mitochondria exposing its C-terminal domain to the matrix space [33]. *Ns*RFK has also been purified as a monomer of 35.5 kDa, with maximal activity at alkaline pH and 55 °C [48]. In general the K_m^{RF} values for the yeast and fungi proteins are low, 0.1–0.2 μM, while larger differences, up to 500-fold, are reported for their K_m^{ATP} values (►Tab. 4.1). Mammalian RFKs have been reported to be soluble cytosolic proteins and have been purified from several tissues [49,50,52]. RFKs purified from rat liver or intestinal mucosa (rRFKl and rRFKim, respectively) are proteins of ~14 kDa, that are similar but not identical, and are able to transform most RF analogues modified at position 8 of the isoalloxazine ring [49]. The gene encoding hRFK has been cloned, overexpressed and the corresponding monomeric protein of ~18.5 kDa purified [50]. Steady-state kinetic analysis, and evaluation of product inhibition, in rRFKl suggested an ordered bi-bi mechanism (►Tab. 4.1), in which RF binds first, followed by ATP, then ADP is released followed by FMN [53]. The crystal structure of the apo form has been determined for the *Schizosaccharomyces pombe* RFK (*Sp*RFK) [54], while several complexes of this enzyme, as well as of hRFK, in complexes with the ligands have also been reported (ADP, ADP:Mg^{2+}, FMN:ADP, FMN:ADP:Mg^{2+}) [50,54,55] (►Fig. 4.3A). The RFK structures represent a novel

Tab. 4.1: Summary of published steady-state kinetic parameters for the RFK activity (RF + ATP → FMN + ADP) in enzymes from different organisms

	k_{cat} (min^{-1})	K_m^{RF} (μM)	K_i (μM)	k_{cat}/K_m^{RF} (min^{-1}μM^{-1})	k_{cat} (min^{-1})	K_m^{ATP} (μM)	k_{cat}/K_m^{ATP} (min^{-1}μM^{-1})	
*Pg*RFK		0.1				6.7		[47]
*Nc*RFK		0.12				0.21		[48]
rRFKl		11.4				3.7		[53]
rRFKim						5		[49]
hRFK	18/30[a]	117/36[a]		0.15/0.83				[101]
*Ca*FADS	<302[b]	<13[b]	4.0	23.2	68[b, c]	13.7[b, c]	4.93[c]	[71]
*Ca*FADS	24.6	<<<0.01	0.08	–		89.3	0.28	[25]
*Bs*FADS		<0.1				6.5		[30]
*Bs*RFK		0.18				112		[69]
*At*FMN/FHy	25.2	0.01		2520	26.4	2.0	13.2	[41]
*Nt*RFK		0.32[d]						[27]

[a] The first value of the pair corresponds to the reaction in the absence of sodium dithionite, while the second corresponds to the same parameter in the presence of 24 mM sodium dithionite
[b] Inhibition by substrate prevented the determination of true parameters and these correspond to apparent ones; ${}^{app}k_{cat}$ and ${}^{app}K_m$
[c] Parameters estimated using a RF concentration exhibiting ~80% of the experimentally detected maximal activity
[d] Parameter corresponding to $S_{0.5}$

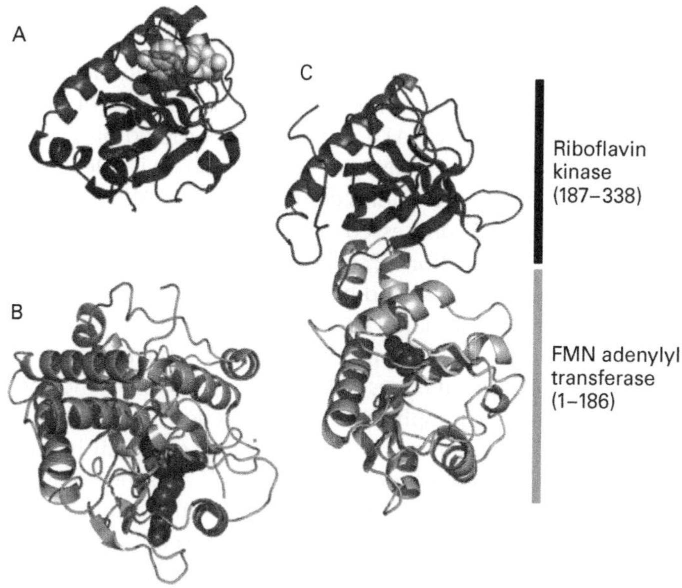

Fig. 4.3: Cartoon representation of the crystal structures of the enzymes involved in the production of FMN and FAD in different organisms. (A) hRFK in complex with ADP:Mg^{2+} (light grey spheres on the front) and FMN (dark spheres at the back) products of the reaction (pdb 1Q9S). (B) *Cg*FMNAT in complex with FAD:Mg^{2+} (dark grey spheres) (pdb 3G6K). (C) *Ca*FADS (pdb 2X0K). The RFK and FMNAT modules are represented in dark and light grey, respectively. At the FMNAT module the PP_i product of the reaction is also included as dark grey spheres.

family of phosphoryl transferring enzymes that utilizes a unique nucleotide-binding site. These structures revealed a six-stranded antiparallel beta barrel core, clasped on one side by two C-terminal helices that display an L-like shape. The opposite side of the β-barrel serves as a platform for substrate binding as demonstrated by complexes with the products of the reaction, ADP:Mg^{2+} and FMN. The binding site of ADP:Mg^{2+} defines a novel nucleotide-binding motif that encompasses a loop, a 3_{10} helix, and a reverse turn followed by a short β-strand that contains the PTAN consensus motif. This active site loop forms an arch, with ATP and RF binding at the opposite side. The phosphoryl transfer appears to occur through the hole underneath the arch [50]. The structure of the RFK:ADP:Mg^{2+}:FMN complex shows drastic conformational changes associated with flavin binding, primarily at the so-called Flap I and Flap II loop regions (▶Fig. 4.3A). The bound FMN molecule interacts with the enzyme extensively and is well-ordered. Residues from Flap II interact with Flap I and shield the FMN molecule from the solvent, whereas one of the FMN phosphate oxygens and the Asn36 (belonging to the PTAN conserved motif) side-chain carbonyl coordinate the Mg^{2+} ion. This structure also indicates that the proposed catalytic base, Glu86, is well-positioned for activation of the O5′ hydroxyl of RF for the phosphoryl transfer reaction. The unusual metal-binding site involves, in addition to the ADP phosphates, the strictly conserved Thr and Asn from the PTAN motif. These structural data are also in agreement with the ordered kinetic mechanism of RFK previously proposed [54,55].

Proteins with FMNAT activity have been characterized from different species, and this activity has been shown to reside in both cytosol and mitochondria [46,53]. Differences in protein thermal stability, molecular mass, and specific activity provided evidence for the presence of different isoforms in mitochondria and cytosol [43]. Early kinetic analysis using rat liver FMNAT (rFMNATI) suggested an ordered bi-bi mechanism in which ATP binds to enzyme prior to FMN, and pyrophosphate is released before FAD [53]. A number of *in vitro* and *in vivo* studies confirmed that hFADS1, but not hFADS2, is localized in mitochondria and is processed into a lower molecular mass protein once it reaches this subcellular localization [23]. Recombinant hFADS1 has a ${}^{app}K_m^{FMN}$ of 1.3 μM, is strongly inhibited by FMN, and has a strict requirement for Mg^{2+}. hFADS2 has also been purified as a recombinant His-tagged protein of ~57 kDa with pI ~5.6–6.0. The purified enzyme binds the FAD product very tightly, but noncovalently. Under turnover conditions, the enzyme catalyzes FAD assembly from ATP and FMN and, at a much lower rate, the reverse FAD pyrophosphorolytic reaction (FAD pyrophosphorylase activity, FADpp) to produce FMN and ATP in the presence of PP_i [36]. This reaction is also catalyzed by *Sc*FMNAT, but not by rFMNATI [56,57]. Absence of $MgCl_2$, as well as the presence of PP_i or GTP, prevents FAD synthesis, while lack of PP_i prevents the pyrophosphorolytic process. $MgCl_2$ has also been shown to be essential for activities in other recombinant hFADS preparations [34,35] and for the enzymes from rat, plants and yeast [27,56]. In general, kinetic parameters thus far reported for the FMNAT enzymes from yeasts and animals are in the same order of magnitude (▶Tab. 4.2). However,

Tab. 4.2: Summary of published steady-state kinetic parameters for the FMNAT (FMN + ATP → FAD + PPi) activity in enzymes from different organisms

	k_{cat} (min^{-1})	K_m^{FMN} (μM)	K_m^{ATP} (μM)	k_{cat}/K_m^{FMN} (min^{-1}μM^{-1})	k_{cat}/K_m^{ATP} (min^{-1}μM^{-1})	K_i^{FMN} (μM)	K_i^{ATP} (μM)	
rFMNATI	9	9.1	71	0.98	0.13	0.75	1.3	[53]
hFADS1		1.3						[23]
hFADS2	4.7	0.35	15.3	13.4	0.31			[36]
hFADS2	0.2	0.36		3.33				[34]
hFADS2	1.2/4.8[a]	109/68[a]		0.01/0.07[a]				[101]
*Cg*FADS	5.22	0.76	10.7	20	1.4			[40]
*Ca*FADS	17.0	1.2	35.7	14.3	0.48			[71]
*Ca*FADS	18.0	1.0	37.0	18.0				[25]
*Bs*FADS		0.38						[30]
*At*RibF1	0.65	18.9	11.1	0.03	0.06	0.85	18.7	[28]
*At*RibF2	0.77	20.8	13.1	0.07	0.06	0.92	13.02	[28]

[a] The first value of the pair corresponds to the reaction in the absence of sodium dithionite, while the second corresponds to the same parameter in the presence of 24 mM sodium dithionite

important differences have been reported for different preparations of the same protein, such as hFADS2, suggesting different production, isolation procedures or components in the assay reaction mixture might alter the efficiency of the reaction. FAD release appears to be the rate-limiting step of the whole catalytic cycle. Therefore, it is proposed that the interaction of the enzyme with an acceptor apoprotein, and probably also an adequate redox state or some accessory proteins, must be required to complete the cycle. This suggests that the process leading to FAD synthesis and delivery to client apoproteins may be tightly controlled [36]. Three-dimensional structures for eukaryotic FMNAT have been reported for two yeast enzymes: the FMNAT from the *Candida glabrata* pathogen (*Cg*FMNAT; pdb 3FWK) and the FAD1 isoform from *S. cerevisiae* (*Sc*FMNAT; pdb 2WSI) (►Fig. 4.3B) [40,58]. These structures confirm eukaryotic FMNAT as a member of the PAPS superfamily, revealing a novel flavin-binding mode and a unique enzyme-bound FAD conformation. *Cg*FMNAT structures have been reported that capture the enzyme in different complexes with its ligands, providing models for different active-site states prior to and after catalysis [40]. The structure of *Sc*FMNAT in complex with the FAD product also shows important conformational differences in the two motifs involved in binding the phosphate moieties of FAD as compared to the *C. glabrata* orthologue. This suggests that the loop involved in binding of phosphates is dynamic and undergoes substantial conformational changes during the enzyme catalytic cycle. Although no structure has been reported yet for an FMNAT from mammals, hFADS is predicted to include a molybdo-pterin binding domain at the N-terminus that is not present in available structures from other species. This raises several interesting questions about the role of this domain in animals [36].

In addition to the enzymes involved in FMN and FAD synthesis, eukaryotic cells contain FMN-hydrolyzing enzymes in the cytosol and in the mitochondrial intermembrane space [59–63]. In the *S. cerevisiae* mitochondrial intermembrane space, FAD pyrophosphatase (FADase) (EC 3.6.1.18) and FHy (EC 3.1.3.2) activities, which catalyze the FAD + $H_2O \rightarrow$ FMN + AMP and FMN + $H_2O \rightarrow$ RF + P_i reactions, respectively, have been shown to provide RF [61]. Moreover, monofunctional enzymes bearing FMNAT might also exhibit FADpp activity, as mentioned above [36,57]. In the case of hFADS2, the kinetic parameters for this activity have been reported with k_{cat}, K_m^{FAD} and K_m^{PPi} values of 0.31 min^{-1}, <0.1 μM and 82 μM, respectively [36]. Considering calculated RF, FMN and FAD concentrations in different organelles, it has been proposed that mitochondria might play a major role in regulating the flavin pool, and that all of these enzymes will have a key role in regulating flavin homeostasis [63]. Therefore, the structure-based investigation of the kinetic properties of these enzymes should offer insights into the regulatory mechanisms of flavin homeostasis in eukaryotic organisms. Flavin traffic across the mitochondrial membranes also occurs and it is another key aspect in the metabolism of flavins. RF uptake by the mitochondria is proposed to occur independent of flavin efflux, while export of the synthesized FMN and FAD has been related to the existence of an RF/FAD (or FMN) antiporter [26,45,46,64]. The occurrence of an RF/FAD cycle was proposed to account for RF uptake in mitochondria biogenesis and RF recovery in mitochondrial flavoprotein degradation [26]. In *S. cerevisiae* a nuclear gene, *FLX1*, was evidenced to encode the tripartite repeat structure of FLX1, a protein involved in flavin transport that is a member of a protein family consisting of mitochondrial substrate and nucleotide carriers [65].

4.4 FMN and FAD metabolism in bacteria depends on a bifunctional enzyme

Purification to homogeneity of type I FADS from *Bacillus subtilis* (*Bs*FADS) and *Ca*FADS confirmed that a single polypeptide of moderate molecular weight catalyzed the RFK and FMNAT activities in these organisms [29,30,66–68]. The FMNAT activity was shown to be reversible in both *Ca*FADS and *Bs*FADS, producing FMN and ATP from FAD, indicating both enzymes have a FADpp activity [25,30]. Surprisingly, *Bs*FADS has a strict specificity for reduced flavins in all reactions, while *Ca*FADS acts on the oxidized forms. Although bacterial RFK and FMNAT activities are generally encoded in a single protein, some bacilli, *e.g., B. subtilis,* also have a monofunctional RFK that may play a regulatory or supplementary role in flavin cofactor biosynthetic pathways and also has a strict specificity for reduced RF [69].

Investigations of the steady-state kinetic properties of *Ca*FADS described two catalytic cycles of a bi-substrate sequential ordered mechanism for the initial transformation of RF into FMN, and then of FMN into FAD [25]. It was also concluded that RF binds prior to ATP, while ADP is released before FMN during the RFK activity [25], in a mechanism similar to that described above for monofunctional RFKs [55]. In the FMNAT reaction, FMN was proposed to bind after ATP, and PP_i to be released before FAD. An initial hypothetical model proposed a single binding site for the isoalloxazine and/or ribityl portions of the flavins for the RFK and FMNAT processes, but two independent ATP binding sites [25]. Later binding analysis of substrates and products to *Ca*FADS allowed determination of ATP binding parameters for two independent sites in the presence of Mg^{2+}, but for only a single site in its absence [70]. A single binding site was detected for FMN and FAD, but RF showed two binding sites. The presence of adenine nucleotides induces the appearance of a second FMN binding site, allowing the determination of the dissociation constants for FMN at two independent binding sites. These data proved for the first time that *Ca*FADS has two independent flavin binding sites, as well as two independent ATP binding sites. Moreover, they indicated that binding of the adenine nucleotide favors flavin binding at the RFK site of the enzyme [70].

Recently, the kinetic properties of *Ca*FADS have been revisited. RF concentrations above ~5 µM lead to considerable substrate inhibition of the RFK activity of the enzyme (►Tab. 4.1) [71,72]. The kinetic dependences on RF indicate the formation of dead-end combinations of RF to give a non-productive enzyme-substrate/product complex. So far, the nature of this complex is not clear. It might be a *Ca*FADS:ADP:Mg^{2+}:FMN or a *Ca*FADS:ADP:Mg^{2+}:RF complex, since formation of these type of complexes has positive cooperativity in *Ca*FADS [70]. On the other hand, excess RF might induce incorporation of two RF molecules in the RFK active site, either preventing the binding of the second substrate, ATP, or the catalytically competent allocation of, at least, one of the RF molecules. Unfortunately, the similarity between K_i and the apparent K_m^{RF} (►Tab. 4.1) indicates a very strong RF inhibitory effect that prevents determination of the real cause of the observed inhibition at the molecular level [72]. The RF inhibitory effect might also have implications in the interpretation of the bi-bi substrate RFK catalytic mechanism previously proposed, but further work should be done to clarify this particular point. The K_m^{RF} and K_m^{ATP} obtained for the RFK activity are in the range of the reported binding parameters. However, K_m^{RF} and k_{cat} are both considerably larger than previously reported, while K_m^{ATP} is smaller [25]. Such differences might be due to the different buffer composition, particularly to the Mg^{2+} concentration, as well as to the influence of the large inhibitory effect induced by RF. Dependence of the steady-state rates for the FMNAT activity of *Ca*FADS on both the FMN and ATP concentrations exhibited a Michaelis-Menten saturation profile with kinetic parameters determined in different laboratories being similar (►Tab. 4.2) [25,72]. K_m^{FMN} and K_m^{ATP} are slightly larger than the values reported for their corresponding binding affinities [70]. Finally, the reported kinetic parameters for the RFK and FMNAT activities of *Ca*FADS and *Bs*FADS suggest the prokaryotic enzyme is more efficient than the corresponding proteins involved in those functions in humans (►Tab. 4.2).

FADS crystallographic structures have been reported from *Thermotoga maritima* (*Tm*-FADS, pdb 1MRZ), *Corynebacterium ammoniagenes* (pdb 2X0K) and *Streptococcus pneumoniae* (*Spn*FADS, pdb 3PO1). All these structures confirmed that prokaryotic type I FADS folds in two almost independent modules, which correspond to those predicted from sequence alignment, each one putatively containing an independent site for flavin and ATP binding (in agreement with the binding studies) (►Fig. 4.3C) [73,74]. The N-terminal module (residues 1–186 *Ca*FADS, FMNAT module) forms an α/β dinucleotide binding domain with a typical Rossmann fold topology, and ends in a small subdomain built by a β-hairpin and two short α-helices. As predicted from sequence analysis this module does not present structural homology with eukaryotic FMNATs, but shows an important structural homology with enzymes of the NT superfamily [74]. The molecular surface of the crystal structure of *Ca*FADS in complex with PP_i shows a large open cavity in the FMNAT module whose central portion is positively charged and contributes to stabilization of the PP_i product. Comparison of this cavity with those of some NTs in complex with substrates/products allowed identification of putative binding sites for substrates and products for the FMNAT reaction [38,74]. Main differences in the *Ca*FADS structure from other proteins of its family are located in the β1n strand (absent in *Tm*FADS and NT homologues), as well as in the loops L1n, L4n (a 3_{10} helix in *Spn*FADS), and L6n (an α-helix in *Spn*FADS). Although the FMNAT module is proposed to catalyze the adenylylation of FMN, its independent production has shown that it might not be self-sufficient to carry out this process [38,74].

The C-terminal module (187–338 *Ca*FADS, RFK module) folds into a globular domain consisting of a β-barrel with six antiparallel β-strands, a terminal α-helix perpendicularly oriented to the barrel, and seven loops connecting them (►Fig. 4.3). As envisaged from sequence homology with monofunctional RFKs, it has a high structural homology with these proteins and it has been shown to catalyze the RFK activity when individually expressed [38,54,70,74]. The main difference from other members of the family is an insertion of 12 residues present in the L3c (L232–V246) loop of *Ca*FADS that is only present in the *Corynebacterium* and *Mycobacterium* species. The consensus PTAN motif, conserved in sequence and structure in the RFK module of type I FADS and in RFK families, is proposed to stabilize the metal ion and the phosphate groups of ATP:Mg^{2+} during RFK catalysis [50,54,55,74,75]. Substrate binding sites have been modeled in the RFK module of *Ca*FADS taking into account the structures reported for monofunctional RFKs. Surprisingly, the RFK module in the *Ca*FADS structure has an ATP binding site closed by several loops, some that are absent in homologous monofunctional RFKs (as *Sp*RFK and hRFK) and others that have different conformations [50,54,55] (►Fig. 4.3). In particular, the predicted binding locus of the adenine moiety of ATP is blocked by T208 [50,54], and the arrangement of the 207-PTAN-210 consensus motif is inverted when compared with that exhibited by monofunctional RFKs. Thus, the organization of the Pro-Thr peptide bond results in the nitrogen and carbonyl groups oriented in opposite directions than observed in monofunctional RFKs and, therefore, the T208 side-chain is too far away to interact with the isoalloxazine or ribityl chain, or even the phosphates of ATP. Conformational changes of the loop at the end of the PTAN motif (known as FlapI in RFKs and L1c in *Ca*FADS) are proposed to be critical during the catalytic cycle of monofunctional enzymes [50]. Recent site-directed mutagenesis data for T208 and N210 in *Ca*FADS have shown that both residues are critical for correct formation of the RFK active site geometry for substrate binding and catalysis, suggesting concerted conformational changes in the RFK module of *Ca*FADS during its activity [71]. These observations indicate that the architecture of the *Ca*FADS active site must suffer different and additional conformational changes to those proposed for monofunctional RFKs to make the ATP binding cavity accessible to this ligand, and to situate T208 at the RFK catalytic site.

The *Ca*FADS structure is proposed to be organized in a dimer of trimers [74]. In the trimer, the three protomers are connected in a head-tail disposition (►Fig. 4.4), with the flavin site in the FMNAT module, which is highly accessible to the solvent in

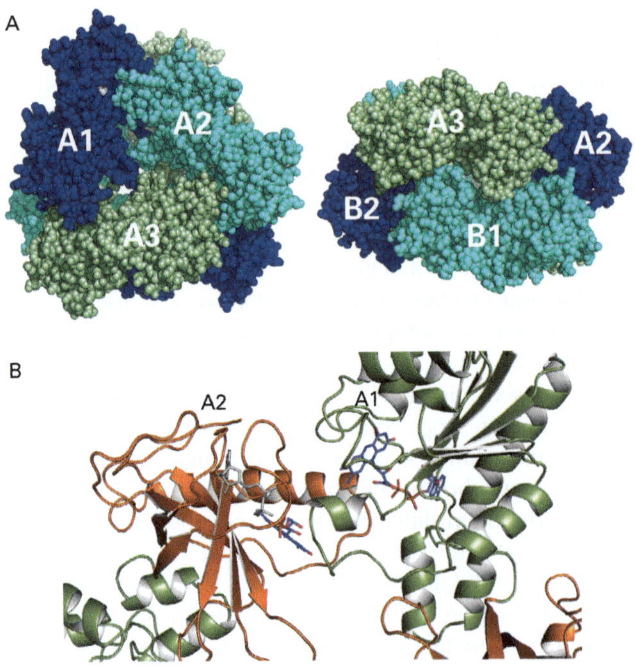

Fig. 4.4: Oligomeric structure of *Ca*FADS. (A) Perpendicular orientations of the dimer of trimers A_3:B_3 where A and B correspond to identical protomers but incluced within each one of the trimers. (B) Detail of the head-tail disposition between two protomers in the trimer. Ligands modeled as described [38].

the monomer, instead facing the RFK site of the neighboring protomer. Data from our group indicate that mutations at the RFK site modulate the binding parameters and the catalytic efficiency at the FMNAT active site of *Ca*FADS, supporting the hypothesis that the dimer of trimers must play a functional role during catalysis and that channeling between the RFK site and the FMNAT site of different protomers might take place [71]. As described for several NTs that also stabilize hexameric organizations [76–78], the *Ca*FADS dimer of trimers might contribute to regulation of flavin homeostasis. This might be an alternative to the cellular compartmentalization of different isoforms found in eukaryotes to maintain flavin homeostasis. In solution, oligomeric *Ca*FADS ensembles have been detected together with the monomeric form [70,74], and further work is underway to characterize the factors inducing their formation and stoichiometry, their putative functional role and production under physiological conditions, and whether they correspond to the oligomers observed in the crystal structure.

Finally, a number of pathogenic/opportunistic pathogenic prokaryotes are not able to synthesize RF *de novo* (*e.g., Listeria monocytogenes* or *Streptococcus pyogenes*) and must obtain it from the environment [5,79]. In addition to these, many other gram-positive bacteria seem to be able to import RF. Flavin importers have been identified and characterized in *B. subtilis* (the proton-RF symporter YpaA), *Corynebacterium glutamicum* (the energy-independent RF facilitator RibM) and some other bacteria [80–83].

4.5 FMN and FAD metabolism in plants

In plants, flavins are also required in many metabolic processes in chloroplasts, mitochondria and cytosol. RF only appears to be synthesized in chloroplasts, but, as noted above, bioinformatics approaches suggest that plants have sequence homologs of both monofunctional and bifunctional enzymes with RFK and FMNAT activities in different subcellular localizations [37,41]. Although different activities involved in the biosynthesis, hydrolysis, and transport of flavin nucleotides have been detected in different plant organelles, not all of the proteins involved in these transformations have been identified at the gene level yet, fewer have been biochemically characterized, and no structural data for any of these proteins are available. Therefore, much further work is required to fully understand the role of all the proteins involved in flavin metabolism in plants [28,41,84,85].

At the cytosolic level, RFK, FHy, and FMNAT activities have been reported [28,41]. In *Arabidopsis thaliana,* the enzyme homologs of monofunctional RFK, *At*FMN/FHy, showed both an RFK activity and an FHy activity, the latter of which catalyzes dephosphorylation of FMN to RF. Characterization of the purified enzyme, upon heterologous expression in *E. coli,* indicated that Tween 20 enhanced both activities, while the FHy activity greatly decreased in the absence of Mg^{2+} [41]. This later observation is expected for phosphatases of the Mg^{2+}-dependent HAD family. The C-terminal (*At*FMN) and N-terminal (*At*FHy) modules of *At*FMN/FHy can be physically separated with little change in their kinetic parameters (▶Tab. 4.1), indicating that the RFK and FHy activities reside, respectively, in each one of the modules [41]. The FHy module is an effective catalyst at neutral pH (k_{cat} and K_m^{FMN}, 105 min^{-1} and 0.67 μM, respectively), in agreement with cytosolic localization, and in contrast with previously known FMN hydrolyzing acid phosphatases [42,60]. Genomic evidence additionally suggests that this bifunctional enzyme might have originated early in plant evolution, since no evidence of a monofunctional RFK without the FHy sequence has been identified [41,86]. In *A. thaliana,* the existence of a putative cytosolic FMNAT (*At*FADS) homologous with monofunctional enzymes responsible for this activity has also been proposed, but little is known about its molecular properties [41].

Chloroplasts have been evidenced as having, RFK, FHy, FMNAT and FADase activities [28,87,88]. Despite the presence of RFK activity in percoll-isolated pea chloroplasts and the ability of pea chloroplasts to synthesize FMN from imported RF [28], the inability to find candidate sequences for organellar RFK in plant databases suggests that enzymes with little or no sequence similarity to known RFK must exist in plants. Plant chloroplasts have been shown to be unique among eukaryotes in that they have sequence homologs of FADS previously found only in prokaryotes [28,37]. These enzymes differ from the prokaryotic proteins in that they are shorter at the C-terminus and lack the RFK activity. The two plant-type FADS isoforms encoded by the *A. thaliana* genome (*At*RibF1 and *At*RibF2), both with N-terminal plastid targeting sequences, have been heterologously expressed in *E. coli.* Both proteins exhibit FMNAT activity, but excess FMN and ATP also inhibit this activity [28]. Kinetic parameters of *At*RibF1 and *At*RibF2 for the two substrates were similar (▶Tab. 4.2). These enzymes have lower k_{cat} values than those of the bacterial and mammalian enzymes, their K_m^{FMN} values are higher than those of bacterial enzymes, and they show similar catalytic efficiencies to the mammalian enzymes but lower than those of the prokaryotic enzymes. Remote similarity of the C-terminal module of plant-like FADS has been found with the C-terminal module of other families of pyrophosphohydrolases. Moreover, a remote sequence consensus of this protein region was found with serine/threonine phosphatases 2C and members of the hydrolase superfamily. These observations suggest that the C-terminal module of plant-like FADS could have a function. Preliminary experiments with *Glycine max* plant-like FADS indicate that its C-terminal module might have a hydrolytic activity that needs further characterization [37]. Recently, an FHy activity was also detected in pea chloroplasts [89]. The gene encoding this protein in *A. thaliana, At*cpFHy1, has been cloned and expressed in *E. coli,* allowing its biochemical characterization. *At*cpFHy1 hydrolyzes FMN most effectively at pH 6.5–7.0 and at ~40 °C, while cytosolic FHy has optimum activity at lower pH values and higher temperatures. Moreover, the determined values of K_m^{FMN}, k_{cat} and catalytic efficiency for *At*cpFHy1 are ~21-fold lower, ~3-fold higher and ~63-fold lower, respectively, than those for the cytosolic enzyme. *At*cpFHy1 localizes in the stroma of chloroplasts. Finally, the chloroplast *At*cpFHy1 and the cytosolic FHy have different protein sequences and distinct evolutionary origins, although both belong to the HAD superfamily [89]. Twenty-seven Nudix hydrolase genes have been found in *Arabidopsis* that encode proteins able to hydrolyze nucleoside diphosphates [87]. Among them, *At*NUDX23, appears to be the sole enzyme with FADase activity, corresponding to the activity detected in chloroplasts [28]. Recently, the transcript levels of *At*NUDX23, as well as of genes involved in flavin metabolism (*e.g., At*FADS, *At*RibF1, *At*RibF2, and *At*FMN/FHy), have been shown to significantly increase under continuous light. These studies propose a negative feedback regulation through the hydrolysis of FAD by *At*NUDX23 that, together with changes in the light-dark environment, contribute to the fine-tuning of intracellular flavin levels through regulation of the balance of FAD and FMN in plastids and, therefore, of the flavin homeostasis in plant cells [90].

Early studies detected RFK activity in plant mitochondria [91], and later, FHy and FADase activities were also detected [28]. However, only recently was it clearly shown that the synthesis of FAD occurs in this cellular compartment [27]. Mitochondria from *Nicotiana tabacum* were shown to take up RF, and to convert it into FMN and FAD by monofunctional RFK and FMNAT present in mitochondria. Both enzymes were activated by $MgCl_2$ and their activities found to be physically separated in the mitochondrial matrix and membrane fractions, respectively [27]. The sigmoidal shape of the saturation dependence of the FMN synthesis rate on RF concentration allowed determination of a $S_{0.5}$ value for *Nt*RFK in the same order of magnitude as the K_m^{RF} measured for the RFK partially purified from the plant *Solanum nigrum* [91], and one order of magnitude higher than the K_m^{RF} value of the bifunctional *At*FMN/FHy (►Tab. 4.1) [41]. A detailed kinetic study of the mitochondrial *Nt*FMNAT activity was prevented by the rapid conversion of FMN to RF, stimulated by $MgCl_2$, and catalyzed by the FHy activity present in mitochondria. High FHy activities were also detected in mitochondrial extracts from pea, preventing previous detection of FMNAT activity in plant mitochondria [28]. Nevertheless, preliminary data on *Nt*FMNAT are consistent with K_m^{FMN} values determined for the monofunctional recombinant *At*RibF1 and *At*RibF2 [28]. Mitochondrial FMNAT activity is estimated to represent only 3% of the total activity in the protoplasts, while this activity is maximal in chloroplasts.

All together these data suggest a cross-talk between chloroplast, cytosol and mitochondria during flavin cofactor biosynthesis, hydrolysis, and transport, allowing the following adaptation of the model initially proposed by Sandoval *et al.* [27,28]: RF is synthesized in plastids; FMN is synthesized and hydrolyzed in chloroplasts, mitochondria, and the cytosol; and FAD is synthesized in chloroplasts, mitochondria and the cytosol, and hydrolyzed in plastids and mitochondria. Additionally, RF is exported from chloroplasts to the cytosol, and mitochondria import RF from the cytosol via carrier-mediated systems that operate at different concentration ranges and have different uptake efficiencies. Then FAD produced in plant mitochondria can also be exported to the cytosol via a putative FAD exporter [27]. This is in line with the existence of both mitochondrial FAD synthesis and FAD exporter pathways in plants, as already reported for mammalian and yeast mitochondria [26,46,64]. At present, no candidate gene encoding any mitochondrial RF transporter has been identified. In contrast, sequence searches revealed the presence in plants of more than 30 sequence homologs of the mitochondrial FAD exporter FLX1 from *S. cerevisiae* [64,65].

4.6 Conclusions and future research directions

Given the large number of flavoproteins and flavoenzymes involved in the cellular metabolism of all types of organisms, deficiency of RFK or FMNAT activities will prevent the assembly of essential holoproteins, causing accumulation of the corresponding apoproteins unable to carry out their functions, and, therefore, inducing metabolic alterations closely linked to physiological or pathological states. In mammals, RF deficiency has been related to vascular diseases, complications in pregnancy and postnatal development, cataracts, and deficiency in the activities of flavin-metabolizing enzymes [1,15,22–24]. Respiratory infections also alter RF metabolism by leading to its excessive urinary lost [3]. Changes in the expression levels of RFK and FMNAT activities have also been connected with several cellular stresses, suggesting these proteins have additional and diverse cellular functions. Thus, patients with amyotrophic lateral sclerosis show a decrease in the expression of both proteins [92], while RFK overexpression up-regulates the cellular levels of FMN, FAD and glutathione, increases the activity of the cellular antioxidant system, and promotes cell growth and survival in prostate cancer [93]. Moreover, RFK is also proposed to contribute to the assembly of protein complexes involved in signal transduction related to cellular defense processes dependent on tumor necrosis factor [94,95]. Finally, flavin cofactors themselves might play regulatory roles. This includes, for example, the modulation of protein conformational stability induced by FAD that explains the beneficial RF therapy in patients suffering from acyl-CoA dehydrogenase deficiency or from pathologies related to malfunctioning of respiratory chain complex II or frataxin [17–19,96]. Among this group we can also include the FMN riboswitches in bacteria. They are predominantly located in 5′-unstranslated regions of mRNAs and represent gene regulatory systems that consist of an FMN-responsive aptamer and an overlapping expression platform [97]. The importance of all these processes shows the potential interest for the better understanding of the mechanisms involved in the metabolism of flavins.

The differential molecular characteristics of enzymes that carry out the same chemistry for synthesis of FMN and FAD in different organisms, or even organelles, are also of interest. These differences have identified prokaryotic FADS as a potential drug target for the development of inhibitors endowed with anti-pathogenic activity [40,74,98,99]. Thus, selective inhibition of prokaryotic FADS by drugs with no inhibitory activity toward the eukaryote homologues may be a feasible treatment of diseases produced by pathogens. Prokaryotic FADS catalyzes several complex reactions, which provide numerous opportunities for the rational design of transition state analogue inhibitors. Prokaryotic FADS and the enzymes exhibiting FMNAT activity in eukaryotes utilize different sets of active-site residues to carry out the same chemistry, making the substrate binding region in the FMNAT active site a possible target for the development of FADS inhibitors. Structural comparisons also detect important differences between the conformational mechanism for the RFK activity in the RFK module of type I FADS and that in mammalian RFKs, pointing at the loops responsible for such differences as additional sites for the differential inhibition of this activity in prokaryotes versus eukaryotes. Finally, the quaternary organization proposed in some prokaryotic FADSs might be another leading structure to target for development of inhibitors of the FADS activities.

Cellular homeostasis of flavin coenzymes is achieved by an intricate interplay of the different enzymes in the pathway responding to regulators, together with differential organelle compartmentation, and altered susceptibility of holo- and apo-flavoproteins to proteolytic digestion [32,42,100]. Despite the vital importance of FMN and FAD in metabolism, many of the factors that affect the homeostasis of these cofactors within living cells are still incompletely understood. The presence of enzymes that synthesize flavin nucleotides in organelles and the cytosol is not surprising because it ensures that the individual subcellular compartments have the ability to independently regulate flavin content in response to specific metabolic needs. It is more surprising to find an abundance of hydrolytic enzymes, which might have different functions. The released RF can be exported where it is needed, or could be degraded to avoid photochemical sensitization.

Despite the increase in understanding of flavin metabolism in recent years, many questions are still open. Their answer will surely provide knowledge that will have interesting applications.

4.7 Acknowledgments

This work has been supported by the Spanish Ministry of Science and Innovation [BIO2010-14983 to M.M.]. The author thanks Drs. A. Serrano, S. Frago, B. Herguedas, M. Martínez-Júlvez and A. Velázquez-Campoy for their contribution along the last years in the characterization of *Ca*FADS.

4.8 References

[1] Lee SS, McCormick DB. Effect of riboflavin status on hepatic activities of flavin-metabolizing enzymes in rats. J Nutr 1983;113:2274–9.
[2] Lee SS, McCormick DB. Thyroid hormone regulation of flavocoenzyme biosynthesis. Arch Biochem Biophys 1985;237:197–201.
[3] Brijlal S, Lakshmi AV, Bamji MS, Suresh P. Flavin metabolism during respiratory infection in mice. Br J Nutr 1996;76:453–62.
[4] Lee SS, Ye JH, Jones DP, McCormick DB. Correlation of H_2O_2 production and liver catalase during riboflavin deficiency and repletion in mammals. Biochem Biophys Res Commun 1983;117:788–93.
[5] Fischer M, Bacher A. Biosynthesis of flavocoenzymes. Nat Prod Rep 2005;22:324–50.
[6] Volk R, Bacher A. Biosynthesis of riboflavin. Studies on the mechanism of L-3,4-dihydroxy-2-butanone 4-phosphate synthase. J Biol Chem 1991;266:20610–8.
[7] Bacher A. Biosynthesis of flavins. In: Müller F. ed. Chemistry and Biochemistry of Flavoproteins. Boca Raton, FL: CRC Press, 1991:215–59.
[8] Bacher A, Ladenstein R. The lumazine synthase/riboflavin synthase complex of *Bacillus subtilis*. In: Müller F. ed. Chemistry and Biochemistry of Flavoproteins. Boca Raton, FL: CRC Press, 1991:293–316.
[9] Eisenreich W, Schwarzkopf B, Bacher A. Biosynthesis of nucleotides, flavins, and deazaflavins in *Methanobacterium thermoautotrophicum*. J Biol Chem 1991;266:9622–31.
[10] Ghisla S, Massey V. Mechanisms of flavoprotein-catalyzed reactions. Eur J Biochem 1989;181:1–17.
[11] Massey V. Introduction: flavoprotein structure and mechanism. FASEB J 1995;9:473–5.
[12] Massey V. The chemical and biological versatility of riboflavin. Biochem Soc Trans 2000;28:283–96.
[13] Mansoorabadi SO, Thibodeaux CJ, Liu HW. The diverse roles of flavin coenzymes--nature's most versatile thespians. J Org Chem 2007;72:6329–42.
[14] Macheroux P, Kappes B, Ealick SE. Flavogenomics-a genomic and structural view of flavin-dependent proteins. Febs J 2011;278:2625–34.
[15] Depeint F, Bruce WR, Shangari N, Mehta R, O'Brien PJ. Mitochondrial function and toxicity: role of B vitamins on the one-carbon transfer pathways. Chem Biol Interact 2006;163:113–32.
[16] Manthey KC, Rodriguez-Melendez R, Hoi JT, Zempleni J. Riboflavin deficiency causes protein and DNA damage in HepG2 cells, triggering arrest in G1 phase of the cell cycle. J Nutr Biochem 2006;17:250–6.
[17] Henriques BJ, Rodrigues JV, Olsen RK, Bross P, Gomes CM. Role of flavinylation in a mild variant of multiple acyl-CoA dehydrogenation deficiency: a molecular rationale for the effects of riboflavin supplementation. J Biol Chem 2009;284:4222–9.
[18] Gonzalez-Cabo P, Ros S, Palau F. Flavin adenine dinucleotide rescues the phenotype of frataxin deficiency. PLoS One 2010;5:e8872.
[19] Gregersen N, Andresen BS, Pedersen CB, Olsen RK, Corydon TJ, Bross P. Mitochondrial fatty acid oxidation defects-remaining challenges. J Inherit Metab Dis 2008;31:643–57.
[20] Giancaspero TA, Wait R, Boles E, Barile M. Succinate dehydrogenase flavoprotein subunit expression in *Saccharomyces cerevisiae*-involvement of the mitochondrial FAD transporter, Flx1p. Febs J 2008;275:1103–17.

[21] Bacher A. Riboflavin kinase and FAD syntehtase. In: Müller F. ed. Chemistry and Biochemistry of Flavoproteins. Boca Raton, FL: CRC Press, 1991:349–70.
[22] Powers HJ. Riboflavin (vitamin B_2) and health. Am J Clin Nutr 2003;77:1352–60.
[23] Torchetti EM, Brizio C, Colella M, Galluccio M, Giancaspero TA, Indiveri C, Roberti M, Barile M. Mitochondrial localization of human FAD synthetase isoform 1. Mitochondrion 2010;10:263–73.
[24] Yates CA, Evans GS, Pearson T, Powers HJ. Absence of luminal riboflavin disturbs early postnatal development of the gastrointestinal tract. Dig Dis Sci 2003;48:1159–64.
[25] Efimov I, Kuusk V, Zhang X, McIntire WS. Proposed steady-state kinetic mechanism for *Corynebacterium ammoniagenes* FAD synthetase produced by *Escherichia coli*. Biochemistry 1998;37:9716–23.
[26] Barile M, Brizio C, Valenti D, De Virgilio C, Passarella S. The riboflavin/FAD cycle in rat liver mitochondria. Eur J Biochem 2000;267:4888–900.
[27] Giancaspero TA, Locato V, de Pinto MC, De Gara L, Barile M. The occurrence of riboflavin kinase and FAD synthetase ensures FAD synthesis in tobacco mitochondria and maintenance of cellular redox status. Febs J 2009;276:219–31.
[28] Sandoval FJ, Zhang Y, Roje S. Flavin nucleotide metabolism in plants: monofunctional enzymes synthesize FAD in plastids. J Biol Chem 2008;283:30890–900.
[29] Manstein DJ, Pai EF. Purification and characterization of FAD synthetase from *Brevibacterium ammoniagenes*. J Biol Chem 1986;261:16169–73.
[30] Kearney EB, Goldenberg J, Lipsick J, Perl M. Flavokinase and FAD synthetase from *Bacillus subtilis* specific for reduced flavins. J Biol Chem 1979;254:9551–7.
[31] Mashhadi Z, Xu H, Grochowski LL, White RH. Archaeal RibL: a new FAD synthetase that is air sensitive. Biochemistry 2010;49:8748–55.
[32] Wu M, Repetto B, Glerum DM, Tzagoloff A. Cloning and characterization of FAD1, the structural gene for flavin adenine dinucleotide synthetase of *Saccharomyces cerevisiae*. Mol Cell Biol 1995;15:264–71.
[33] Santos MA, Jimenez A, Revuelta JL. Molecular characterization of FMN1, the structural gene for the monofunctional flavokinase of *Saccharomyces cerevisiae*. J Biol Chem 2000;275: 28618–24.
[34] Galluccio M, Brizio C, Torchetti EM, Ferranti P, Gianazza E, Indiveri C, Barile M. Over-expression in *Escherichia coli*, purification and characterization of isoform 2 of human FAD synthetase. Protein Expr Purif 2007;52:175–81.
[35] Brizio C, Galluccio M, Wait R, Torchetti EM, Bafunno V, Accardi R, Gianazza E, Indiveri C, Barile M. Over-expression in *Escherichia coli* and characterization of two recombinant isoforms of human FAD synthetase. Biochem Biophys Res Commun 2006;344:1008–16.
[36] Torchetti EM, Bonomi F, Galluccio M, Gianazza E, Giancaspero TA, Lametti S, Indiveri C, Barile M. Human FAD synthase (isoform 2): a component of the machinery that delivers FAD to apo-flavoproteins. Febs J 2011;278:4434–49.
[37] Yruela I, Arilla-Luna S, Medina M, Contreras-Moreira B. Evolutionary divergence of chloroplasts FAD synthetase proteins. BMC Evol Biol 2010;10:311.
[38] Frago S, Martínez-Júlvez M, Serrano A, Medina M. Structural analysis of FAD synthetase from *Corynebacterium ammoniagenes*. BMC Microbiol 2008;8:160.
[39] Krupa A, Sandhya K, Srinivasan N, Jonnalagadda S. A conserved domain in prokaryotic bifunctional FAD synthetases can potentially catalyze nucleotide transfer. Trends Biochem Sci 2003;28:9–12.
[40] Huerta C, Borek D, Machius M, Grishin NV, Zhang H. Structure and mechanism of a eukaryotic FMN adenylyltransferase. J Mol Biol 2009;389:388–400.
[41] Sandoval FJ, Roje S. An FMN hydrolase is fused to a riboflavin kinase homolog in plants. J Biol Chem 2005;280:38337–45.
[42] Mitsuda H, Tsuge H, Tomozawa Y, Kawai F. Multiplicity of acid phosphatase catalyzing FMN hydrolysis in spinach leaves. J Vitaminol (Kyoto) 1970;16:52–7.

[43] McCormick DB. Two interconnected B vitamins: riboflavin and pyridoxine. Physiol Rev 1989;69:1170–98.
[44] de Luca C, Kaplan NO. Flavin adenine dinucleotide synthesis in animal tissues. Biochim Biophys Acta 1958;30:6–11.
[45] Barile M, Passarella S, Bertoldi A, Quagliariello E. Flavin adenine dinucleotide synthesis in isolated rat liver mitochondria caused by imported flavin mononucleotide. Arch Biochem Biophys 1993;305:442–7.
[46] Pallotta ML, Brizio C, Fratianni A, De Virgilio C, Barile M, Passarella S. *Saccharomyces cerevisiae* mitochondria can synthesise FMN and FAD from externally added riboflavin and export them to the extramitochondrial phase. FEBS Lett 1998;428:245–9.
[47] Kashchenko VE, Shavlovskii GM. Purification and properties of the riboflavin kinase of the yeast *Pichia guilliermondii*. Biokhimiia 1976;41:376–83.
[48] Rajeswari SR, Jonnalagadda VS, Jonnalagadda S. Purification and characterization of flavokinase from *Neurospora crassa*. Indian J Biochem Biophys 1999;36:137–42.
[49] Kasai S, Nakano H, Maeda K, Matsui K. Purification, properties, and function of flavokinase from rat intestinal mucosa. J Biochem 1990;107:298–303.
[50] Karthikeyan S, Zhou Q, Mseeh F, Grishin NV, Osterman AL, Zhang H. Crystal structure of human riboflavin kinase reveals a beta barrel fold and a novel active site arch. Structure 2003;11:265–73.
[51] Faiura LR, Kashchenko VS. Thermostable riboflavin kinase in the yeast *Pichia guilliermondii*. Ukr Biokhim Zh 1997;69:21–5.
[52] Merrill AH, Jr., McCormick DB. Affinity chromatographic purification and properties of flavokinase (ATP:riboflavin 5′-phosphotransferase) from rat liver. J Biol Chem 1980;255:1335–8.
[53] Yamada Y, Merrill AH, Jr., McCormick DB. Probable reaction mechanisms of flavokinase and FAD synthetase from rat liver. Arch Biochem Biophys 1990;278:125–30.
[54] Bauer S, Kemter K, Bacher A, Huber R, Fischer M, Steinbacher S. Crystal structure of *Schizosaccharomyces pombe* riboflavin kinase reveals a novel ATP and riboflavin-binding fold. J Mol Biol 2003;326:1463–73.
[55] Karthikeyan S, Zhou Q, Osterman AL, Zhang H. Ligand binding-induced conformational changes in riboflavin kinase: structural basis for the ordered mechanism. Biochemistry 2003;42:12532–8.
[56] Oka M, McCormick DB. Complete purification and general characterization of FAD synthetase from rat liver. J Biol Chem 1987;262:7418–22.
[57] Schrecker AW, Kornberg A. Reversible enzymatic synthesis of flavin-adenine dinucleotide. J Biol Chem 1950;182:795–803.
[58] Leulliot N, Blondeau K, Keller J, Ulryck N, Quevillon-Cheruel S, van Tilbeurgh H. Crystal structure of yeast FAD synthetase (Fad1) in complex with FAD. J Mol Biol 2010;398:641–6.
[59] Fuchs KR, Shekels LL, Bernlohr DA. Analysis of the ACP1 gene product: classification as an FMN phosphatase. Biochem Biophys Res Commun 1992;189:1598–605.
[60] Granjeiro JM, Ferreira CV, Juca MB, Taga EM, Aoyama H. Bovine kidney low molecular weight acid phosphatase: FMN-dependent kinetics. Biochem Mol Biol Int 1997;41:1201–8.
[61] Barile M, Brizio C, De Virgilio C, Delfine S, Quagliariello E, Passarella S. Flavin adenine dinucleotide and flavin mononucleotide metabolism in rat liver--the occurrence of FAD pyrophosphatase and FMN phosphohydrolase in isolated mitochondria. Eur J Biochem 1997;249: 777–85.
[62] Lee RS, Ford HC. Purification and characterization of 5′-nucleotidase/FAD pyrophosphatase from human placenta. Methods Enzymol 1997;280:424–36.
[63] Pallotta ML. Evidence for the presence of a FAD pyrophosphatase and a FMN phosphohydrolase in yeast mitochondria: a possible role in flavin homeostasis. Yeast 2011;28:693–705.
[64] Bafunno V, Giancaspero TA, Brizio C, Bufano D, Passarella S, Boles E, Barile M. Riboflavin uptake and FAD synthesis in *Saccharomyces cerevisiae* mitochondria: involvement of the Flx1p carrier in FAD export. J Biol Chem 2004;279:95–102.

[65] Tzagoloff A, Jang J, Glerum DM, Wu M. FLX1 codes for a carrier protein involved in maintaining a proper balance of flavin nucleotides in yeast mitochondria. J Biol Chem 1996;271:7392–7.
[66] Nakagawa S, Igarashi A, Ohta T, Hagihara T, Fujio T, Aisaka K. Nucleotide sequence of the FAD synthetase gene from *Corynebacterium ammoniagenes* and its expression in *Escherichia coli*. Biosci Biotechnol Biochem 1995;59:694–702.
[67] Hagihara T, Fujio T, Aisaka K. Cloning of FAD synthetase gene from *Corynebacterium ammoniagenes* and its application to FAD and FMN production. Appl Microbiol Biotechnol 1995;42:724–9.
[68] Nakagawa S, Hagihara T, Fujio T, Aisaka K. Metaphosphate-dependent phosphorylation of riboflavin to FMN by *Corynebacterium ammoniagenes*. Appl Microbiol Biotechnol 1995;42: 724–9.
[69] Solovieva IM, Tarasov KV, Perumov DA. Main physicochemical features of monofunctional flavokinase from *Bacillus subtilis*. Biochemistry (Mosc) 2003;68:177–81.
[70] Frago S, Velázquez-Campoy A, Medina M. The puzzle of ligand binding to *Corynebacterium ammoniagenes* FAD synthetase. J Biol Chem 2009;284:6610–9.
[71] Serrano A, Frago S, Herguedas B, Martínez-Júlvez M, Velázquez-Campoy A, Medina M. Key residues at the riboflavin kinase catalytic site of the bifunctional Riboflavin kinase/FMN adenylyltransferase from *Corynebacterium ammoniagenes*. Cell Biochem Biophys 2012; doi: 10.1007/s12013-012-9403-9.
[72] Serrano A, Ferreira P, Martínez-Júlvez M, Medina M. The prokaryotic FAD Synthetase family: a potential drug target. Current Pharm Design 2012; *In press*.
[73] Wang W, Kim R, Yokota H, Kim SH. Crystal structure of flavin binding to FAD synthetase of *Thermotoga maritima*. Proteins 2005;58:246–8.
[74] Herguedas B, Martínez-Júlvez M, Frago S, Medina M, Hermoso JA. Oligomeric state in the crystal structure of modular FAD synthetase provides insights in its sequential catalysis in prokaryotes. J Mol Biol 2010;400:218–30.
[75] Cheek S, Ginalski K, Zhang H, Grishin NV. A comprehensive update of the sequence and structure classification of kinases. BMC Struct Biol 2005;5:6.
[76] Izard T, Geerlof A. The crystal structure of a novel bacterial adenylyltransferase reveals half of sites reactivity. EMBO J 1999;18:2021–30.
[77] Garavaglia S, D'Angelo I, Emanuelli M, Carnevali F, Pierella F, Magni G, Rizzi M. Structure of human NMN adenylyltransferase. A key nuclear enzyme for NAD homeostasis. J Biol Chem 2002;277:8524–30.
[78] Zhai RG, Rizzi M, Garavaglia S. Nicotinamide/nicotinic acid mononucleotide adenylyltransferase, new insights into an ancient enzyme. Cell Mol Life Sci 2009;66:2805–18.
[79] Abbas CA, Sibirny AA. Genetic control of biosynthesis and transport of riboflavin and flavin nucleotides and construction of robust biotechnological producers. Microbiol Mol Biol Rev 2011;75:321–60.
[80] Cecchini G, Perl M, Lipsick J, Singer TP, Kearney EB. Transport and binding of riboflavin by *Bacillus subtilis*. J Biol Chem 1979;254:7295–301.
[81] Hemberger S, Pedrolli DB, Stolz J, Vogl C, Lehmann M, Mack M. RibM from *Streptomyces davawensis* is a riboflavin/roseoflavin transporter and may be useful for the optimization of riboflavin production strains. BMC Biotechnol 2011;11:119.
[82] Vogl C, Grill S, Schilling O, Stulke J, Mack M, Stolz J. Characterization of riboflavin (vitamin B_2) transport proteins from *Bacillus subtilis* and *Corynebacterium glutamicum*. J Bacteriol 2007;189:7367–75.
[83] Grill S, Yamaguchi H, Wagner H, Zwahlen L, Kusch U, Mack M. Identification and characterization of two *Streptomyces davawensis* riboflavin biosynthesis gene clusters. Arch Microbiol 2007;188:377–87.
[84] Fischer M, Romisch W, Saller S, Illarionov B, Richter G, Rohdich F, Eisenreich W, Bacher A. Evolution of vitamin B_2 biosynthesis: structural and functional similarity between pyrimidine deaminases of eubacterial and plant origin. J Biol Chem 2004;279:36299–308.
[85] Roje S. Vitamin B biosynthesis in plants. Phytochemistry 2007;68:1904–21.

[86] Sobhanaditya J, Rao NA. Plant flavokinase. Affinity-chromatographic procedure for the purification of the enzyme from mung-bean (Phaseolus aureus) seeds and conformational changes on its interaction with orthophosphate. Biochem J 1981;197:227–32.
[87] Ogawa T, Yoshimura K, Miyake H, Ishikawa K, Ito D, Tanabe N, Shigeoka S. Molecular characterization of organelle-type Nudix hydrolases in Arabidopsis. Plant Physiol 2008;148: 1412–24.
[88] Ravindranath SD, Rao NA. Nucleotidases in plants. 3. Effect of metabolites on the enzyme hydrolyzing flavine adenine dinucleotide (FAD) from Phaseolus radiatus. Arch Biochem Biophys 1969;133: 54–9.
[89] Rawat R, Sandoval FJ, Wei Z, Winkler R, Roje S. An FMN hydrolase of the haloacid dehalogenase superfamily is active in plant chloroplasts. J Biol Chem 2011;286:42091–8.
[90] Maruta T, Yoshimoto T, Ito D, Ogawa T, Tamoi M, Yoshimura K, Shigeoka S. An Arabidopsis FAD pyrophosphohydrolase, AtNUDX23, is involved in the flavin homeostasis. Plant Cell Physiol 2012;doi:10.1093/pcp/pcs054
[91] Sadasivam S, Shanmugasundaram ER. Studies on the flavokinase of Solanum nigrum L. Enzymologia 1966;31:203–8.
[92] Lin J, Diamanduros A, Chowdhury SA, Scelsa S, Latov N, Sadiq SA. Specific electron transport chain abnormalities in amyotrophic lateral sclerosis. J Neurol 2009;256:774–82.
[93] Hirano G, Izumi H, Yasuniwa Y, Shimajiri S, Ke-Yong W, Sasagiri Y, Kusaba H, Matsumoto K, Hasegawa T, Akimoto M, Akashi K, Kohno K. Involvement of riboflavin kinase expression in cellular sensitivity against cisplatin. Int J Oncol 2011;38:893–902.
[94] Yazdanpanah B, Wiegmann K, Tchikov V, Krut O, Pongratz C, Schramm M, Kleinridders A, Wunderlich T, Kashkar H, Utermohlen O, Bruning JC, Schutze S, Kronke M. Riboflavin kinase couples TNF receptor 1 to NADPH oxidase. Nature 2009;460:1159–63.
[95] Park KJ, Lee CH, Kim A, Jeong KJ, Kim CH, Kim YS. Death receptors 4 and 5 activate Nox1 NADPH oxidase through riboflavin kinase to induce reactive oxygen species-mediated apoptotic cell death. J Biol Chem 2012;287:3313–25.
[96] Bugiani M, Lamantea E, Invernizzi F, Moroni I, Bizzi A, Zeviani M, Uziel G. Effects of riboflavin in children with complex II deficiency. Brain Dev 2006;28:576–81.
[97] Winkler WC, Breaker RR. Regulation of bacterial gene expression by riboswitches. Annu Rev Microbiol 2005;59:487–517.
[98] Spencer R, Fisher J, Walsh C. Preparation, characterization, and chemical properties of the flavin coenzyme analogues 5-deazariboflavin, 5-deazariboflavin 5′-phosphate, and 5-deazariboflavin 5′-diphosphate, leads to 5′-adenosine ester. Biochemistry 1976;15:1043–53.
[99] Gerdes SY, Scholle MD, D'Souza M, Bernal A, Baev MV, Farrell M, Kurnasov OV, Daugherty MD, Mseeh F, Polanuyer BM, Campbell JW, Anantha S, Shatalin KY, Chowdhury SA, Fonstein MY, Osterman AL. From genetic footprinting to antimicrobial drug targets: examples in cofactor biosynthetic pathways. J Bacteriol 2002;184:4555–72.
[100] Forster C, Santos MA, Ruffert S, Kramer R, Revuelta JL. Physiological consequence of disruption of the VMA1 gene in the riboflavin overproducer *Ashbya gossypii.* J Biol Chem 1999;274:9442–8.
[101] Pedrolli DB, Nakanishi S, Barile M, Mansurova M, Carmona EC, Lux A, Gartner W, Mack M. The antibiotics roseoflavin and 8-demethyl-8-amino-riboflavin from *Streptomyces davawensis* are metabolized by human flavokinase and human FAD synthetase. Biochem Pharmacol 2011;82:1853–9.

4.9 Abbreviations

RF	Riboflavin
FMN	flavin mononucleotide
FAD	flavin adenine dinucleotide

RFK	ATP:riboflavin kinase
FMNAT	ATP:FMN adenylyltransferase
FADS	bifunctional FAD synthetase
PAPS	3′-phosphoadenosine 5′-phosphosulfate reductase-like family
hFADS1 and hFADS2	isoforms 1 and 2 of the human enzyme catalyzing the FMNAT reaction
CaFADS	Corynebacterium ammoniagenes FADS
type 1 and 2 FADS	isoforms 1 and 2 of prokaryotic FADS
NT	ucleotidyltransferase
PgRFK	Pichia guilliermondii RFK
ScRFK	S. cerevisiae RFK
NsRFK	Neurospora crassa RFK
rRFK	rat RFK
hRFK	humans RFK
SpRFK	Schizosaccharomyces pombe RFK
rRFKl	rat liver RFK
rRFKim	rat intestinal mucosa RFK
rFMNATl	rat liver FMNAT
FADpp	FAD pyrophosphorylase activity
PPi	pyrophosphate
CgFMNAT	Candida glabrata FMNAT
*Sc*FMNAT	*S. cerevisiae* FAD1 isoform
FADase	FAD pyrophosphatase
FHy	FMN hydrolase
BsFADS	Bacillus subtilis FADS
TmFADS	Thermotoga maritime FADS
SpnFADS	Streptococcus pneumoniae FADS
*At*FMN/FHy	*Arabidopsis thaliana* enzyme with RFK and FHy activities
AtcpFHy1	Arabidopsis thaliana FHy
*At*RibF1 and *At*RibF2	*Arabidopsis thaliana* plant-type FADS isoforms 1 and 2

5 Mechanisms of bacterial luciferase and related flavin reductases

Shiao-Chun Tu

Abstract

Bacterial luciferase and NAD(P)H:flavin oxidoreductase (flavin reductase) together are a member of the flavin-dependent two-component monooxygenase family, unique in their ability to emit light as a reaction product. This report focuses on mechanistic aspects of bacterial luciferase and flavin reductases, individually and as a functional enzyme couple. For luciferase, proposed reaction mechanisms and intermediates are evaluated. Several important mechanistic issues are discussed. For flavin reductases from luminous bacteria, their reaction mechanisms as individual enzymes are described. Moreover, mechanisms of $FMNH_2$ transfer from reductase to luciferase are discussed along with a summary of supporting evidence. A brief description of reduced flavin transfers by other two-component monooxygenases is also presented.

5.1 Introduction

Bacterial luciferase (hereafter referred to as luciferase) catalyzes the oxidation of reduced FMN ($FMNH_2$) and a long-chain aliphatic aldehyde (RCHO) by oxygen to generate FMN, aliphatic carboxylic acid (RCOOH), water and light (λ_{max} ~490 nm). Luciferase, being a uni-functional monooxygenase, has no bound FMN prosthetic group and cannot catalyze the reduction of FMN. The required $FMNH_2$ is supplied *in vivo* by NAD(P)H-flavin oxidoreductase (flavin reductase). Together, they are a functionally coupled two-component monooxygenase system.

The luciferase-flavin reductase two-component monooxygenase is unique in its light-emitting activity. Cell physiology of luminous bacteria, *lux* operon structure and regulation, actions of autoinducer or quorum-sensing inducer, applications as sensors, and associated fluorescent proteins have all been extensively studied. In addition, mechanisms and structures of luciferase and flavin reductases have been the subjects for several reviews [1–5]. This article focuses on a number of major mechanistic properties of luciferase and flavin reductases, as individual enzymes and as a two-component monooxygenase couple. Functional couplings by a few other two-component monooxygenases are also briefly described.

5.2 Luciferase mechanism overview

►Fig. 5.1 depicts a chemical mechanism of luciferase that is most supported by experimental evidence. Luciferase-bound reduced FMN (intermediate I) reacts with oxygen to form a 4a-hydroperoxy-FMNH intermediate II (HF-OOH). A subsequent reaction of II with an aliphatic aldehyde forms the 4a-peroxyhemiacetal-FMNH intermediate III (HF-OOCH(OH)R). The decay of III forms a carboxylic acid and excited 4a-hydroxy-FMNH (HF-OH* or intermediate IV*) following a Chemically Initiated Electron Exchange (CIEEL) or, for short, Electron Transfer (ET) mechanism. This mechanism involves the proposed formation of a caged pair of carboxylic acid radical anion R-C(OH)$O^{-\bullet}$ and a 4a-hydroxyFMNH radical cation (HF-$OH^{+\bullet}$ or intermediate $IV^{+\bullet}$) from III. A subsequent 1-e transfer from R-C(OH)$O^{-\bullet}$ to HF-$OH^{+\bullet}$ produces RCOOH and HF-OH* (IV*). Relaxation of HF-OH* to ground state HF-OH (IV) is coupled to light emission. With RCOOH bound to luciferase, HF-OH breaks down to FMN and H_2O. Subsequently, a slow release of RCOOH is followed by the dissociation of FMN to complete a single cycle of the luciferase reaction [6].

II has been isolated at –20 °C in 50% ethylene glycol-phosphate [7,8] and at 2–4 °C in aqueous media [9,10]. The isolated II was characterized by absorption and fluorescence spectroscopies [7,9,11] and its proposed 4a-hydroperoxy-FMNH structure was confirmed by an NMR study [12]. II generates FMN and H_2O_2 in a dark decay or reacts, anaerobically [13], with aldehyde to emit luminescence without H_2O_2 formation [8,10]. A synthetic 5-ethyl-4a-hydroperoxy-4a,5-dihydroflavin chemical model was found essentially identical to II in absorption spectrum, and was able to react with aldehyde for chemiluminescence [14]. 4a-Hydroperoxy-flavin turns out to be an essential intermediate for all known flavin-dependent one- or two-component monooxygenases. In all cases, luciferase intermediate II appears to be the best characterized.

Intermediate III has been detected by transient absorption spectroscopy with characteristics consistent with those expected for a flavin 4a-adduct [15]. The proposed mechanism for the formation of HF-OH* from III is discussed below. HF-OH* undergoes a light-emitting relaxation to ground state HF-OH, which has been detected as a transient fluorescent species [16] and, later, isolated and characterized by absorption and fluorescence spectroscopies [17,18].

5.2.1 Mechanism of chemiexcitation

The least understood aspect of the scheme in ►Fig. 5.1 is the formation of HF-OH* from III. Two major mechanisms have been proposed. First, a Baeyer-Villiger (BV) mechanism [19] depicts an abstraction of a proton from the OH in the peroxyhemiacetal group to initiate a step-wise electron- and hydride-transfer to generate RCOOH and excited HF-OH*. Although BV mechanism is well established for many reactions, it is unlikely applicable to luciferase chemiexcitation. First, questionably, the energy released from multiple steps of electron and hydride transfers must be saved and pooled together in order to populate HF-OH to the excited state. Next, cyclohexanone monooxygenase is known to follow a BV mechanism [20] in converting boronic acids to boric acids but bacterial luciferase cannot utilize boronic acids for boric acid formation or light emission [21]. Moreover, as will be discussed below, results of studies using FMN analogs with various

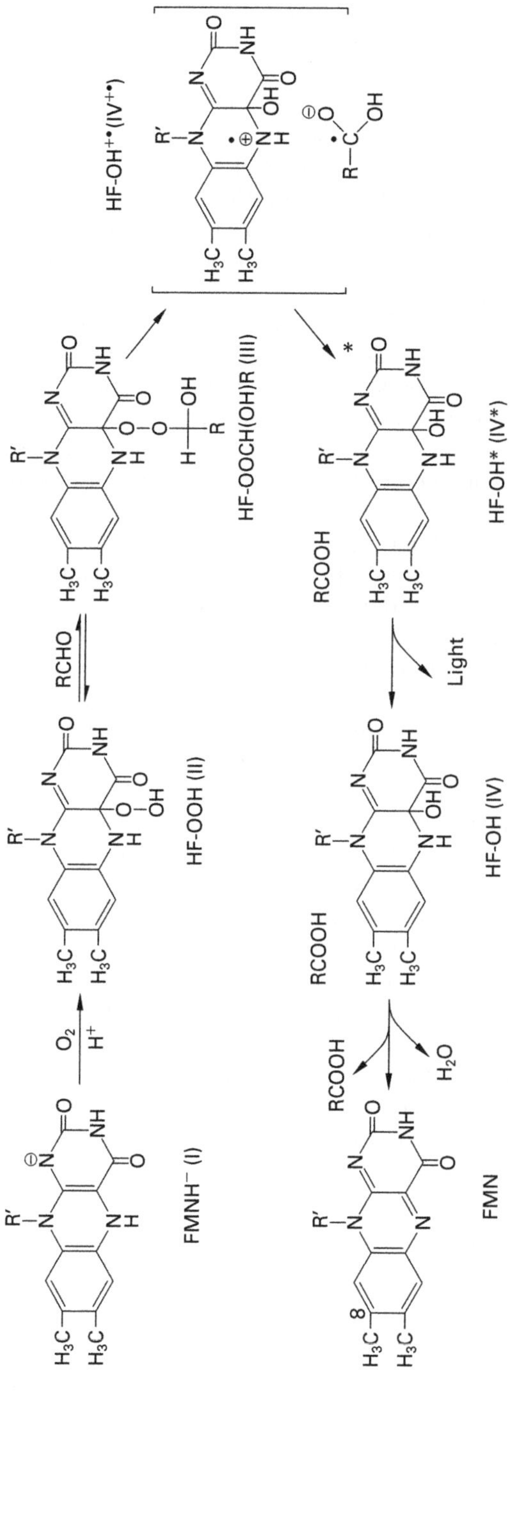

Fig. 5.1: Chemically Initiated Electron Exchange (CIEEL) or Electron Transfer (ET) mechanism of bacterial luciferase-catalyzed bioluminescent reaction.

substitutions at position 8 (see FMN in ▶Fig. 5.1) as substrates for luciferase were also inconsistent with the BV mechanism [22,23].

A second, and now highly favored, mechanism is the CIEEL/ET mechanism (▶Fig. 5.1) [24–27]. Initially, 4a-hydroxy flavin radical cation was a purely hypothesized species. The formation of such flavin radical cations from 4a-hydroxy or methoxy flavins has since been demonstrated by 1-e electrochemical oxidation [28], chemical comproportionation [29], and pulse radiolysis [30]. Based on their redox potentials [28,30], the reduction of 4a-hydroxy flavin radical cation by carboxylic acid radical anion is coupled to the release, in a single step of 1-e transfer, of ~90 kcal/mol [30], more than the 67 kcal/mol needed for the excitation of 4a-hydroxy flavin. Also, chemiluminescence was observed from 1-e reductions of 4a-hydroxy flavin radical cations [31,32]. Moreover, using C8-substituted reduced FMN derivatives as substrates for luciferase, faster luminescence decay rates were expected for flavins with lower (positive) 1-e oxidation potentials for their 4a,5-dihydro forms because of their expected higher tendencies to generate the 4a-hydroxy flavin radical cations. This was indeed observed, in strong support of the CIEEL/ET mechanism but inconsistent with the BV mechanism as mentioned above [22,23].

A minor variation of the CIEEL/ET mechanism involves a dioxirane intermediate, which was first conceived by Cho and Lee [33] and later incorporated into the CIEEL/ET mechanism (▶Fig. 5.2) [34,35]. Steps leading to the formation of III and those from the radical pairs of R-C(OH)O$^{-\bullet}$ and HF-OH$^{+\bullet}$ downward are the same as depicted in the scheme in ▶Fig. 5.1. Following this dioxirane mechanism, HF-OH is generated twice, initially from III by reaction A and, next, from the luminescent relaxation of HF-OH* (not shown in ▶Fig. 5.2). Moreover, reaction B was identified as the rate-limiting step, and the isotope-sensitive step when [1-^{2}H]-aldehyde was used as a substrate. In the latter case, the locations of the deuterium are indicated by circles in the scheme in ▶Fig. 5.2. Since the initial formation of HF-OH is before the rate-limiting reaction B, there should be an accumulation of HF-OH prior to the generation of radical species. However, HF-OH has only been detected and isolated after the light-emitting relaxation of HF-OH*, not before its formation [16–18]. Next, when [1-^{2}H]-aldehyde substrate was used, the observed kinetic isotope effects on the bioluminescence reaction rate were 1.4–1.9 [36,37]. Such an effect is consistent with the original CIEEL/ET mechanism shown in the scheme in ▶Fig. 5.1 for a primary isotope effect with the formation of HF-OH$^{+\bullet}$ and R-C(OH)O$^{-\bullet}$ from III as the isotope-sensitive and partially rate-limiting step. In contrast, with reaction B as the rate-limiting and isotope-sensitive step in the scheme in ▶Fig. 5.2, a secondary deuterium isotope effect is expected. Although not impossible [37], the magnitude of secondary deuterium isotope effects is unlikely as

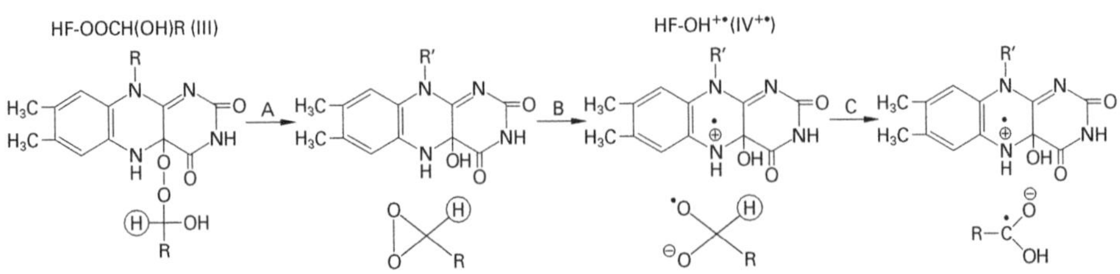

Fig. 5.2: Proposed involvement of a dioxirane intermediate in the bacterial luciferase reaction.

large as 1.4–1.9. The involvement of dioxirane as an obligatory intermediate in luciferase chemiexcitation has, thus far, not been supported by direct experimental findings.

5.2.2 Identity of primary excited state and emitter

Theoretically radical annihilation by 1-e transfer from a radical anion $A^{-\bullet}$ to a radical cation $C^{+\bullet}$ can proceed by two pathways, with one (step a) leading to C* and the other (step b) to A* as the excited state (►Fig. 5.3). Hence, based on the scheme in ►Fig. 5.1, either carboxylic acid (RCOOH) or HF-OH could be converted to the primary excited state (defined as the first excited state formed in the luciferase reaction) by 1-e transfer from R-C(OH)$O^{-\bullet}$ to HF-$OH^{+\bullet}$. The observed 490-nm bioluminescence is consistent with HF-OH* as the primary excited state and the emitter. However, 490-nm bioluminescence can also be observed if RCOOH* is the primary excited species but undergoes a resonance energy transfer to subsequently generate HF-OH* as the emitter. Such possibilities have in fact been voiced [34,38,39].

To resolve this issue, fluorescent cis, trans, trans, cis-9,11,13,15-octadecatetraenoaldehyde (α-parinaraldehyde or α-PAD) and trans, trans, trans, trans-9,11,13,15-octadecatetraenoaldehyde (β-parinaraldehyde or β-PAD) have been synthesized as active substrates for the luciferase reaction [40]. The corresponding reaction products α-parinaric acid (α-PAC) and β-parinaric acid (β-PAC) were each shown to bind at the aldehyde substrate site, and both showed strong fluorescence peaking at 420 nm. With α-PAD or β-PAD as a substrate, only the normal 490-nm bioluminescence was observed without any 420-nm component [40]. The first possibility is that HF-OH* was generated as the primary excited species and the emitter. The second possibility is that α-PAC* or β-PAC* was the primary excited species but generated HF-OH* as an emitter through energy transfer with 100% transfer efficiency. Energy transfer is sensitive to, among other factors, the relative orientation between the donor (in this case, α-PAC* or β-PAC*) and the accepter (in this case HF-OH). Since α-PAC is quite different from β-PAC with respect to stereo structure, it is highly unlikely that both α-PAC* and β-PAC* can transfer energy to HF-OH at 100% efficiency. Therefore, the 490-nm bioluminescence without any 420-nm component using α-PAD or β-PAD as a substrate for luciferase strongly supports HF-OH* as the primary excited species and the emitter.

For a second issue, the quantum yield of the *in vitro* luciferase bioluminescence from the luciferase-bound HF-OH* was found to be 0.1–0.16 [10,41]. However, fluorescence quantum yields of free 5-ethyl-4a-ethoxy-3-methyl-4a,5-dihydroflavin (EtF-OEt) and 5-ethyl-4a-hydroxy-3-methyl-4a,5-dihydroflavin (EtF-OH), as models for HF-OH, were no more than 10^{-5} [31,42]. Moreover, quantum yields of chemiluminescence

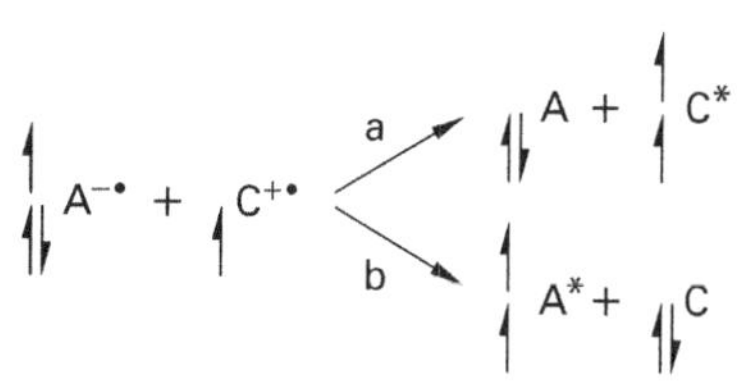

Fig. 5.3: Two pathways for excited state formation by one-electron transfer from an anionic radical to a cationic radical.

by reacting 4a-hydroperoxy flavin models with aldehyde [14] or by 1-e reduction of EtF-OH$^{+\bullet}$ were also in the range of 10^{-4}–10^{-7} [28,29,31]. In view of the huge differences in quantum yields between model systems and the luciferase reaction, the validity of HF-OH* as the luciferase emitter has often been questioned. Luciferase did not bind either EtF-OH or EtF-OEt. Therefore, the possibility that the environment of luciferase active site can greatly enhance the emission intensity of the bound 4a-hydroxy flavin could not be tested at that time.

We have synthesized a new flavin model for HF-OH, namely 5-decyl-4a-hydroxy-FMN (DecylFMN-OH) [43]. Unlike the earlier flavin models, DecylFMN-OH bound to luciferase and an active αC106A mutant with 1:1 molar ratio and K_d of 2.4 and 1.2 μM, respectively, at the luciferase active site as indicated by the inhibition of luciferase activity resulting from DecylFMN-OH binding. The fluorescence of free DecylFMN-OH was very weak but was enhanced several orders of magnitude to 0.08 and 0.05 quantum yields upon binding to luciferase and αC106A, respectively. These findings clearly show that the luciferase active site is suitable for greatly enhancing the emission efficiency of 4a-hydroxyflavin and, therefore, firmly supporting HF-OH* as the emitter of luciferase.

5.2.3 Multiple forms of 4a-hydroperoxy-FMNH intermediate II

Two aspects of intermediate II can benefit from further elaborations. First, immediately after isolation in the dark, II exhibited a weak fluorescence. Upon irradiation within the absorption range such as at 370 nm, the intensity of II fluorescence greatly increased and the peak shifted from 505 to 485 nm (uncorrected for uneven instrument performance) as reported by Balny and Hastings [11] or from 535 to 505 nm (both corrected) as described by Tu [9,44]. Both the initially isolated species (II_x) and the irradiation-transformed species (II_y) were fully active in reacting with aldehyde for bioluminescence [9,11]. Although the work by Balny and Hastings [11] and that by Tu [9,44] clearly described a blue shift of II fluorescence upon irradiation, Lee *et al.* [45] later reported a resolution of a so called "spectral discrepancy" by confirming the blue shift of fluorescence when II_x was exposed to light as described by Balny and Hastings [11] and disputing a "red shift" which they mistakenly attributed to the work by Tu [9,44]. There was no discrepancy in the first place.

What is the difference between II_x and II_y at the molecular level? Possibly II_x was transformed photochemically into II_y. However, a more recent study [46] provided a different answer. II can be isolated in the dark using a luciferase H44A mutant [47]. In the presence of 0.2 M KI, the weakly fluorescent II_x from αH44A was transformed into the highly fluorescent II_y in the dark along with a blue shift of the emission spectrum [46]. Hence, it is likely that such a transformation resulted from a conformational change in luciferase rather than a photochemical reaction.

Unrelated to II_x and II_y, two different forms of II can also be identified by another criterion. Although αH44A and αH44D were extremely weak in bioluminescence, significant amounts of II can be generated by these two luciferase variants [47]. Interestingly, a major and fast-decaying II can be isolated, which was inactive in bioluminescence but exhibited the same absorption spectrum as the II from wild-type luciferase. In addition, a minor and slow-decaying II can be detected by its ability to react with aldehyde for light emission [47]. Since the C(4a)-carbon in HF-OOH is a tetrahedron center, we entertain the possibility that these two forms of II are stereoisomers with respect to the orientation of OOH attachment at position 4a. In consistence with this conjecture is that only one species of II was identified from the wild-type luciferase, which exhibited the same decay kinetics when monitored by absorption or by bioluminescence activity. The active site of wild-type luciferase is well structured to apparently allow just one orientation of OOH attachment to flavin C(4a). With αH44 mutants, a critical change of the luciferase active site occurred. Consequently, the control of II formation was loosened and two stereo-isomers of II were generated, with one active and the other inactive in bioluminescence. Further studies are required to confirm or disprove our hypothesis.

5.2.4 Aldehyde substrate inhibition

Luciferases from *Vibrio harveyi* (formerly MAV or *Benekea harveyi*) [48] and *Photorhabdus luninescens* (formerly *Xenorhabdus luninescens*) [49] can be inhibited, *in vitro*, at high levels of aldehyde. Such an inhibition only occurs when aldehyde is mixed with luciferase prior to $FMNH_2$. Three mechanisms for aldehyde inhibition have been reported. They are referred to as the TS (two-site-sequential), OS (one-site), and TR (two-site-random) mechanisms. The TS mechanism by Holzman and Baldwin [50] was based on sequential binding of an aldehyde substrate and then an aldehyde inhibitor per αβ luciferase. It also depicted a random binding of aldehyde and $FMNH_2$ substrates. Later, Baldwin, Raushel, and coworkers [37] reported the OS mechanism, apparently to replace the TS mechanism. In the OS mechanism, the αβ luciferase has only one aldehyde binding site. However, if luciferase binds aldehyde first, the luciferase:aldehyde complex cannot bind $FMNH_2$ for light emission. Moreover, the dissociation of aldehyde from the inactive luciferase:aldehyde complex is much slower than the autoxidation of $FMNH_2$. Therefore, when a sample containing luciferase:aldehyde complex is mixed with an $FMNH_2$ solution, there will be no or negligible amounts of $FMNH_2$ left to react with any luciferase slowly dissociated from the inactive luciferase:aldehyde complex. We have formulated the TR mechanism [51] with the following critical features: (a) There are two independent aldehyde sites per αβ luciferase for random binding of aldehydes. (b) One aldehyde site has a high affinity and rapid on and off rates for the binding of an aldehyde substrate (A_s). The other aldehyde site binds an aldehyde inhibitor (A_i) weakly and with much slower on and off rates (in comparison with the binding of $FMNH_2$ and A_s). (c) A_s and $FMNH_2$ bind to luciferase in random order, but the A_i site overlaps with the $FMNH_2$ site. Therefore, A_i is a competitive inhibitor against $FMNH_2$ binding. (d) Since $FMNH_2$ undergoes a rapid autoxidation and the release of A_i from luciferase is much slower, dilution of a luciferase:A_i sample with an $FMNH_2$ solution does not generate any additional active luciferase:$FMNH_2$ complex. All aspects as listed above for the TR mechanism have been confirmed experimentally [51].

The TS mechanism and our TR mechanism are both based on two aldehyde sites per luciferase and the random binding of A_s and $FMNH_2$. The TR model requires only random and independent binding of A_s and A_i, for which supporting evidence was reported [51]. In contrast, the TS model relies on sequential binding of A_s and A_i. For this to occur, the two aldehyde sites must be mutually dependent but the possible underlying molecular basis was not discussed. Such a sequential binding is not necessary to account for aldehyde inhibition, and is inconsistent with our experimental findings [51]. The OS model is similar to our TR model with respect to a slow dissociation of aldehyde inhibitor from luciferase in comparison with a fast autoxidation of $FMNH_2$. However, the OS model relies on one aldehyde binding site whereas we have observed two aldehyde sites per luciferase. Following the OS model, once aldehyde binds to luciferase, the luciferase:aldehyde complex is incapable of binding $FMNH_2$ for bioluminescence. If so, when aldehyde is mixed with luciferase prior to $FMNH_2$, higher levels of bioluminescence should not occur at higher levels of aldehyde. This is inconsistent with what has been observed for *in vitro* assays [48,51,52]. At low aldehyde concentrations (*e.g.* ≤5 μM decanal), the binding of A_s dominated over A_i. Upon secondary additions of $FMNH_2$, higher bioluminescence activities were observed at higher aldehyde levels. Only when the aldehyde concentrations were sufficiently high to allow significant A_i binding (*e.g.* >10 μM), further increases in aldehyde concentration then resulted in decreasing luminescence activities. Last, although both OS and our TR mechanisms depicted slow binding and release of aldehyde inhibitor, the corresponding rate constants reported for the OS and TR mechanisms are markedly different. In our chemical relaxation studies, luciferase was first mixed with decanal at several concentrations suitable for A_i binding. Once equilibrium was achieved, the sample was quickly diluted with buffer to allow the time-dependent dissociation of A_i to reach a new equilibrium in aldehyde binding. Time courses of the appearance of A_i-free luciferase were monitored. Results gave rise to chemical relaxation times, in the range of tens of seconds, and the values of $k_{on} = 3.5 \times 10^3\ M^{-1}\ s^{-1}$ and $k_{off} = 0.038\ s^{-1}$ for the decanal inhibitor [51]. The k_{on} and k_{off} for decanal inhibitor binding and release were also determined by numerical fitting of stopped-flow data as $9.1 \times 10^5\ M^{-1}\ s^{-1}$ and $5.8\ s^{-1}$, respectively by Baldwin, Raushel, and coworkers [37,53]. The corresponding chemical relaxation times would be ≤0.1 s, inconsistent with our experimental findings.

5.3 Flavin reductases – general remarks

In the last two decades, reduced flavins generated by flavin reductases have been found to be essential for a wide variety of biological functions. Flavin reductases were initially classified by Nivière, Fontecave, and colleagues [54,55] and later, with their kind agreement, renamed as Class 1 for flavoproteins and Class 2 for non-flavoproteins [3]. The two classes of flavin reductases often show distinct structural and mechanistic properties. Since NADH and NADPH, in general, have different metabolic functions, flavin reductases were also classified as the NADH-preferring flavin reductase D (FRD), the NADPH-preferring flavin reductase P (FRP), and the general flavin reductase (FRG) for those capable of utilizing NADH and NADPH with similar efficiencies [3].

Luciferase must acquire the $FMNH_2$ substrate from flavin reductase donor enzymes *in vivo*. Several species of flavin reductases in luminous bacteria have been identified. These include the FRP (Class 1) [56–58], FRD (Class 2) [56,57,59], and FRG (Class 2) [59] from *V. harveyi*, and FRG (Class 1) from *Vibrio fischeri* (designated FRase I) [60,61]. Genes encoding flavin reductases similar to a Class 2 *Escherichia coli* flavin reductase Fre have also been found in *P. luminescens*, *V. fischeri*, *V. harveyi*, and *Vibrio orientalis* [62]. However, properties of these encoded flavin reductases have not been well characterized. Recently, a flavin reductase encoded by the *luxG* gene of

Photobacterium eiognathi TH1 has been characterized to be of the Class 2-FRD type [63]. The following discussions are limited to flavin reductases from luminous bacteria, with the exception at the end for some comparisons with the flavin reductases in other two-component monooxygenases.

5.3.1 Mechanisms of flavin reductases in single-enzyme reactions

The FRP from *V. harveyi* (FRP_{Vh}) and FRG (FRase 1) from *V. fischeri* (FRG_{Vf}) are both Class 1 flavoproteins with a tightly bound FMN per enzyme monomer, and both exhibit a ping-pong mechanism in single-enzyme reactions [64,65]. NAD(P)H first binds to the reductase holoenzyme and reduces the bound flavin cofactor. $NAD(P)^+$ is released and the oxidized flavin substrate binds to the reduced enzyme. After a transfer of reducing equivalents from the reduced flavin cofactor to the flavin substrate, the reduced flavin product is released to regenerate the original holoenzyme. For FRP_{Vh}, the apoenzyme bound FMN and $FMNH_2$ cofactor tightly with essentially the same K_d [66], supporting the requirement of a ping-pong mechanism for the enzyme to shuttle between oxidized and reduced holoenzyme forms without ever losing the flavin cofactor. We also found that the reduction of FRP_{Vh} FMN cofactor by NADPH was reversible, thus demonstrating that the cofactor reduction is not a committed step for the $FMNH_2$ product formation [67].

The Class 2 FRD and FRG from *V. harveyi* and the newly identified LuxG (Class 2 FRD) from *P. leiognathi* all exhibited the sequential mechanism [56,57,59,68,69]. Both the flavin and NAD(P)H substrates bind to the enzyme to form a ternary complex. The products are formed at the active site and released to regenerate the original enzyme form.

5.3.2 Mechanisms of luciferase:flavin reductase coupled reactions

5.3.2.1 Direct $FMNH_2$ transfer

Free reduced flavins can undergo an autoxidation yielding oxidized flavin and H_2O_2. If acceptor enzymes cannot catch reduced flavins efficiently, there would be a great waist of NAD(P)H for the nonproductive reduction of flavins and formation of harmful active oxygen species. Therefore, mechanisms for the transfer of $FMNH_2$ from flavin reductases to luciferase have emerged as an intriguing research subject. Three possible mechanisms have been considered [3]. The first is by free diffusion. The second one is non-specific reduction of luciferase-bound oxidized flavin by colliding external reductants, including but not limited to NAD(P)H. The resting state luciferase binds FMN weakly. However, FMN generated in the catalytic reaction remains bound to luciferase and is released only after a slow decay of the luciferase:FMN:carboxylic acid ternary complex [6], thus allowing non-specific reductions of luciferase-bound FMN as a possibility. Thus far, there is no definitive evidence to rule out or to quantify these two mechanisms for the supply of $FMNH_2$ to luciferase *in vivo*. The third mechanism is a direct transfer of $FMNH_2$ from flavin reductase to luciferase within a static or transient complex. Supporting evidence for direct $FMNH_2$ transfer already existed when the optimal concentration of FMN for FRG_{Vf} in a coupled reaction with *V. fischeri* luciferase (L_{Vf}) was found much lower than that required for the FRG_{Vf} single-enzyme assay [70]. Later, K_m values for FMN and NAD(P)H were found to be much higher in single-enzyme reactions by FRD_{Vh} and FRP_{Vh} than in *V. harveyi* luciferase (L_{Vh})-coupled reactions [71]. The turnover rates of flavin reductases are ≥2 orders of magnitude faster than that of luciferases mentioned above. As a functional complex, the turnover would be limited by the slow luciferase. Therefore, FMN and NAD(P)H at levels much lower than the K_m for the single-enzyme reaction by flavin reductase were already sufficient to generate $FMNH_2$ at a level suitable for 50% of the V_{max} of the coupled bioluminescence. Consequently, K_m for FMN and NAD(P)H for flavin reductase in the luciferase-coupled reaction became much lower than those in the reductase single-enzyme reaction.

5.3.2.2 Two direct transfer mechanisms

FRP_{Vh} exhibited a ping-pong mechanism in its single-enzyme reaction but a sequential mechanism was observed for FRP_{Vh}:L_{Vh} coupled reactions [64]. Marked decreases in K_m for both flavin and NAD(P)H were also observed for flavin reductases in the reactions by FRP_{Vh}:L_{Vf}, FRG_{Vf}:L_{Vf}, and FRG_{Vf}:L_{Vh} couples [64,72]. However, while the coupled reactions by FRP_{Vh}:L_{Vf} and FRG_{Vf}:L_{Vf} were similar to FRP_{Vh}:L_{Vh} in switching from ping-pong to a sequential mechanism, FRG_{Vf} in the coupled reaction with excess L_{Vh} still retained the ping-pong mechanism [64,72]. These observations were interpreted on the basis of "Direct Transfer of Product" (DTP) and "Direct Transfer of Cofactor" (DTC) mechanisms as shown in ▸Fig. 5.4, in which the subscripts c, s, and p for flavin stand for cofactor, substrate, and product, respectively. Also, E and E:F_c are, respectively, reductase apoenzyme and holoenzyme; L is luciferase; F and FH_2 are, respectively, oxidized and reduced flavin; N and NH are, respectively, $NAD(P)^+$ and NAD(P)H.

The FRG_{Vf}:L_{Vh} DTP reaction followed the same ping-pong mechanism as the FRG_{Vf} single-enzyme reaction except for the last step, in which the reduced flavin product (F_pH_2) was directly transferred to luciferase for bioluminescence. In contrast,the FRP_{Vh}:L_{Vh}, FRP_{Vh}:L_{Vf}, and FRG_{Vf}:L_{Vf} couples followed the DTC pathway, in which NAD(P)H first bound to reductase holoenzyme (E:F_c), and $NAD(P)^+$ was released after the reduction of the flavin cofactor. The reduced flavin cofactor (F_cH_2) was directly transferred to luciferase, leaving reductase as an apoenzyme (E). The exogenously added flavin then bound to the apoenzyme to regenerate the reductase holoenzyme. Starting with the apoenzyme E as shown in the Scheme (▸Fig. 5.4), a sequential mechanism can be identified for the DTC pathway. Since three enzyme couples followed one mechanism and one couple followed the other pathway, the transfer mechanism was apparently governed by the specific pairing of the donor and acceptor enzymes rather than dictated by either flavin reductase or luciferase.

The bound FMN initially in FRP_{Vh} and FRG_{Vf} holoenzymes is a genuine cofactor in their single-enzyme reactions and, for FRG_{Vf}, in the L_{Vh}-coupled reaction. It stays bound to the reductase during catalysis and is recovered in the original form at the end of a catalytic cycle. On the other hand, the FMN initially bound to the reductase is referred to as a cofactor in the DTC pathway only to be consistent with the DTP scheme with respect to the holoenzyme (E:F_c) designation. From the perspective of the DTC mechanism, this flavin should be regarded as a pre-bound substrate, for it was converted to a product and released from the reductase. This is highly unusual and interesting, because

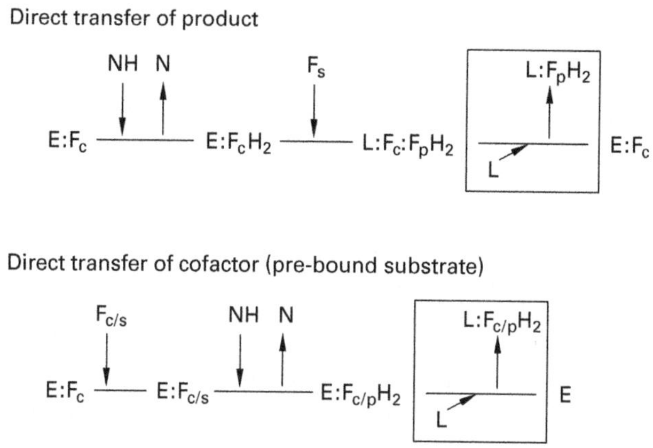

Fig. 5.4: Two pathways for the direct transfer of $FMNH_2$ from flavin reductase to bacterial luciferase.

a flavin prosthetic group in a flavoprotein was found to perform different biochemical functions under different reaction conditions.

According to the Scheme in ▶Fig. 5.4, exogenous flavin when added at high levels should function as a competitive inhibitor against luciferase or a non-competitive (or mixed) inhibitor against NADPH in the DTC coupled reaction but would not be inhibitive for the DTP coupled reaction [64]. The expected inhibition patterns were indeed observed for the three reductase:luciferase couples that exhibited the sequential mechanism whereas the lack of inhibition was found for FRG_{Vf}:L_{Vh} that retained the ping-pong mechanism [64,72]. In a different set of experiments, the FMN cofactor of FRP_{Vh} and FRG_{Vf} was replaced by 2-thioFMN, with retention of substantial reductase activities [64,72]. The reconstituted reductases were then used in the luciferase-coupled reaction in which FMN and NAD(P)H were added as substrates for the flavin reductase. Reduced 2-thioFMN is known to be a poor substrate for the luciferase luminescence reaction. If the reduced cofactor of reductase (2-thio-$FMNH_2$ in this case) is transferred to luciferase, as proposed for the three enzyme couples that showed the sequential mechanism, very little bioluminescence is expected due to the low substrate activity of 2-thio-$FMNH_2$ for luciferase. On the other hand, if the reduced flavin product ($FMNH_2$ in this case) is transferred to luciferase, then normal bioluminescence is expected for FRG_{Vf}:L_{Vh} that retained the ping-pong mechanism. All expected results were observed experimentally [64,72].

For the direct transfer of $FMNH_2$ to occur, flavin reductase and luciferase must form a functional complex. We found that L_{Vh} complexed with FRP_{Vh} only when the latter was in monomeric form [73]. This actually makes great biochemical sense. With a K_d of about 2 μM for FRP_{Vh} monomer-dimer equilibrium [74] and with maximum cellular contents of FRP_{Vh} and L_{Vh} at 3 and 170 μM, respectively, most FRP_{Vh} would be trapped as complexes with L_{Vh} [73]. This would minimize the non-productive formation of free $FMNH_2$ by FRP_{Vh} *in vivo*. Using a fusion protein of FRP_{Vh} and a yellow fluorescent protein (YFP), bioluminescence energy transfers from luciferase to FRP_{Vh}-YFP fusion were observed *in vitro* and *in vivo* [75]. Therefore, FRP_{Vh} and L_{Vh} not only formed a complex *in vitro* and *in vivo*, such a complex was also functional. Since the FRP_{Vh} turnover rate is much faster than that of L_{Vh} and the *in vivo* content of L_{Vh} is much higher than that of FRP_{Vh}, we speculate that the functional complex of L_{Vh} and FRP_{Vh} may be transient in nature. Once direct $FMNH_2$ transfer is achieved, FRP_{Vh} conceivably could dissociate from L_{Vh} during its slow catalytic reaction and seek out other L_{Vh} molecules for complex formation and $FMNH_2$ transfer.

The direct transfer of $FMNH_2$ from FRP_{Vh} to L_{Vh} was also demonstrated by the construction of a fusion protein of FRP_{Vh} and L_{Vh} [76]. Kinetic characteristics for direct $FMNH_2$ transfer were retained by this fusion enzyme. Moreover, at sub-micromolar concentrations, this fusion enzyme exhibited higher bioluminescence than an equimolar mixture of luciferase and reductase with and, especially, without the addition of extra flavin. Without any flavin addition, bioluminescence from the fusion enzyme was not sensitive to high concentrations of oxygen, indicating that $FMNH_2$ formed by the reductase component did not escape from the fusion enzyme. These findings, taken together, strongly indicate that the fusion enzyme mimics a single-component bifunctional monooxygenase.

5.3.2.3 Other considerations on $FMNH_2$ transfer to luciferase *in vivo*

Although direct $FMNH_2$ transfer has gained considerable experimental support, it may not be the sole mechanism for luciferase to acquire $FMNH_2$ *in vivo*. Luciferase would be active in luminescence no matter how $FMNH_2$ is delivered. Therefore, it is not surprising that Fre, the known major flavin reductase in *E. coli*, can support the bioluminescence in *E. coli* cells expressing the cloned luciferase [77]. Currently, little is known about how exactly $FMNH_2$ is acquired by luciferase *in vivo*. By using high concentrations of oxygen as a scavenger of free but not luciferase-bound $FMNH_2$, it was found that, at least *in vitro*, the combination of luciferase and flavin reductase concentrations is important in regulating the relative contributions of free diffusion and direct transfer of $FMNH_2$ for luciferase luminescence [76,78]. Conceivably, both mechanisms, and even the non-specific reduction of luciferase-bound FMN, could all contribute to the acquisition of $FMNH_2$ by luciferase *in vivo*. Their relative contributions could also vary under different physiological conditions.

The reductase LuxG from *P. leiognathi* TH1 is the first characterization of a flavin reductase encoded by a gene (*luxG* in this case) residing within the *lux* operon. It was active in supporting luciferase for bioluminescence [63] but the mechanism of $FMNH_2$ transfer has not been identified. Regulated by the same *lux* operon, luciferase and LuxG would be synthesized in equal molar quantities. If LuxG turns over much faster than luciferase from *P. leiognathi* TH1 (L_{Pl}), as the case for most known reductases and luciferases, then it would seem wasteful to synthesize equal molar quantities of LuxG and L_{Pl} when the former has a much faster turnover rate. Interestingly, L_{Pl} is a fast-reacting type and the k_{cat} for LuxG, at 1.7 s^{-1} [63], happens to be 10-fold or more slower than most flavin reductases from luminous bacteria.

5.3.3 Reduced flavin transfers in two-component monooxygenases in general

Considering the diversity of known flavin-dependent two-component monooxygenases, it is expected that these systems may utilize different mechanisms for reduced flavin transfer. Since no complex of the monooxygenase and reductase was observed during catalysis, free diffusion of reduced flavin was suggested for the StyA:StyB couple from *Pseudomonas* sp. VLB120 [79] and for the *Streptomyces coelicolor* ActVA:ActVB couple [80]. Mechanisms of reduced flavin transfer in the *p*-hydroxyphenylacetate hydroxylase (or 4-hydroxyphenylacetate 3-monooxygenase) system differ according to the organismic sources. For the *E. coli* enzyme system, no reductase:monooxygenase complex was necessary and the monooxygenase can efficiently bind and utilize reduced flavin released by the reductase [81]. A rapid-diffusion process of reduced flavin was also indicated for the same two-component monooxygenase couple from *Acinetobacter baumannii* [82]. In contrast, for the *p*-hydroxyphenylacetate hydroxylase from *Pseudomonas aeruginosa*, the interaction between the monooxygenase and the reductase can facilitate the reduced flavin transfer [83]. The two-component styrene monooxygenase from *Pseudomonas putida* S12 appeared to rely on both free diffusion and direct transfer of flavin [84]. The reductase and the monooxygenase component in the alkanesulfonate monooxygenase system from *E. coli* formed a stable complex [85]. Moreover, the reductase followed an ordered sequential mechanism in single-enzyme reaction but changed to a rapid equilibrium ordered mechanism in the presence of the monooxygenase and its substrate octanesulfonate [86]. A direct transfer of reduced flavin appeared to govern this system. The acrylamine N-oxygenase system from *Pseudomonas fluorescens* was reported to require a complex of the reductase and the monooxygenase for an efficient transfer of reduced flavin [87].

Last but not least, the FMN-containing nitro/flavin reductase (NfrA1) from *Bacillus subtilis* was found to follow a ping-pong mechanism in its single-enzyme reaction by using the bound FMN as a cofactor but switched over to a sequential mechanism in the luciferase-coupled reaction by using the bound FMN as a pre-bound substrate [88]. The bi-functional nature of the bound FMN in FRP_{Vh} and FRG_{Vf} described earlier is apparently not an isolated phenomenon.

5.4 Acknowledgments

SCT is grateful to an endowment of John and Rebecca Moores Professorship and the funding from National Science Council, Taiwan, ROC.

5.5 References

[1] Hastings JW, Potrikus CJ, Gupta SC, Kurfürst M, Makemson JC. Biochemistry and physiology of bioluminescent bacteria. Adv Microb Physiol 1985;26:235–91.

[2] Baldwin TO, Christopher JA, Raushel FM, et al. Structure of bacterial luciferase. Curr Opin Struct Biol 1995;5:798–809.

[3] Tu S-C. Reduced flavin: Donor and acceptor enzymes and mechanisms of channeling. Antioxid Redox Signal 2001;3:881–97.

[4] Tu S-C. Bacterial Bioluminescence: Biochemistry. In: Horspool WM, Lenci F, eds. CRC Handbook of Organic Photochemistry and Photobiology. 2nd ed. Boca Raton, FL, USA: CRC Press, 2004:136/1–17.

[5] Tu S-C. Bacterial Luciferase: Structure-function relationships and protein engineering. In: Viviani VR, Ohmiya Y, eds. Luciferases and Fluorescent Proteins: Principles and Advances in Biotechnology and Bioimaging. Kerala, India: Transworld Research Network, 2007:1–18.

[6] Li Z, Meighen EA. The turnover of bacterial luciferase is limited by a slow decomposition of the ternary enzyme-product complex of luciferase, FMN, and fatty acid. J Biol Chem 1994;269:6640–4.

[7] Hastings JW, Balny C, LePeuch C, Douzou P. Spectral properties of an oxygenated luciferase-flavin intermediate isolated by low-temperature chromatography. Proc Natl Acad Sci USA 1973;70:3468–72.

[8] Hastings JW, Balny C. The oxygenated bacterial luciferase-flavin intermediate. Reaction products via the light and dark pathways. J Biol Chem 1975;250:7288–93.

[9] Tu S-C. Isolation and properties of bacterial luciferase-oxygenated flavin intermediate complexed with long-chain alcohols. Biochemistry 1979;18:5940–5.

[10] Tu S-C. Isolation and properties of bacterial luciferase intermediates containing different oxygenated flavins. J Biol Chem 1982;257:3719–25.

[11] Balny C, Hastings JW. Fluorescence and bioluminescence of bacterial luciferase intermediates. Biochemistry 1975;14:4719–23.

[12] Vervoort J, Muller F, Lee J, van den Berg WAM, Moonen CTW. Identifications of the true carbon-13 nuclear magnetic resonance spectrum of the stable intermediate II in bacterial luciferase. Biochemistry 1986;25:8062–7.

[13] Becvar JE, Tu S-C, Hastings JW. Activity and stability of the luciferase-flavin intermediate. Biochemistry 1978;17:1807–12.

[14] Kemal C, Bruice TC. Simple synthesis of a 4a-hydroperoxy adduct of a 1,5-dihydroflavine: Preliminary studies of a model for bacterial luciferase. Proc Nat Acad Sci USA 1976;73:995–9.

[15] Macheroux P, Ghisla S, Hastings JW. Spectral detection of an intermediate preceding the excited state in the bacterial luciferase reaction. Biochemistry 1993;32:14183–6.

[16] Matheson IBC, Lee J, Müller F. Bacterial bioluminescence: Spectral study of the emitters in the *in vitro* reaction. Proc Natl Acad Sci USA 1981;78:948–52.

[17] Kurfürst M, Ghisla S, Hastings JW. Characterization and postulated structure of the primary emitter in the bacterial luciferase reaction. Proc Natl Acad Sci USA 1984;81:2990–4.

[18] Kurfürst M, Macheroux P, Ghisla S, Hastings JW. Isolation and characterization of the transient, luciferase-bound flavin-4a-hydroxide in the bacterial luciferase reaction. Biochim Biophys Acta 1987;924:104–10.

[19] Eberhard A, Hastings JW. A postulated mechanism for the bioluminescent oxidation of reduced flavin mononucleotide. Biochem Biophys Res Commun 1972;47:348–53.

[20] Branchaud BP, Walsh CT. Functional group diversity in enzymatic oxygenation reactions catalyzed by bacterial flavin-containing cyclohexanone oxygenase. J Am Chem Soc 1985;107:2153–61.

[21] Ahrens M, Macheroux P, Eberhard A, Ghisla S, Branchaud BP, Hastings JW. Boronic acids as mechanistic probes for the bacterial luciferase reaction. Photochem Photobiol 1991;54:295–9.

[22] Macheroux P, Eckstein J, Ghisla S. Studies on the mechanism of bacterial bioluminescence. Evidence compatible with a one electron transfer process and a CIEEL mechanism in the luciferase reaction. In: Edmondson DE, McCormick DB, eds. Flavins and Flavoproteins. Berlin: Water de Gruyter and Co., 1987:613–9.

[23] Eckstein JW, Hastings JW, Ghisla S. Mechanism of bacterial bioluminescence: 4a,5-dihydroflavin analogs as models for luciferase hydroperoxide intermediates and the effect of substitutions at the 8-position of flavin on luciferase kinetics. Biochemistry 1993;32:404–11.

[24] Mager HIX, Addink R. On the role of some flavin adducts as one-electron donors. In: Bray RC, Engel PC, Mayhew SG, eds. Flavins and Flavoproteins. Berlin: Water de Gruyter and Co., 1984:37–40.

[25] Macheroux P, Ghisla S, Kurfürst M, Hastings JW. Studies on the bacterial luciferase reaction: Isotope effects on the light emission. Is a 'CIEEL' mechanism involved? In: Bray RC, Engel PC, Mayhew SG, eds. Flavins and Flavoproteins. Berlin: Water de Gruyter and Co., 1984: 669–72.

[26] Mager HIX, Tu S-C. One-electron transfers in flavin systems: Relevance to the postulated CIEEL mechanism in bacterial bioluminescence. In: Edmondson DE, McCormick DB, eds. Flavins and Flavoproteins. Berlin: Water de Gruyter and Co., 1987:583–92.

[27] Tu S-C, Mager HIX, Shao R, Cho KW, Xi L. Mechanisms of bacterial luciferase and aromatic hydroxylases. In: Curti B, Ronchi S, Zanetti G, eds. Flavins and Flavoproteins. Berlin: Walter de Gruyter and Co., 1991:253–60.

[28] Mager HIX, Sazou D, Liu YH, Tu S-C, Kadish KM. Reversible one-electron generation of 4^a,5-substituted flavin radical cations: Models for a postulated key intermediate in bacterial bioluminescence. J Am Chem Soc 1988;110:3759–62.

[29] Mager HIX, Tu S-C. Spontaneous formation of flavin radicals in aqueous solution by comproportionation of a flavinium cation and a flavin pseudobase. Tetrahedron 1988;44:5669–74.

[30] Merényi G, Lind J, Mager HIX, Tu S-C. Properties of 4a-hydroxy-4a,5-dihydroflavin radicals in relation to bacterial bioluminescence. J Phys Chem 1992;96:10528–33.

[31] Kaaret TW, Bruice TC. Electrochemical luminescence with N(5)-ethyl-4a-hydroxy-3-methyl-4a,5-dihydrolumiflavin. The mechanism of bacterial luciferase. Photochem Photobiol 1990;51:629–33.

[32] Tu S-C. Oxygenated flavin intermediates of bacterial luciferase and flavoprotein aromatic hydroxylases: Enzymology and chemical models. In: Baumstark AL, ed. Advances in Oxygenated Processes. Greenwich: Jai Press, 1991:115–40.

[33] Cho K-W, Lee H-J. Nature of high energy intermediate involved in the bioluminescence reaction catalyzed by bacterial luciferase. Korean Biochem J 1984;17:1–9.

[34] Raushel FM, Baldwin TO. Proposed mechanism for the bacterial bioluminescence reaction involving a dioxirane intermediate. Biochem Biophys Res Commun 1989;164:1137–42.

[35] Francisco WA, Abu-Soud HM, DelMonte AJ, Singleton DA, Baldwin TO, Raushel FM. Deuterium kinetic isotope effects and the mechanism of the bacterial luciferase reaction. Biochemistry 1998;37:2596–606.

[36] Shannon P, Presswood RB, Spencer R, Becvar JE, Hastings JW, Walsh C. A study of deuterium isotope effects on the bacterial bioluminescence reaction. In: Singer TP, Ondarza RN, eds. Mechanisms of Oxidizing Enzymes. Amsterdam: Elsevier/North Holland, 1978:69–78.

[37] Francisco WA, Abu-Soud HM, Baldwin TO, Raushel FM. Interaction of bacterial luciferase with aldehyde substrates and inhibitors. J Biol Chem 1993;268:24734–41.

[38] Ziegler MM, Baldwin TO. Biochemistry of bacterial bioluminescence. Curr Top Bioenerg 1981;12:65–113.

[39] Lee J, Matheson IBC, Müller F, O'Kane DJ, Vervoort J, Visser AJWG. The mechanism of bacterial bioluminescence. In: Müller F, ed. Chemistry and Biochemistry of Flavins and Flavoenzymes. Orlando: CRC Press, 1991:109–51.

[40] Cho KW, Tu S-C, Shao R. Fluorescent polyene aliphatics as spectroscopic and mechanistic probes for bacterial luciferase: Evidence against carbonyl product from aldehyde as the primary excited species. Photochem Photobiol 1993;57:396–402.

[41] Hastings JW, Nealson KH. Bacterial bioluminescence. Ann Rev Microbiol 1977;31:549–95.

[42] Ghisla S, Massey V, Lhoste J-M, Mayhew SG. Fluorescence and optical characteristics of reduced flavins and flavoproteins. Biochemistry 1974;14:589–97.

[43] Lei B, Ding Q, Tu S-C. Identity of the emitter in the bacterial luciferase luminescence reaction: Binding and fluorescence quantum yield studies of 5-decyl-4a-hydroxy-4a,5-dihydroriboflavin-5'-phosphate as a model. Biochemistry 2004;43:15975–82.

[44] Tu S-C. Bacterial luciferase 4a-hydroperoxyflavin intermediates: stabilization, isolation, and properties. Methods Enzymol 1986;133:128–39.

[45] Lee J, O'Kane DJ, Gibson BG. Dynamic fluorescence properties of bacterial luciferase intermediates. Biochemistry 1988;27:4862–70.
[46] Huang S, Tu S-C. Effects of iodide on the fluorescence and activity of the hydroperoxyflavin intermediate of *Vibrio harveyi* luciferase. Photochem Photobiol 2005;81:425–30.
[47] Xin X, Xi L, Tu S-C. Functional consequences of site-directed mutation of conserved histidyl residues of the bacterial luciferase a subunit. Biochemistry 1991;30:11255–62.
[48] Hastings JW, Weber K, Friedland J, Eberhard A, Mitchell GW, Gunsalus A. Structurally distinct bacterial luciferases. Biochemistry 1969;8:4681–9.
[49] Xi L, Cho KW, Tu S-C. Cloning and nucleotide sequences of *lux* genes and characterization of luciferase of *Xenorhabdus luminescens* from a human wound. J Bacteriol 1991;173: 1399–405.
[50] Holzman TF, Baldwin TO. Reversible inhibition of the bacterial luciferase catalyzed bioluminescence reaction by aldehyde substrate: Kinetic mechanism and ligand effects. Biochemistry 1983;22:2838–46.
[51] Lei B, Cho KW, Tu S-C. Mechanism of aldehyde inhibition of *Vibrio harveyi* luciferase. Identification of two aldehyde sites and relationship between aldehyde and flavin binding. J Biol Chem 1994;269:5612–8.
[52] Meighen EA, MacKenzie RE. Flavine specificity of enzyme-substrate intermediates in the bacterial bioluminescent reaction. Structural requirements of the flavine side chain. Biochemistry 1973;12:1482–91.
[53] Abu-Soud HM, Clark AC, Francisco WA, Baldwin TO, Raushel FM. Kinetic destabilization of the hydroperoxy flavin intermediate by site-directed modification of the reactive thiol in bacterial luciferase. J Biol Chem 1993;268:7699–706.
[54] Nivière V, Vanoni MA, Zanetti G, Fontecave M. Reaction of NAD(P)H:flavin oxidoreductase from *Escherichia coli* with NADPH and riboflavin: Identification of intermediates. Biochemistry 1998;37:11879–87.
[55] Nivière V, Fieschi F, Décout J-L, Fontecave M. The NAD(P)H:flavin oxidoreductase from *Escherichia coli*. Evidence for a new mode of binding for reduced pyridine nucleotides. J Biol Chem 1999;274:18252–60.
[56] Gerlo E, Charlier J. Identification of NADH-specific and NADPH-specific FMN reductases in *Beneckea harveyi*. Eur J Biochem 1975;57:461–7.
[57] Jablonski E, DeLuca M. Purification and properties of the NADH and NADPH specific FMN oxidoreductases from *Beneckea harveyi*. Biochemistry 1977;16:2932–6.
[58] Lei B, Liu M, Huang S, Tu S-C. *Vibrio harveyi* NADPH-flavin oxidoreductase: cloning, sequencing and overexpression of the gene and purification and characterization of the cloned enzyme. J Bacteriol 1994;176:3552–8.
[59] Watanabe H, Hastings JW. Specificities and properties of three reduced pyridine nucleotide-flavin mononucleotide reductases coupling to bacterial luciferase. Mol Cell Biochem 1982;44:181–7.
[60] Inouye S. NAD(P)H-flavin oxidoreductase from the bioluminescent bacterium, *Vibrio fischeri* ATCC 7744, is a flavoprotein. FEBS Lett 1994;347:163–8.
[61] Zenno S, Saigo K, Kanoh H, Inouye S. Identification of the gene encoding the major NAD(P)H-flavin oxidoreductase of the bioluminescent bacterium *Vibrio fischeri* ATCC 7744. J Bacteriol 1994;176:3536–43.
[62] Zenno S, Saigo K. Identification of the genes encoding NAD(P)H-flavin oxidoreductases that are similar in sequence to *Escherichia coli* Fre in four species of luminous bacteria: *Photorhabdus luminescens, Vibrio fischeri, Vibrio harveyi,* and *Vibrio orientalis.* J Bacteriol 1994;176:3544–51.
[63] Nijvipakul S, Wongratana J, Suadee C, Entsch B, Ballou DP, Chaiyen P. LuxG is a functioning flavin reductase for bacterial luminescence. J Bacteriol 2008;190:1531–8.
[64] Lei B, Tu S-C. Mechanism of reduced flavin transfer from *Vibrio harveyi* NADPH-FMN oxidoreductase to luciferase. Biochemistry 1998;37:14623–9.

[65] Tang CK, Jeffers CE, Nichols JC, Tu S-C. Flavin specificity and subunit interaction of *Vibrio fischeri* general NAD(P)H-flavin oxidoreductase FRG/FRase I. Arch Biochem Biophys 2001;392:110–6.

[66] Li X, Chow DC, Tu SC. Thermodynamic analysis of the binding of oxidized and reduced FMN cofactor to *Vibrio harveyi* NADPH-FMN oxidoreductase FRP apoenzyme. Biochemistry 2006;45:14781–7.

[67] Lei B, Wang H, Yu Y, Tu S-C. Redox potential and equilibria in the reductive half-reaction of *Vibrio harveyi* NADPH-FMN oxidoreductase. Biochemistry 2005;44:261–7.

[68] Michaliszyn GA, Wing SS, Meighen EA. Purification and properties of a NAD(P)H:flavin oxidoreductase from the luminous bacterium, *Beneckea harveyi*. J Biol Chem 1977;252:7495–9.

[69] Nijvipakul S, Ballou DP, Chaiyen P. Reduction kinetics of a flavin oxidoreductase LuxG from Photobacterium leiognathi (TH1): half-sites reactivity. Biochemistry 2010;49:9241–8.

[70] Duane W, Hastings JW. Flavin mononucleotide reductase of luminous bacteria. Mol Cell Biochem 1975;6:53–64.

[71] Jablonski E, DeLuca M. Studies of the control of luminescence in *Beneckea harveyi*: properties of the NADH and NADPH:FMN oxidoreductases. Biochemistry 1978;17:672–8.

[72] Jeffers CE, Tu S-C. Differential transfers of reduced flavin cofactor and product by bacterial flavin reductase to luciferase. Biochemistry 2001;40:1749–54.

[73] Jeffers CE, Nichols JC, Tu S-C. Complex formation between *Vibrio harveyi* luciferase and monomeric NADPH:FMN oxidoreductase. Biochemistry 2003;42:529–34.

[74] Liu M, Lei B, Ding Q, Lee JC, Tu S-C. *Vibrio harveyi* NADPH:FMN oxidoreductase: preparation and characterization of the apoenzyme and monomer-dimer equilibrium. Arch Biochem Biophys 1997;337:89–95.

[75] Low JC, Tu S-C. Energy transfer evidence for *in vitro* and *in vivo* complexes of *Vibrio harveyi* flavin reductase P and luciferase. Photochem Photobiol 2003;77:446–52.

[76] Jawanda N, Ahmed K, Tu SC. *Vibrio harveyi* flavin reductase--luciferase fusion protein mimics a single-component bifunctional monooxygenase. Biochemistry 2008;47:368–77.

[77] Campbell ZT, Baldwin TO. Fre Is the Major Flavin Reductase Supporting Bioluminescence from *Vibrio harveyi* Luciferase in *Escherichia coli*. J Biol Chem 2009;284:8322–8.

[78] Li X, Tu SC. Activity coupling of *Vibrio harveyi* luciferase and flavin reductase (FRP): oxygen as a probe. Arch Biochem Biophys 2006;454:26–31.

[79] Otto K, Hofstetter K, Rothlisberger M, Witholt B, Schmid A. Biochemical characterization of StyAB from *Pseudomonas* sp. strain VLB120 as a two-component flavin-diffusible monooxygenase. J Bacteriol 2004;186:5292–302.

[80] Valton J, Mathevon C, Fontecave M, Niviere V, Ballou DP. Mechanism and regulation of the Two-component FMN-dependent monooxygenase ActVA-ActVB from *Streptomyces coelicolor*. J Biol Chem 2008;283:10287–96.

[81] Louie TM, Xie XS, Xun L. Coordinated production and utilization of $FADH_2$ by NAD(P)H-flavin oxidoreductase and 4-hydroxyphenylacetate 3-monooxygenase. Biochemistry 2003;42:7509–17.

[82] Sucharitakul J, Phongsak T, Entsch B, Svasti J, Chaiyen P, Ballou DP. Kinetics of a two-component *p*-hydroxyphenylacetate hydroxylase explain how reduced flavin is transferred from the reductase to the oxygenase. Biochemistry 2007;46:8611–23.

[83] Chakraborty S, Ortiz-Maldonado M, Entsch B, Ballou DP. Studies on the mechanism of *p*-hydroxyphenylacetate 3-hydroxylase from *Pseudomonas aeruginosa*: a system composed of a small flavin reductase and a large flavin-dependent oxygenase. Biochemistry 2010;49:372–85.

[84] Kantz A, Chin F, Nallamothu N, Nguyen T, Gassner GT. Mechanism of flavin transfer and oxygen activation by the two-component flavoenzyme styrene monooxygenase. Arch Biochem Biophys 2005;442:102–16.

[85] Abdurachim K, Ellis HR. Detection of protein-protein interactions in the alkanesulfonate monooxygenase system from *Escherichia coli*. J Bacteriol 2006;188:8153–9.

[86] Gao B, Ellis HR. Altered mechanism of the alkanesulfonate FMN reductase with the monooxygenase enzyme. Biochem Biophys Res Commun 2005;331:1137–45.
[87] Lee JK, Zhao H. Identification and characterization of the flavin:NADH reductase (PrnF) involved in a novel two-component arylamine oxygenase. J Bacteriol 2007;189:8556–63.
[88] Zenno S, Kobori T, Tanokura M, Saigo K. Purification and characterization of NfrA1, a *Bacillus subtilis* nitro/flavin reductase capable of interacting with the bacterial luciferase. Biosci Biotechnol Biochem 1998;62:1978–87.

6 Amine and amino acid oxidases and dehydrogenases

Paul F. Fitzpatrick

Abstract

A large number of flavoproteins oxidize carbon-nitrogen bonds. Most of these enzymes can be grouped into three structural families, the D-amino acid oxidase family, the monoamine oxidase family, and the trimethylamine dehydrogenase family. While the overall folds of the D-amino acid oxidase and monoamine oxidase families diverge, there is remarkable similarity in the orientation of the reactive carbon-nitrogen bond to the flavin within the active site, consistent with a common mechanism; less structural information is available regarding trimethylamine dehydrogenase and related enzymes. The mechanisms that have been considered for amine oxidation by these enzymes include carbanion, radical, nucleophilic, and hydride transfer mechanisms. There is growing consensus that D-amino acid oxidase and related enzymes utilize hydride transfer and that at least some of the members of the monoamine oxidase family do also. Given the structural conservation within the monoamine oxidase family, it is likely that all members of this family utilize the same mechanism. While there is less mechanistic information regarding the trimethylamine dehydrogenase family, mechanistic studies to date can also be accommodated by a hydride transfer mechanism.

6.1 Introduction

Flavoenzymes capable of oxidizing a carbon-nitrogen single bond by removal of a hydride equivalent to form a carbon-nitrogen double bond include several structural families with a wide range of substrate specificities. While the individual reactions catalyzed by these enzymes include oxidation of primary, secondary, and tertiary amines and of amino acids, the essential reaction in all cases is the same, the oxidation of an amine to an imine, as shown in ▶Fig. 6.1, with the electrons being transferred to the flavin cofactor. The imine product is typically released from the enzyme, where it can be hydrolyzed to the corresponding aldehyde or ketone and free amine; exceptions to this occur when the substrate is cyclic, so that the imine is stable to hydrolysis. The subsequent reoxidation of the reduced flavin occurs by a variety of mechanisms. In the case of the oxidases, the reduced flavin reacts directly with oxygen to yield hydrogen peroxide and the oxidized flavin. This also likely occurs with the amino acid monooxygenases, in which the

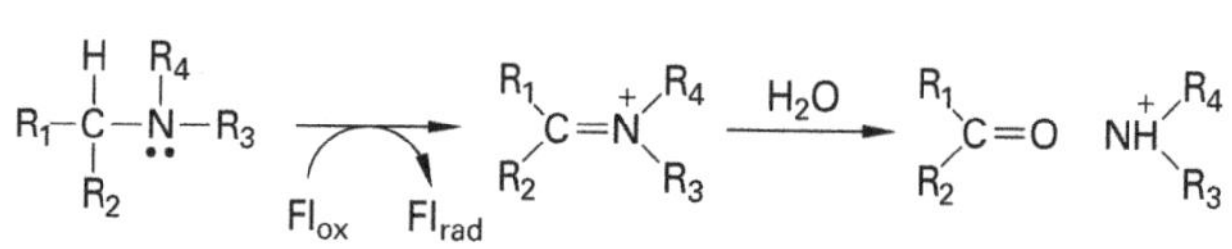

Fig. 6.1: The general reaction of amine oxidation by flavoproteins.

reaction of the reduced enzyme-imine complex with oxygen yields an amide and CO_2 as products. In the case of the dehydrogenases, the electrons are typically transferred from the reduced flavin to an additional cofactor, such as an iron-sulfur center, or to another protein. The focus of the present review will be on the common amine oxidation reaction that makes up the reductive half reaction of these enzymes. Details of the subsequent oxidative half reactions are discussed in greater detail in the chapters on oxygen activation by flavoproteins and on specific oxidases and dehydrogenases. Here, the individual enzymes are grouped together based on common protein structures rather than common substrates to facilitate discussion of mechanistic proposals and reaction specificities.

6.2 D-Amino acid oxidase and related enzymes

Kidney D-amino acid oxidase (DAAO) was one of the first enzymes shown to contain FAD as a cofactor [1] and subsequently has been a model enzyme for understanding amine-oxidizing flavoproteins in general. Structurally, DAAO is a member of a family of enzymes that also includes NikD, sarcosine oxidase, D-arginine dehydrogenase, glycine oxidase, L-proline dehydrogenase, and N,N-dimethylglycine oxidase [2] (►Fig. 6.2). In all cases the substrates are α-amino acids, and a bond between the α-nitrogen and an adjacent carbon is oxidized. However, in the cases where the physiological substrate contains a secondary or tertiary nitrogen, such as sarcosine, proline, or N,N-dimethylglycine, the bond to the α-carbon is not oxidized; instead the bond between the nitrogen and a carbon of one of the other substituents on the nitrogen is oxidized. The reaction specificity is determined by the protein rather than the substrate, in that DAAO oxidizes sarcosine (N-methylglycine) to methylamine and glyoxylate [3], whereas sarcosine oxidase oxidizes sarcosine to glycine and formaldehyde, as shown in ►Fig. 6.3 [4]. Comparison of the available structures of these proteins with ligands or substrates bound provides a likely explanation for the differences in the identity of the bond being oxidized. ►Fig. 6.4 shows an overlay of the active site of DAAO with D-alanine bound [5] with that of sarcosine oxidase with the inhibitor N,N-dimethylglycine bound [6]. The two enzymes bind the substrates with opposite orientations relative to the flavin. The different orientations result in the α-carbon and the amine nitrogen in DAAO being nearly superimposable on a methyl carbon and the amine nitrogen in sarcosine oxidase. This difference in the orientations of the substrate is seen in structures of the other members of this family of flavin amine oxidases.

Given the common structures of these enzymes it is reasonable to assume they use very similar mechanisms to oxidize their amine substrates. An early mechanistic proposal was that the reaction was initiated by removal of the substrate α-proton to form a carbanion (►Fig. 6.5A). This was based on the effects of ring-substitution on the kinetics for a series of substituted phenylglycines in which the correlation of the electron-donating ability of the substituent with the V_{max} value gave a ρ value of 5.4 [7]. However, for most of the substrates in this series the V_{max} value was insensitive to the substituent on the aromatic ring of the substrate due to substrate oxidation not being rate-limiting [8]. An alternative mechanism for flavoprotein-catalyzed dehydrogenation reactions was proposed by Hamilton [9] at about the same time. In this proposal, oxidation of amines or alcohols is initiated by attack of the heteroatom on the C4a position of the flavin (►Fig. 6.5B); this mechanism also required an active site base. The subsequent finding that DAAO catalyzes the elimination

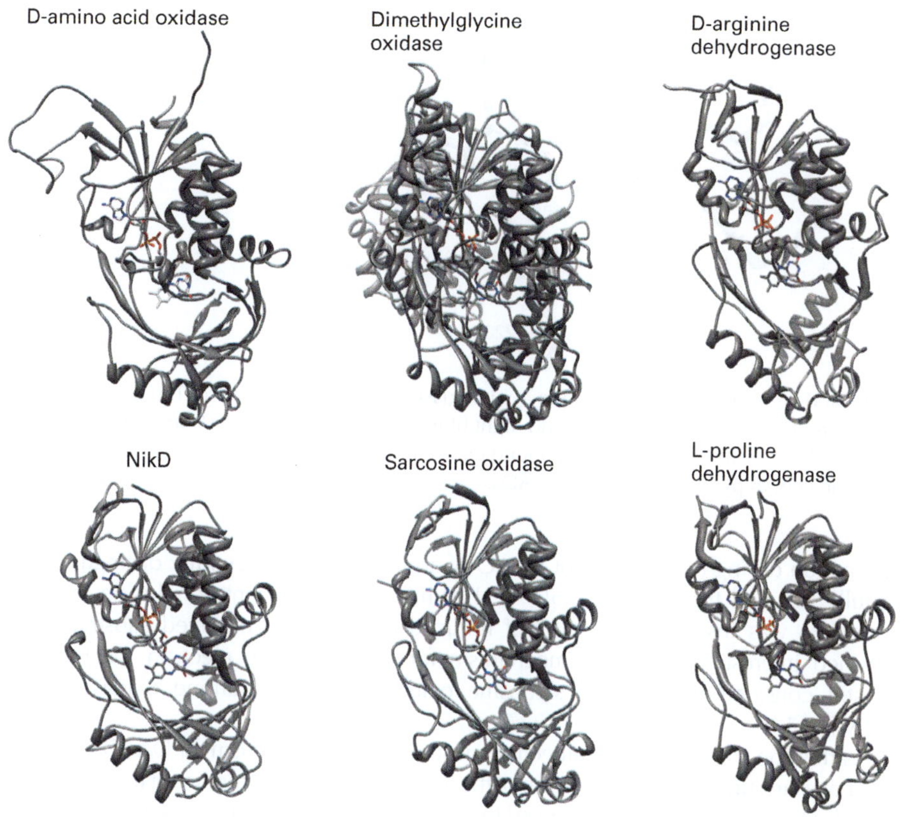

Fig. 6.2: Three-dimensional structures of members of the D-amino acid oxidase family of amine-oxidizing flavoproteins. The structures were drawn using Chimera [73] and the following pdb files: *R. gracilis* D-amino acid oxidase, 1C0P; *Arthrobacter globiformis* dimethylglycine oxidase, 1PJ5; *Pseudomonas aeruginosa* D-arginine dehydrogenase, 2NYE; *Streptomyces tendae* NikD, 2OLN; *Bacillus sp. B-0618* sarcosine oxidase, 1EL5; *Pyrococcus horikoshii* L-proline dehydrogenase, 1Y56.

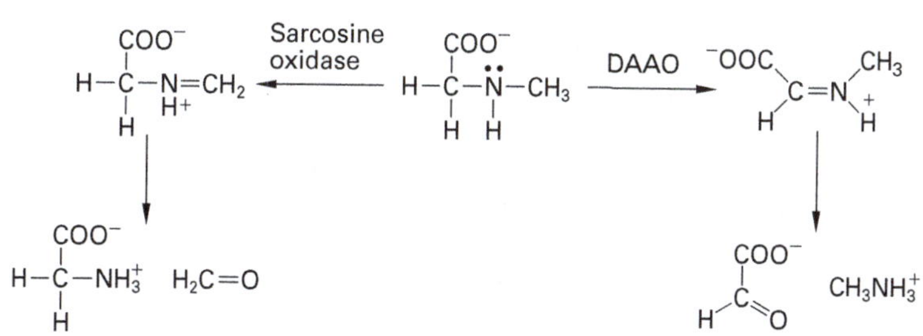

Fig. 6.3: Products of sarcosine oxidation by D-amino acid oxidase and sarcosine oxidase.

of chloride from 3-Cl-D-alanine [10] was taken as strong evidence for a carbanion intermediate. A further possibility, oxidation of the amine by single electron steps with intermediate formation of the flavin semiquinone and a substrate radical, found support in

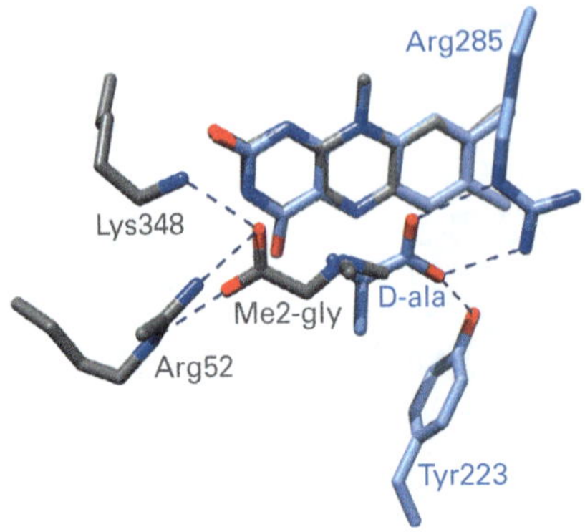

Fig. 6.4: Binding of substrates in D-amino acid oxidase and sarcosine oxidase. The active sites of *R. gracilis* DAAO with D-alanine bound (light blue carbons, pdb file 1COP) and *Bacillus* sarcosine oxidase with dimethylglycine bound (gray carbons, pdb file 1EL5) were overlaid by aligning the central pyrazine rings of the FADs.

the formation during the reaction of DAAO of an intermediate spectrally distinguishable from fully reduced flavin (▶Fig. 6.5C) [11]. While the subsequent conclusion that this intermediate was instead a complex of reduced enzyme and oxidized substrate [12] diminished interest in a radical mechanism for DAAO, single electron mechanisms were still considered viable for sarcosine oxidase and other amine oxidases fairly recently [13,14].

The carbanion mechanism remained the most-accepted mechanism for DAAO [15] for a number of years, until the determination of the three-dimensional structure of the pig kidney enzyme by X-ray crystallography showed that the enzyme's active site lacks a base capable of removing a proton from the substrate [16]. This led to consideration of a hydride transfer mechanism (▶Fig. 6.5D) as an alternative possibility. Subsequent mechanistic studies further supported a hydride transfer mechanism. In contrast to kidney DAAO, the yeast enzyme exhibits a ρ value of 0.73 for the effect of substituents in the ring of phenylglycine on the rate constant for flavin reduction [17], consistent with little charge development in the transition state for amine oxidation. The pig kidney enzyme exhibits an inverse ^{15}N kinetic isotope effect on amino acid oxidation [18]; this establishes that rehybridization of the amine nitrogen occurs in the same transition state as cleavage of the carbon-hydrogen bond, a result only consistent with hydride transfer among the mechanisms shown in ▶Fig. 6.5. At present, the hydride transfer mechanism is the most generally accepted mechanism for oxidation of amino acids by DAAO and similar flavoproteins.

The protonation state of the amino acid substrate must also be considered in discussing mechanisms of amine oxidation. At neutral pH the amino groups of amines and amino acids are protonated, so that oxidation necessitates the removal of a proton from the charged amino group. This deprotonation could either occur nonenzymatically or involve an active site residue. The k_{cat}/K_m-pH profiles for both pig kidney [19] and yeast DAAO [20] show a pK_a of ~9 for a moiety in the free substrate or enzyme that must be unprotonated for catalysis; this can be assigned to the amino group of the substrate in solution if the anionic amino acid is the substrate. The effect of pH on ^{15}N isotope effects is also consistent with the anionic form of the amino acid rather than the zwitterion being the substrate [18]. Finally, there is no

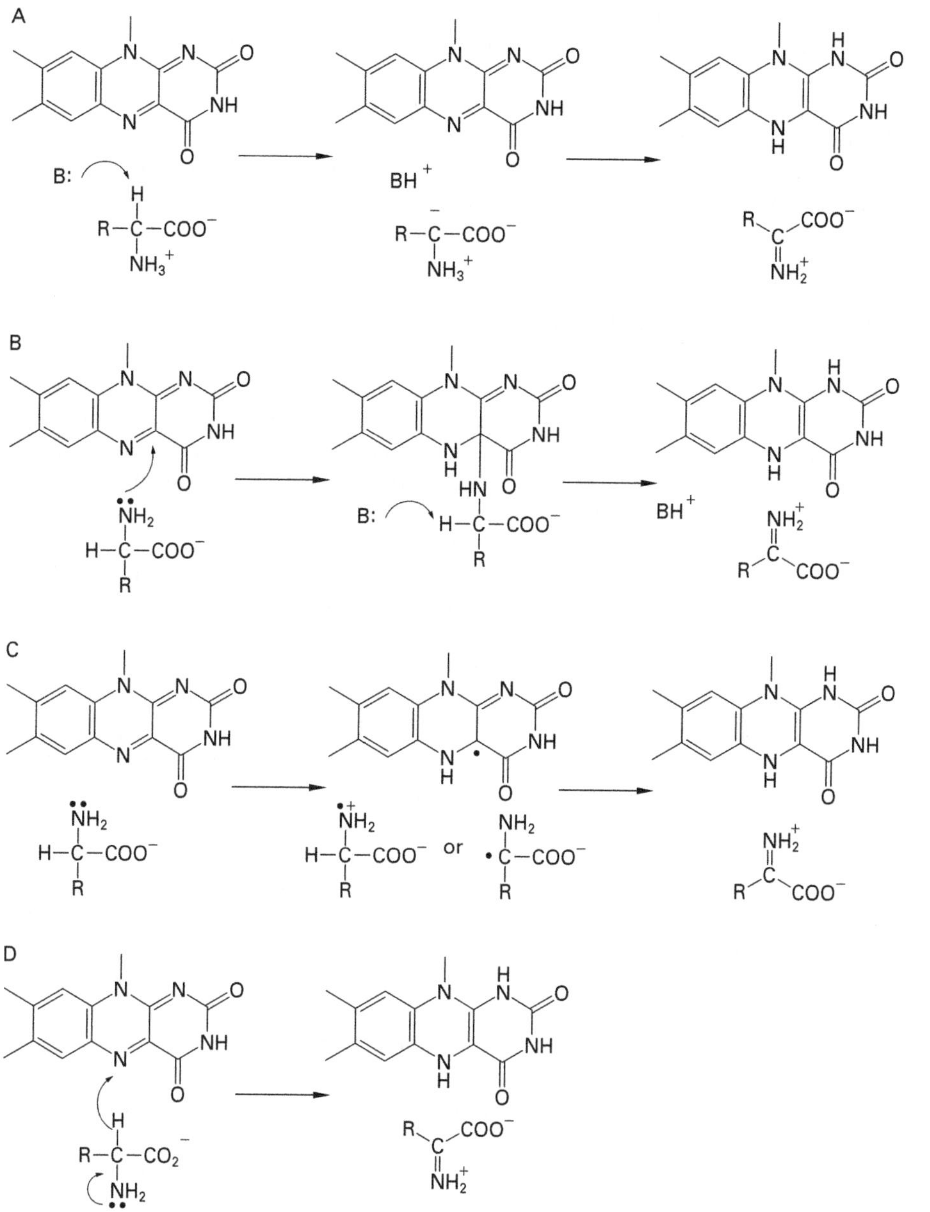

Fig. 6.5: Proposed mechanisms for oxidation of amino acids by flavoproteins: A, carbanion mechanism; B, nucleophilic addition to the flavin C4a position; C, radical mechanism; D, hydride transfer mechanism.

solvent isotope effect on the reductive half reaction once the effects of D_2O on pK_a values are corrected for [20,21], establishing that the amine is already neutral when hydride transfer occurs. All of these results are consistent with deprotonation of the substrate amino group not being enzyme-catalyzed in the case of DAAO. A similar conclusion has been reached for sarcosine oxidase [22]. The need for the unprotonated amine is not necessarily very unfavorable for the reaction. DAAO, like a number of flavoprotein oxidases, is found in peroxisomes [23], where the pH is between 8 and 8.5 [24]. Under these conditions ~10% of the amino groups of most amino acids would be unprotonated, so that an active site base would not be essential to provide the anionic amino acid for the reaction.

The structural and mechanistic results summarized here consistently support hydride transfer from the neutral amine as the mechanism of oxidation of amino acids by the DAAO family of flavoproteins. This is the mechanism shown in ▶Fig. 6.5D.

6.3 Monoamine oxidase and related enzymes

The group of flavoprotein amine oxidases with folds similar to that of monoamine oxidase (MAO) is much larger than that with the same fold as DAAO (▶Fig. 6.6). This family also oxidizes a wider range of substrates. Monoamines, amino acids, and polyamines are all substrates for different members of this family. In addition, the substrates can be primary, secondary, or tertiary amines. While in general the substrates are small molecules, the native substrate for LSD1 is a histone and the minimal substrate for the enzyme is a peptide [25].

There are several structural features common to members of the MAO family. In general, the active sites of these enzymes contain two hydrophobic residues that lie on opposite sides of the substrate amino group. ▶Tab. 6.1 lists the respective residues for a number of these enzymes, and ▶Fig. 6.7 illustrates their positions relative to the FAD. As is clear, these are generally aromatic residues that have the face of the aromatic ring oriented toward the substrate. Consequently, the two tyrosines in MAO have been referred to as an aromatic cage [26,27]. The two residues do not appear to be equally critical for amine oxidation. Mutagenesis of Tyr435 in MAO B to other aromatic or hydrophobic residues or of Tyr444 in MAO A results in significant decreases in the k_{cat}/K_m values for amine substrates [27,28]. In contrast, the kinetic parameters for the Y407F mutant of MAO A are within 5-fold of the wild-type values. Tyr430 of mouse polyamine oxidase corresponds to Tyr407 in MAO; the Y430F mutation similarly decreases the k_{cat}/K_m and k_{cat} values by only ~7-fold [29]. Furthermore, as shown in ▶Tab. 6.1, several MAO family members contain only one aromatic residue in the active site rather than two. The different mechanistic proposals for MAO can yield different roles for these two residues. One possible role independent of the specific mechanism is to provide a π-cation interaction to stabilize the positive charge that must develop on the substrate nitrogen during the reaction. As is the case with the DAAO family, the effects of pH on kinetic parameters have established that the neutral amine is the substrate for MAO and related enzymes [30,31].

A second conserved structural feature in this enzyme family is an active site lysine that forms a water-mediated hydrogen bond with the flavin N5 (▶Fig. 6.8). This lysine is on the opposite face of the isoalloxazine ring from the substrate, making it unlikely that it directly interacts with the substrate. The effects of mutating this residue have been quite variable. Mutation of this residue in maize PAO, Lys300, decreases the rate constant for flavin reduction by three orders of magnitude [32] and the rate constant for oxidation by 30-fold [33]. The structure of the mutant enzyme suggests that the mutation results in substrate binding in a nonreactive fashion [33]. Mutation of the corresponding residue in LSD1, Lys661, is reported to reduce the enzyme activity below detectable levels [34]. In contrast, mutation of this residue in mouse PAO, Lys315, has no effect on the reductive half-reaction, but slows flavin reoxidation ~20-fold. The variability of

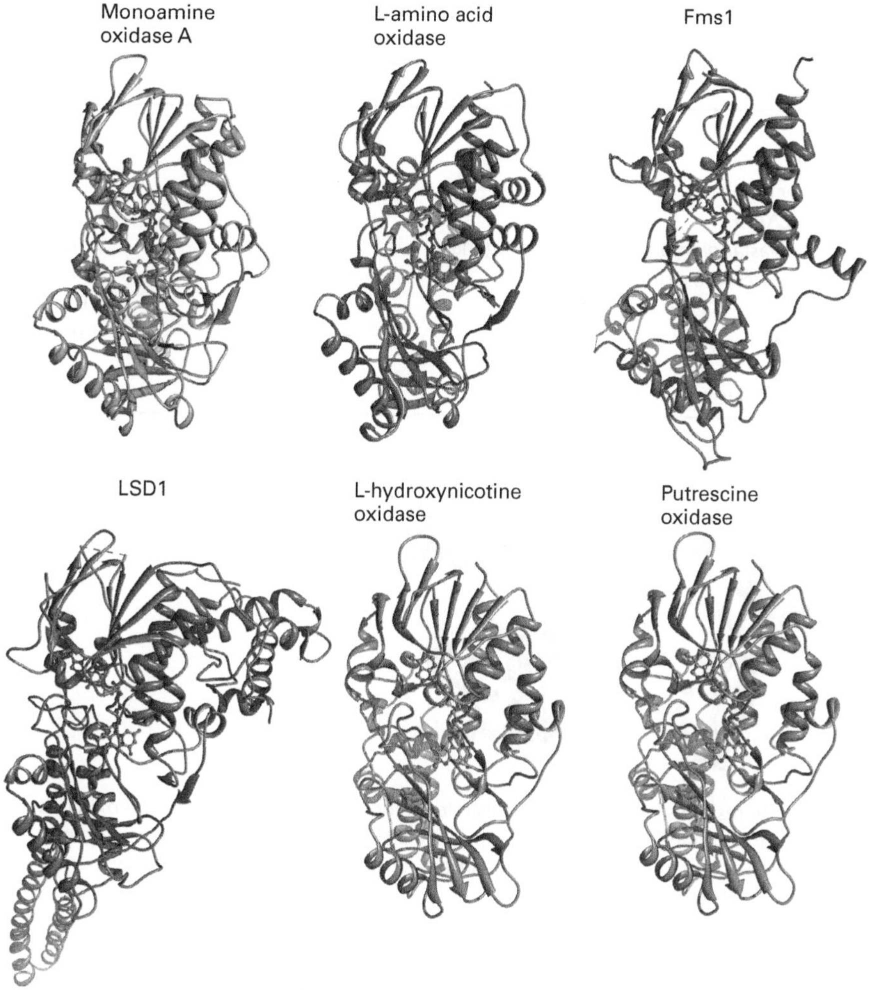

Fig. 6.6: Representative structures of members of the monoamine oxidase structural family of amine-oxidizing flavoproteins: human monoamine oxidase A, pdb file 2Z5X; *Calloselasma rhodostoma* L-amino acid oxidase, pdb file 2IID; *Saccharomyces cerevisiae* fms1, pdb file 1yy5; human LSD1, pdb file 2H94; *Arthrobacter nicotinovorans* L-hydroxynicotine oxidase, pdb file 3NH3; *Rhodococcus erythropolis* putrescine oxidase, pdb file 2YG3.

these effects suggests that this lysine plays a significant role in maintaining the proper active site structure by positioning the flavin and also modulates the flavin reactivity.

The MAO family includes the L-amino acid oxidases (LAAOs). As noted when the first structure of an LAAO became available, although the overall structure of DAAO and LAAO differ, their active sites are effectively mirror images of one another [35]. This is illustrated in ▶Fig. 6.9 in which the interactions of D-alanine in the active site of DAAO are shown along with the interactions of L-phenylalanine in the active site of LAAO. The two amino acid substrates bind above the flavin in opposite orientations. In both cases their carboxylates interact with arginine and tyrosine residues. These arginine residues lie on opposite sides of the active site, consistent with the opposite

Tab. 6.1: Residues bracketing the substrate amine in members of the monoamine oxidase structural family

Protein		
human MAO A	Tyr407	Tyr444
human MAO B	Tyr398	Tyr435
Aspergillus niger MAO	Trp430	Phe466
C. rhodostoma LAAO	Ile430	Trp465
Pseudomonas savastanoi tryptophan monooxygenase	Phe476	Trp519
R. erythropolis putrescine oxidase	Tyr395	His432
human Fms1	Tyr450	Cys488
S. cerevisiae LSD1	Tyr761	Thr810
maize PAO	Phe403	Tyr439
A. nicotinovorans L-HO-nicotine oxidase	Trp371	Tyr407

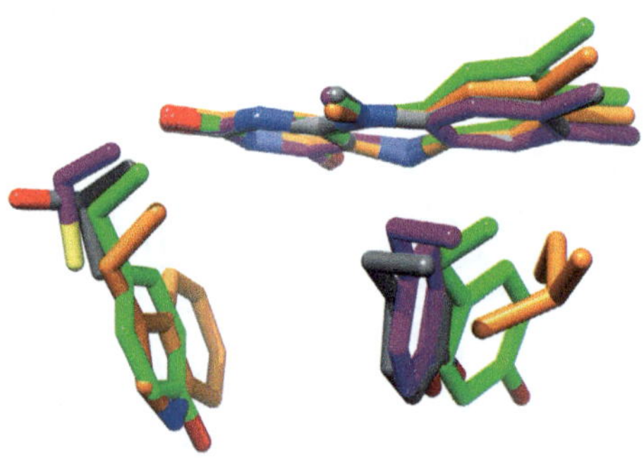

Fig. 6.7: Active site residues bracketing the substrate amino moiety in selected amine oxidases: green carbons, human monoamine oxidase A, pdb file 2Z5X; orange carbons, *C. rhodostoma* L-amino acid oxidase, pdb file 2IID; magenta carbons, *S. cerevisiae* fms1, pdb file 1yy5; gray carbons, human LSD1, pdb file 2H94.

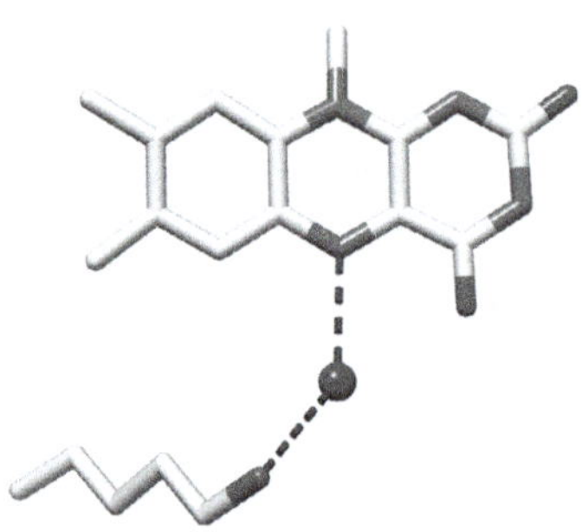

Fig. 6.8: The conserved active site lysine in the monoamine oxidase family. The FAD, lysine, and water are from a structure of L-6-HO-nicotine oxidase, pdb file 3NH3.

orientations of the substrates, but the phenolic oxygens of the two tyrosines are virtually superimposable. The carbon-nitrogen bonds that are oxidized in the two substrates also occupy almost the same location above the flavin ring. The similarity of the active sites is consistent with the two enzymes utilizing the same mechanism. LAAO, like DAAO,

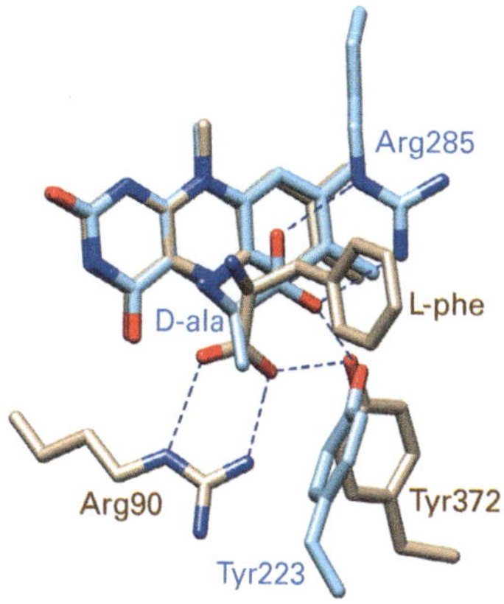

Fig. 6.9: Binding of substrates in D-amino acid oxidase and L-amino acid oxidase. The active sites of *R. gracilis* D-amino acid oxidase with D-alanine bound (light blue carbons, pdb file 1COP) and *C. rhodostoma* L-amino acid oxidase with phenylalanine bound (tan carbons, pdb file 2IID) were overlaid by aligning the central pyrazine rings of the FADs

lacks a conserved active site base that could accept a proton from the substrate, making a carbanion mechanism unlikely and suggesting that the amino group of the amino acid substrate must be deprotonated before binding. Solution studies of LAAO are also consistent with similar mechanisms for the two enzymes. The k_{cat}/K_m-pH profiles for different LAAOs show that a moiety with a pK_a value close to that of the amino group of the substrate must be deprotonated for activity [31,36], there is no solvent isotope effect on the reductive half-reaction of LAAOs, and the ^{15}N kinetic isotope effects for LAAO and DAAO are similar [37]. Thus, if DAAO utilizes a hydride transfer mechanism, it is reasonable to conclude that LAAOs do also.

Compared to the mechanistic proposals for LAAOs, which were based mostly on precedent from DAAO, a wider variety of mechanisms have been proposed for MAO and related enzymes. Amines containing cyclopropyl groups, such as phenylcyclopropylamine, inactivate MAO [38] and LSD1 [39]. In light of the propensity of cyclopropyl rings alpha to a radical to open to form a highly reactive primary carbon radical and the observation that inactivation of MAO is accompanied by attachment of the inhibitor to the protein, Silverman and coworkers [40,41] proposed that the enzyme oxidized amines in single electron steps (►Fig. 6.10A). Alternative mechanisms involving transient formation of amino acid radicals in the active site were also proposed [41]. The proposal that amine oxidation by MAO involved transient formation of amino acid radicals gained some support from reports that stable tyrosyl radicals could be detected in MAO [42,43]. However, there are a number of other problems with a single electron transfer mechanism for these enzymes. The single electron reduction potential of amines is well over 1 V, making single electron oxidation by a flavin with a reduction potential of 200 mV or less very unfavorable energetically. Transient radical intermediates are not seen during the oxidation of amines by flavoprotein oxidases [44–47]. While Tyr407 has been identified as the most likely site of the radical seen in MAO [42], mutagenesis of this residue does not dramatically affect the enzyme's activity. Finally, inactivation of MAO B by phenylcyclopropylamine results in modification of Cys365 [48], but this residue is far from the active site and unlikely to be involved in the normal oxidation reaction.

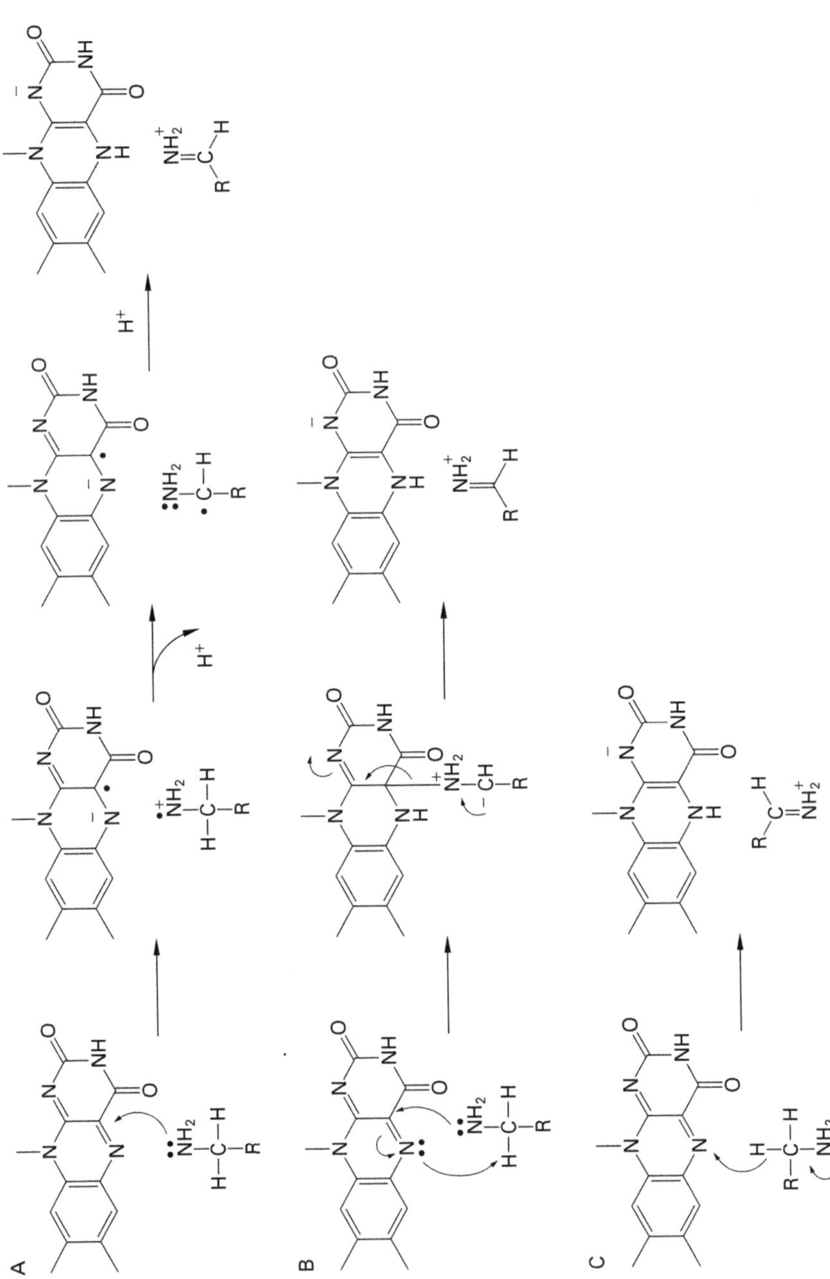

Fig. 6.10: Mechanisms proposed for oxidation of amines by monoamine oxidase: A, radical mechanism; B, nucleophilic addition to the flavin C4a position; C, hydride transfer mechanism.

An alternative mechanism proposed by Edmondson and coworkers [49] is shown in ▶Fig. 6.10B. The mechanism was based on studies of the reaction of benzylamine with lumiflavin [50], but it has clear similarities to the earlier proposal of Hamilton shown in ▶Fig. 6.5B. The mechanism was proposed to explain the finding that the k_{cat} value for oxidation of a series of para-substituted benzyl amines by MAO A decreases as the substituent becomes more electron donating, for a ρ value of 1.8. The positive ρ value was taken as evidence for removal of the substrate hydrogen as a proton. In light of the absence of a conserved base in the MAO active site, the flavin N5 was proposed to be the base. However, oxidation of substituted phenethylamines by MAO A [51] or of benzylamines by human MAO B [52] shows no effect of the ring substituent, for a ρ value of ~0. A more recent study of the effects of ring substituents on the rate constant for reduction of mouse PAO by N,N′-dibenzyl-1,4-diaminobutanes gave a ρ value of –0.6 at the pH optimum [53]. The variation in the ρ value for essentially identical reactions makes it difficult to draw definitive conclusions regarding the mechanisms of these enzymes from such analyses. While a large positive ρ value is most consistent with development of some negative charge in the transition state, a ρ value of 1.7 has been reported for the pyridine-nucleotide dependent enzyme xylose reductase, which necessarily utilizes a hydride transfer mechanism [54], and even DAAO exhibits a positive ρ value [17].

As is the case with DAAO, ^{15}N kinetic isotope effects have been used as mechanistic probes for members of the MAO structural family. The first value to be determined was for tryptophan monooxygenase, an LAAO. This enzyme exhibits an ^{15}N isotope effect on the k_{cat}/K_m value for the slow substrate alanine of 0.992 [37], similar to the values for DAAO [18] and N-methyltryptophan oxidase [55] and consistent with similar transitions states for all three enzymes. Subsequently, a value of 0.985 was reported for MAO B with benzylamine as substrate [56]. The ^{15}N effect increased to 0.990 when deuterated benzylamine was the substrate. This change with deuteration was taken as evidence for the nucleophilic addition reaction of ▶Fig. 6.10B, in that a concerted mechanism such as hydride transfer would be expected to show no change or a more inverse effect with the deuterated substrate [57]. However, *ab initio* calculations for the oxidation of dimethylamine by a flavin indicate that the mechanism of Figure 6.5B would yield ^{15}N isotope effects significantly greater than unity, while radical and hydride transfer mechanisms would both yield isotope effects less than one [55]. The decrease in the isotope effect for MAO upon deuteration is also significantly less than would be expected given the large deuterium effect of ~5 for MAO [52] and the expectation that the ^{15}N effect is on a step after an effectively irreversible CH bond cleavage step rather than before it. An alternative explanation for the ^{15}N isotope effects on MAO is that the mechanism is the same as those of LAAO and DAAO, hydride transfer, but that CH bond cleavage and the rehybridization of the nitrogen are slightly asynchronous in MAO B with benzylamine as substrate. If rehybridization of the nitrogen is slightly ahead of hydride transfer, negative charge would develop at the reacting carbon, consistent with the observed ρ value.

Recent structural studies of L-6-hydroxynicotine oxidase (6HLNO) provide further insight into the mechanisms of this enzyme family. This enzyme catalyzes the oxidation of 6-HO-L-nicotine to form 6-HO-pseudooxynicotine (▶Fig. 6.11). The product of the reaction was proposed to be 6-HO-N-methylmyosmine based on the ability of cells containing the enzyme to convert this compound to 6-HO-pseudooxynicotine [58], but the actual reaction product has never been isolated. The structure of 6HLNO shows that it is a member of

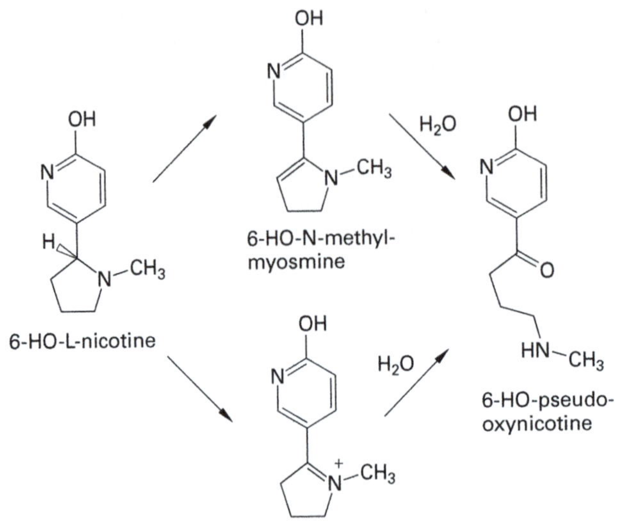

Fig. 6.11: The L-HO-nicotine oxidase reaction.

the MAO family, with the conserved lysine and pair of aromatic residues in the active site discussed above (►Tab. 6.1) [59]. Given this structure and the lack of precedent for oxidation of a carbon-carbon bond by any member of the MAO family, a more likely possibility is that the enzyme catalyzes the oxidation of a carbon-nitrogen bond in its substrate to form an imine, as shown in the bottom path in ►Fig. 6.11. Such an imine would readily open to form 6-HO-pseudooxynicotine. The orientation of the amine substrate relative to the flavin in 6HLNO has been established by structures of both reduced and oxidized 6HLNO with bound substrate. ►Fig. 6.12 portrays an overlay of the FAD and bound L-6-HO-nicotine in 6HLNO with the FAD and bound D-alanine

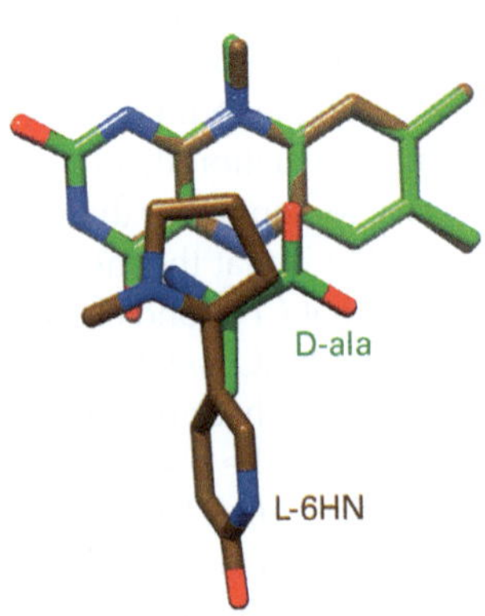

Fig. 6.12: Binding of substrates in D-amino acid oxidase and L-HO-nicotine oxidase. The active site of *A. nicotinovorans* L-HO-nicotine oxidase with L-HO-nicotine (6LHN) bound (brown carbons, pdb file 3NG7) and that of *R. gracilis* DAAO with D-alanine bound (green carbons, pdb file 1C0P) were overlaid by aligning the central pyrazine rings of the FAD in the two structures.

in DAAO. This figure shows that the orientations of the carbon-nitrogen bonds in the two substrates relative to the FAD are essentially identical. In both cases the reactive carbons of the substrate are positioned 3.4–3.5 Å from the flavin N5, appropriate positioning for hydride transfer [60]. Thus, the structural data for 6HLNO is consistent with a mechanism similar to that of DAAO.

Clearly, the mechanism of oxidation of amines by members of the MAO family remains a matter of some argument. Still, comparison of the LAAO and 6HLNO active sites with that of DAAO establishes that the relative orientation of the reactive carbon-nitrogen bond to the flavin is essentially identical in all three cases. This conservation of active site geometry strongly suggests that both the DAAO and the MAO families of flavin amine oxidases use the same mechanism. In light of the preponderance of evidence for a hydride transfer mechanism for DAAO, a similar hydride transfer mechanism seems most likely for the MAO family.

6.4 Trimethylamine dehydrogenase

Trimethylamine dehydrogenase (TMDH), dimethylamine dehydrogenase, and histamine dehydrogenase (HDH) make up a smaller family of amine-oxidizing flavoproteins [61]. The FMN is covalently bound to a cysteinyl residue through a 6-S-linkage. In addition to the flavin, these enzymes contain a 4Fe-4S center and a tightly bound ADP (▶Fig. 6.13). Reoxidation of the reduced flavin formed upon amine oxidation occurs by transfer of electrons first to the 4Fe-4S center and then to an electron transferring flavoprotein [62]. The role of the bound ADP, which is in a separate domain from the redox-active cofactors [63], is not known.

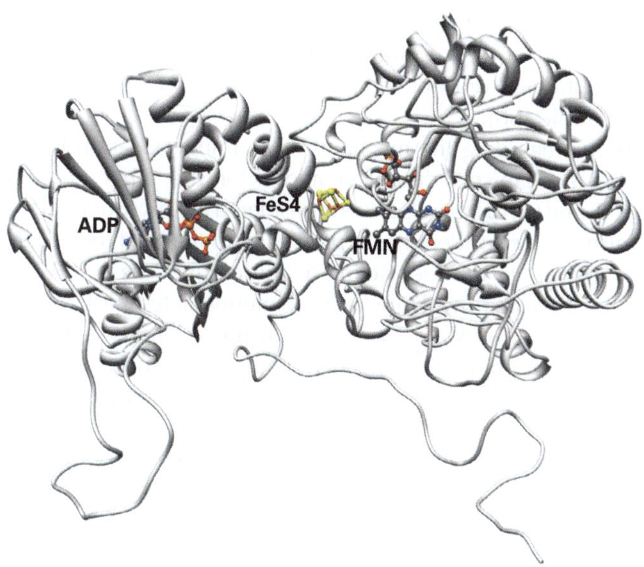

Fig. 6.13: Structure of trimethylamine dehydrogenase from *Methylophilus methylotrophus*, pdb file 1DJN, showing the FMN, iron-sulfur center, and bound ADP.

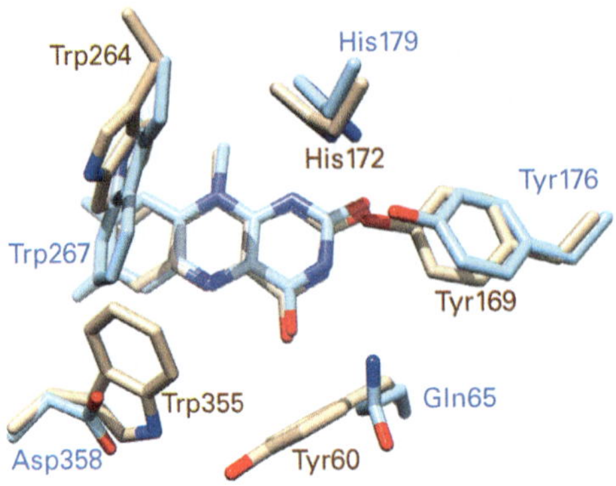

Fig. 6.14: The active sites of *M. methylotrophus* trimethylamine dehydrogenase (light brown carbons, pdb file 1DJN) and *Nocarioides simplex* histamine dehydrogenase (light blue carbons, pdb file 3K30) were overlaid by aligning the central pyrazine rings of the FAD in the two structures.

The active sites of TMDH and HDH are compared in ▶Fig. 6.14. No structures are available of substrates or inhibitors bound in the active site of either enzyme, although models have been proposed [64,65]. In TMDH, the trimethylamine substrate is proposed to be bound as the neutral amine [66] in an aromatic bowl made up of Tyr60, Trp264, and Trp355 [67]. Of these, only Trp264 is conserved in all three enzymes, while Tyr60 is replaced by a glutamine in both HDH and dimethylamine dehydrogenase, and Trp355 is replaced by Asp358 in HDH. The conserved tryptophan may stabilize the positive charge on the amine in the transition state by a π-cation interaction in a similar fashion to the role of the aromatic residues listed in ▶Tab. 6.1 for the MAO structural family. The conserved His172 in TMDAH had been proposed to act as an active site base to form the carbanion of the substrate, but the finding that mutagenesis of this residue in TMDH to glutamine did not significantly alter the rate constant for amine oxidation ruled out this role [68].

There is no consensus regarding the mechanism of amine oxidation by TMDH and related enzymes. The mechanisms discussed above for the other amine oxidizing flavoproteins (▶Figs. 6.5 and 6.10) have also been considered for this structural family. Thus, a carbanion mechanism was considered as a possibility until site-directed mutagenesis ruled out any of the active site residues acting as the required base [69,70]. This led to consideration of mechanisms involving radical intermediates [70] or nucleophilic addition to the flavin [14,63,66]. A drawback to both possibilities was the assumption that the cationic form of the amine was considered to be the substrate for both TMDH [64] and HDH [65,71]. The subsequent conclusion from pH and isotope studies that the neutral base is the substrate for TMDH [66,72] alleviated this concern, although the inability to assay HDH above pH 9.3 has precluded establishing the protonation state of the substrate for that enzyme [61]. Despite the overwhelming evidence for hydride transfer as the mechanism for the majority of amine-oxidizing flavoproteins, there does not appear to have been significant consideration of this possibility for the TMDH family [14]. However, computational analyses of the oxidation of dimethylamine by flavin indicate that the hydride transfer pathway would be of significantly lower energy than either a radical or nucleophilic reaction [55]; the significance of this observation is tempered by the significant uncertainties in calculating energetics with present *ab initio* methods. Still, hydride transfer has to be considered a viable mechanism for the TMDH family.

6.5 Conclusions

Fundamental to our growing understanding of the mechanism of amine oxidation by flavoproteins described here has been the combination of detailed structures of multiple enzymes catalyzing these reactions with increasingly powerful mechanistic studies of representative enzymes. The structural studies of both the DAAO and the MAO structural families show remarkable similarities in the positioning of substrates relative to the flavin in the active site, and the mechanistic studies indicate that the transition states for the reactions are also quite similar. While we have less information regarding the smaller TMDH family, there is no indication that these enzymes diverge from the patterns seen with the others. Although the mechanisms of both MAO and TMDH remain a matter of some disagreement, it is possible to make the argument that all the enzymes discussed here utilize similar hydride transfer mechanisms. The three families thus represent an example of convergent evolution on a common solution for a common biochemical problem.

6.6 Acknowledgments

I would like to thank the collaborators and graduate students and postdoctoral fellows from my laboratory who have contributed to our understanding of flavoprotein mechanisms over the years. The research on this project in the Fitzpatrick laboratory has been supported by NIH grant GM 58698.

6.7 References

[1] Warburg O, Christian W. Isolation of the prosthetic group of the d-amino acid oxidase. Biochem Zeit 1938;298:150–68.
[2] Carrell CJ, Bruckner RC, Venci D, Zhao G, Jorns MS, Mathews FS. NikD, an unusual amino acid oxidase essential for nikkomycin biosynthesis: structures of closed and open forms at 1.15 and 1.90 Å resolution. Structure 2007;15:928–41.
[3] Naoi M, Yagi K. Oxidation of sarcosine and N-alkyl derivatives of glycine by D-amino-acid oxidase. Biochim Biophys Acta 1976;438:61–70.
[4] Wagner MA, Jorns MS. Monomeric sarcosine oxidase: 2. Kinetic studies with sarcosine, alternate substrates, and a substrate analogue. Biochemistry 2000;39:8825–9.
[5] Umhau S, Pollegioni L, Molla G, Diederichs K, Welte W, Pilone MS, Ghisla S. The X-ray structure of D-amino acid oxidase at very high resolution identifies the chemical mechanism of flavin-dependent substrate dehydrogenation. Proc Natl Acad Sci USA 2000;97:12463–8.
[6] Wagner MA, Trickey P, Chen Z-W, Mathews FS, Jorns MS. Monomeric sarcosine oxidase: 1. Flavin reactivity and active site binding determinants. Biochemistry 2000;39:8813–24.
[7] Neims AH, De Luca DC, Hellerman L. Studies on crystalline D-amino acid oxidase. III. Substrate specificity and σ-ρ relationship. Biochemistry 1966;5:203–13.

[8] Fitzpatrick PF, Massey V. The kinetic mechanism of D-amino acid oxidase with D-α-aminobutyrate as substrate: Effect of enzyme concentration on the kinetics. J Biol Chem 1982;257:12916–23.
[9] Brown LE, Hamilton GA. Some model reactions and a general mechanism for flavoenzyme-catalyzed dehydrogenations. J Am Chem Soc 1970;92:7225–7.
[10] Walsh CT, Schonbrunn A, Abeles RH. Studies on the mechanism of action of D-amino acid oxidase. Evidence for removal of substrate α-hydrogen as a proton. J Biol Chem 1971;246: 6855–66.
[11] Massey V, Gibson QH. Role of semiquinones in flavoprotein catalysis. Fed Proc 1964;23:18–29.
[12] Massey V, Ghisla S. Role of charge-transfer interactions in flavoprotein catalysis. Ann N Y Acad Sci 1974;227:446–65.
[13] Zhao G, Jorns MS. Spectral and kinetic characterization of the Michaelis charge transfer complex in monomeric sarcosine oxidase. Biochemistry 2006;45:5985–92.
[14] Scrutton NS. Chemical aspects of amine oxidation by flavoprotein enzymes. Nat Prod Rep 2004;21:722–30.
[15] Fitzpatrick PF. Carbanion versus hydride transfer mechanisms in flavoprotein-catalyzed dehydrogenations. Bioorg Chem 2004;32:125–39.
[16] Mattevi A, Vanoni MA, Todone F, Rizzi M, Teplyakov A, Coda A, Bolognesi M, Curti B. Crystal structure of D-amino acid oxidase: A case of active site mirror-image convergent evolution with flavocytochrome b_2. Proc Natl Acad Sci USA 1996;93:7496–501.
[17] Pollegioni L, Blodig W, Ghisla S. On the mechanism of D-amino acid oxidase. Structure/linear free energy correlations and deuterium kinetic isotope effects using substituted phenylgylcines. J Biol Chem 1997;272:4924–34.
[18] Kurtz KA, Rishavy MA, Cleland WW, Fitzpatrick PF. Nitrogen isotope effects as probes of the mechanism of D-amino acid oxidase. J Am Chem Soc 2000;122:12896–7.
[19] Denu JM, Fitzpatrick PF. pH and kinetic isotope effects on the reductive half-reaction of D-amino acid oxidase. Biochemistry 1992;31:8207–15.
[20] Harris CM, Pollegioni L, Ghisla S. pH and kinetic isotope effects in D-amino acid oxidase catalysis: Evidence for a concerted mechanism in substrate dehydrogenation via hydride transfer. Eur J Biochem 2001;268:5504–20.
[21] Denu JM, Fitzpatrick PF. Intrinsic primary, secondary, and solvent kinetic isotope effects on the reductive half-reaction of D-amino acid oxidase: Evidence against a concerted mechanism. Biochemistry 1994;33:4001–7.
[22] Zhao G, Jorns MS. Ionization of zwitterionic amine substrates bound to monomeric sarcosine oxidase. Biochemistry 2005;44:16866–74.
[23] de Duve C, Baudhuin P. Peroxisomes (microbodies and related particles). Physiol Rev 1966;46:323–57.
[24] Dansen TB, Wirtz KWA, Wanders RJA, Pap EHW. Peroxisomes in human fibroblasts have a basic pH. Nat Cell Biol 1999;2:51–3.
[25] Forneris F, Binda C, Vanoni MA, Battaglioli E, Mattevi A. Human histone demethylase LSD1 reads the histone code. J Biol Chem 2005;280:41360–5.
[26] Binda C, Newton-Vinson P, Hubalek F, Edmondson DE, Mattevi A. Structure of human monoamine oxidase B, a drug target for the treatment of neurological disorders. Nat Struct Biol 2002;9:22–6.
[27] Akyüz MA, Erdem SS, Edmondson DE. The aromatic cage in the active site of monoamine oxidase B: effect on the structural and electronic properties of bound benzylamine and p-nitrobenzylamine. J Neural Transm 2007;114:693–8.
[28] Nandigama RK, Miller JR, Edmondson D. Loss of serotonin oxidation as a component of the altered substrate specificity in the Y444F mutant of recombinant human liver MAO A. Biochemistry 2001;40:14839–46.

[29] Royo M, Fitzpatrick PF. Mechanistic studies of mouse polyamine oxidase with N1,N12–bis-ethylspermine as a substrate. Biochemistry 2005;44:7079–84.
[30] Dunn RV, Marshall KR, Munro AW, Scrutton NS. The pH dependence of kinetic isotope effects in monoamine oxidase A indicates stabilization of the neutral amine in the enzyme-substrate complex. FEBS Journal 2008;275:3850–8.
[31] Emanuele JJ, Jr., Fitzpatrick PF. Mechanistic studies of the flavoprotein tryptophan 2-monooxygenase. 2. pH and kinetic isotope effects. Biochemistry 1995;34:3716–23.
[32] Polticelli F, Basran J, Faso C, Cona A, Minervini G, Angelini R, Federico R, Scrutton NS, Tavladoraki P. Lys300 plays a major role in the catalytic mechanism of maize polyamine oxidase. Biochemistry 2005;44:16108–20.
[33] Fiorillo A, Federico R, Polticelli F, Boffi A, Mazzei F, Di Fusco M, Ilari A, Tavladoraki P. The structure of maize polyamine oxidase K300M mutant in complex with the natural substrates provides a snapshot of the catalytic mechanism of polyamine oxidation. FEBS Journal 2011;278:809–21.
[34] Lee MB, Winder C, Cooch H, Shiekhattar R. An essential role for CoREST in nucleosomal histone3 lysine4 demethylation. Nature 2005;437:432–5.
[35] Pawelek PD, Cheah J, Coulombe R, Macheroux P, Ghisla S, Vrielink A. The structure of L-amino acid oxidase reveals the substrate trajectory into an enantiomerically conserved active site. EMBO J 2000;19:4204–15.
[36] Porter DJT, Bright HJ. Interpretation of the pH dependence of flavin reduction in the L-amino acid oxidase reaction. J Biol Chem 1980;255:2969–75.
[37] Ralph EC, Anderson, MA, Cleland, WW, Fitzpatrick, PF. Mechanistic studies of the flavoenzyme tryptophan 2-monooxygenase: Deuterium and ^{15}N kinetic isotope effects on alanine oxidation by an L-amino acid oxidase. Biochemistry 2006;45:15844–52.
[38] Paech C, Salach JI, Singer TP. Suicide inactivation of monoamine oxidase by trans-phenylcyclopropylamine. J Biol Chem 1980;255:2700–4.
[39] Schmidt DMZ, McCafferty DG. trans-2-Phenylcyclopropylamine is a mechanism-based inactivator of the histone demethylase LSD1. Biochemistry 2007;46:4408–16.
[40] Silverman RB, Hoffman SJ, Catus WB. A mechanism for mitochondrial monoamine oxidase catalyzed amine oxidation. J Am Chem Soc 1980;102:7126–8.
[41] Silverman RB. Radical ideas about monoamine oxidase. Acc Chem Res 1995;28:335–42.
[42] Rigby SE, Hynson RM, Ramsay RR, Munro AW, Scrutton NS. A stable tyrosyl radical in monoamine oxidase A. J Biol Chem 2005;280:4627–31.
[43] Dunn RV, Munro AW, Turner NJ, Rigby SEJ, Scrutton NS. Tyrosyl radical formation and propagation in flavin dependent monoamine oxidases. ChemBioChem 2010;11:1228–31.
[44] Miller JR, Edmondson DE, Grissom CB. Mechanistic probes of monoamine oxidase B catalysis: Rapid-scan stopped flow and magnetic field independence of the reductive half-reaction. J Am Chem Soc 1995;117:7830–1.
[45] Gaweska H, Henderson Pozzi M, Schmidt DMZ, McCafferty DG, Fitzpatrick PF. Use of pH and kinetic isotope effects to establish chemistry as rate-limiting in oxidation of a peptide substrate by LSD1. Biochemistry 2009;48:5440–5.
[46] Henderson Pozzi M, Gawandi V, Fitzpatrick PF. pH Dependence of a mammalian polyamine oxidase: insights into substrate specificity and the role of lysine 315. Biochemistry 2009;48:1508–16.
[47] Adachi MS, Juarez PR, Fitzpatrick PF. Mechanistic studies of human spermine oxidase: kinetic mechanism and pH effects. Biochemistry 2010;49:386–92.
[48] Zhong B, Silverman RB. Identification of the active site cysteine in bovine liver monoamine oxidase B. J Am Chem Soc 1997;119:6690–1.
[49] Miller JR, Edmondson DE. Structure-activity relationships in the oxidation of para-substituted benzylamine analogues by recombinant human liver monoamine oxidase A. Biochemistry 1999;38:13670–83.

[50] Kim J-M, Bogdan MA, Mariano PS. Mechanistic analysis of the 3-methyllumiflavin-promoted oxidative deamination of benzylamine. A potential model for monoamine oxidase catalysis. J Am Chem Soc 1993;115:10591–5.
[51] Nandigama RK, Edmondson DE. Structure-activity relations in the oxidation of phenethylamine analogues by recombinant human liver monoamine oxidase A. Biochemistry 2000;39:15258–65.
[52] Walker MC, Edmondson DE. Structure-activity relationships in the oxidation of benzylamine analogues by bovine liver mitochondrial monoamine oxidase B. Biochemistry 1994;33:7088–98.
[53] Henderson Pozzi M, Gawandi V, Fitzpatrick PF. Mechanistic studies of para-substituted N,N,-dibenzyl-1,4-diaminobutanes as substrates for a mammalian polyamine oxidase. Biochemistry 2009;48:12305–13.
[54] Kratzer R, Kavanagh KL, Wilson DK, Nidetzky B. Studies of the enzymic mechanism of *Candida tenuis* xylose reductase (AKR 2B5): X-ray structure and catalytic reaction profile for the H113A mutant. Biochemistry 2004;43:4944–54.
[55] Ralph EC, Hirschi JS, Anderson MA, Cleland WW, Singleton DA, Fitzpatrick PF. Insights into the mechanism of flavoprotein-catalyzed amine oxidation from nitrogen isotope effects on the reaction of N-methyltryptophan oxidase. Biochemistry 2007;46:7655–64.
[56] MacMillar S, Edmondson DE, Matsson O. Nitrogen kinetic isotope effects for the monoamine oxidase B-catalyzed oxidation of benzylamine and (1,1-$^{2}H_2$)benzylamine: nitrogen rehybridization and CH bond cleavage are not concerted. J Am Chem Soc 2011;133:12319–21.
[57] O'Leary MH. Multiple isotope effects on enzyme-catalyzed reactions. Ann Rev Biochem 1989;58:377–401.
[58] Gries FA, Decker K, Bruehmueller M. Decomposition of nicotine by bacterial enzymes. V. The oxidation of 1-6-hydroxynicotine to Î³-methylaminopropyl 6-hydroxy-3-pyridyl ketone. Hoppe-Seyler's Z Physiol Chem 1961;325:229–41.
[59] Kachalova GS, Bourenkov GP, Mengesdorf T, Schenk S, Maun HR, Burghammer M, Riekel C, Decker K, Bartunik HD. Crystal Structure Analysis of Free and Substrate-Bound 6-Hydroxy-l-Nicotine Oxidase from *Arthrobacter nicotinovorans.* J Mol Biol 2010;396:785–99.
[60] Kachalova G, Decker K, Holt A, Bartunik HD. Crystallographic snapshots of the complete reaction cycle of nicotine degradation by an amine oxidase of the monoamine oxidase (MAO) family. Proc Natl Acad Sci USA 2011;108:4800–5.
[61] Limburg J, Mure M, Klinman JP. Cloning and characterization of histamine dehydrogenase from *Nocardioides simplex*. Arch Biochem Biophys 2005;436:8–22.
[62] Steenkamp DJ, Beinert H. Mechanistic studies on the dehydrogenases of methylotrophic bacteria. 2. Kinetic studies on the intramolecular electron transfer in trimethylamine and dimethylamine dehydrogenase. Biochem J 1982;207:241–52.
[63] Trickey P, Basran J, Lian L-Y, Chen Z-w, Barton JD, Sutcliffe MJ, Scrutton NS, Mathews FS. Structural and biochemical characterization of recombinant wild type and a C30A mutant of trimethylamine dehydrogenase from *Methylophilus methylotrophus* (sp. W3A1). Biochemistry 2000;39:7678–88.
[64] Raine AR, Yang CC, Packman LC, White SA, Mathews FS, Scrutton NS. Protein recognition of ammonium cations using side-chain aromatics: a structural variation for secondary ammonium ligands. Protein Sci 1995;4: 2625–8.
[65] Reed T, Lushington GH, Xia Y, Hirakawa H, Travis DM, Mure M, Scott EE, Limburg J. Crystal Structure of Histamine Dehydrogenase from *Nocardioides simplex*. J Biol Chem 2010;285:25782–91.
[66] Basran J, Sutcliffe MJ, Scrutton NS. Optimizing the Michaelis complex of trimethylamine dehydrogenase: identification of interactions that perturb the ionization of substrate and facilitate catalysis with trimethylamine base. J Biol Chem 2001;276:42887–92.

[67] Bellamy HD, Lim LW, Mathews FS, Dunham WR. Studies of crystalline trimethylamine dehydrogenase in three oxidation states and in the presence of substrate and inhibitor. J Biol Chem 1989;264:11887–92.

[68] Basran J, Sutcliffe MJ, Hille R, Scrutton NS. Reductive half-reaction of the H172Q mutant of trimethylamine dehydrogenase: evidence against a carbanion mechanism and assignment of kinetically influential ionizations in the enzyme-substrate complex. Biochem J 1999;341: 307–14.

[69] Basran J, Jang M-H, Sutcliffe MJ, Hille R, Scrutton NS. The role of Tyr-169 of trimethylamine dehydrogenase in substrate oxidation and magnetic interaction between FMN cofactor and the 4Fe/4S center. J Biol Chem 1999;274:13155–61.

[70] Jang MH, Basran, J, Scrutton, N S, Hille, R. The reaction of trimethylamine dehydrogenase with trimethylamine. J Biol Chem 1999;274:13147–54.

[71] Fujieda N, Satoh A, Tsuse N, Kano K, Ikeda T. 6-S-Cysteinyl flavin mononucleotide-containing histamine dehydrogenase from *Nocardioides simplex*: molecular cloning, sequencing, overexpression, and characterization of redox centers of enzyme. Biochemistry 2004;43:10800–8.

[72] Basran J, Sutcliffe MJ, Scrutton NS. Deuterium isotope effects during carbon-hydrogen bond cleavage by trimethylamine dehydrogenase. Implications for mechanism and vibrationally assisted hydrogen tunneling in wild-type and mutant enzymes. J Biol Chem 2001;276:24581–7.

[73] Pettersen EF, Goddard TD, Huang CC, Couch GS, Greenblatt DM, Meng EC, Ferrin TE. UCSF Chimera—A visualization system for exploratory research and analysis. J Comput Chem 2004;25:1605–12.

7 Monoamine oxidases A and B: membrane-bound flavoenzymes of medical importance

Dale E. Edmondson, Claudia Binda and Andrea Mattevi

Abstract

Monoamine oxidases A and B (MAO A and B) are flavoenzymes containing covalent (8α-S-Cysteinyl) FAD cofactors that have important physiological functions in catalyzing the oxidation of amine neurotransmitters. Both enzymes are bound to the outer mitochondrial membrane via trans-membrane α-helices located at their respective C-termini. Their three-dimensional structures have been determined over the past decade and have led to a more definitive understanding of their functions at a molecular level. In spite of sequence conservation (70% identity) and of similar overall topology, the two enzymes have structurally different active site cavities to account for observed differences in substrate and inhibitor interactions. Highly specific reversible and irreversible inhibitors have been identified for MAO B and our understanding of this enzyme exceeds current detailed insights of MAO A. An extensive literature exists on both enzymes and this review summarizes the molecular properties of both purified and membrane-bound forms of the enzymes. Recent developments have shown the enzymes are topologically situated on distinct faces of the mitochondrial outer membrane and that they exhibit quite different behaviors when *p*-substituted benzylamine analogues are used as mechanistic probes. The impact of these differential structural and functional properties on their biological roles remain open questions in the field.

7.1 Introduction

Monoamine Oxidases A and B (MAO A, MAO B) serve an important physiological function in the degradation and control of levels of amine neurotransmitters (Dopamine, Serotonin, Norepinephrine) (▶Fig. 7.1A). In addition, these enzymes degrade ingested amines (*e.g.*, Tyramine), which could serve as false neurotransmitters. The prominence of monoamine oxidases (MAOs) in the flavoenzyme field and in pharmacology arose from initial findings in the 1960's that arylalkyl hydrazines, originally used to treat tuberculosis patients, exhibited a mood-elevating response by the irreversible inhibition of MAO. This finding sparked considerable investigations into various MAO inhibitors by both academic and pharmaceutical laboratories to develop drugs that could be used as anti-depressants and active research in this field continues [1].

Demonstration that MAO is a flavoenzyme came from publications in Singer's laboratory [2,3] that showed the FAD cofactor is covalently bound to the enzyme via an 8α-S-cysteinyl linkage to the isoalloxazine ring (▶Fig. 7.1B). MAO was the first example

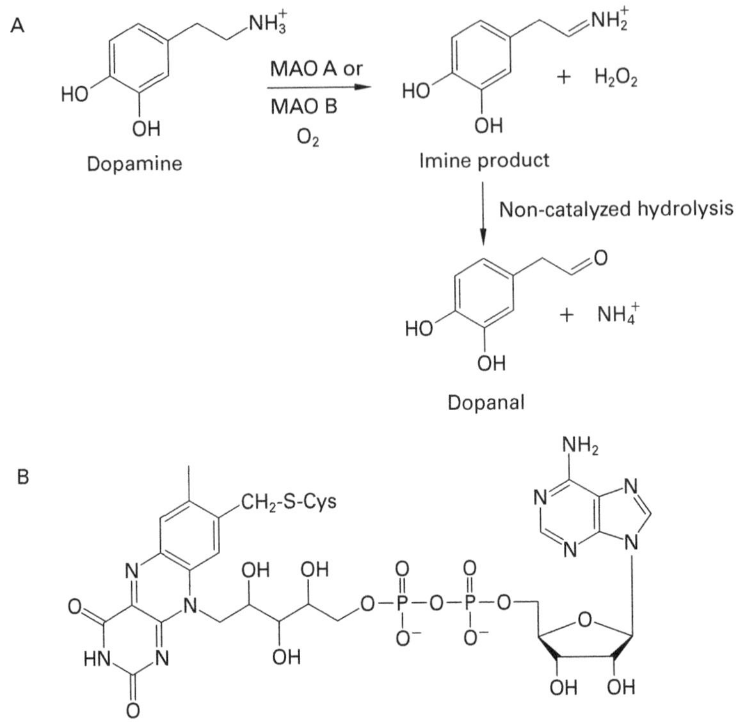

Fig. 7.1: (A) Oxidation of dopamine catalyzed by monoamine oxidases A and B and the reaction products. (B) Structure of 8α-S-cyteinyl FAD, which is the cofactor found in human and rat MAO A and MAO B and the cofactor for zebrafish MAO.

of such a covalent linkage although several other flavoenzymes have since been found to also contain cysteinyl flavins [4].

The presence of two types of MAO (MAO A and MAO B) was suggested from inhibition patterns with the acetylenic inhibitor deprenyl [5], which shows a higher affinity for the B form than for the A form. Unequivocal demonstration that the two forms are separate gene products came from delineations of their respective gene sequences [6] and location of those genes on the human X-chromosome [7]. MAO A and MAO B exhibit ~70% sequence identity and have been postulated to result from an evolutionary gene duplication event since teleosts contain a single gene for MAO [8] whose functional properties more closely resemble those of MAO A rather than MAO B [9]. MAO A is expressed early during development in mammals then followed later by MAO B expression [10]. Further work has identified both enzymes are bound to the mitochondrial outer membrane in all tissues although the tissue distribution of MAO A and MAO B differ among mammals investigated [11].

Since the MAO-catalyzed reaction results in reduction of O_2 to form H_2O_2 (▶Fig. 7.1A), they are classified as flavoprotein oxidases. Their mitochondrial localization and the finding that MAO B expression increases ~4-fold in neuronal tissues in an age-dependent manner [12] and MAO A expression increases ~9-fold in cardiac cells

[13] suggests they may play a role in age-related neuronal disease such as Parkinson's Disease and in age-related cardiomyopathies which are both thought to be due to the formation of reactive oxygen species. Therefore, both enzymes are the focus of continued investigations for their roles in these age-related diseases.

7.2 Structural studies of MAO A and MAO B

Since they are membrane-associated enzymes, the structural elucidations of MAO A and MAO B constituted a considerable challenge. A breakthrough in this endeavor came with the development of expression systems for the production and purification of both human MAO B and MAO A in reagent quantities. Expression of the individual recombinant proteins in *Pichia pastoris* was found to provide high levels of both human enzymes [14,15] as well as rat MAO A [16], MAO B [17], and zebrafish MAO [18]. Expression and purification of the rat and zebrafish MAOs allow for comparative functional studies with the human enzymes to facilitate the interpretation of results from animal models for drug development. Prior to this work, bovine liver MAO B [19] was used for MAO B studies and either human placenta MAO A [20] or *S. cerevisiae*-expressed MAO A [21] were used as sources for MAO A studies. Attempts by this laboratory to express high levels of human MAO B in *S. cerevisiae* were unsuccessful although low levels of recombinant enzyme could be expressed [22]. Commercial sources of membrane-bound human MAO A and MAO B are derived from expression of the human genes in the baculovirus/insect cell system.

Human MAO B was the first of the two enzymes to be crystallized and the structure determined [23]. The enzyme crystallizes as a dimer and exhibits a "*p*-hydroxybenzoate hydroxylase fold" (▶Fig. 7.2A). The covalent FAD is in an extended conformation and is situated at one end of a bipartite cavity that includes an entrance cavity and a substrate cavity separated by an Ile199 'gate' (▶Fig. 7.2B). Ile199 is conserved among most mammalian MAO B sequences with the exception of bovine MAO B where a Phe is present. This difference in gating residues is proposed to account for differences in inhibitor specificities observed on comparison of bovine and human MAO B [24,25]. The structure also shows the *re* face of the flavin ring to be the site for interaction(s) with the substrate and to be surrounded by an "aromatic cage" structure consisting of Tyr398 and Tyr435 (▶Fig. 7.2B). The flavin ring conformation is unusual in that it is bent, deviating ~30° from planarity whereas the isoalloxazine ring is usually in a planar conformation both in its free form and when bound in other flavoenzyme oxidases. Another unusual source of strain is the *cis* amide bond linking Tyr398 and Cys397 (that forms the covalent flavin linkage), which allows the side chain ring of Tyr398 to occupy its position as part of the "aromatic cage" in the catalytic site. A similar strain and aromatic cage is also present in MAO A.

Structural data on MAO B show the enzyme is bound to the outer mitochondrial membrane *via* a trans-membrane helix situated at the carboxyl terminus (▶Fig. 7.2A). The last 20 residues of this helix are too disordered to be visible in the structure. A similar C-terminal trans-membrane helix is observed in human and rat MAO A [26,27] (▶Fig. 7.3A,B). The Cα fold of human MAO A is similar to that of MAO B as is the covalent FAD conformation, and aromatic cage structure. A major difference between the two is that the catalytic site in human MAO A is a monopartite structure (▶Fig. 7.3C) rather than the dipartite structure in MAO B (▶Fig. 7.2B).

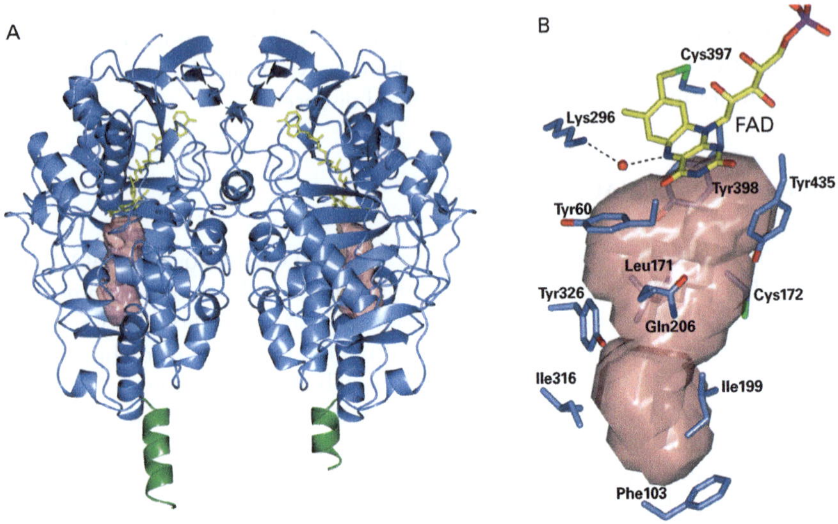

Fig. 7.2: (A) Structural representation of human Monoamine Oxidase B. The protein structure is represented in light blue, the protein chains representing the membrane binding domain in green and the dipartite active site cavity as pink transparent surface. The FAD moiety is depicted in yellow, with nitrogen, oxygen, sulfur and phosphor atoms in blue, red, green and magenta, respectively. The 8α-S-cysteinyl covalent linkage between FAD and Cys397 is shown. (B) Close-up representation of the dipartite active site cavity in human Monoamine Oxidase B. The labeled amino acid residues (color code as for FAD except carbon atoms in light blue) in the cavity include the "gating" residues Ile199 and Tyr326.

Of interest, human MAO A crystallizes as a monomer [27,28] and rat MAO A crystallizes as a dimer (as does MAO B) (►Fig. 7.3A,B). This result raised the interesting question as to whether there is any functional importance of this difference in oligomeric forms between human MAO A and MAO B. One suggestion [29] was that replacement of Glu151 in rat MAO A by a Lys in human MAO A was the source of this difference in oligomeric forms. Pulsed EPR DEER studies [30] were conducted with soluble and membrane-bound MAO A and MAO B [30] using TEMPO-spin labeled pargyline analogues to provide distance information. Distances determined for MAO A and for MAO B were found to be consistent with dimeric structures for both enzymes in the mitochondrial outer membrane. Purified MAO B is fully dimeric whereas purified human or rat MAO A are ~50% dimeric. These data show that human MAO A readily dissociates into monomers on liberation from the membrane as does rat MAO A. The human MAO A monomer appears to crystallize more readily whereas it is the dimeric form of the rat enzyme that favors crystallization. These data also demonstrate that the structures of the membrane-bound forms of MAO A and MAO B exhibit distances identical with those found in their respective crystalline forms.

One added benefit from studies of the TEMPO-substituted pargylines as active site MAO probes is that they are unable to cross the outer membrane in intact mitochondria [31]. Using these inhibitors as probes in conjunction with protease inactivation experiments, the membrane topologies of human and rat MAO A and B were investigated. Human MAO A

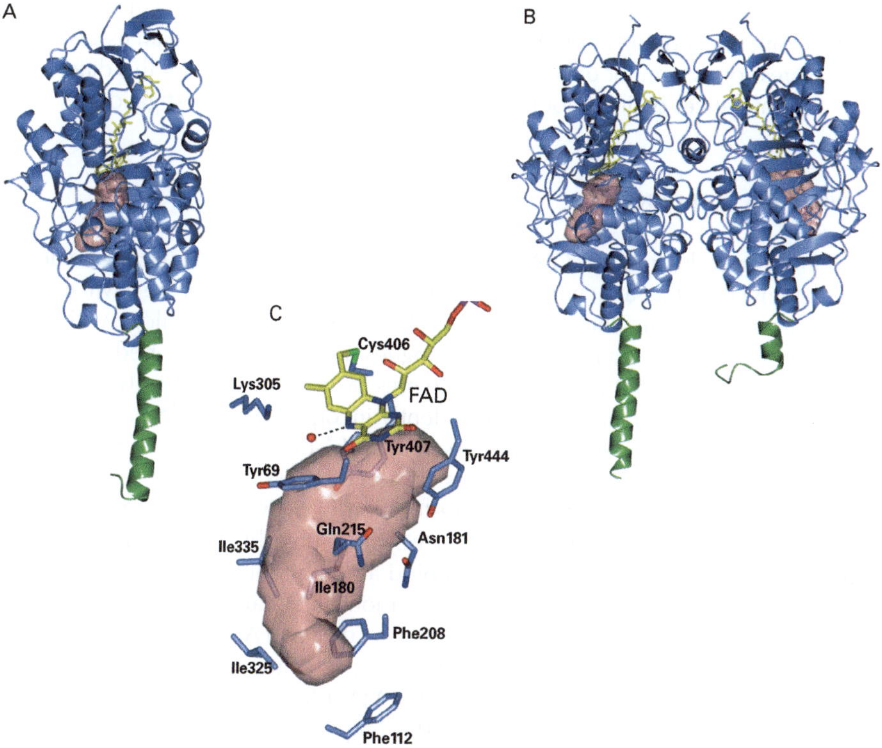

Fig. 7.3: Structural representations of human (A) and rat (B) Monoamine Oxidase A. The coloring scheme is as described in ►Fig. 7.2. (C) Close-up representation of active site cavity of human MAO A. Color code is as in ►Fig. 7.2B.

is found to be on the cytosolic face in intact human placental mitochondria but situated on the inter-membrane face of *Pichia pastoris* mitochondria [32]. Studies of rat liver MAO showed MAO B to be situated on the inter-membrane face and MAO A also to be on the cytosolic face (►Fig. 7.4). In the yeast recombinant system, the topology of the expressed

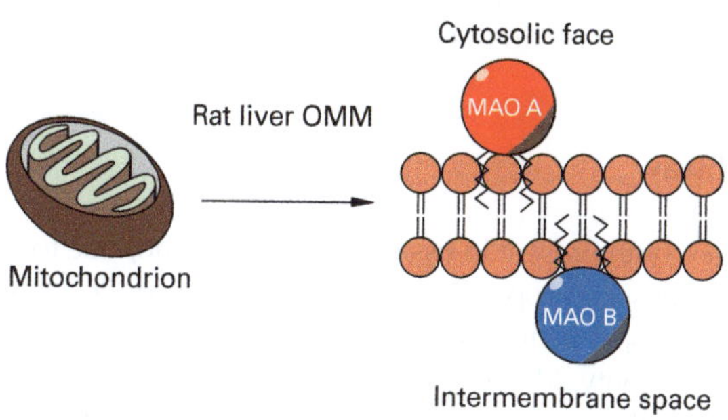

Fig. 7.4: Depiction of the topological orientations of MAO A and of MAO B in the outer mitochondrial membrane in rat liver.

rat enzymes were the same as observed for the expressed human enzymes. These studies demonstrate differing topologies of MAO A and MAO B in both mammalian tissue and in the yeast expression system for unknown reasons. The molecular basis for this differential topology and influences on inhibitor interactions is under ongoing investigation.

7.3 Flavin cofactor properties

As stated in the introduction, MAO A and MAO B both contain covalent FAD cofactors where a cysteinyl thioether attachment to the 8α-methylene position of the flavin occurs. The mechanism for this attachment appears to be autocatalytic by analogy with other flavoenzyme systems investigated. Flavin analogues are able to be inserted covalently into either MAO A or MAO B [33] and the enzymes found to be catalytically active. The only flavin analogue tested that was not covalently incorporated is 5-deaza FAD. MAO A with Cys406 mutated to an Ala and expressed in a *rib5*-deficient strain of *S. cerevisiae* [34] is no longer able to bind FAD analogues covalently but is capable of reconstitution of catalytic activity on their addition. No restoration of activity was observed on the addition of either FMN or riboflavin. The apoenzyme is unstable to solubilization from its membrane environment; however it does bind a number of FAD analogues with K_d values in the 50–100 nM range. These studies demonstrate that covalent binding is not essential for MAO activity but that the presence of a covalent FAD stabilizes the enzyme structure permitting solubilization in detergents and purification.

Unlike other flavoenzyme oxidases [35], neither MAO A nor MAO B reacts with sulfite at the N(5) position of their respective flavins. This property is the result of both enzymes possessing very hydrophobic active sites (see ▶Figs. 7.2B and 7.3C), which are expected to lower their affinities for binding the anionic sulfite near their flavins. Both MAO A and MAO B exhibit anionic flavin semiquinones on reductive titrations with sodium dithionite as shown for other flavoprotein oxidases [35]. Attempts to measure the redox potentials of the flavins in MAO A and MAO B using xanthine oxidase mediated reduction [36] resulted in potentials around –200 mV, which is close to that of free FAD and lower than the measured potential for 8α-S-cysteinylriboflavin (–169 mV) [37]. Using spectroelectrochemical titrations, Newton-Vinson and Edmondson measured the 1-e^- potentials for human MAO B to be +43 for the E_{ox}/E_{sq} couple and +37 mV for the E_{sq}/E_{hq} couple [38] which are closer to the potentials measured for other flavoenzyme oxidases [39]. This higher potential is also thermodynamically consistent with the observation that O_2 is reduced directly to H_2O_2 (E_m = +695 mV) rather than to $O_2^{-\bullet}$ (E_m = –137 mV) [40].

7.4 Catalytic reaction pathway

MAO A and MAO B usually are both present in tissue preparations and, due to their membrane association and similar physical properties, are not readily separated from one another. Since MAO A has a higher specificity than MAO B for serotonin, this substrate has been used to estimate MAO A activity. MAO B has a higher specificity for phenethylamine or benzylamine so these are the substrates of choice for estimating MAO B activity. Membrane particles do not lend themselves well to UV/vis absorption spectral studies due to light scattering and therefore, radiochemical procedures have been extensively used where the apolar aldehyde product is separated from the polar amine substrate by solvent extraction and product formation quantified by scintillation counting. Tipton and Singer have pointed out pitfalls in this assay procedure [40] and most current assay procedures utilize tyramine as substrate and the fluorometric detection of H_2O_2 with the horseradish peroxidase-coupled Amplex Red assay [41]. This assay is convenient and sensitive but does have limitations. It cannot be used with catechol-type substrates such as serotonin or dopamine since these compounds also are substrates for horseradish peroxidase. Thus, caution is needed when using this assay procedure to measure MAO inhibition since any inhibitor with catechol or aniline functionalities may also react with the peroxidase. Another caution to be observed when using the fluorescence assay is that oxidation of the resorufin product to a non-fluorescent resazurin by horseradish peroxidase or to other non-fluorescent species has also been observed when initial rate determinations of more than 3–4 min duration are used [43]. In principle, resorufin could function as a reversible inhibitor of either MAO A or MAO B since these enzymes tend to bind aromatic compounds such as Methylene Blue [43]. The fact that no reports of any resorufin inhibition of MAO have been noted in publications may reflect the low concentration of product formed in initial rate assays.

One major difference between MAO A and MAO B is the magnitude of their respective K_M values for O_2. MAO A exhibits a K_M for O_2 of ~10 μM whereas MAO B exhibits a value near air saturation (~240 μM). Therefore, under normal assay conditions, MAO A is saturated with O_2 whereas MAO B is only half-saturated. Temperature effects on MAO B activity can therefore exhibit large changes in rates since the $[O_2]$ in aqueous solutions is highly temperature dependent. Extensive kinetic work on bovine MAO B [44,45] has shown the enzyme follows a ternary complex (*i.e.*, enzyme-product-O_2) mechanistic pathway using benzylamine (or its *para*-substituted analogues) as substrate (▶Fig. 7.5). In contrast, phenethylamine oxidation functions via a classical "ping-pong"

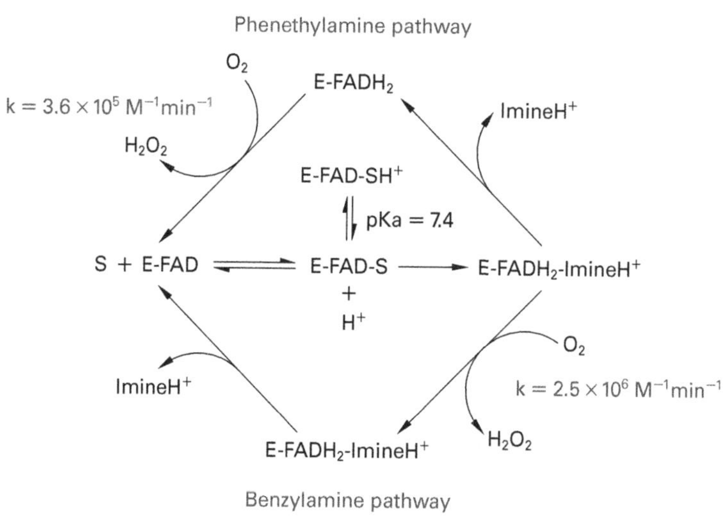

Fig. 7.5: Proposed mechanistic pathway for benzylamine oxidation (bottom loop) and for phenethylamine oxidation (top loop) for MAO B catalysis. The measured second-order rate constants for O_2 oxidation of the free, reduced enzyme and for the reduced enzyme-protonated imine complex are shown in red lettering.

or binary complex pathway where the imine product rapidly dissociates from the reduced enzyme which then reacts with O_2. With the benzylamine analogues, O_2 reacts with the imine-reduced enzyme complex to form H_2O_2 at a rate ~10-fold faster than that observed with the free, reduced form of the enzyme. Most of the available data with MAO A support the enzyme following a ternary complex mechanism as well since the presence of substrate enhances the rate of oxidation of the reduced form of the enzyme [46]. Little is known regarding the molecular basis for enhancement of the rate of O_2 oxidation by either protonated imine product (MAO B) or by substrate (MAO A). The most reasonable explanation is the electrostatic stabilization of the formation of the anionic superoxide intermediate (resulting from the single electron reduction of O_2) as found to be the role of a His residue in glucose oxidase [47].

Direct evidence for formation of the protonated imine as the initial product formed at a rate identical with that of flavin reduction and subsequent release into the medium is found with studies on the MAO B oxidation of *para*-dimethylamino-benzylamine [48]. This analogue exhibits spectral properties that allow detection of the protonated form of the imine as reaction product and permits measurement of its rate of hydrolysis, which occurs non-enzymatically in the solvent rather than in the catalytic site. Similar results are also found with MAO A showing that imine hydrolysis occurs after its release from the enzyme. Suggestions in the literature [49] that Cys and/or Glu residues in MAO A or MAO B may have a catalytic role in water activation for imine hydrolysis are not supported experimentally and therefore constitute pure speculation.

The current steady state and stopped flow kinetic data support the pathway for MAO catalysis as shown in ▶Fig. 7.5. Similar pathways have been shown for other flavoprotein oxidases as well. It is well known that D-amino acid oxidase follows either a binary or ternary complex mechanism depending on the nature of the amino acid substrate. Other more complex mechanisms such as that suggested by Ramsay *et al.* [50] require comment. In their published scheme, it is suggested that a possible cause for finding deviations from Michaelis-Menten behavior in MAO B (and apparently not in MAO A, in contrast to their prior work) is due to displacement of imine product from the reduced enzyme by substrate resulting in the differential expression of steady state parameters (V_{max} and K_M values) for both the oxidized and reduced enzyme species. These deviations from Michaelis-Menten behavior were observed using the Amplex-Red coupled assay with fluorescent resorufin product detection using 96-well plate reader assays which allow for a large number of assays to improve the errors commonly observed in the measurement of rates in single catalytic assays. What is not considered in these studies is the inclusion of error from formation of non-fluorescent resorufin products [42] or the possibility of any substrate inhibition. Extensive data supporting reoxidation of the reduced enzyme-imine complex by O_2 [44,45] in benzylamine oxidation assays would argue that very little if any free reduced MAO B is available in the steady state to bind to substrate. A better case could be made for such catalytic complexes involving the free reduced form of MAO B interacting with the substrate using phenethylamine since free, reduced enzyme occurs as an intermediate during catalysis [44]. Unpublished work in our laboratory on MAO B shows substrate (Benzylamine) inhibition at pH 9.0 where ~50% of the amine is in its deprotonated form. Thus, MAO B is susceptible to substrate inhibition, which becomes more observable at high pH values and provides a further complication in the interpretation of MAO B steady state kinetic data. Our working hypothesis is that MAO B is particularly susceptible to substrate inhibition since a second molecule of substrate could readily bind in the entrance cavity on deprotonation thus inhibiting release of the imine product from the substrate cavity. Such behavior is not observed with MAO A and probably reflects the monopartite nature of its active site cavity. No substrate inhibition is observed with the Ile199Ala-Tyr326Ala double mutant of MAO B [51] in which the gating residues separating the two cavities have been altered to result in an "open" single cavity structure, therefore providing additional evidence for the role of the dipartite cavity structure in substrate inhibition.

7.5 Mechanism of C-H bond cleavage and flavin reduction

Work from a number of laboratories has demonstrated that MAO A and MAO B oxidation of primary amines is stereospecific with cleavage of the *pro-R* C-H bond [52–54]. This specificity contrasts with the Cu-TOPA plasma amine oxidase in which abstraction of either the *R* or *S* hydrogen can occur [55]. These data support the view that the amine substrate is bound to MAO A and MAO B in a similar manner with respect to the flavin cofactor. Stereospecificity is also observed in the effectiveness of the acetylenic inhibitors *L*-Deprenyl or *R*-(+)-Rasagiline with MAO B [56].

The mechanism of amine oxidation and flavin reduction in MAO has been a controversial topic with three possibilities suggested in the literature. Silverman and co-workers have suggested the initial step to be a 1-electron oxidation of the amine by the flavin to form an aminium cation radical and a flavin radical (*i.e.*, a single electron transfer or SET mechanism) [57] (▶Fig. 7.6). Formation of this intermediate species should increase the acidity of the α-C-H bond sufficiently such that H^+ abstraction by a basic amino acid residue could occur with subsequent electron donation from the amine radical to the flavin radical forming the reduced flavin and imine product. The major experimental support for this mechanism is the observation of cyclopropyl ring opening of the mechanism-based inhibitor tranylcypromine concomitant with formation of a flavin C(4a) adduct. Radical mechanisms have also been suggested by Scrutton's laboratory [58] but modified to involve electron transfer to one of the tyrosyl residues involved in the aromatic cage in front of the flavin. Supporting evidence for a single electron transfer mechanism is weak and a number of experiments argue against it. The 1-electron redox potential of the MAO flavin is considerably lower (+40 mV) than that for the amine 1-electron couple (+1.0–1.5 V) and constitutes a strong thermodynamic argument against the SET mechanism. No effect of magnetic field on the reduction rate of bovine MAO B is observed [59] as might be expected for the reaction of a radical pair. Rapid scan stopped flow experiments with substrate analogues where C-H bond cleavage is rate limiting (large ^{D}k values) provide no spectral evidence for any flavin semiquinone intermediates. Thus, there is no strong experimental evidence to support a SET mechanism for either MAO A or MAO B catalysis.

With the demonstration that D-amino acid oxidase follows a hydride ion transfer mechanism for flavin reduction [60], Fitzpatrick and co-workers have argued that flavin-dependent amine oxidases (including MAO) function by hydride ion transfer steps since no spectral intermediates are observed in a number of flavoenzymes catalyzing amine oxidation [61]. Mechanistic experiments on MAO A show Hammett plots with positive ρ values both at pH 7.5 and at 9.0 [62,63] (and is also observed with zebrafish MAO

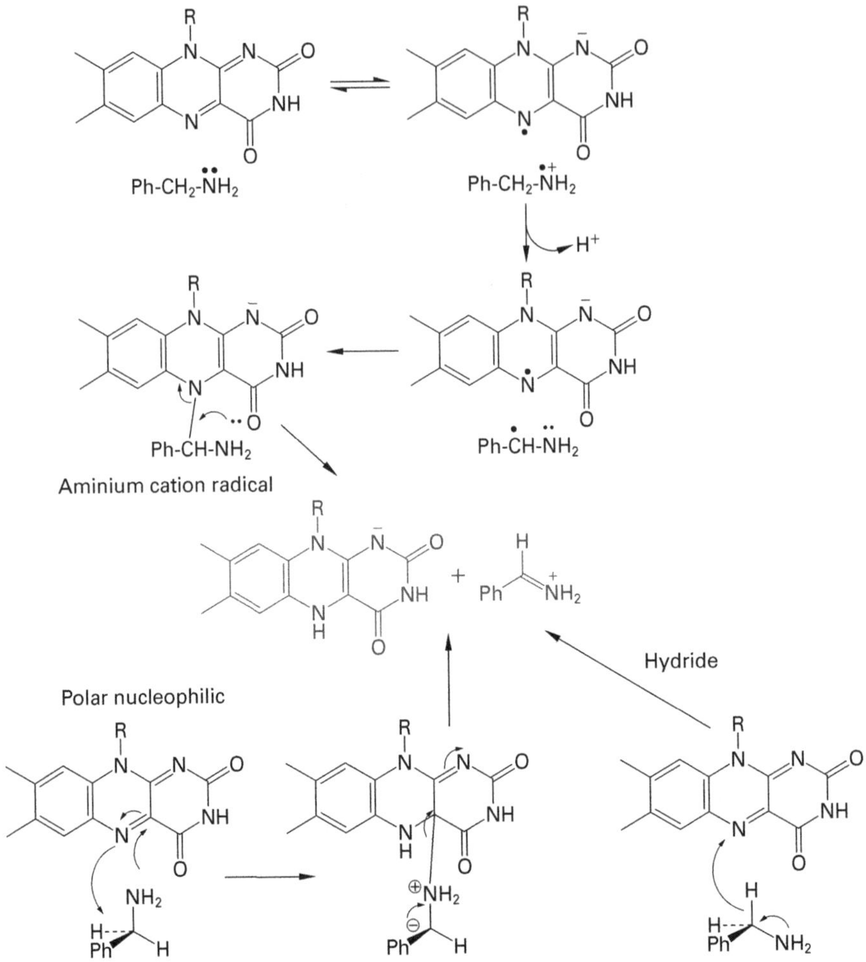

Fig. 7.6: Composite depiction of proposed mechanisms for C-H bond cleavage in the reductive half reaction in Monoamine Oxidase catalysis.

[9]) which constitute strong arguments against a hydride transfer since model system oxidation of benzylamine analogues by hydride acceptors [64] show negative ρ values. Bovine MAO B oxidation of benzylamine analogues show no contributions of electronic effects from *para*-substituents [45] and has been interpreted [62] as being due to the conformation of the bound substrate in the catalytic site preventing orbital overlap for transmission of electronic effects. Recent ^{15}N kinetic isotope effect measurements of human MAO B oxidation of ^{1}H and ^{2}H benzylamine analogues [65] show C-H bond cleavage is not concerted with the change in bond order to N in the transition state, which argues that a Hydride ion mechanism for C-H bond cleavage, if it does occur, is not a concerted step in MAO B catalysis.

The available mechanistic data support a polar nucleophilic mechanism for MAO A catalysis [62]. This mechanism is an adaptation of Hamilton's original suggestion but differs in that following nucleophilic attack of the amine lone pair at the C(4a) position of the flavin ring, the abstracting base is the basic N(5) nitrogen (▶Fig. 7.6) rather than a basic amino acid residue since structural data do not support the participation of an active site basic residue in either MAO A or MAO B. Macheroux and coworkers have estimated a pK_a for the N(5)-H of model flavin hydroquinones to be ~25 from NMR H-exchange kinetic experiments [66], and Bordwell has calculated pK_a values for benzylic C-H bonds that are in the same range [67]. Such a mechanism would be consistent with the Hammett data for MAO A and for zebrafish MAO. To date, no spectral evidence for the intermediate formation of a C(4a) adduct with the flavin has been demonstrated with either MAO A or MAO B. It remains for future mechanistic work to further test whether this mechanism is operative in MAO B since it is generally assumed that MAO A and MAO B operate by identical mechanisms. Whether this assumption is correct can only be defined by detailed comparative mechanistic experiments.

7.6 Reaction with O_2 to form H_2O_2

As required for definition as flavoprotein oxidases, both MAO A and MAO B react with O_2 to form H_2O_2 with no evidence for any release of $O_2^{-\bullet}$. Two mechanistic pathways for this reaction have been suggested as shown in ▶Fig. 7.7. MAO A exhibits a high affinity for O_2 with a K_M value of ~10 µM whereas MAO B exhibits a K_M for O_2 of ~240 µM. Despite these differences in K_M values, the reduced forms of both enzymes react with O_2 in a second order reaction with rate constants of ~5×10^5 M^{-1}-min^{-1} [45,46]. Acceleration of the rate of oxidation (~10-fold) of the reduced enzyme is observed when the imine product is bound in bovine MAO B [45] and in the presence of substrate in MAO A [46]. By analogy with the role of a His in glucose oxidase [47] that stabilizes the initial formation of a radical pair of superoxide anion and flavin semiquinone, the presence of imine or amine in the active sites of MAO B or MAO A could perform a similar stabilization resulting in enhanced rates. Both MAO A and MAO B contain Lys residues H-bonded through a water molecule to the N(5) of the flavin. This structure is also seen in a number of flavoprotein oxidases and may have an important role in the reaction. Jorns and colleagues found a dramatic decrease in O_2 reactivity of monomeric sarcosine oxidase [68] on mutagenesis of this Lys to a Met. Similar experiments with other flavoprotein oxidases have resulted in mixed results [69] and, therefore, no general role of this Lys residue in flavoprotein oxidases can be formulated regarding its role in the reaction with O_2. This situation may reflect differing mechanisms for the reaction of O_2 with the reduced flavin in different flavoproteins with some involving formation of a C4a peroxy adduct intermediate while others may proceed via a proton coupled electron transfer reaction (▶Fig. 7.7) in the formation of hydrogen peroxide from the superoxide-flavin radial pair intermediate that is generally accepted in the field.

7.7 Biological and pharmacological significance of MAO A and MAO B

Although MAO A and MAO B have been extensively investigated, new information continues to come forth regarding their biological and pharmacological importance in humans and human disease states. It is beyond the scope of this review to discuss this

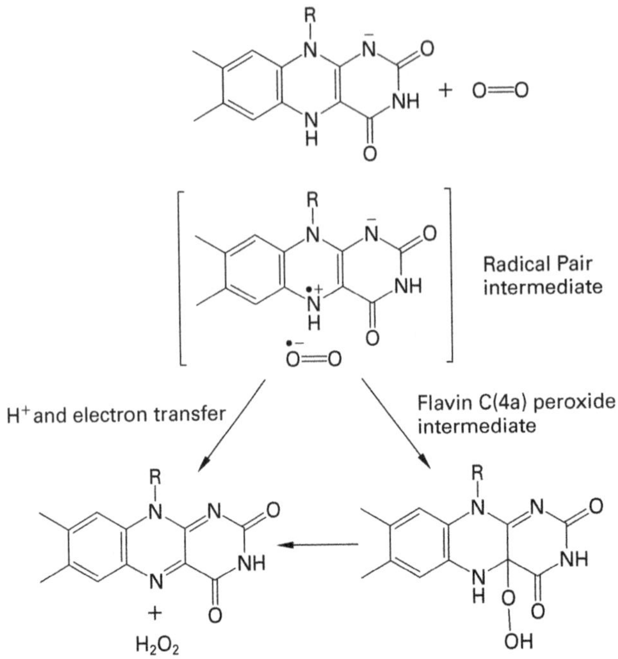

Fig. 7.7: Proposed general mechanisms for oxygen reaction of reduced flavin coenzymes in flavoprotein oxidase catalysis leading to the production of hydrogen peroxide.

aspect of MAO in any detail. The most salient aspects to consider by the interested reader is that the tissue specific expression of MAO is altered along development, with MAO B levels increasing 4-fold in neuronal tissue while cardiac levels of MAO A increase 9-fold on aging. Since MAO A and B play important roles in the catabolism of neurotransmitters and produce hydrogen peroxide as a source of reactive oxygen species, their roles in age-dependent degenerative diseases remain an integral component in continued research on the etiology of diseases including Parkinson's Disease and heart disease. This is in addition to their better known involvement in disease states such as depression.

7.8 Acknowledgements

This work has been supported by NIH GM29433, Cariplo Foundation (2008-3148) and PRIN09 (MIUR).

7.9 References

[1] Youdim MBH, Edmondson DE, Tipton KF. The therapeutic potential of monoamine oxidase inhibitors. Nat Revs Neurosci 2006;7:295–309.

[2] Kearney EB, Salach JI, Walker WH, Seng RL, Kenney W, Zeszotek E, Singer TP. The covalently-bound flavin of hepatic monoamine oxidase. 1. Isolation and sequence of a flavin peptide and evidence for binding at the 8 α-position. Eur J Biochem 1971;24:321–7.

[3] Nagy J, Salach JI. Identity of the active site peptide fragments from human "A"-form and the bovine "B" form of monoamine oxidase. Arch Biochem Biophys 1981;208:388–94.
[4] Heuts DPH, Scrutton NS, McIntire WS, Fraaje MW. What's in a covalent bond? On the role and formation of covalently bound flavin cofactors. FEBS J 2009;276:3405–27.
[5] Bach AW, Lan NC, Johnson DL, Abell CW, Bembenek ME, Kwan SW, Seeburg PH, Shih JC. cDNA Cloning of human liver monoamine oxidase A and B: Molecular differences in enzymatic properties. Proc Natl Acad Sci USA 1988;85:4934–38.
[6] Grimsby J, Chen K, Wang LJ, Lan NC, Shih JC. Human MAO A and B genes exhibit identical exon intron organization. Proc Natl Acad Sci USA 1991;88:3647–3641.
[7] Setini A, Pierucci F, Senatori O, Nicotra A. Molecular characterization of monoamine oxidase in zebrafish (*Danio rerio*). Comp Biochem Phys B 2005;140:153–61.
[8] Aldeco M, Arslan BK, Edmondson DE. Catalytic and inhibitor binding properties of zebrafish monoamine oxidase (zMAO): Comparisons with human MAO A and MAO B. Comp Biochem Physiol Part B. 2011;159:78–83.
[9] Strolin-Benedetti M, Dostert P, Tipton KF. Developmental aspects of the monoamine degrading enzyme monoamine oxidase. Dev Pharmac Ther 1992;18:191–200.
[10] Sura J, Kettler R, Da Prada M, Richard JG. Quantitative enzyme radioautography with 3H-Ro 41-1049 and 3H-Ro 19-6327 in vitro: Localization and abundance of MAO A and MAO B in rat CNS, peripheral organs, and human brain. J Neurosci 1992;12:1977–99.
[11] Fowler CJ, Wiberg A, Oreland L, Marcusson J, Winblad B. The effect of age on the activity and molecular properties of human brain monoamine oxidase. J Neural Transm. 1980;49:1–20.
[12] Maurel A, Hernandez C, Kunduzova O, Bompart G, Cambon C, Parini A, Frances B. Age-dependent increase in hydrogen peroxide production by cardiac monoamine oxidase A in rats. Am J Physiol-Heart Circul Physiol 2003;284:H1460–67.
[13] Newton-Vinson P, Hubalek F, Edmondson DE. High-level expression of human liver monoamine oxidase B in *Pichia pastoris*. Prot Exp Purif 2000;20:334–45.
[14] Li M, Hubalek F, Newton-Vinson P, Edmondson DE. High level expression of human liver monoamine oxidase A in *Pichia pastoris*. Comparison with the enzyme expressed in *Saccharomyes cerevisae*. Prot Exp Purif 2002;24:152–62.
[15] Upadhyay A, Edmondson DE. Characterization of detergent purified recombinant rat liver monoamine oxidase B expressed in *Pichia pastoris*" Prot Expr Purif 2008;59:349–56.
[16] Wang J, Edmondson DE. High-level expression and purification of rat monoamine oxidase A (MAO A) in *Pichia pastoris*: Comparison with human MAO A. Prot Expr Purif 2010; 70:211–17.
[17] Arslan BK, Edmondson DE. Expression of zebrafish (*Danio rerio*) monoamine oxidase (MAO) in *Pichia pastoris*: Purification and comparison with human MAO A and MAO B. Prot Expr Purif 2010;70:290–7.
[18] Binda C, Newton-Vinson P, Hubalek F, Edmondson DE, Mattevi A. Structure of human monoamine oxidase B, a drug target for the treatment of neurological disorders. Nat Struct Biol 2002;9:22–6.
[19] Hubalek F, Binda C, Khalil A, Li M, Mattevi A, Castagnoli N, Edmondson DE. Demonstration of isoleucine 199 as a structural determinant for the selective inhibition of human monoamine oxidase B by specific reversible inhibitors. J Biol Chem 2005;280:15761–6.
[20] Kruger MP, Mazouz F, Ramsay RR, Milcent R, Singer TP. Dramatic differences in the susceptibility of monoamine oxidase B to a group of powerful inhibitors. Biochem Biophys Res Commun 1995;206:556–62.
[21] Ma J, Yoshimura M, Yamashita E, Nakagawa A, Ito A, Tsukihara T. Structure of rat monoamine oxidase A and its specific recognitions for substrates and inhibitors. J Mol Biol 2004;338: 103–14.
[22] Binda C, Li M, Hubalek F, Restelli N, Edmondson DE, Mattevi A. Insights into the mode of inhibition of human mitochondrial monoamine oxidase B from high-resolution crystal structures. Proc Natl Acad Sci USA 2003;100:9750–5.

[23] Son SY, Ma J, Kondou Y, Yoshimura M, Yamashita E, Tsukihara T. Structure of human monoamine oxidase A at 2.2-angstrom resolution: the control of opening the entry for substrates/inhibitors. Proc Natl Acad Sci USA 2008;105:5739–44.

[24] Andres AM, Soldevila M, Navarro A, Kidd KK, Oliva B, Bertranpetit J. Positive selection in MAO A gene is human exclusive: determination of the putative amino acid change selected in the human lineage. Human Genetics 2004;115:377–86.

[25] Upadhyay AK, Borbat PP, Wang J, Freed JH, Edmondson DE. Determination of the oligomeric states of human and rat monoamine oxidases in the outer mitochondrial membrane and octyl beta-D-glucopyranoside micelles using pulsed dipolar electron spin resonance spectroscopy. Biochemistry 2008;47:1554–66.

[26] Upadhyay AK, Edmondson DE. Development of spin-labeled pargyline analogues as specific inhibitors of human monoamine oxidase A and B. Biochemistry 2009;48:3928–35.

[27] Salach JI. Monoamine oxidase from beef liver mitochondria: Simplified isolation procedure, properties, and determination of its cysteinyl flavin Arch Biochem Biophys 1979;192:128–37.

[28] Weyler W, Salach JI. Purification and properties of mitochondrial monoamine oxidase A from human placenta. J Biol Chem 1985;260:13199–207.

[29] Weyler W, Titlow CC, Salach JI. Catalytically active monoamine oxidase Type A from human liver expressed in *Saccharomyces cerevisiae* contain covalent FAD. Biochim Biophys Res Commun 1990;173:1205–11.

[30] Urban P, Andersen JK, Hsu H-P, Pompon D. Comparative membrane locations and activities of human monoamine oxidases expressed in yeast. FEBS Lett 1991;286:142–6.

[31] Wang J, Edmondson DE. Topological probes of monoamine oxidases A and B in rat liver mitochondria: Inhibition by TEMPO-substituted pargyline analogues and inactivation by proteolysis. Biochemistry 2011;50:2499–505.

[32] Miller JR, Edmondson DE. Influence of flavin analogue structure on the catalytic activities and flavinylation reactions of recombinant human liver monoamine oxidases A and B. J Biol Chem 1999;274:23515–35.

[33] Nandigama RK, Edmondson DE. Influence of FAD structure on its binding and activity with the C406A mutant of recombinant human liver monoamine oxidase A. J Biol Chem 2000;275:20527–32.

[34] Massey V, Müller F, Feldberg R, Schuman M, Sullivan PA, Howell LG, Mayhew SG, Matthews RG, Foust GP. The reactivity of flavoproteins with sulfite. Possible relevance to the problem of oxygen reactivity. J Biol Chem 1969;244:3999–4006.

[35] Ramsay RR, Sablin SO, Singer TP. redox properties of the flavin cofactor of monoamine oxidases A and B and their relationship to the kinetic mechanism. Brain Res 1995;106:33–9.

[36] Edmondson DE, Singer TP. Oxidation-reduction properties of the 8α-substituted flavins. J Biol Chem 1973;284:8144–49.

[37] Newton-Vinson P, Edmondson DE. High-level expression, structural, kinetic , and redox characterization of recombinant human liver monoamine oxidase B. In: Ghisla S, Kroneck P, Macheroux P, Sund H. eds. Flavins and Flavoproteins. Berlin: Agency for Scientific Publications, 1999:431–8.

[38] Stankovich M. Redox properties of flavins and flavoproteins. In: Müller F. ed. Chemistry and Biochemistry of Flavoenzymes Vol. 1. Boca Raton, FL, USA:, CRC Press, 1991:401–25.

[39] Petlicki J, van de Ven TGM. The equilibrium between the oxidation of hydrogen peroxide by oxygen and the dismutation of peroxyl or superoxide radicals in aqueous solutions in contact with oxygen. J Chem Soc Faraday Trans 1998;94:2763–67.

[40] Tipton KF, Singer TP. The radiochemical assay for monoamine oxidase activity. Problems and pitfalls. Biochem Pharmacol 1993;46:1311–16.

[41] Zhou M, Panchuk-Voloshina N. A one-step fluorometric method for the continuous measurement of monoamine oxidase activity. Anal Biochem 1997;253:169–74.

[42] Towne V, Will M, Oswald B, Zhao Q. Complexities in horse-radish peroxidase-catalyzed oxidation of dihydroxyphenoxine derivatives: appropriate ranges for pH values and hydrogen peroxide concentrations in quantitative analysis. Anal Biochem 2004;334:290–96.
[43] Ramsay RR, Dunford C, Gillman PK. Methylene blue and serotonin toxicity: inhibition of monoamine oxidase A (MAO A) confirms a theoretical prediction. Br J Pharm 2007;152: 946–51.
[44] Husain M, Edmondson DE, Singer TP. Kinetic studies on the catalytic mechanism of liver monoamine oxidase, Biochemistry 1982;21:595–600.
[45] Walker MC, Edmondson DE. Structure-activity relationships in the oxidation of benzylamine analogues by bovine liver mitochondrial monoamine oxidase B. Biochemistry 1994;33:7088–98.
[46] Ramsay RR. Kinetic mechanism of monoamine oxidase-A. Biochemistry 1991;30:4624–29.
[47] Roth JP, Klinman JP. Catalysis of electron transfer during activation of O_2 by the flavoprotein glucose oxidase Proc Natl Acad Sci USA 2003;100:62–67.
[48] Edmondson DE, Bhattacharyya AK, Walker MC. Spectral and kinetic studies of imine product formation in the oxidation of *p*-(N,N-dimethylamino)benzylamine analogues by monoamine oxidase B. Biochemistry 1993;32:5196–202.
[49] Kachalova G, Decker K, Holt A, Bartunik HD. Crystallographic snapshots of the complete reaction cycle of nicotine degradation by an amine oxidase of the monoamine oxidase (MAO) family. Proc Natl Acad Sci USA 2011;108:4800–05.
[50] Ramsay RR, Olivieri A, Holt A. An improved approach to steady state analysis of monoamine oxidases. J Neural Transm 2011;118:1003–19.
[51] Milczek E, Binda C, Rovida S, Mattevi A, Edmondson DE. The "gating" residues Ile199 and Tyr326 in human monoamine oxidase B function in substrate and inhibitor recognition. FEBS J 2011;278:4860–69.
[52] Yu PH, Bailey BA, Durden DA, Boulton AA. Stereospecific deuterium substitution at the alpha-carbon position of dopamine and its effect on oxidative deamination catalyzed by MAO-A and MAO-B from different tissues. Biochem Pharmacol 1986;35:1027–36.
[53] Belleau B, Moran J. Deuterium isotope effects in relation to chemical mechanism of monoamine oxidase. Ann N Y Acad Sci 1963;107:822–39.
[54] Li, Min, Ph.D. Dissertation, Emory University, 2005.
[55] Farnum MF, Klinman JP. Stereochemical probes of bovine plasma amine oxidase: Evidence for mirror image processing and a *syn* abstraction of hydrogens from C-1 and C-2 of dopamine. Biochemistry 1986;25:6028–36.
[56] Youdim MBH, Weinstock M. Molecular basis of neuroprotective activities of rasagiline and the anti-Alzheimer drug TV 3326 [(N-propargyl-(3R)-aminoindane-5-yl-ethylmethycarbamate]. Cell Molec Neurosci 2002;21:555–73.
[57] Lu X, Rodriguez M, Ji H, Silverman RB, Vintem PB, Ramsay RR. Irreversible inactivation of mitochondrial monoamine oxidases. In: Chapman S, Perham R, Scrutton N, eds. Flavins and Flavoproteins. Berlin: Agency for Scientific Publications, 2002:817–30.
[58] Dunn R, Munro AW, Turner NJ, Rigby SEJ, Scrutton N. Tyrosyl radical formation and propogation in flavin dependent monoamine oxidases. ChemBiochem 2010;11:1228–31.
[59] Miller JR, Edmondson DE, Grissom CB. Mechanistic probes of monoamine oxidase B catalysis: Rapid-scan stopped-flow and magnetic field independence of the reductive half reaction. J Am Chem Soc 1995;117:7830–1.
[60] Harris CM, Pollegioni L, Ghisla S. pH and kinetic isotope effects in D-amino acid oxidase catalysis - Evidence for a concerted mechanism in substrate dehydrogenation via hydride transfer, Eur J Biochem 2001;268:5504–20.
[61] Fitzpatrick PF. Oxidation of amines by flavoproteins. Arch Biochem Biophys 2010;493: 13–25.

[62] Miller JR, Edmondson DE. Structure-activity relationships in the oxidation of para-substituted benzylamine analogues by recombinant human liver monoamine oxidase A. Biochemistry 1999;38:13670–83.

[63] Wang J, Edmondson DE. 2H-Kinetic isotope effects and pH dependence of catalysis as mechanistic probes of rat monoamine oxidase A: Comparisons with the human enzyme. Biochemistry 2011;50:7710–17.

[64] Dubey R, Kothari S, Banerji KK. Kinetics and mechanism of the oxidation of substituted benzylamines by hexamethylenetetramine-bromine. J Phys Org Chem 2002;15:103–107.

[65] MacMillar S, Edmondson DE, Matsson O. Nitrogen kinetic isotope effects for the monoamine oxidase catalyzed oxidation of benzylamine and (1,1-2H2)-benzylamine: nitogen rehybridization and C-H bond cleavage are not concerted. J Am Chem Soc 2011;133:12319–21.

[66] Macheroux P, Ghisla S, Sanner C, Ruterjans H, Müller F. Reduced flavin: NMR investigation of N5-H exchange mechanism, estimation of ionisation constants and assessment of properties as biological catalyst, BMC Biochem 2005;6:26.

[67] Bordwell FG, Cheng JP, Satish AV, Twyman CL. Acidities and homolytic bond-dissociation energies (Bdes) of benzyl-type C-H bonds in sterically congested substrates, J Org Chem 1992;57:6542–6.

[68] Zhao GH, Bruckner RC, Jorns MS. Identification of the oxygen activation site in monomeric sarcosine oxidase: Role of Lys265 in catalysis, Biochemistry 2008;47:9124–35.

[69] McDonald CA, Fagan RL, Collard F, Monnier VM, Palfey BA. Oxygen reactivity in flavoenzymes: Context matters. J Am Chem Soc 2011;133:16809–11.

8 Choline oxidase and related systems

Giovanni Gadda

Abstract

Choline oxidase catalyzes the four-electron oxidation of choline to glycine betaine, with betaine aldehyde as intermediate. Molecular oxygen acts as electron acceptor for the oxidation of both the alcohol substrate and the aldehyde intermediate. Choline and its oxidation product are central metabolites in biological processes, with importance for health and disease and in biotechnological and medical applications. The enzyme from *Arthrobacter globiformis* has been characterized in its biochemical, structural, kinetic and mechanistic properties, disclosing structural-functional insights on the mechanism of alcohol oxidation and oxygen activation for reaction with the reduced flavin. Site-directed mutagenesis and X-ray crystallography have been used to complement mechanistic and biochemical investigations, thereby allowing for an understanding at the molecular level of the roles carried out in substrate binding, alcohol oxidation or reaction with oxygen by a number of active site residues, including Ser101, Glu312, His351, Val464, His466, and Asn 510. Flavin covalent attachment to the protein has been demonstrated to be important for the hydride transfer reaction resulting in alcohol oxidation. Knowledge of the biochemistry of choline oxidase is applicable to choline dehydrogenase, for which stable preparations of enzyme in purified and active form have been difficult to obtain due to its association with membranes.

8.1 Introduction

Choline oxidase (E.C. 1.1.3.17, choline-oxygen 1-oxidoreductase) catalyzes the oxidation of choline to glycine betaine, with molecular oxygen (O_2) as electron acceptor [1,2]. The reaction occurs through two subsequent flavin-mediated oxidations, in which the alcohol substrate is oxidized to betaine aldehyde and the (hydrated) aldehyde intermediate to the carbon acid (▶Fig. 8.1) [1,2]. Most of the biochemical knowledge on the enzyme derives from mechanistic and structural studies on recombinant choline oxidase from *Arthrobacter globiformis* [3,4]. The enzymes from *A. nicotianae* and *Alcaligenes* sp. have also been purified and characterized [5–9]. A membrane-associated enzyme that reacts poorly with O_2, choline dehydrogenase, has been partially characterized in rat [10–13], *Pseudomonas* sp. [14], *Halomonas elongata* [1], and described in oysters [15] and bacteria, including *Escherichia coli* [16,17] and *Sinorhizobium meliloti* [18]. However, difficulty in obtaining choline dehydrogenase in purified, stable and active form has hampered biochemical studies [14,19]. As of May 2012, close to

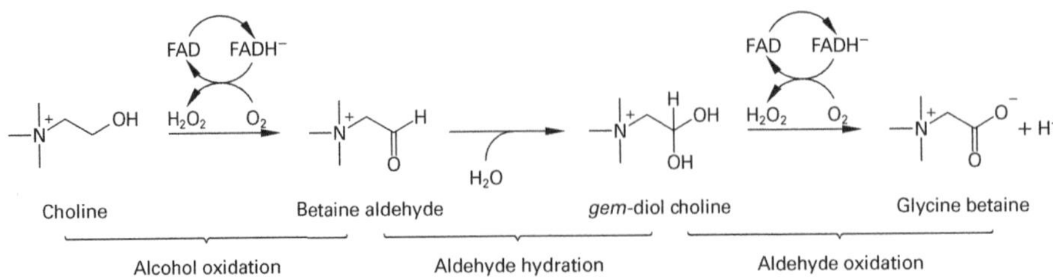

Fig. 8.1: The two-step oxidation of choline as catalyzed by choline oxidase.

6000 ORFs from a myriad of organisms have been annotated as "choline dehydrogenase" or "choline oxidase", including a variety of human pathogenic bacteria, underscoring the relevance of the enzyme in biological processes.

8.1.1 Glucose-methanol-choline enzyme oxidoreductase superfamily

Choline oxidase has been grouped in the Glucose-Methanol-Choline oxidoreductase (GMC) enzyme superfamily, initially on the basis of amino acid sequence alignment [20]. This superfamily includes glucose oxidase, cholesterol oxidase, methanol oxidase, pyranose 2-oxidase, aryl-alcohol oxidase, cellobiose dehydrogenase, and a number of other flavin-dependent enzymes. These enzymes utilize FAD as a cofactor for catalysis and oxidize a plethora of alcohols, which are structurally unrelated. GMC enzymes share limited sequence similarity in their substrate binding domains, in accordance with the variety of different substrates they oxidize. In contrast, their overall crystal structures appear alike and reveal highly conserved catalytic sites (▶Fig. 8.2). In a number of GMC oxidoreductase members, including choline oxidase, pyranose 2-oxidase and aryl-alcohol oxidase, mechanistic studies with kinetic isotope effects have established that alcohol oxidation occurs *via* the transfer of a hydride ion from the alcohol substrate to the enzyme-bound flavin [21–23]. Based on the similar active site architectures and the available mechanistic data on three enzyme members it is plausible to expect that members of the GMC enzyme superfamily utilize similar catalytic mechanisms for alcohol oxidation.

8.1.2 Choline, glycine betaine and choline-oxidizing enzymes in biotechnology and medicine

Choline and its oxidation product glycine betaine are central molecules in biological processes, with importance for health and disease [24,25]. In humans, choline is an essential nutrient [26], which is required for the biosynthesis, structural integrity and signaling functions of cellular membranes and that affects lipid transport from liver through VLDL assembly and secretion [27]. The only *de novo* source of choline other than diet is the methylation of phosphatidylethanolamine catalyzed by phosphatidylethanolamine-*N*-methyltransferase in the liver and its subsequent enzyme-catalyzed hydrolysis [28]. Choline metabolism is linked to the metabolism of methionine and folate and to the biosynthesis of acetylcholine and phosphatidylcholine [27,29]. Betaine homocysteine methyltransferase utilizes betaine, either from the diet or derived from choline oxidation, as a methyl group donor for the conversion of homocysteine to methionine [30].

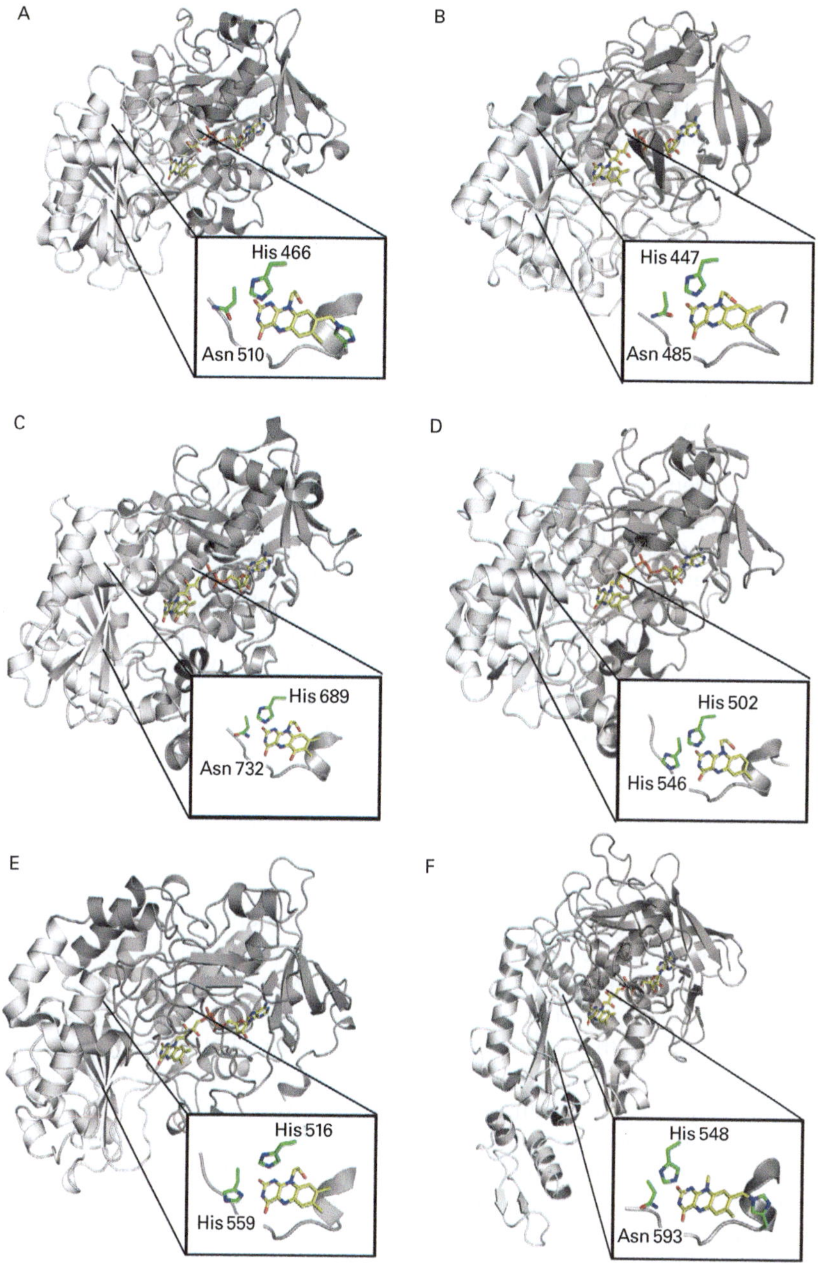

Fig. 8.2: Overall structures and conserved active site residues of GMC oxidoreductase enzyme superfamily members. (A) *Arthrobacter globiformis* choline oxidase (Ser101Ala) PDB 3NNE, (B) *Brevibacterium sterolicum* cholesterol oxidase PDB 1COY, (C) *Phanerochaete chrysosporium* cellobiose dehydrogenase PDB 1KDG, (D) *Plerotus eryngii* aryl-alcohol oxidase PDB 3FIM, (E) *Aspergillus niger* glucose oxidase PDB 1CF3, and (F) *Trametes ochracea* pyranose 2-oxidase PDB 2IGK (F).

Hypomethylation of DNA has been demonstrated in rats fed a diet devoid of choline, which results in decreased concentrations of methionine and *S*-adenosylmethionine [31,32]. This has implications for gene mutation and expression, cell proliferation and genome stability [33–41]. Choline deficiency has also been shown to alter histone methylation and increase lymphocyte apoptosis and DNA damage [42,43]. In Alzheimer's disease choline deficiency has been linked to elevated plasma homocysteine [44]. Glycine betaine is the primary source of methyl groups in the liver and kidney for the biosynthesis of methionine from homocysteine by betaine homocysteine S-methyltransferase [45,46]; it is also a ubiquitous osmoprotectant in mammals, plants and microorganisms, which is accumulated in most tissues or cells to assist cell volume regulation [17,25,47,48]. Glycine betaine has been demonstrated to be an important osmoprotectant in human pathogenic bacteria, such as *Pseudomonas aeruginosa, Escherichia coli, Staphylococcus aureus* and *Listeria monocytogenes* [17,49–52]. In *Mycobacterium tuberculosis*, glycine betaine has been shown to initiate growth in human macrophages [53]. Due to the importance of choline and glycine betaine in biological processes, choline oxidase and choline dehydrogenase have received considerable attention over recent years in biotechnology and medicine.

Two major biotechnological applications of choline oxidase, and to a lesser extent choline dehydrogenase, are in the construction of amperometric biosensors for the detection of acetylcholine in biological fluids [54] and the genetic engineering of increased tolerance towards abiotic stress in economically important crops [47,55]. The former couples the reactions of immobilized acetylcholinesterase and choline oxidase for the rapid, noninvasive, *in situ* detection of fluctuating extracellular levels of acetylcholine in normal and pathological conditions [56]. The latter involves the creation of transgenic plants that lack the ability to naturally accumulate glycine betaine, such as rice, maize, rapeseed, mustard green, kaki, cotton, tomato, tobacco or potato, resulting in enhanced tolerance toward abiotic stresses, such as hypersalinity, chilling, freezing, heat, strong light, drought, oxidative stress or combinations thereof [47].

Clinically, the gene coding for choline dehydrogenase was found to be overexpressed in subjects with favorable outcome in a study on a cohort of patients with early-stage estrogen receptor-positive breast cancer that were treated with tamoxifen [57] (choline dehydrogenase originally reported as Genbank accession number AI240933). A subsequent study demonstrated that in breast cancer the gene for choline dehydrogenase is regulated by estrogen and this can be used as a prognostic biomarker [58]. Abnormal choline metabolism has been emerging as a common pathway in almost every cancer type studied by bench-to-bedside magnetic resonance spectroscopy evaluating total choline-containing compounds [59,60]. While high cell proliferation in tumors was initially proposed to explain the high levels of choline-containing compounds [61], studies on non-malignant breast or prostate epithelial cells treated with growth hormones established that aberrant choline metabolism was associated with malignant transformation in cancer [62]. Choline deficiency has also been shown to induce L to M2 transition in pyruvate kinase isoenzymes in hepatocarcinogenesis in rats [63]. Among several single nucleotide polymorphisms that occur frequently in the human gene for choline dehydrogenase, rs12676 has been associated with increased susceptibility to dietary choline limitations, which results in steatosis and muscle cell damage [64], and increased risk of breast cancer [65]. More recently, the same choline dehydrogenase polymorphism has been demonstrated to be a functional variation associated with altered sperm motility patterns and altered mitochondrial morphology in human sperm associated with sterility [66]. A choline dehydrogenase knockout mouse model displayed impaired fertility due to decreased sperm mobility, with abnormal mitochondrial morphology in testis, liver, kidney, and heart [67].

8.2 Choline oxidase

8.2.1 Three-dimensional structure

The three-dimensional structure of choline oxidase from *A. globiformis* in the free form devoid of ligand has been elucidated by X-ray crystallography and refined to a resolution of 1.86 Å [68]. The enzyme is present in the crystal as a homodimer, in agreement with solution determinations under native conditions using gel-electrophoresis and gel filtration chromatography [69]. The dimer interface includes two sets of six identical inter-subunit electrostatic contacts clustered on the outer edge of the interface. The overall protein fold resembles that of other GMC enzymes (►Fig. 8.2), with a typical *p*-hydroxybenzoate hydroxylase fold [70] comprising both substrate (residues 160–200 and 312–463) and flavin (residues 1–159, 201–311 and 464–527) binding domains. A long loop (residues 64–95) with amphipathic character covers the active site, which is typical of members of the GMC enzyme superfamily. Molecular dynamics demonstrates very limited movement of the loop during a 60 ns simulation, consistent with the loop not being involved in substrate access to the active site [71]. In agreement with limited movement of the loop the average *B*-factor of the loop residues is 20 Å^2 and the loop is well-defined in the crystal structure [68].

The isoalloxazine ring of the FAD is buried within the protein, secluded from bulk solvent and not visible from the protein surface (►Fig. 8.3A). The flavin is covalently linked through its C8M atom to the Nε2 atom of His99 (►Fig. 8.3B). An interesting feature of the flavin is that it is not planar, but has a 120° "butterfly" angle between the dimethylbenzene and piperazine rings that is consistent with the flavin C4a atom being *sp*^{3} hybridized. Electron density maps further suggest that the flavin C4a atom is involved in a covalent linkage with either a single atom or a diatomic molecule [68]. Single-crystal spectroscopy and density functional theory calculations have provided evidence strongly suggesting that a flavin C4a-OH or C4a-OO(H) adduct forms in an X-ray-dependent fashion under the cryogenic conditions used in the experiment [72]. After forming in the crystal, the flavin-adduct would not decay at cryogenic temperatures due to an insufficient proton inventory being present in the enzyme devoid of ligands [72]. While no evidence for the transient formation of the flavin C4a-adduct with oxygen in solution is available, a C4a-adduct with oxygen has been reported for pyranose 2-oxidase, another member of the GMC enzyme superfamily [73]. Interestingly, in the X-ray structure of the choline oxidase variant with Ser101 replaced with alanine, which was cocrystallized with the product analogue acetate, the flavin does not participate in a C4a-adduct with O_2 but is instead planar (►Fig. 8.3C) [74]. In the active site of the Ser101Ala enzyme, a water molecule is 3.8 Å from the flavin C4a atom, where O_2 is predicted to react with the reduced flavin by enhanced-statistics molecular dynamics simulations in other flavin-dependent enzymes [75].

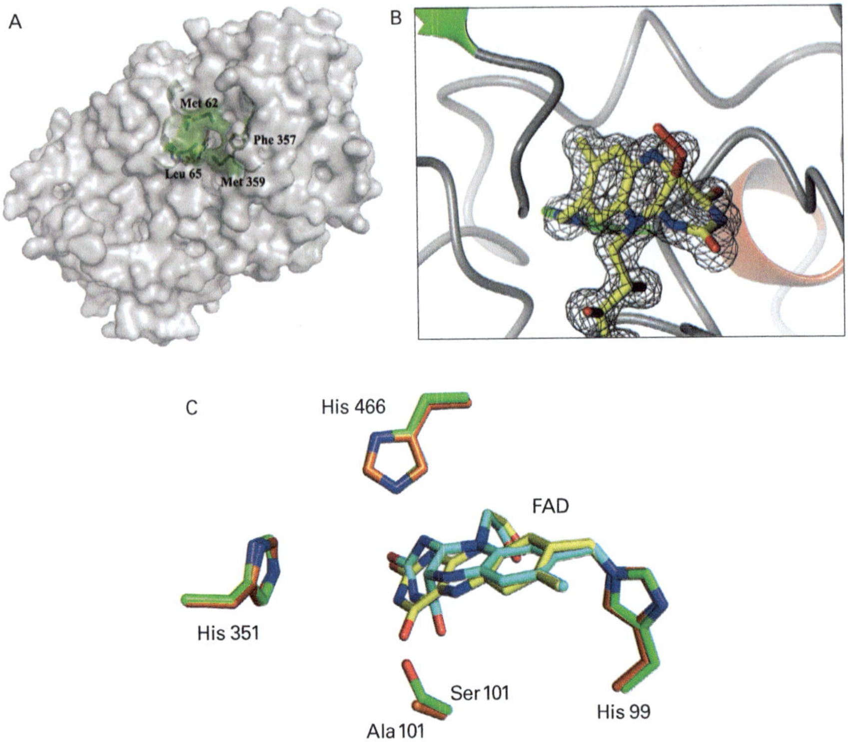

Fig. 8.3: (A) The hydrophobic cluster on the surface of choline oxidase (PDB 3NNE), proposed to act as gate for substrate access to the active site. The overall structure is represented as transparent grey surface, the hydrophobic residues (Met62, Leu65, Phe357 and Met359) in green and FAD in yellow. (B) Active site of choline oxidase wild type (PDB entry 2JBV) with distorted flavin: FAD in yellow and oxygen adduct in red. (C) Comparison of the active site of choline oxidase wild type (green; PDB 2JBV) and Ser101Ala enzyme (orange; PDB 3NNE). FAD is shown in cyano for the wild type and yellow for the Ser101Ala enzyme.

The active site cavity is completely secluded from the bulk solvent and has an approximate volume of 125 Å^3, which is slightly larger than the volume of choline (*e.g.*, 93 Å^3) [68]. This likely contributes to the optimal positioning of the substrate for catalysis in the enzyme-substrate complex and to the high specificity of the enzyme for choline (and betaine aldehyde). The cavity is decorated with both polar and hydrophobic residues. An aromatic cage composed of Trp331, Phe357 and Val464 is topped by Glu312 and constitutes the binding site for the positively charged trimethylammonium group of choline via π-cation and electrostatic interactions (►Fig. 8.4). The side chains of His351 and His466 are located closer to the isoalloxazine of the flavin, with the latter at 4.2 Å from both the flavin N1 and C4a atoms (►Fig. 8.2). Ser101 is less than 4 Å from the N5 atom of the flavin, whereas the side chain of Asn510 is sandwiched between the imidazole side chains of His351 (3 Å) and His466 (4 Å) and at 4.7 Å from the flavin O2 and N3 atoms. Through site-directed mutagenesis the role in catalysis of several active site residues has been elucidated, as illustrated in later sections of this Chapter.

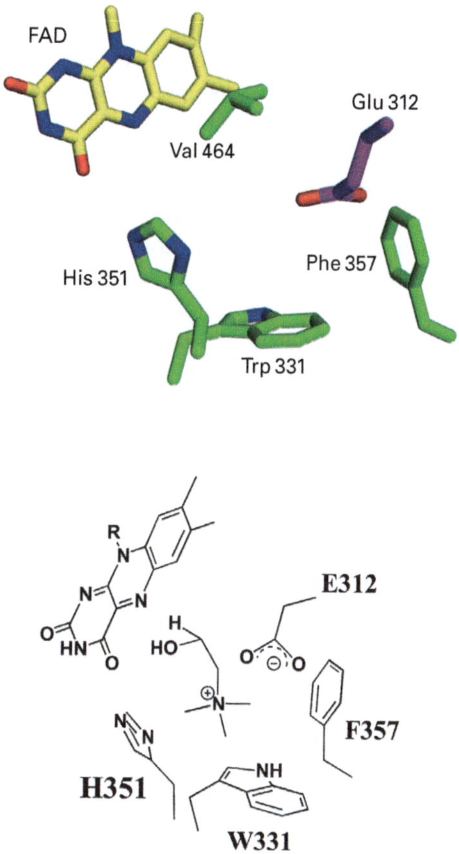

Fig. 8.4: The trimethylammonium-binding site in the active site of choline oxidase (PDB 3NNE; Ser101Ala enzyme). FAD is shown in yellow, Glu312 in magenta, His351, Phe357 and Trp331 in green.

The X-ray structures of two active site variants of choline oxidase engineered through site-directed mutagenesis with Ser101 or Val464 replaced with alanine are available to resolutions of 2.2 Å [74,76]. Both mutant enzymes maintain an overall fold that is identical to that of the wild type enzyme, with average rmsd values ≤0.67 Å for 527 topologically equivalent Cα atoms in each subunit upon superimposition with the wild type enzyme [74,76]. No differences are seen in the relative location of the flavin and the active site residues, with all the residue side chains maintaining the same position and orientation in the wild type and mutant enzymes. In contrast to the wild type enzyme, in both mutant enzymes the flavin C4a atom is sp^2 hybridized and no flavin-adduct is present, with the isoalloxazine slightly bent along the N5-N10 axis defining a 172° angle between the dimethylbenzene and pyrimidine rings (▶Fig. 8.3C) [74,76]. In all three structures the last 16–20 amino acids at the C-terminal end are not observable crystallographically, consistent with high mobility of the protein C-terminus [68,74,76].

8.2.2 Biophysical properties

Choline oxidase is a homodimer of 120 kDa, as indicated by both solution and crystallographic studies [68,69]. Each subunit contains FAD covalently linked to the protein through its C8 methyl and the Nε2 atom of His99 (▶Fig. 8.3B) [68]. The UV-visible absorbance spectrum of the oxidized enzyme shows peaks at 359 and 452 nm, consistent with covalent attachment of the flavin to the protein moiety.

The anionic hydroquinone, with typical absorbance maximum at 356 nm (pH 8.0), is obtained upon anaerobic reduction of the enzyme with choline [77]. This reduced enzyme species is stabilized between pH 6.0 and 10.0 [78], consistent with the flavin N1 atom not being accessible to bulk solvent or having a pK_a value lower than 6.7 for flavins in solution. The enzyme has high affinity for sulfite (K_d ~40 μM, at pH 7.0 and 15 °C), although it reacts slowly (k_{on} = 0.04 $M^{-1}s^{-1}$) [77], probably due to constrained access to the active site of the negatively charged sulfite (choline, in contrast, is positively charged). Upon reduction of choline oxidase with dithionite, the anionic flavin semiquinone is obtained [77,78]. In the crystal structure of the enzyme two features are likely responsible for the stabilization of the negative charge on the flavin N1-C2=O atoms: the *N*-terminal end of a long α-helix (residues 512–527) and the side chain of His466 at <4 Å [68,74,76]. Replacement of His466 with aspartate yields the neutral hydroquinone upon dithionite reduction, without formation of a flavin semiquinone, consistent with the importance of His466 for charge stabilization on the reduced flavin [78].

At pH 8.0, the anionic flavin semiquinone is unusually stable in the presence of O_2 [69]. The enzyme containing flavin semiquinone is inert in turnover, as indicated by similar k_{cat} and $k_{cat}/K_{choline}$ values for choline or betaine aldehyde expressed per active site oxidized flavin content of enzyme samples with flavin in semiquinone/oxidized ratios of 0/100 and 63/37 [77]. The nonreactive flavin semiquinone of choline oxidase can be slowly oxidized by prolonged incubation at pH 6.0, likely through a pH-dependent conformational change of the enzyme [69]. In an unrelated study, it was established that a localized and reversible conformational change induced by freezing at low pH is associated with the complete and reversible loss of catalytic activity in choline oxidase [79]. Slow reactivation of the enzyme occurs in a pH-dependent fashion, being facilitated at pH ≥ 6.5, implicating incorrect ionization(s) of protein side chains that are not directly accessible to bulk solvent as being associated with the reversible conversion of inactive and active enzyme species [79]. A catalytically incompetent form of enzyme in slow equilibrium with an active species with catalytic properties in the reductive half-reaction similar to the wild type was also obtained through site-directed mutagenesis upon replacing Val464 with either alanine or threonine [80].

In the enzyme devoid of ligand, the reduction potentials of the enzyme-bound flavin are amongst the highest reported for flavin-dependent enzymes, with $E_{ox/sq,7}$ = +211 mV and $E_{sq/red,7}$ = –65 mV [78]. With glycine betaine bound at the active site, the enzyme loses the ability to stabilize thermodynamically the flavin semiquinone, displaying a midpoint reduction potential $E_{ox/red,7}$ of +132 mV for the oxidized/reduced flavin couple [81]. The elevated reduction potentials are partly due to the covalent attachment of the flavin to the protein [82] and the presence of the positively charged imidazolium of His466, as shown by mutagenesis studies with replacement of His466 with alanine or aspartate [78,81]. Other factors likely further contribute to the +330 mV perturbation of the midpoint reduction potential of the flavin from the value of –207 mV in solution to +132 mV in the active site of the enzyme.

8.2.3 Substrate specificity and inhibitors

Choline oxidase is active on choline and betaine aldehyde [69,77], with $^{app}k_{cat}$ and $^{app}K_m$ values under atmospheric O_2 concentration in the 11–15 s^{-1} and 0.6–2.3 mM (at pH 7.0 and 25 °C) [69,77]. The trimethylammonium group of the substrate, with its positive charge and geometry, is the key determinant for substrate binding, as indicated by studies with substrate and product analogs and site-directed mutagenesis [68,83,84]. Both the k_{cat}/K_m and K_{is} values for substituted substrates and products decrease with a logarithmic pattern with decreasing number of methyl groups on the amine moiety, allowing for an estimate of ~1.7 kJ mol^{-1} for the contribution of each methyl group to binding [84]. Lack of solvent viscosity effects, pH effects and steady-state kinetics with 3,3-dimethyl-butanol, an isosteric analogue of choline devoid of positive charge, allow a K_d value of ~55 mM to be estimated [83]. This value is ~180-times larger than the kinetically determined K_d value for choline using a stopped-flow spectrophotometer, allowing for an estimate of ~13 kJ mol^{-1} for the contribution of the substrate positive charge to binding [83]. This value agrees well with the value of ~15 kJ mol^{-1} for the ionic contribution to substrate binding in the active site of the enzyme estimated from the comparison of the K_d values for choline determined for the wild type enzyme and a mutant enzyme where Glu312 is replaced with glutamine through side-directed mutagenesis [68].

Through molecular dynamics, a cluster of hydrophobic residues located on the solvent accessible surface of the enzyme, Met62, Leu65, Phe357 and Met359, has been proposed to form a gate that allows access of choline to and egress of glycine betaine from the active site cavity [71]. The opening and closing of the gate occurs in the 10–20 ns timescale, consistent with product release not limiting the overall turnover of the enzyme [21,71]. The electrostatic potential of the enzyme has been proposed to control the diffusion of the substrate to the active site through the hydrophobic gate based on computational studies [71].

A number of amines variably substituted with methyl and ethyl groups are competitive inhibitors of choline oxidase with respect to choline, with the best inhibitor being trimethylamine with a K_{is} value of 2.4 mM at pH 6.5 and 25 °C [84]. Glycine betaine inhibits the enzyme competitively with respect to choline in a pH-dependent fashion, with a limiting K_{is} of 12.5 mM at low pH that increases with increasing pH, with a thermodynamic pK_a of 7.5 [77].

Steady-state kinetics and NMR spectroscopy with betaine aldehyde and its isosteric analogue 3,3-dimethylbutyraldehyde have established that choline oxidase utilizes the hydrated species of the aldehyde as substrate [85]. In contrast, the enzyme is inhibited by the non-hydrated form of the aldehyde, which is present in solution in negligible amounts at equilibrium with betaine aldehyde but it is the predominant form with 3,3-dimethylbutyraldehyde [85]. From a chemical standpoint hydrated betaine aldehyde is equivalent to a substituted form of choline carrying an extra hydroxyl on the α-carbon, *i.e.*, *gem*-diol choline. Thus, it is conceivable that the enzyme uses the

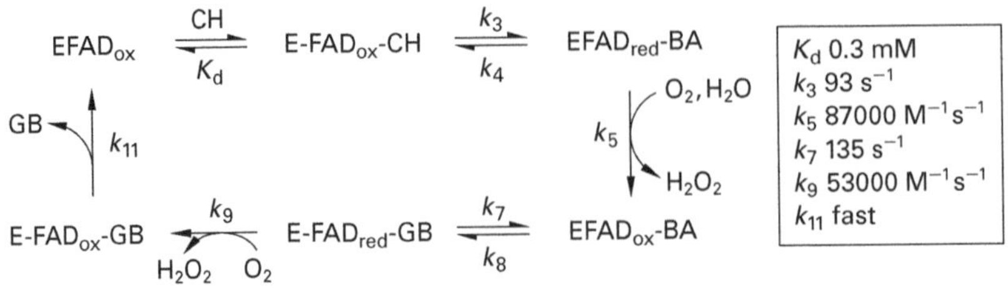

Fig. 8.5: A minimal steady-state kinetic mechanism of choline oxidase: E, enzyme; FAD_{ox}, oxidized flavin; FAD_{red}, reduced flavin; CH, choline; BA, betaine aldehyde; GB, glycine betaine. The rate constants are for pH 10.0 (pH independent region for kinetic steps).

same chemical mechanism for the oxidation of the alcohol substrate and the aldehyde intermediate (*see below*).

8.2.4 Steady-state kinetic mechanism

The steady-state kinetic mechanism of choline oxidase with choline as substrate is shown in ▶Fig. 8.5, and is valid throughout the accessible pH range in which the enzyme is stable, *i.e.*, pH 5.0 to 10.0 [1,77]. Colorimetric assays of betaine aldehyde release when the enzyme turns over with choline, steady-state and anaerobic rapid reaction kinetics with choline or betaine aldehyde and substrate kinetic isotope effects with 1,2-[$^4H_2$]-choline establish that during turnover with choline betaine aldehyde is a reaction intermediate that predominantly remains bound in the active site [1,21,77,86]. The order of the kinetic steps involving substrate binding, catalysis and product release is not changed upon substituting several of the active site residues by mutagenesis, *i.e.*, His99 [87], Ser101 [74], Glu312 [68], His351 [88], Val464 [80], His466 [81], and Asn510 [89]. Similarly, analogues of choline, such as *N,N*-dimethylamine, *N*-methylamine, and 3,3-dimethyl-butanol, follow the same steady-state kinetic mechanism [83,84]. When betaine aldehyde is substrate, O_2 reaction with the reduced flavin occurs prior to the release of the reaction product glycine betaine [1,77]. Thus, in all cases flavin oxidation occurs in the enzyme-product complex rather than in the free enzyme devoid of ligand.

The overall turnover of the enzyme with choline is limited by the two chemical steps of flavin reduction (k_3 and k_7 in ▶Fig. 8.5) between pH 6.5 and 10.0, as suggested by steady-state and anaerobic rapid reaction kinetics with choline or betaine aldehyde and substrate kinetic isotope effects with 1,2-[$^4H_2$]-choline [21,86]. The overall turnover with betaine aldehyde instead is limited exclusively by the chemical step of flavin reduction in which the aldehyde is oxidized to the carbon acid (k_{cat} of 133 s^{-1} and k_{red} of 135 s^{-1}; pH 10 and 25 °C) [21].

8.2.5 Chemical mechanism for alcohol oxidation

The chemical mechanism for alcohol oxidation has been studied in choline oxidase with substrate and solvent kinetic isotope effects [21], solvent viscosity [21], pH [21,77,86,90] and temperature effects [87,91,92] on the kinetic parameters and site-directed mutagenesis [68,74,76,81,87–89,93,94].

Alcohol oxidation begins with the enzyme-catalyzed abstraction of the substrate hydroxyl proton by a catalytic base with pK_a of 7.5, as indicated by pH and kinetic isotope effects on the $k_{cat}/K_{choline}$ and k_{cat} values with choline as substrate for the wild type enzyme [21,77,86]. Inhibition studies with glycine betaine and kinetic isotope effects establish the pK_a of 7.5 as being an intrinsic value, with a low commitment to catalysis when the enzyme is saturated with O_2 [21,77,91].

Unequivocal identification of the amino acid group responsible for this reaction has not been obtained, since individual replacement of each of the three active site histidine residues with alanine (His351 and His466) or asparagine (His99) yields mutant enzymes still displaying pH profiles for $k_{cat}/K_{choline}$ consistent with the presence of a base [81,87,88]. However, the presence of water acting as surrogate base in the His351Ala or His466Ala enzymes cannot be ruled out, and more studies are required. In the GMC member aryl-alcohol oxidase QM/MM and site-directed mutagenesis studies are consistent with His502, equivalent to His466 in choline oxidase, acting as catalytic base in the alcohol oxidation reaction [95]. Similarly, site-directed mutagenesis investigations on pyranose 2-oxidase have suggested that the equivalent histidine (*i.e.*, His548) may act as catalytic base in that enzyme [96]. Irrespective of the identity of the base in choline oxidase, proton abstraction yields an activated alkoxide intermediate that is stabilized in the active site of the enzyme through multiple electrostatic interactions with the side chains of Glu312 [68] and His466 [81], and hydrogen bonding interactions with the side chains of Ser101 [93], His351 [88] and Ans510 [89] (▶Fig. 8.6). These interactions, along with the covalent linkage of the flavin to His99 [87], collectively contribute to preorganization of the enzyme-substrate complex that results in subsequent quantum mechanical tunneling of the hydride ion from the α-carbon of the activated substrate to the flavin N5 atom (*see below*) [4,91]. The formation of the alkoxide intermediate in the wild type enzyme is kinetically fast (≥1900 s^{-1}), as suggested by the lack of a solvent kinetic isotope effects on $k_{cat}/K_{choline}$ and k_{red} values determined in steady-state and reductive half-reaction [21,93]. Interestingly, in a number of mutant enzymes substituted at Ser101 the rate constant for proton abstraction decreases to values that are measurable by stopped-flow time-resolved spectroscopy, with values between 40 s^{-1} and 125 s^{-1} [93]. In these mutant enzymes the rate constant for flavin reduction shows a logarithmic decrease spanning four orders of magnitude with increasing hydrophobicity of the side chain at position 101 (Ser > Thr > Cys > Val) [93]. This suggests that the hydrophilic character of a serine residue close to the flavin C(4a) and N(5) atoms is important for the hydride transfer reaction to the flavin.

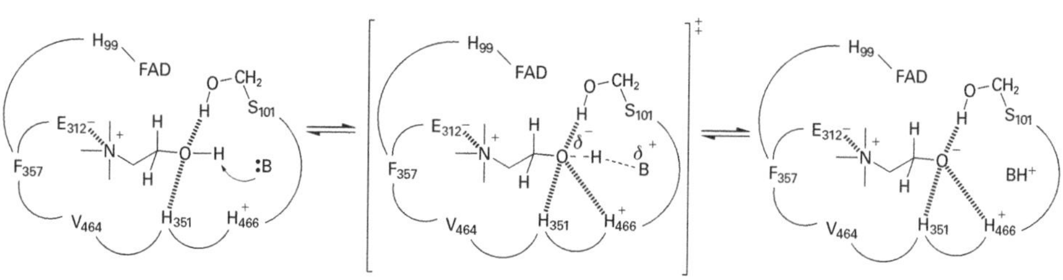

Fig. 8.6: The mechanism for alcohol activation catalyzed by choline oxidase.

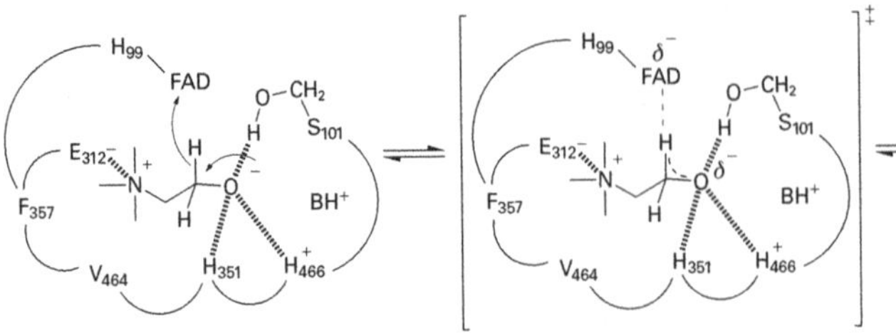

Fig. 8.7: The mechanism for flavin reduction catalyzed by choline oxidase.

Flavin reduction occurs via hydride transfer from the α-carbon of the activated substrate to the flavin N5 atom, yielding betaine aldehyde (►Fig. 8.7) [21]. This reaction occurs in a stepwise fashion with respect to the proton abstraction reaction producing the alkoxide species, as demonstrated by substrate and deuterium kinetic isotope effects on the steady-state kinetic and rapid reaction parameters for the reductive half-reaction of choline oxidase with choline as substrate [21]. Flavin reduction with choline is partially rate-limiting for overall turnover of the enzyme, along with the second flavin reduction linked to the oxidation of the aldehyde intermediate to glycine betaine, with pH-independent rate constants of 93 s^{-1} and 135 s^{-1} (at pH 10 and 25 °C) [21]. Temperature dependences of the $k_{cat}/K_{choline}$ and its associated kinetic isotope effect establish that the hydride transfer occurs via quantum mechanical tunneling within a highly pre-organized enzyme-substrate complex, with minimal independent movements of the alkoxide α-carbon acting as hydride donor and the flavin N5 atom acting as hydride acceptor [91]. Characteristic of such a tunneling reaction in a highly preorganized enzyme-substrate complex is the temperature independent $^{D}(k_{cat}/K_{choline})$ with a value of ~11 between 15 and 40 °C, the large isotope effect of 14 on the Eyring preexponential factors with choline and 1,2-[$^{2}H_4$]-choline (A_H'/A_D'), and the similar enthalpies of activation ($\Delta H^{‡}$) for the hydride transfer reaction with choline and 1,2-[$^{2}H_4$]-choline with finite values of ~18 kJ/mol [91]. A number of electrostatic and hydrogen bonding interactions between choline alkoxide and the side chains of several amino acid residues in the active site of the enzyme are responsible for the highly preorganized enzyme-substrate complex (*see above*) [68,80,81,88,89,93], including the flavin covalent attachment to the protein moiety [87]. In agreement with the importance of the flavin covalent attachment to the protein for preorganization of the enzyme-substrate, the hydride transfer reaction in the His99Asn enzyme (in which the cofactor is tightly but not covalently bound to the protein) requires significant sampling of the reactive configuration that results in the tunneling reaction [87]. This is suggested by the A_H'/A_D' of 1, the large isotope effect on the energy of activation $\Delta E^{‡}$ with protiated and deuterated choline, the dependence on temperature of the $^{D}k_{red}$ value and the different $\Delta H^{‡}$ values for cleavage of the CH and CD bonds of the substrate [87]. Similar disruption of enzyme-substrate complex preorganization is observed upon replacing the main site of substrate binding in the active site Glu312 with aspartate [92]. The latter conservative substitution also represents an interesting example of how substitution of residues strictly involved in substrate binding, but not catalysis, may impact catalysis with minimal effects on binding.

The hydride tunneling reaction of choline oxidase has been also investigated at sub-saturating concentrations of O_2, in which the reaction is reversible, by looking at how temperature affects pH and kinetic isotope effects [90]. After accounting for the kinetic complexity ensuing from the reversible catalytic regime, analyses of the $k_{cat}/K_{choline}$ and $^{D}(k_{cat}/K_{choline})$ values indicate that the quantum mechanical nature of the reaction is maintained in the reversible regime [90]. Interestingly, an enthalpically unfavorable internal equilibrium of the enzyme-substrate complex occurring prior to the cleavage of the substrate CH bond is evident from a comparison of the thermodynamic and kinetic parameters for the hydride transfer reaction under irreversible and reversible catalytic regimes [90]. Such an internal equilibrium is likely required to achieve preorganization in the enzyme-substrate complex for the subsequent tunneling reaction of the hydride ion. Substrate deuterium kinetic isotope effects in steady-state kinetics have disclosed the presence of an internal equilibrium of enzyme-substrate complexes in a mutant form of choline oxidase with Glu312 replaced with alanine [68]. Solvent viscosity and kinetic isotope effects on the reductive half-reaction of mutant enzymes with Val464 replaced with threonine or alanine also disclosed the presence of an enzyme-substrate isomerization occurring prior to hydride transfer [80]. With this mutant enzyme the isomerization is likely associated with the reaction of proton abstraction from the hydroxyl group of the substrate [80].

8.2.6 Chemical mechanism for aldehyde oxidation

The hydrated form of betaine aldehyde, *i.e.*, *gem*-diol choline, is the species that is oxidized to glycine betaine in the second oxidation reaction catalyzed by choline oxidase, as suggested by kinetic and spectrophotometric data with betaine aldehyde and 3,3-dimethylbutyraldehyde as substrate [85]. NMR spectroscopy has established that betaine aldehyde is predominantly in *gem*-diol form in solution [85], whereas kinetics data have demonstrated that betaine aldehyde is a good substrate for the enzyme with a pH-independent, maximal rate of anaerobic flavin reduction slightly faster than with choline (135 s^{-1} vs. 93 s^{-1} at pH 10) [21]. As expected from the similar chemical structures of choline and *gem*-diol choline, the enzyme shows similar K_d values for choline and betaine aldehyde at pH 8.0 (~2 mM) and 10.0 (~0.4 mM) [21,85]. In contrast, the substrate analogue 3,3-dimethylbutyraldehyde is present in solution at equilibrium predominantly in the aldehydic form, with only ~20% being hydrated [85]. This yields reduction of the enzyme-bound flavin only for the fraction of enzyme that binds the hydrated form, with the majority of the enzyme being inhibited because the aldehyde form of 3,3-dimethylbutyraldehyde is not oxidized and acts as a competitive inhibitor with respect to choline [85]. The proposed mechanism for the second oxidation reaction catalyzed by choline oxidase is illustrated in ▶Fig. 8.8. A catalytic base with an apparent pK_a of 6.7 under atmospheric O_2 conditions is required for the deprotonation reaction that abstracts the hydroxyl proton of the hydrated aldehyde, as suggested by the pH profile of the $^{app}(k_{cat}/K_{betaine\ aldehyde})$ value [85]. Since the enzyme retains the aldehyde intermediate in the active site during turnover with choline [1], it is likely that the same

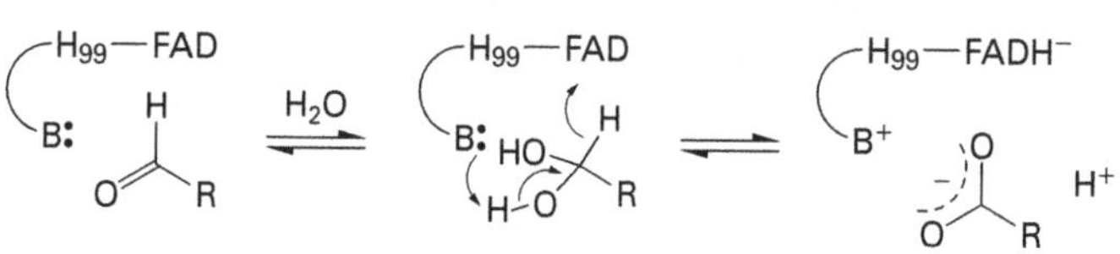

Fig. 8.8: The mechanism for oxidation of betaine aldehyde to glycine betaine catalyzed by choline oxidase.

group acts as base in the first and second oxidation reactions. The observed absence of deuterium solvent kinetic isotope effects on the k_{cat} value with choline is consistent with the hydration reaction producing *gem*-diol choline during turnover of the enzyme with choline being kinetically fast [21,85]. With betaine aldehyde as substrate for the enzyme, the relative timing of OH and CH bond cleavage and the possible mechanism of hydride transfer through quantum tunneling have not been investigated.

8.2.7 Oxygen activation for reaction with reduced flavin

Despite the presence of three histidine residues in the active site, it is the non-ionizable positive charge harbored on the substrate/product of the enzymatic reaction that plays the predominant role in the oxidative half-reaction(s) catalyzed by choline oxidase (for a recent review see [3]). Evidence for this conclusion comes from steady-state kinetics with choline or a substrate analogue devoid of charge, and site-directed mutagenesis [3,81,83,84,87,88]. Steady-state kinetics established that the reaction of O_2 with the reduced flavin occurs with the reaction product bound in the active site of the enzyme (▶Fig. 8.5). A decrease of 80-fold in the k_{cat}/K_{oxygen} value is seen when 3,3-dimethyl-butanol, a substrate analogue devoid of positive charge, is used as substrate for the enzyme instead of choline [83]. Since the analogue is isosteric with choline, the effect appears primarily due to the lack of charge. Individual replacements of His351 and His466 with alanine and His99 with asparagine yield mutant enzymes with k_{cat}/K_{oxygen} values ≤1.5-times lower than the wild type enzyme, consistent with each histidine residue not being important for O_2 reaction with the reduced enzyme [81,87,88]. In agreement with the non-ionizable quaternary amine of the substrate/product of the enzymatic reaction being important for O_2 reaction with the flavin, the wild type enzyme shows a k_{cat}/K_{oxygen} value that is independent of pH between 6.0 and 10.0 [77].

Localization of O_2 to the reaction site is aided by the presence of a nonpolar aminoacyl side chain, Val464, located in proximity (3.8 Å) to the flavin C4a atom, as suggested by site-directed mutagenesis [76]. This side chain is spatially adjacent to His466 (3.8 Å), which is fully conserved in the GMC enzyme superfamily, at 3.5 Å from a water molecule that is identified in the X-ray structure of the Ser101Ala mutant enzyme [76]. The position of the water molecule with respect to the flavin is identical to that where O_2 is predicted to react with the reduced flavin based on enhanced-statistics molecular dynamics on other flavin-dependent oxidases and monooxygenases [75]. Upon replacing Val464 with either alanine, which allows for an enlarged space proximal to the flavin C4a atom, or threonine, which increases polarity of the flavin microenvironment, k_{cat}/K_{oxygen} values ~1,500 $M^{-1}s^{-1}$ are determined [76]. Such values are ~50-fold lower than the k_{cat}/K_{oxygen} of 90,000 $M^{-1}s^{-1}$ determined for the wild type enzyme, consistent with Val464 being important in the oxidative half-reaction. Interestingly, a hydrophobic side chain in similar position is present in several flavoprotein oxidases, including glucose oxidase, monomeric sarcosine oxidase, cholesterol oxidase, pyranose 2-oxidase, monoamine oxidase A and B, polyamine oxidase, D-amino acid oxidase and glycolate oxidase, suggesting that O_2 localization by a nonpolar site proximal to the flavin reaction site is a general feature in flavoprotein oxidases [3]. Nonpolar cavities that encapsulate and desolvate O_2 have been identified crystallographically in flavoprotein monooxygenases [75]. In these enzymes, highly reactive C4a-OO(H) adducts with the flavin are typical reaction intermediates that must be stabilized and prevented from reacting unproductively in order to hydroxylate substrates. In contrast, the requirement for a nonpolar site encapsulated in a cavity is less stringent in oxidases since any reaction intermediate on the path from O_2 to hydrogen peroxide is likely to be destabilized, rather than stabilized, in order to avoid leakage of reactive oxygen species from the active site of the enzyme [3].

8.3 Choline dehydrogenase

Knowledge of the biochemistry of choline dehydrogenase has lagged behind that for choline oxidase, primarily due to the difficulty of obtaining stable preparations of enzyme in purified and active form [13,14,19]. In 1980 Tsuge *et al.* noted in a study on the purification of rat liver choline dehydrogenase that "All of the characteristics reported thus far have been obtained using a relatively crude preparation and a highly purified preparation is an urgent priority in the field" [13]. Unfortunately, this statement remains valid in 2012 with most of the studies published over the past 30 years focusing on the molecular characterization of the gene coding the enzyme [18,97] and its biotechnological application for engineering stress tolerance in plants [98], while only a few studies are available on the biochemistry of partially purified enzymes from *Pseudomonas* and *Halomonas elongata* [14,19].

Choline dehydrogenase is membrane-associated to the plasma membrane in bacteria and the mitochondria in eukaryotic cells, although the structural features governing such association are not known [11,14,19,99]. The enzyme can oxidize either choline or betaine aldehyde [11,19] and it reacts poorly with O_2, although phenazine methosulfate is preferred as electron acceptor [19]. The enzyme extracted from rat liver mitochondria has also been shown to utilize cytochrome *c* as electron acceptor [11]. Based on the association of the enzyme with the membrane, ubiquinone has been proposed to act as electron acceptor *in vivo*, but definitive evidence is not available [13]. The presence of flavin in (partially) purified preparations of choline dehydrogenase from *Pseudomonas*, *H. elongata*, and rat liver [11,13,14,19] has not been established conclusively, although an FAD-binding consensus sequence of the type GXGXXG is present at the N-terminal end of the protein from these cellular sources. Comparison of the amino acid sequence of choline dehydrogenase from various cellular sources with the amino acid sequence of *A. globiformis* choline oxidase demonstrates that active site residues of the oxidase that play important roles in substrate binding and catalysis are fully conserved in the dehydrogenase, *e.g.*, Glu312, His351, His466 and Asn510 (numbering of choline oxidase). Thus, in the absence of biochemical data it is only possible to postulate that what has been learned at the molecular level on the biochemistry and mechanism of action of choline oxidase may be applicable to the dehydrogenase.

8.4 Thiamine oxidase/dehydrogenase

Thiamine oxidase catalyzes the four-electron, two-step oxidation of thiamine, an alcohol, to thiamine acetic acid, a carbon acid [100,101]. The reaction is flavin-mediated and involves formation of an enzyme-bound thiamine aldehyde intermediate that is not released to the bulk solvent [102]. In these respects, thiamine oxidase/dehydrogenase can be grouped with choline oxidase and choline dehydrogenase, since they catalyze similar oxidations of alcohols to carbon acids through enzyme-bound aldehyde intermediates.

From a biochemical standpoint, thiamine oxidase/dehydrogenase has been characterized in some of its basic features, demonstrating that the enzyme has high affinity for sulfite, with K_d of 117 μM, and forms a red anionic semiquinone upon anaerobic photoreduction with EDTA [100]. O_2 acts as electron acceptor yielding stoichiometric amounts of hydrogen peroxide without release of superoxide [100]. Thiamine oxidase/dehydrogenase was the first flavoprotein in which an 8α-N(1)-histidyl-FAD linkage was identified [101,103,104]. The midpoint reduction potentials of the enzyme-bound flavin have been determined at pH 7.2 and equal $E_{ox}/E_{sq,7.2}$ = +80 mV and $E_{sq}/E_{red,7.2}$ = +30 mV [100].

8.5 Conclusions

In summary, choline oxidase has been extensively characterized in its biochemical, structural, kinetic, and mechanistic properties. These studies have revealed structural-functional details on how the substrate access to and bind at the active site of the enzyme, the highly preorganized nature of the enzyme-substrate complex, the chemical mechanism for alcohol oxidation and, to a lesser extent, aldehyde oxidation, and the nature of the group(s) that participate in the activation of O_2 for reaction with the enzyme-bound flavin. Knowledge of the biochemistry of choline oxidase is applicable to choline dehydrogenase, for which difficulty of obtaining stable preparations of enzyme in purified and active form has hampered its studies.

8.6 Acknowledgements

This work was supported in part by Grants PRF 37351-G4 and PRF 47363-AC4 from the American Chemical Society, a Research Initiation Grant and a Quality Improvement Fund from Georgia State University, a CAREER Award MCB-0545712 and NSF MCB-1121695 from the National Science Foundation, and a number of Molecular Basis of Diseases Fellowships and Roland E. McNair Post-Baccalaureate Achievement Program to graduate and undergraduate students who have worked on this project. I thank the valuable experimental and intellectual contributions of all of the students and postdoctoral research associates who have worked or currently work in my research group, as well those of my collaborators and colleagues. A heartfelt "Thank you" goes to Ms. Francesca Salvi for the preparation of the artwork and figures.

8.7 References

[1] Gadda G. Kinetic mechanism of choline oxidase from Arthrobacter globiformis. Biochim Biophys Acta 2003;1646:112–8.
[2] Ikuta S, Imamura S, Misaki H, Horiuti Y. Purification and characterization of choline oxidase from Arthrobacter globiformis. J Biochem 1977;82:1741–9.
[3] Gadda G. Oxygen activation in flavoprotein oxidases: the importance of being positive. Biochemistry 2012;51:2662–9.
[4] Gadda G. Hydride transfer made easy in the reaction of alcohol oxidation catalyzed by flavin-dependent oxidases. Biochemistry 2008;47:13745–53.

[5] Ribitsch D, Winkler S, Gruber K, et al. Engineering of choline oxidase from Arthrobacter nicotianae for potential use as biological bleach in detergents. Appl Microbiol Biotechnol 2010;87:1743–52.
[6] Ribitsch D, Karl W, Wehrschutz-Sigl E, et al. Heterologous expression and characterization of choline oxidase from the soil bacterium Arthrobacter nicotianae. Appl Microbiol Biotechnol 2009;81:875–86.
[7] Hekmat A, Saboury AA, Moosavi-Movahedi AA, Ghourchian H, Ahmad F. Effects of pH on the activity and structure of choline oxidase from Alcaligenes species. Acta Biochim Pol 2008;55:549–57.
[8] Tavakoli H, Ghourchian H, Moosavi-Movahedi AA, Chilaka FC. Effects of paraoxon and ethylparathion on choline oxidase from Alcaligenes species: inhibition and denaturation. Int J Biol Macromol 2005;36:318–23.
[9] Ohta-Fukuyama M, Miyake Y, Emi S, Yamano T. Identification and properties of the prosthetic group of choline oxidase from Alcaligenes sp. J Biochem 1980;88:197–203.
[10] Ebisuzaki K, Williams JN, Jr. Preparation and partial purification of soluble choline dehydrogenase from liver mitochondria. Biochem J 1955;60:644–6.
[11] Rendina G, Singer TP. Studies on choline dehydrogenase. I. Extraction in soluble form, assay, and some properties of the enzyme. J Biol Chem 1959;234:1605–10.
[12] Barrett MC, Dawson AP. The reaction of choline dehydrogenase with some electron acceptors. Biochem J 1975;151:677–83.
[13] Tsuge H, Nakano Y, Onishi H, Futamura Y, Ohashi K. A novel purification and some properties of rat liver mitochondrial choline dehydrogenase. Biochim Biophys Acta 1980;614:274–84.
[14] Russell R, Scopes RK. Use of hydrophobic chromatography for purification of the membrane-located choline dehydrogenase from a Pseudomonas strain. Bioseparation 1994;4:279–84.
[15] Perrino LA, Pierce SK. Choline dehydrogenase kinetics contribute to glycine betaine regulation differences in chesapeake bay and atlantic oysters. J Exp Zool 2000;286:250–61.
[16] Andresen PA, Kaasen I, Styrvold OB, Boulnois G, Strom AR. Molecular cloning, physical mapping and expression of the bet genes governing the osmoregulatory choline-glycine betaine pathway of Escherichia coli. J Gen Microbiol 1988;134:1737–46.
[17] Landfald B, Strom AR. Choline-glycine betaine pathway confers a high level of osmotic tolerance in Escherichia coli. J Bacteriol 1986;165:849–55.
[18] Pocard JA, Vincent N, Boncompagni E, Smith LT, Poggi MC, Le Rudulier D. Molecular characterization of the bet genes encoding glycine betaine synthesis in Sinorhizobium meliloti 102F34. Microbiology 1997;143 (Pt 4):1369–79.
[19] Gadda G, McAllister-Wilkins EE. Cloning, expression, and purification of choline dehydrogenase from the moderate halophile Halomonas elongata. Appl Environ Microbiol 2003;69:2126–32.
[20] Cavener DR. GMC oxidoreductases. A newly defined family of homologous proteins with diverse catalytic activities. J Mol Biol 1992;223:811–4.
[21] Fan F, Gadda G. On the catalytic mechanism of choline oxidase. J Am Chem Soc 2005;127:2067–74.
[22] Ferreira P, Hernandez-Ortega A, Herguedas B, Martinez AT, Medina M. Aryl-alcohol oxidase involved in lignin degradation: a mechanistic study based on steady and pre-steady-state kinetics and primary and solvent isotope effects with two alcohol substrates. J Biol Chem 2009;284:24840–7.
[23] Sucharitakul J, Wongnate T, Chaiyen P. Kinetic isotope effects on the noncovalent flavin mutant protein of pyranose 2-oxidase reveal insights into the flavin reduction mechanism. Biochemistry 2010;49:3753–65.
[24] Ueland PM. Choline and betaine in health and disease. J Inherit Metab Dis 2011;34:3–15.
[25] Lever M, Slow S. The clinical significance of betaine, an osmolyte with a key role in methyl group metabolism. Clin Biochem 2010;43:732–44.

[26] Institute of Medicine, and National Academy of Sciences USA (1998) Choline. In: Dietary reference intakes for folate, thiamin, riboflavin, niacin, vitamin B12, panthothenic acid, biotin, and choline Vol 1. Washington DC: National Academy Press; 1998:309–422.
[27] Zeisel SH, Blusztajn JK. Choline and human nutrition. Annu Rev Nutr 1994;14:269–96.
[28] Blusztajn JK, Zeisel SH, Wurtman RJ. Developmental changes in the activity of phosphatidylethanolamine N-methyltransferases in rat brain. Biochem J 1985;232:505–11.
[29] Zeisel SH. Gene response elements, genetic polymorphisms and epigenetics influence the human dietary requirement for choline. IUBMB Life 2007;59:380–7.
[30] Sunden SL, Renduchintala MS, Park EI, Miklasz SD, Garrow TA. Betaine-homocysteine methyltransferase expression in porcine and human tissues and chromosomal localization of the human gene. Arch Biochem Biophys 1997;345:171–4.
[31] Locker J, Reddy TV, Lombardi B. DNA methylation and hepatocarcinogenesis in rats fed a choline-devoid diet. Carcinogenesis 1986;7:1309–12.
[32] Tsujiuchi T, Tsutsumi M, Sasaki Y, Takahama M, Konishi Y. Hypomethylation of CpG sites and c-myc gene overexpression in hepatocellular carcinomas, but not hyperplastic nodules, induced by a choline-deficient L-amino acid-defined diet in rats. Jpn J Cancer Res 1999;90:909–13.
[33] Holliday R, Grigg GW. DNA methylation and mutation. Mutat Res 1993;285:61–7.
[34] Jaenisch R. DNA methylation and imprinting: why bother? Trends Genet 1997;13:323–9.
[35] Robertson KD, Jones PA. Dynamic interrelationships between DNA replication, methylation, and repair. Am J Hum Genet 1997;61:1220–4.
[36] Zingg JM, Jones PA. Genetic and epigenetic aspects of DNA methylation on genome expression, evolution, mutation and carcinogenesis. Carcinogenesis 1997;18:869–82.
[37] Jones PA, Gonzalgo ML. Altered DNA methylation and genome instability: a new pathway to cancer? Proc Natl Acad Sci U S A 1997;94:2103–5.
[38] Niculescu MD, Craciunescu CN, Zeisel SH. Dietary choline deficiency alters global and gene-specific DNA methylation in the developing hippocampus of mouse fetal brains. FASEB J 2006;20:43–9.
[39] Niculescu MD, Craciunescu CN, Zeisel SH. Gene expression profiling of choline-deprived neural precursor cells isolated from mouse brain. Brain Res Mol Brain Res 2005;134:309–22.
[40] Niculescu MD, Yamamuro Y, Zeisel SH. Choline availability modulates human neuroblastoma cell proliferation and alters the methylation of the promoter region of the cyclin-dependent kinase inhibitor 3 gene. J Neurochem 2004;89:1252–9.
[41] Niculescu MD, Zeisel SH. Diet, methyl donors and DNA methylation: interactions between dietary folate, methionine and choline. J Nutr 2002;132:2333S–5S.
[42] Mehedint MG, Niculescu MD, Craciunescu CN, Zeisel SH. Choline deficiency alters global histone methylation and epigenetic marking at the Re1 site of the calbindin 1 gene. FASEB J 2010;24:184–95.
[43] da Costa KA, Niculescu MD, Craciunescu CN, Fischer LM, Zeisel SH. Choline deficiency increases lymphocyte apoptosis and DNA damage in humans. Am J Clin Nutr 2006;84:88–94.
[44] Vafai SB, Stock JB. Protein phosphatase 2A methylation: a link between elevated plasma homocysteine and Alzheimer's Disease. FEBS Lett 2002;518:1–4.
[45] Pajares MA, Perez-Sala D. Betaine homocysteine S-methyltransferase: just a regulator of homocysteine metabolism? Cell Mol Life Sci 2006;63:2792–803.
[46] Zeisel SH. Choline: critical role during fetal development and dietary requirements in adults. Annu Rev Nutr 2006;26:229–50.
[47] Chen TH, Murata N. Glycinebetaine: an effective protectant against abiotic stress in plants. Trends Plant Sci 2008;13:499–505.
[48] Graham JE, Wilkinson BJ. Staphylococcus aureus osmoregulation: roles for choline, glycine betaine, proline, and taurine. J Bacteriol 1992;174:2711–6.

[49] Kunin CM, Rudy J. Effect of NaCl-induced osmotic stress on intracellular concentrations of glycine betaine and potassium in Escherichia coli, Enterococcus faecalis, and staphylococci. J Lab Clin Med 1991;118:217–24.

[50] Lucht JM, Bremer E. Adaptation of Escherichia coli to high osmolarity environments: osmoregulation of the high-affinity glycine betaine transport system proU. FEMS Microbiol Rev 1994;14:3–20.

[51] Peddie BA, Lever M, Randall K, Chambers ST. Osmoprotective activity, urea protection, and accumulation of hydrophilic betaines in Escherichia coli and Staphylococcus aureus. Antonie Van Leeuwenhoek 1999;75:183–9.

[52] Wemekamp-Kamphuis HH, Wouters JA, Sleator RD, Gahan CG, Hill C, Abee T. Multiple deletions of the osmolyte transporters BetL, Gbu, and OpuC of Listeria monocytogenes affect virulence and growth at high osmolarity. Appl Environ Microbiol 2002;68:4710–6.

[53] Price CT, Bukka A, Cynamon M, Graham JE. Glycine betaine uptake by the ProXVWZ ABC transporter contributes to the ability of Mycobacterium tuberculosis to initiate growth in human macrophages. J Bacteriol 2008;190:3955–61.

[54] Ozdemir M, Arslan F, Arslan H. An amperometric biosensor for choline determination prepared from choline oxidase immobilized in polypyrrole-polyvinylsulfonate film. Artif Cells Blood Substit Immobil Biotechnol 2012.

[55] Chen TH, Murata N. Glycinebetaine protects plants against abiotic stress: mechanisms and biotechnological applications. Plant Cell Environ 2011;34:1–20.

[56] Khan A, Ab Ghani S. Multienzyme microbiosensor based on electropolymerized o-phenylenediamine for simultaneous in vitro determination of acetylcholine and choline. Biosens Bioelectron 2012;31:433–8.

[57] Ma XJ, Wang Z, Ryan PD, et al. A two-gene expression ratio predicts clinical outcome in breast cancer patients treated with tamoxifen. Cancer Cell 2004;5:607–16.

[58] Wang Z, Dahiya S, Provencher H, et al. The prognostic biomarkers HOXB13, IL17BR, and CHDH are regulated by estrogen in breast cancer. Clin Cancer Res 2007;13:6327–34.

[59] Glunde K, Bhujwalla ZM, Ronen SM. Choline metabolism in malignant transformation. Nat Rev Cancer 2011;11:835–48.

[60] Glunde K, Jiang L, Moestue SA, Gribbestad IS. MRS and MRSI guidance in molecular medicine: targeting and monitoring of choline and glucose metabolism in cancer. NMR Biomed 2011;24:673–90.

[61] Daly PF, Lyon RC, Faustino PJ, Cohen JS. Phospholipid metabolism in cancer cells monitored by 31P NMR spectroscopy. J Biol Chem 1987;262:14875–8.

[62] Aboagye EO, Bhujwalla ZM. Malignant transformation alters membrane choline phospholipid metabolism of human mammary epithelial cells. Cancer Res 1999;59:80–4.

[63] Hacker HJ, Steinberg P, Bannasch P. Pyruvate kinase isoenzyme shift from L-type to M2-type is a late event in hepatocarcinogenesis induced in rats by a choline-deficient/DL-ethionine-supplemented diet. Carcinogenesis 1998;19:99–107.

[64] da Costa KA, Kozyreva OG, Song J, Galanko JA, Fischer LM, Zeisel SH. Common genetic polymorphisms affect the human requirement for the nutrient choline. FASEB J 2006;20: 1336–44.

[65] Xu X, Gammon MD, Zeisel SH, et al. Choline metabolism and risk of breast cancer in a population-based study. FASEB J 2008;22:2045–52.

[66] Johnson AR, Lao S, Wang T, Galanko JA, Zeisel SH. Choline Dehydrogenase Polymorphism rs12676 Is a Functional Variation and Is Associated with Changes in Human Sperm Cell Function. PLoS One 2012;7:e36047.

[67] Johnson AR, Craciunescu CN, Guo Z, et al. Deletion of murine choline dehydrogenase results in diminished sperm motility. FASEB J 2010;24:2752–61.

[68] Quaye O, Lountos GT, Fan F, Orville AM, Gadda G. Role of Glu312 in binding and positioning of the substrate for the hydride transfer reaction in choline oxidase. Biochemistry 2008;47:243–56.

[69] Fan F, Ghanem M, Gadda G. Cloning, sequence analysis, and purification of choline oxidase from Arthrobacter globiformis: a bacterial enzyme involved in osmotic stress tolerance. Arch Biochem Biophys 2004;421:149–58.

[70] Fraaije MW, Mattevi A. Flavoenzymes: diverse catalysts with recurrent features. Trends Biochem Sci 2000;25:126–32.

[71] Xin Y, Gadda G, Hamelberg D. The cluster of hydrophobic residues controls the entrance to the active site of choline oxidase. Biochemistry 2009;48:9599–605.

[72] Orville AM, Lountos GT, Finnegan S, Gadda G, Prabhakar R. Crystallographic, spectroscopic, and computational analysis of a flavin C4a-oxygen adduct in choline oxidase. Biochemistry 2009;48:720–8.

[73] Sucharitakul J, Prongjit M, Haltrich D, Chaiyen P. Detection of a C4a-hydroperoxyflavin intermediate in the reaction of a flavoprotein oxidase. Biochemistry 2008;47:8485–90.

[74] Finnegan S, Yuan H, Wang YF, Orville AM, Weber IT, Gadda G. Structural and kinetic studies on the Ser101Ala variant of choline oxidase: catalysis by compromise. Arch Biochem Biophys 2010;501:207–13.

[75] Baron R, Riley C, Chenprakhon P, et al. Multiple pathways guide oxygen diffusion into flavoenzyme active sites. Proc Natl Acad Sci U S A 2009;106:10603–8.

[76] Finnegan S, Agniswamy J, Weber IT, Gadda G. Role of valine 464 in the flavin oxidation reaction catalyzed by choline oxidase. Biochemistry 2010;49:2952–61.

[77] Ghanem M, Fan F, Francis K, Gadda G. Spectroscopic and kinetic properties of recombinant choline oxidase from Arthrobacter globiformis. Biochemistry 2003;42:15179–88.

[78] Ghanem M, Gadda G. Effects of reversing the protein positive charge in the proximity of the flavin N(1) locus of choline oxidase. Biochemistry 2006;45:3437–47.

[79] Hoang JV, Gadda G. Trapping choline oxidase in a nonfunctional conformation by freezing at low pH. Proteins 2007;66:611–20.

[80] Finnegan S, Gadda G. Substitution of an active site valine uncovers a kinetically slow equilibrium between competent and incompetent forms of choline oxidase. Biochemistry 2008;47:13850–61.

[81] Ghanem M, Gadda G. On the catalytic role of the conserved active site residue His466 of choline oxidase. Biochemistry 2005;44:893–904.

[82] Heuts DP, Scrutton NS, McIntire WS, Fraaije MW. What's in a covalent bond? On the role and formation of covalently bound flavin cofactors. FEBS J 2009;276:3405–27.

[83] Gadda G, Fan F, Hoang JV. On the contribution of the positively charged headgroup of choline to substrate binding and catalysis in the reaction catalyzed by choline oxidase. Arch Biochem Biophys 2006;451:182–7.

[84] Gadda G, Powell NL, Menon P. The trimethylammonium headgroup of choline is a major determinant for substrate binding and specificity in choline oxidase. Arch Biochem Biophys 2004;430:264–73.

[85] Fan F, Germann MW, Gadda G. Mechanistic studies of choline oxidase with betaine aldehyde and its isosteric analogue 3,3-dimethylbutyraldehyde. Biochemistry 2006;45:1979–86.

[86] Gadda G. pH and deuterium kinetic isotope effects studies on the oxidation of choline to betaine-aldehyde catalyzed by choline oxidase. Biochim Biophys Acta 2003;1650:4–9.

[87] Quaye O, Cowins S, Gadda G. Contribution of flavin covalent linkage with histidine 99 to the reaction catalyzed by choline oxidase. J Biol Chem 2009;284:16990–7.

[88] Rungsrisuriyachai K, Gadda G. On the role of histidine 351 in the reaction of alcohol oxidation catalyzed by choline oxidase. Biochemistry 2008;47:6762–9.

[89] Rungsrisuriyachai K, Gadda G. Role of asparagine 510 in the relative timing of substrate bond cleavages in the reaction catalyzed by choline oxidase. Biochemistry 2010;49:2483–90.

[90] Fan F, Gadda G. An internal equilibrium preorganizes the enzyme-substrate complex for hydride tunneling in choline oxidase. Biochemistry 2007;46:6402–8.

[91] Fan F, Gadda G. Oxygen- and temperature-dependent kinetic isotope effects in choline oxidase: correlating reversible hydride transfer with environmentally enhanced tunneling. J Am Chem Soc 2005;127:17954–61.

[92] Quaye O, Gadda G. Effect of a conservative mutation of an active site residue involved in substrate binding on the hydride tunneling reaction catalyzed by choline oxidase. Arch Biochem Biophys 2009;489:10–4.
[93] Yuan H, Gadda G. Importance of a serine proximal to the C(4a) and N(5) flavin atoms for hydride transfer in choline oxidase. Biochemistry 2011;50:770–9.
[94] Quaye O, Nguyen T, Gannavaram S, Pennati A, Gadda G. Rescuing of the hydride transfer reaction in the Glu312Asp variant of choline oxidase by a substrate analogue. Arch Biochem Biophys 2010;499:1–5.
[95] Hernandez-Ortega A, Borrelli K, Ferreira P, Medina M, Martinez AT, Guallar V. Substrate diffusion and oxidation in GMC oxidoreductases: an experimental and computational study on fungal aryl-alcohol oxidase. Biochem J 2011;436:341–50.
[96] Wongnate T, Sucharitakul J, Chaiyen P. Identification of a catalytic base for sugar oxidation in the pyranose 2-oxidase reaction. Chembiochem 2011;12:2577–86.
[97] Canovas D, Vargas C, Kneip S, et al. Genes for the synthesis of the osmoprotectant glycine betaine from choline in the moderately halophilic bacterium Halomonas elongata DSM 3043, USA. Microbiology 2000;146 (Pt 2):455–63.
[98] Yilmaz JL, Bulow L. Enhanced stress tolerance in Escherichia coli and Nicotiana tabacum expressing a betaine aldehyde dehydrogenase/choline dehydrogenase fusion protein. Biotechnol Prog 2002;18:1176–82.
[99] Rendina G, Singer TP. Solubilization and assay of choline dehydrogenase. Biochim Biophys Acta 1958;30:441–2.
[100] Gomez-Moreno C, Choy M, Edmondson DE. Purification and properties of the bacterial flavoprotein: thiamin dehydrogenase. J Biol Chem 1979;254:7630–5.
[101] Kenney WC, Edmondson DE, Seng RL. Identification of the covalently bound flavin of thiamin dehydrogenase. J Biol Chem 1976;251:5386–90.
[102] Gomez-Moreno C, Edmondson DE. Evidence for an aldehyde intermediate in the catalytic mechanism of thiamine oxidase. Arch Biochem Biophys 1985;239:46–52.
[103] Kenney WC, Edmondson DE, Singer TP. A novel form of covalently bound flavin from thiamine dehydrogenase. Biochem Biophys Res Commun 1974;57:106–11.
[104] Edmondson DE, Kenney WC. Identification and properties of 8alpha-(N(1)-histidyl)-riboflavin: the flavin component of thiamine dehydrogenase and beta-cyclopiazonate oxidocyclase. Biochem Biophys Res Commun 1976;68:242–8.

9 Pyranose oxidases

Thanyaporn Wongnate and Pimchai Chaiyen

Abstract

Pyranose oxidases catalyze the oxidation of aldopyranoses by using molecular oxygen as an electron acceptor to yield the corresponding keto-aldoses and hydrogen peroxide. Typical enzymes of this type include glucose 1-oxidase (GO) and pyranose 2-oxidase (P2O). These enzymes belong to the glucose-methanol-choline (GMC) oxidase superfamily and contain flavin adenine dinucleotide (FAD) as a cofactor. GO catalyzes regiospecific oxidation at the C1 position whereas P2O oxidizes pyranoses at the C2 position. Site-directed mutagenesis, kinetic isotope effects, pH-dependent and transient kinetics studies have shown that His548 in P2O is a catalytic base which abstracts D-glucose C2-OH to form an alkoxide intermediate prior to a hydride transfer from D-glucose to flavin. P2O is the first flavoprotein oxidase for which C4a-hydroperoxyflavin has been detected as an intermediate during the oxidative half-reaction. Kinetic isotope effect studies on the oxidative half-reaction of P2O clearly indicate that H_2O_2 elimination from the C4a-hydroperoxyflavin is controlled by a single proton transfer from the flavin N(5) to the peroxide leaving group. Currently, it is not clear whether a C4a-hydroperoxyflavin is also a common intermediate in other oxidases but it is not stable enough for kinetic detection, or the active site of P2O is an exception among oxidases for which the intermediate can be stabilized.

9.1 Introduction

Pyranose oxidases are flavoenzyme oxidases that catalyze the oxidation of aldopyranoses by using molecular oxygen as an electron acceptor to yield the corresponding keto-aldoses and hydrogen peroxide [1,2]. These enzymes contain flavin adenine dinucleotide (FAD) as a cofactor and belong to the glucose-methanol-choline (GMC) oxidase superfamily. Similar to other flavoenzyme oxidases, their reactions consist of two half-reactions. The first half-reaction is a reductive half-reaction in which a hydride equivalent from a pyranose substrate is transferred to FAD to generate the reduced flavin cofactor. The second half-reaction is an oxidative half-reaction, in which two electrons from the reduced FAD are transferred to O_2 to generate H_2O_2 and oxidized flavin. The most well understood pyranose oxidases are glucose 1-oxidase (GO) [1,3,4] and pyranose 2-oxidase (P2O) [5–8]. GO has long been studied and its reaction mechanisms have been investigated since the 1960s. The current chapter focuses on the mechanistic and structural knowledge of a more recently discovered enzyme, P2O. The reduction mechanism of GO and future prospects for studying both enzymes are also discussed.

9.2 Pyranose 2-oxidase (EC 1.13.10)

Pyranose 2-oxidase (P2O) catalyzes the oxidation of several aldopyranoses by O_2 at the C2 position to yield the corresponding 2-keto-aldoses and H_2O_2 [2,9] (▶Scheme 9.1).

The first part of the reaction is a reductive half-reaction in which two electrons are transferred as a hydride equivalent from a sugar substrate to the flavin cofactor to generate reduced FAD and the corresponding 2-keto-sugar (1) [6,7]. For the following oxidative half-reaction, two electrons are transferred from the reduced flavin to O_2 to form H_2O_2 (2) [7,10] (▶Scheme 9.2). P2O is unique among flavoenzyme oxidases for the formation of C4a-hydroperoxyflavin intermediate in the oxidative half-reaction [10] (discussed in further detail below). The reaction mechanism of P2O is a ping-pong bi-bi mechanism similar to other flavoprotein oxidoreductases [7]. An exception to this was found in the T169A variant, in which a hydrogen bond between Thr169 and N5 of the FAD is absent. The kinetic mechanism of this T169A mutant involves the formation of a ternary complex [8].

9.2.1 Importance and applications

White-rot fungi produce a variety of oxidases including P2O [11]. The role of these oxidases is thought to be the generation of H_2O_2 as an essential co-substrate for the lignin degrading enzyme. P2O was identified from many fungal species including *Phanerochaete chrysosporium* [12], *Phelebiosis gigantean* [13], *Plaeurotus ostreatus* [14], *Polyporus obtusus* [15], and *Trametes multicolor* [2]. In some species of white rot fungi, P2O is involved in a secondary metabolic pathway that produces antibiotics such as β-pyrone-cortalcerone which helps protect the fungi from bacterial attack [16,17]. P2O converts D-glucose to 2-keto-D-glucose, which is an intermediate for the production of β-pyrone-cortalcerone.

P2O is a key biocatalyst in several biotechnological applications including the biotransformation of carbohydrates such as D-glucose and D-galactose to generate 2-keto sugars that can be further reduced at the C1 position to yield D-fructose and D-tagatose, respectively [18]. P2O has recently been proposed for use in biofuel cell applications [19]. Several variants, especially those at positions L537 and E542 have been shown to be improved thermostability compared to the wild-type enzyme [19].

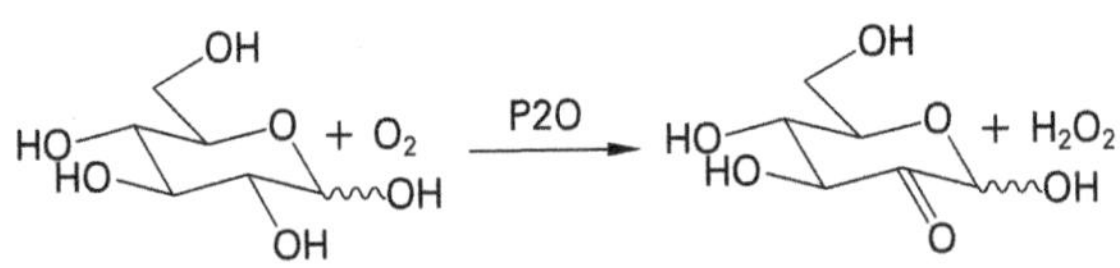

Scheme 9.1: Reaction of pyranose 2-oxidase using D-glucose as a substrate.

$$\text{E-FAD} + \text{D-glucose} \longrightarrow \text{E-FADH}_2 + \text{2-keto-D-glucose} \quad (1)$$

$$\text{E-FADH}_2 + O_2 \longrightarrow \text{E-FAD} + H_2O_2 \quad (2)$$

Scheme 9.2: Reductive (1) and oxidative (2) half-reactions of P2O

9.2.2 General biochemical and biophysical properties of P2O

P2O from *Trametes multicolor* was isolated, purified, and biochemically and biophysically characterized. This enzyme is a homotetrameric flavoprotein with a native molecular mass of 270 kDa [2]. Each subunit contains one FAD covalently linked via the C8 methyl group of FAD to the $N^{\varepsilon 2}$ of the His167 residue [9,20]. The flavin covalent linkage increases the flavin redox potential by ~50 mV [21] and the reduction rate constant by ~22-fold [6]. Interactions around the flavin N(1)/N(5) positions with positively charged or H-bonding residues are required for the biosynthesis of the flavin covalent linkage because mutations which perturb interactions around the positions N(1) and N(5) such as substitution mutations of His548 and Thr169 resulted in incomplete or no flavinylation [5,22] (►Tab. 9.1).

The formation of a flavin-sulfite adduct is a typical characteristic found in flavoenzyme oxidases including P2O. Addition of sodium sulfite at various concentrations to solutions of P2O (WT), H167A, T169A, H167A/H548N, H167A/H548R and N593H resulted in the formation of the flavin N(5)-sulfite adduct [5,22]. The formation of an adduct implies that the negative charge developed at the flavin N(1) and N(5) loci in the P2O active site can be accommodated and stabilized by the residues in the vicinity, similar to other flavoprotein oxidases. This conclusion is confirmed by the aberrant flavin-sulfite adduct formed in the H167A/H548A mutant in which the interactions from both residues His167 and His548 were altered. The K_d value of the flavin N(5)-sulfite adduct complex of H167A/H548N (K_d >2 M) is >4000-fold higher than the value of the wild-type enzyme (K_d = 0.51 mM) and >50-fold higher than that of the H167A mutant (K_d = 37.5 mM) [22], implying that Asn cannot substitute His548 in this function. In contrast, the K_d value of the flavin N(5)-sulfite adduct complex of the H167A/H548R mutant is 0.98 mM, which is only ~2-fold higher than the value for the wild-type enzyme (K_d = 0.51 mM) [5]. For the T169A mutant in which the H-bond between the flavin N5 to Thr169 residue was removed, the K_d value for the formation of the sulfite adduct is 181 mM which is ~355-fold higher than the value of the wild-type enzyme [22]. These results suggest that the stabilizing interactions of His548 and Thr169 with the flavin-sulfite adduct are key for the formation of the adduct in P2O. A summary of the K_d values of the flavin-sulfite adduct formation is summarized in ►Tab. 9.2.

Tab. 9.1: Percentage of covalently linked FAD found in P2O wild-type and mutants [5,22]

Enzyme	Covalently linked (%)	Enzyme	Covalently linked (%)
Wild-type	98 ± 2	T169A	17 ± 3
T169S	95 ± 3	H548A	63 ± 3
T169N	95 ± 2	H548N	42 ± 4
T169G	91 ± 4	H548R	85 ± 3
N593H	97 ± 1	H548S	49 ± 2
H167A	9 ± 2	H548D	26 ± 1

Tab. 9.2: Summary of K_d values for the P2O-sulfite complexes and the reduction potential values of wild-type and mutant P2O [5,8,21,22]

Enzyme	K_d of P2O-sulfite adducts (mM)	E_m^o(mV)
Wild-type	0.51	–105 ± 1
H167A	37.5	–150 ± 2
T169A	181	–197 ± 3
H167A/H548A	ND	–190 ± 2
H167A/H548N	>2000	–217 ± 2
H167A/H548R	0.98	+17 ± 3

ND, not determined. The K_d value could not be determined because the flavin N(5)-sulfite adduct did not form.

The reduction potential values (E_m^o) of WT and various mutants of P2O were measured (►Tab. 9.2). The E_m^o of the H167A/H548A, and H167A/H548N mutants are in the range of ~ –200 mV, which is ~100 mV lower than that of the wild type and ~50 mV lower than that of the H167A mutant [5,21] (►Tab. 9.2). The decrease in redox potential values in these mutants is due to the absence of the flavin covalent linkage and the absence of a positive charge provided by the His548 residue in the vicinity of the flavin N(1). For the H167A/H548R mutant, the midpoint potential value is rather high (+17 mV) [5] (►Tab. 9.2). This result emphasizes the importance of having a positive charge close to the flavin N1/N5 locus in order to maintain high reduction potential values.

9.2.3 Structural studies on P2O

Crystal structures of P2O from *T. multicolor* indicate that the overall structure of P2O is similar to other enzymes in the GMC oxidoreductase superfamily [9,21]. Other members of the GMC family include glucose 1-oxidase from *Aspergillus niger* [23], cholesterol oxidase from *Streptomyces* [24], and choline oxidase from *Arthrobacter globiformis* [25]. These enzymes also contain a conserved His-Asn pair, except for glucose 1-oxidase, in which the asparagine residue is replaced by histidine (►Fig. 9.1) [3]. According to the crystal structure of P2O, His548 and Asn593 are the conserved residues positioned near the isoalloxazine ring and sugar-binding sites of the enzyme [9,21].

P2O structures in several conformations have been reported and the data reveal heterogeneity in the conformations *i.e.* the open, semi-open and closed conformations. The former two conformations were proposed to be relevant for flavin reduction while the closed conformation was thought to be relevant for flavin oxidation [9,21,26,27]. The closed conformation refers to a structure in which a flexible loop (residues 452–457) closes off the active site and is found in the crystal structure of P2O from *Trametes multicolor* in complex with acetate (a competitive inhibitor) solved at 1.8 Å resolution (►Fig. 9.2A) [9]. In this structure, the carboxylate carbon atom of acetate is positioned 3.3 Å below the flavin N(5) atom and the O^{γ} of Thr169 forms a H-bond with the flavin N(5). The open conformation refers to the conformation in which the flexible

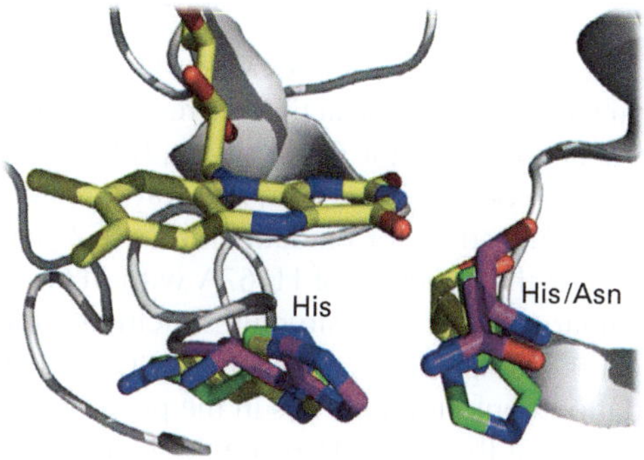

Fig. 9.1: Comparison of the active site of P2O with related GMC enzymes. Active sites of P2O and other enzymes in the GMC family were superimposed with respect to the flavin ring. The figure shows the alignment of the conserved His-Asn residues of P2O (yellow; PDB code 1TTO), Cellobiose dehydrogenase (magenta; PDB code 1KDG), and the His-His pair of GO (green; PDB code 1GAL).

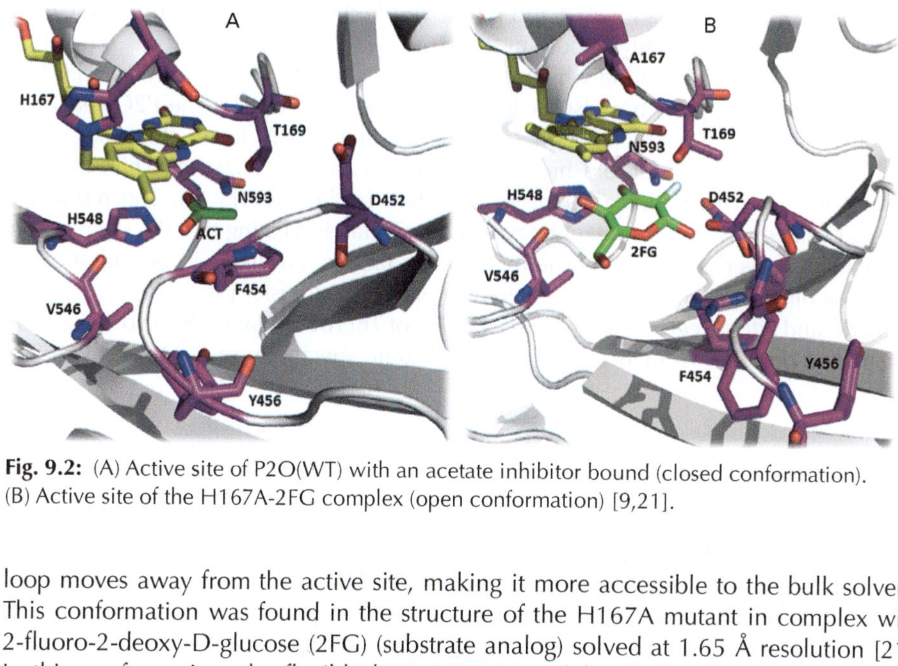

Fig. 9.2: (A) Active site of P2O(WT) with an acetate inhibitor bound (closed conformation). (B) Active site of the H167A-2FG complex (open conformation) [9,21].

loop moves away from the active site, making it more accessible to the bulk solvent. This conformation was found in the structure of the H167A mutant in complex with 2-fluoro-2-deoxy-D-glucose (2FG) (substrate analog) solved at 1.65 Å resolution [21]. In this conformation, the flexible loop is positioned far away from the active site in order to accommodate the binding of a monosaccharide, which would otherwise not be possible in the closed conformation (▸Fig. 9.2B). Recently, the crystal structure of H167A in complex with 3-fluoro-3-deoxy-D-glucose (3FG) was solved, which demonstrated that the flexible loop assumes a position half way between the open and closed conformations [27]. This structure is referred to the semi-open conformation and is more optimal for interactions with the sugar substrate (see below). These data demonstrate the heterogeneity in the local environment of the flavin, and was recently confirmed by fluorescence life-time measurements [28,29].

9.2.4 Substrate recognition

The recently reported crystal structure of H167A in complex with 3FG at 1.35 Å resolution showed interactions of the sugar substrate with the active site residues which may be key for the regio-specific oxidation of D-glucose at the C2-position [27]. In the previously reported structure [21], the C2-position of 2FG is oriented far away (greater than 4 Å) from His548 and flavin N(5), which is not ideal for oxidation at the C2 position, but favors oxidation at C3 (▸Fig. 9.2B). For the structure of H167A with 3FG, the overall structure of the complex is very similar to other P2O structures including the H167A-2FG complex; however bound sugar is oriented toward oxidation at C2. A major difference between the C2 and C3 oxidation conformations lies in the position of a flexible loop (residues 451–461) that forms part of the substrate recognition site (▸Fig. 9.3). In the H167A-2FG complex (C3-oxidation), the substrate recognition loop swings further away from the active site to assume the open conformation while in the C2 oxidation complex (H167A-3FG), the gating segment (residues 454–456) swings inward towards the active site to provide interactions for substrate binding *i.e.* Asp452 forms H-bonds with glucose O4 while Tyr456 forms H-bonds with glucose O6 (green; ▸Fig. 9.3) [27]. Mutagenesis of the residues Phe454 and Tyr456 resulted in inactive enzymes, indicating that this region is functionally important for P2O [27].

Asn593, a conserved residue which is located close to the D-glucose C2 and C3 in the H167A-3FG complex, has also been shown to be important for the P2O reaction, possibly by providing interactions for substrate binding. The K_d for D-glucose binding to the N593H mutant is 19-fold higher than that for the WT enzyme, while the rate constant for flavin reduction decreased ~114-fold in the N593H [5]. The reduction potential values of WT and mutant enzymes are similar, indicating that the decrease in the catalytic power is not due to thermodynamic factors, but likely to perturbation of D-glucose binding which affects the reactivity of hydride transfer [5].

Kinetic and site-directed mutagenesis studies of Thr169 mutants [8] also show that mutation at position 169 can broaden the substrate specificity of P2O such that the

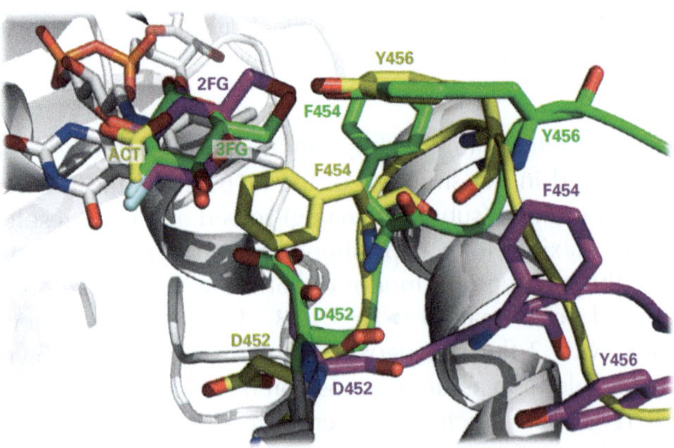

Fig. 9.3: Overlay of the three states of the substrate loop: closed (yellow; PDB 1TTO) [9], semi-open $C2_{ox}$ (green; PDB 3PL8) [27] and open $C3_{ox}$ (magenta; PDB 2IGO) [21]. The key residues in the substrate recognition loop Asp452, Phe454 and Tyr456 are labeled and their side chains are shown.

enzyme can efficiently oxidize D-galactose. For T169S, T169N, and T169G, the overall turnover with D-galactose is faster than that of WT enzyme with D-galactose due to the increase of k_{red} [8]. P2O binds D-galactose in a one-step binding process, which is different from the binding of D-glucose. Interestingly, the oxidation of D-glucose is not significantly affected from changing Thr169 to Asn or Ser. D-glucose binds to WT, T169S, and T169N with the same K_d (45–47 mM), and the hydride transfer rate constants (k_{red}) of these enzymes are similar (15.3–9.7 s^{-1} at 4 °C). In contrast, the k_{red} of T169G with D-glucose (0.7 s^{-1}, 4 °C) is significantly less than that of wild type enzyme, and the k_{red} of D-glucose oxidation is severely affected for T169A (k_{red} of 0.03 s^{-1} at 25 °C). Based on the crystal structures, H-bonding to the flavin N(5) in the T169A mutant is absent, and this interaction is proposed to be important for efficient flavin reduction in P2O [8].

9.2.5 Flavin reduction (sugar oxidation) mechanism

Using transient kinetics and isotope effects, considerable progress towards the understanding of flavin reduction in P2O has been achieved. Investigation into the mechanism of the reductive half-reaction of P2O (WT) using D-glucose and 2-*d*-D-glucose at pH 7.0 indicates that D-glucose binds to the enzyme in a 2-step process (►Scheme 9.3). The first step is the formation of an initial complex of the enzyme and substrate (E-Fl_{ox}:G*) without any change in absorption, while the second step is an isomerization of the enzyme-substrate complex to form an active Michaelis complex (E-Fl_{ox}:G) which can be detected by an increase in absorbance at 395 nm [7]. The isomerization step showed an inverse isotope effect (k_{2H}/k_{2D} of ~0.60), implying that the C2-H bond of D-glucose is more rigid in the E-Fl_{ox}:G complex than in the free form [7]. When the covalent linkage between His167 and FAD was removed in the H167A mutant, multiple steps were observed for D-glucose binding [6]. The multi-step binding process of D-glucose may be related to the presence of multiple sugar binding modes such as those observed in the crystal structures of H167A-2FG (E/S-$C3_{ox}$, open conformation) [21] and the H167A-3FG (E/S-$C2_{ox}$, semi-open conformation) complexes [27]. D-glucose may initially bind to the open conformation. The subsequent isomerization step may involve rearrangement of the initial complex into a semi-open E/S- $C2_{ox}$ state which is more suitable for D-glucose oxidation at the C2-position.

Pre-steady-state kinetic studies using 2-*d*-D-glucose as a substrate indicated a large primary kinetic isotope effect (k_{3H}/k_{3D} = 8.84) on the flavin reduction step [7], while no additional kinetic isotope effect was observed when all deuterated D-glucose (1,2,3,4,5,6,6-d_7-D-glucose) was used in the reactions of P2O(H167A) enzyme [6]. These results clearly indicate that P2O regio-specifically oxidizes the C-H moiety of D-glucose at the C2 position. Pre-steady-state and steady-state kinetics have also shown that flavin reduction is the rate-limiting step in the overall reactions of P2O (WT) and the H167A mutant [6,7].

Rapid equilibrium

$$\text{E-Fl}_{ox} + \text{G} \underset{K_d}{\rightleftharpoons} \text{E-Fl}_{ox}\text{:G*} \underset{k_{-2}}{\overset{k_2}{\rightleftharpoons}} \text{E-Fl}_{ox}\text{:G} \overset{k_3}{\rightleftharpoons} \text{E-Fl}_{red}\text{:P} \overset{k_4}{\rightleftharpoons} \text{E-Fl}_{red} + \text{G}$$

Scheme 9.3: Kinetic mechanism of the reductive half-reaction of P2O [6,7]

9.2.6 Catalytic base for sugar oxidation in the P2O reaction

Studies of solvent kinetic isotope effects on transient kinetics of P2O (WT) and the H167A mutant showed no solvent kinetic isotope effect on the flavin reduction step, suggesting that the deprotonation of the C2-OH bond is not immediately followed by the hydride transfer from the C2-H bond [6]. If the C2-OH deprotonation occurs simultaneously with the hydride transfer, the reaction performed in D_2O should have had a lower rate constant due to the deprotonation of C2-OD instead of C2-OH. The data showing no solvent kinetic isotope effect indicated that a D-glucose alkoxide intermediate exists prior to the transfer of a hydride equivalent from the C2-H group to the flavin N(5). Therefore, a catalytic base to facilitate the formation of D-glucose alkoxide must be present.

The use of site-directed mutagenesis to study the reductive half-reaction of P2O at various pH values suggest that the conserved His548 is a catalytic base that is important for the deprotonation of D-glucose in P2O [5]. Mutations of His548 to various amino acid such as in H548A, H548N, H548S, H548D, H167A/H548A, H167A/H548N, H167A/H548S and H167A/H548D resulted in inactive enzymes at all pH values. Only the mutation to Arg in the H167A/H548R mutant resulted in an active enzyme. (Note that the double mutation with H167A in the H167A/H548 mutants was carried out to obtain homogeneous non-covalently linked FAD because the single mutation at His548 resulted in mixtures of non-covalently linked and covalently linked FAD (see previous discussion). The observed rate constant for flavin reduction of H167A/H548R (0.018 s^{-1}) at pH 7.0 is ~2670-fold lower than that for WT (48 s^{-1}) and ~220-fold lower than that for H167A (3.97 s^{-1}). A key piece of evidence for the mechanism of sugar oxidation by P2O comes from the finding that the reduction rate constant of H167A/H548R increases ~360-fold upon increasing the pH (▶Fig. 9.4, ▶Tab. 9.3) and

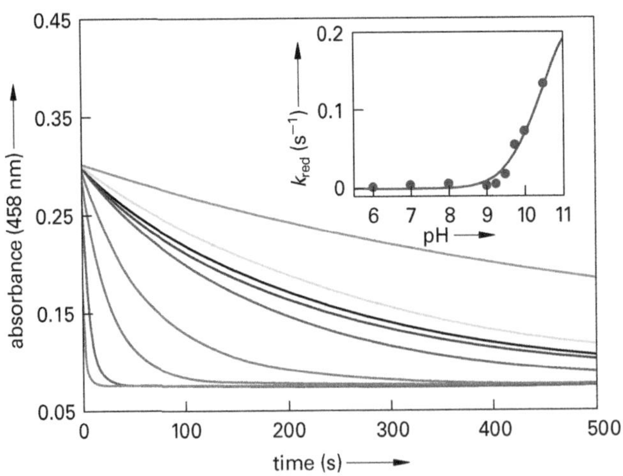

Fig. 9.4: Reactions of the oxidized H167A/H548R at various pH values. The flavin reduction of the H167A/H548R mutant by D-glucose is faster when the pH is increased (traces from left to right [5]). The inset shows a plot of rate constants for the flavin reduction *versus* pH, consistent with a curve associated with a pK_a value >10.1 [5].

Tab. 9.3: Summary of the reduction rate constants for the reactions of oxidized H167A/H548R mutant with D-glucose at various pH values [5]

pH	Reduction rate constants (s^{-1}) by D-glucose at 25 °C
6.0	0.006 ± 0.001
7.0	0.018 ± 0.001
8.0	0.057 ± 0.004
9.5	0.63 ± 0.03
10.25	1.60 ± 0.06
10.5	2.13 ± 0.03

is consistent with a pK_a value of >10.1 (Inset in ▶Fig. 9.4) [5]. This pK_a value is in the range of Arg in aqueous solution (~12.4). These data suggest that Arg548 must be present in the unprotonated form in order to act as a good catalytic base, implying a similar scenario for His548 in the wild-type enzyme. Because the flavin reduction rate constants of P2O (WT) and the H167A mutant are pH-independent between pH 5.5–10.5, the data imply that the His548 residue in the wild-type enzyme must be unprotonated (neutral) throughout this pH range. Therefore, the flavin reduction mechanism of P2O is consistent with a mechanism in which the unprotonated form of His548 acts as a catalytic base to abstract a proton at the C2-OH of D-glucose to generate an alkoxide intermediate prior to the hydride transfer step (▶Scheme 9.4) [5]. The resulting protonated His548 may also contribute to stabilization of the alkoxide intermediate (▶Scheme 9.4).

9.2.7 Detection of a C4a-hydroperoxyflavin intermediate in the reaction of P2O

The most interesting catalytic feature of P2O is that the enzyme stabilizes the formation of C4a-hydroperoxyflavin as an intermediate during the oxidative half-reaction. C4a-hydroperoxyflavin is a common intermediate for flavin-dependent monooxygenases, but as of yet has only been detected in one flavoprotein oxidase, P2O, under natural turnover conditions [10]. The reduced P2O reacts with O_2 with a forward rate constant of 5.8×10^4 M^{-1} s^{-1} and a reverse rate constant of 2 s^{-1}, resulting in the formation of a C4a-hydroperoxyflavin intermediate which decays with a rate constant of 18 s^{-1} (▶Scheme 9.5) [10]. C4a-hydroperoxyflavin also forms during the oxidative half-reaction of the H167A mutant with similar rate constants as those for the WT enzyme, implying that the covalent linkage is not an important factor for the formation of the intermediate [6].

The ability of P2O to stabilize C4a-hydroperoxyflavin has been proposed to be due to the local environment around the flavin C4a/N5 area [8,10]. A hydrophobic cavity formed at the *re*-face of the flavin ring in the closed conformation may help in protecting the intermediate from external solvent. The position and geometry of Thr169 in relation to the flavin N5/O4 locus is also required for stabilization of the C4a-hydroperoxyflavin because mutation of Thr169 to other residues (Ala, Ser,

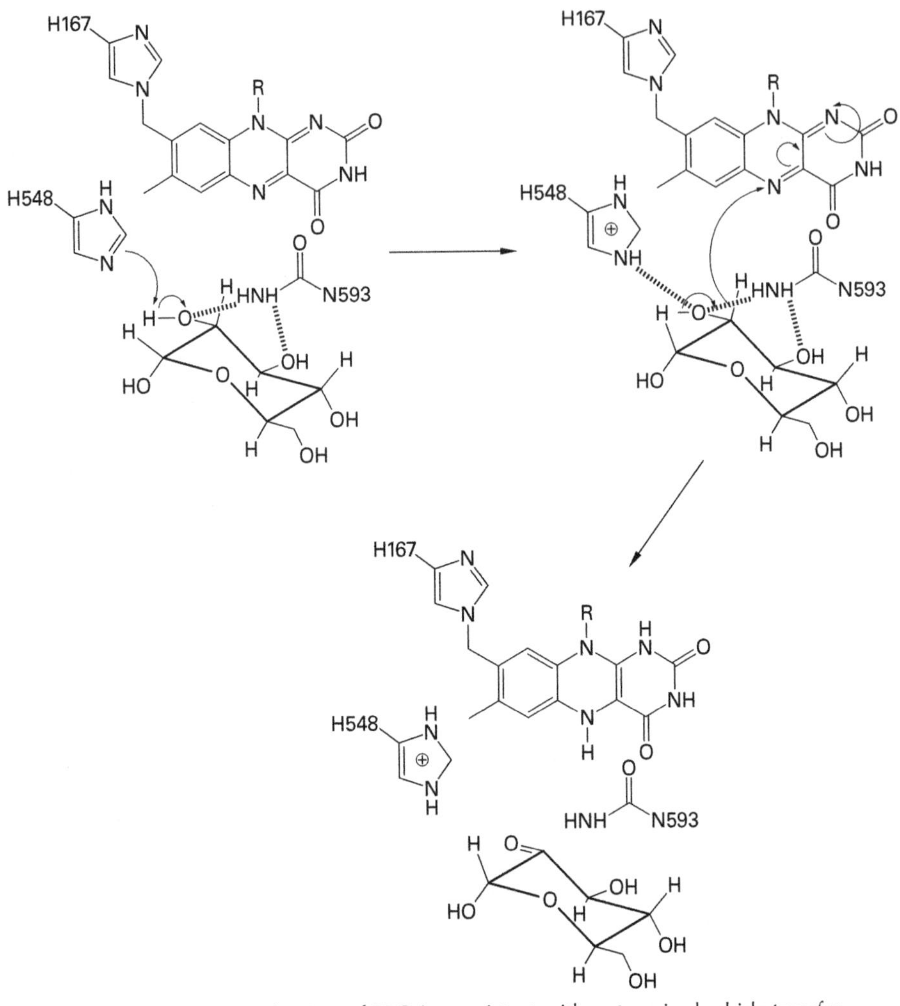

Scheme 9.4: Reaction mechanism of P2O is consistent with a stepwise hydride transfer mechanism [5].

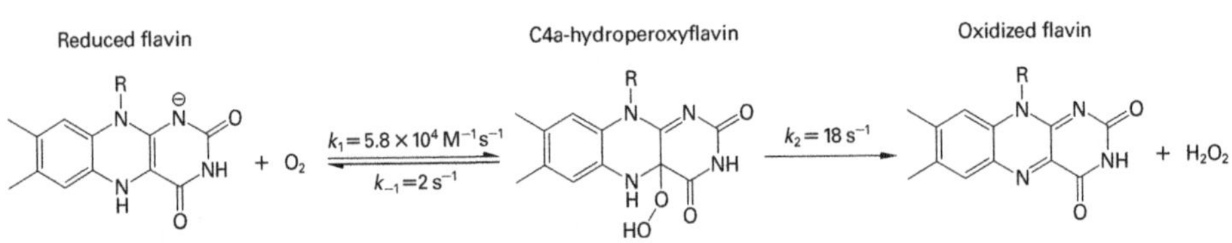

Scheme 9.5: The reaction of reduced P2O with oxygen [10].

Asn, Gly) results in failure of the enzyme to stabilize the intermediate [8]. Currently, we speculate that H-bonding at the flavin N5 is important for stabilization of the intermediate (see below).

9.2.8 The mechanism of H_2O_2 elimination from C4a-hydroperoxyflavin

As a unique oxidase that stabilizes C4a-hydroperoxyflavin, the P2O enzyme was used as a model to investigate the mechanism of H_2O_2 elimination from C4a-hydroperoxyflavin. Recent investigations using transient kinetics and kinetic isotope effects have revealed that breakage of the flavin N(5) H-bond controls the overall process of H_2O_2 elimination from C4a-hydroperoxyflavin [30]. The results (►Tab. 9.4) show that D_2O had negligible effects on the bimolecular rate constant for C4a-hydroperoxyflavin formation (k_1 in ►Scheme 9.5) and the reverse step (k_{-1} in ►Scheme 9.5). The currently accepted mechanism of the reaction between reduced flavin and oxygen predicts that the first step of the reaction involves a one-electron transfer to form a radical pair of flavin semiquinone and superoxide radical, which rapidly collapses to form C4a-peroxyflavin [31]. While for P2O, a net transfer of one proton is required for the subsequent formation of C4a-hydroperoxyflavin, the results do not identify any solvent kinetic isotope effect (SKIE) on this step. These data indicate that the protonation process to form C4a-hydroperoxyflavin is rapid and is not the rate-limiting step for the formation of the intermediate. In contrast, the H_2O_2 elimination step from the intermediate (k_2 in ►Scheme 9.5) was shown to have a SKIE of 2.8 (►Tab. 9.4). The proton inventory for this step shows a linear relationship, suggesting that a single proton transfer process causes the SKIE for the H_2O_2 elimination step [30].

The key finding that identified the factor responsible for controlling the H_2O_2 elimination process came from a creative specific labeling experiment. Taking advantage of the slow exchange at N(5) in the reduced oxidase, P2O was reduced by 2-*d*-D-glucose to generate the reduced enzyme specifically labeled at the flavin N(5)-position. Double-mixing and single-mixing stopped-flow experiments performed in H_2O buffer indicated that the N(5)-deuterated reduced flavin generates similar kinetic isotope effects as those found for the experiments performed with the enzyme pre-equilibrated in D_2O buffer. These results suggest that the N-H(D) cleavage at the flavin N(5) position is responsible for the solvent kinetic isotope effect and is the proton-in-flight that is transferred during the transition state. The mechanism of H_2O_2 elimination from C4a-hydroperoxyflavin is consistent with a single proton transfer from the flavin N(5) to the peroxide leaving group, possibly via the formation of an intra-molecular hydrogen bridge [30] (►Scheme 9.6). This finding is consistent with the previous site-directed mutagenesis studies in which perturbation of H-bonding at the flavin N(5) in Thr169 mutants of P2O resulted in no intermediate formation. Similar results were also observed for the mutation disrupting the H-bonding at the flavin N(5) in the oxygenase component of *p*-hydroxyphenylacetate hydroxylase [32].

Tab. 9.4: Effects of the solvent isotope on the kinetics of the P2O oxidative half-reaction at 4 °C, pH 7.0 [30]

Parameters	**H_2O buffer**	**D_2O buffer**
k_1 (M^{-1} s^{-1})	$6.5 \pm 0.3 \times 10^4$	$6.5 \pm 0.3 \times 10^4$
k_{-1} (s^{-1})	2 ± 0.1	2 ± 0.1
k_2 (s^{-1})	18 ± 1	6.4 ± 0.3

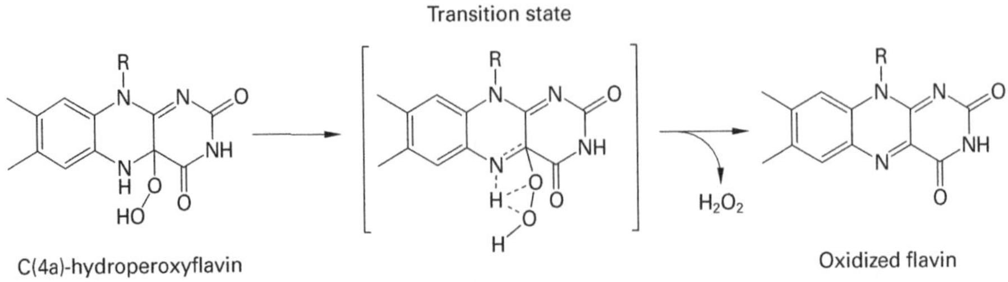

Scheme 9.6: Proposed reaction mechanism for the H_2O_2 elimination from C4a-hydroperoxyflavin in P2O. Dotted lines represent hydrogen bonds that may form between the flavin N(5) proton and the distal and proximal oxygen during the transition state [30].

9.3 Glucose 1-oxidase (EC. 1.1.3.4)

Glucose 1-oxidase (GO) catalyzes the oxidation of D-glucose specifically at the C1 position to yield glucono-lactone with the concomitant reduction of the enzyme bound flavin. The reduced enzyme is oxidized by O_2 to yield H_2O_2 and oxidized flavin (►Scheme 9.7).

9.3.1 Biochemical properties and application of GO

GO from *Aspergillus niger* and *Penicillium amagasakiene* have been investigated with regards to the biochemical, structural and mechanistic aspects [1,3,33]. GO from *Aspergillus niger* was cloned and expressed in yeast. This enzyme is a homodimeric FAD-bound enzyme and is glycosylated (both *N*- and *O*-glycosylated). The native molecular mass of the enzyme varies from 130 to 325 kDa depending on the degree of glycosylation [34]. Each subunit carries one molecule of a non-covalently bound FAD cofactor [3,23,35]. Removal of the glycosylated sugars from the enzyme has no effect on the enzyme activity or stability [36,37]. However, studies of kinetic isotope effects showed that less glycosylation in GO results in a greater degree of quantum mechanical tunneling in the reductive half-reaction [34]. Similar to P2O, GO is a member of the glucose-methanol-choline (GMC) family of oxidoreductases that oxidize alcohol moieties [38]. GO is one of the most widely used enzymes in biosensor applications such as in glucose-sensing electrodes [39,40]. Transfer of electrons between the active site and the electrode can be facilitated by various electron mediators. Mediators commonly used are metal-ligand

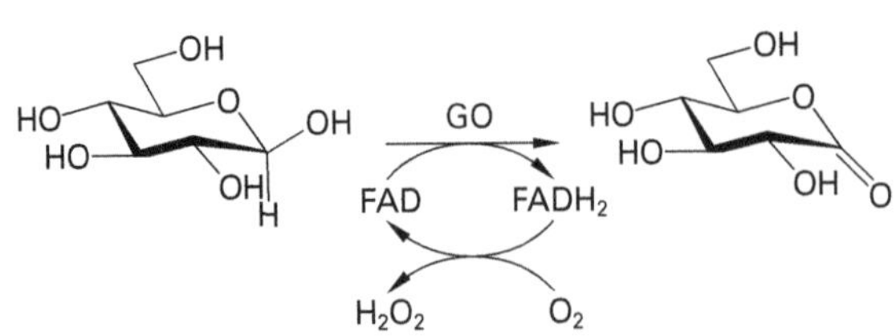

Scheme 9.7: Reaction catalyzed by glucose 1-oxidase.

complexes of iron, cobalt, manganese and chromium, which generally mediate one electron at a time [41].

9.3.2 Flavin reduction of GO

GO can use various sugars, derivatives of D-glucose, and α-hydroxycarbonyl compounds such as dihydroxyacetone, glyceraldehyde, phenacyl alcohol and furoin as substrates [1,42]. Among these, D-glucose is the most effective substrate based on the k_{cat} value (1150 s^{-1} at pH 5.6, 25 °C) [1]. Steady-state kinetics of the oxidation of glucose by molecular oxygen or other electron acceptors in GO can be described by a ping-pong mechanism [1]. Stopped-flow studies on the reductive half-reaction using D-glucose as a substrate indicate that the rate constant of the flavin reduction is very high and is dependent on substrate concentration in a second-order fashion. The reactions with D-mannose, D-xylose, and D-galactose also show similar results as for D-glucose but with much lower rate constants [1]. Therefore, a Michaelis (enzyme-glucose) complex in GO could not be detected based on these data. When a deuterated glucose (1-*d*-D-glucose) was used, the reaction showed a large primary kinetic isotope effect and a saturating rate constant of 67 s^{-1} at 3 °C, indicating that an enzyme-glucose complex exists during the reductive half-reaction [33].

The crystal structure of GO from *Aspergillus niger* at 2.3 Å resolution was solved with no ligand bound and D-glucose was modeled into the active site [23]. The crystallographic data have identified three amino acids located near the reaction center (▶Fig. 9.5): glutamic acid (Glu412), and two histidine conserved residues, His516 and His559. His559 forms H–bonds to N^{ε} and O^{ε} of Glu412. Glu412 is partially buried inside the protein molecule. The side chain of His516 is more flexible and more exposed to the solvent. Although the distances between the C1-hydroxyl proton of the modeled glucose and the nitrogen atoms of His516 and His559 are similar, the His516 has been proposed to be the catalytic base due to its lower pK_a value according to computations [43]. Therefore, the mechanism proposed for the reductive half-reaction of GO is described as a removal of a proton from the C1-hydroxyl group of D-glucose by His516 which is in concert with a direct hydride transfer from the C1 position of glucose to the N(5) position of FAD [3,44].

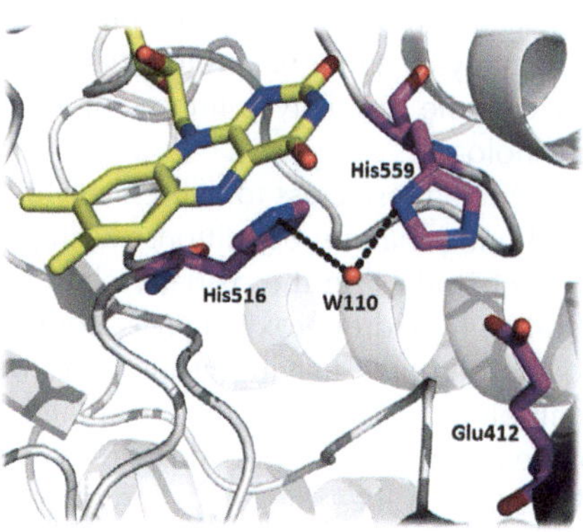

Fig. 9.5: Active site residues in glucose 1-oxidase [23].

9.3.3 Oxidative half-reaction of GO

For the oxidative half-reaction of GO, several two-electron acceptors such as benzoquinones, naphthoquinones and molecular oxygen are good substrates. One-electron acceptors, such as potassium ferricyanide, ferrocene salts, nitroxide radicals, and organic metal complexes are also efficient substrates for GO. The flavin re-oxidation mechanism presumably occurs via two single electron transfer steps as in other flavoenzyme oxidases because no C4a-hydroperoxyflavin intermediate is detected during turnover [1]. However, C4a-hydroperoxyflavin was detected in GO when pulse radiolysis was used to generate flavin semiquinone and superoxide radical in order to initiate intermediate formation [45]. The reduction of O_2 to H_2O_2 involves the uptake of a proton that is thought to be derived from the active site histidine (►Fig. 9.5). Site-directed mutagenesis studies reveal that the pH dependence of the oxidation rate is caused by protonation of a highly conserved histidine in the active site. The activity of the enzyme is optimal at low pH. Furthermore, the H516A mutant has rates that resemble the uncatalyzed reaction [46]. GO was used as a model for investigating the mechanism of oxygen activation by a flavoenzyme.

9.4 Conclusions and future prospects

In conclusion, the reaction of GO and P2O are of the ping-pong type where the sugar product leaves prior to the oxygen reaction. It has been shown that both enzymes oxidize a broad range of sugar substrates. Glucose oxidase catalyzes regio-specific oxidation at the C1 position, whereas pyranose 2-oxidase oxidizes at the C2 position. On the basis of solvent kinetic isotope effect studies on the reductive half-reaction of both enzymes, it appears that the deprotonation of the hydroxyl proton of D-glucose does not limit any steps in the reductive half-reaction. In recent years, considerable understanding towards the mechanism of sugar oxidation in P2O has been achieved while there was not much mechanistic investigation of GO. The mechanism of these enzymes has been proposed to follow a hydride transfer mechanism. An active site base (His548 in P2O and His516 in GO) is proposed to abstract a proton from a hydroxyl group before the transfer of a hydride equivalent from a sugar substrate to the flavin N(5) occurs. Besides applications in the food industry, P2O has become an attractive candidate for biocatalysis applications such as converting sugars into more valuable products. In order to improve the performance of P2O in biocatalysis, a rational protein design in order to broaden substrate specificity and modify enzyme reactivity is important for improving the applications of this enzyme in biotechnology.

P2O is the first enzyme in the flavoprotein oxidase class for which a C4a-hydroperoxyflavin intermediate has been detected under natural conditions during the oxidative half-reaction. The kinetic isotope effect studies on the oxidative half-reaction of P2O indicate that H_2O_2 elimination from C4a-hydroperoxyflavin is controlled by a proton transfer from the flavin N(5) to the peroxide leaving group. The mechanism might involve the formation of an intramolecular H-bridge that facilitates the H_2O_2 elimination process. Currently, it is not clear whether a C4a-hydroperoxyflavin is also a common intermediate in other oxidases but its life-time is not long enough for kinetic detection. Biochemical and biophysical factors which allow the formation of the

C4a-hydroperoxyflavin intermediate in P2O are not understood and are surely important for understanding how flavoenzyme active sites fine-tune their flavin environment for reacting with oxygen.

9.5 References

[1] Gibson QH, Swoboda BE, Massey V. Kinetics and mechanism of action of glucose oxidase. J Biol Chem 1964;239:3927–34.

[2] Leitner C, Volc J, Haltrich D. Purification and characterization of pyranose oxidase from the white rot fungus Trametes multicolor. Appl Environ Microbiol 2001;67:3636–36.

[3] Wohlfahrt G, Witt S, Hendle J, Schomburg D, Kalisz HM, Hecht HJ. 1.8 and 1.9 Å resolution structures of the *Penicillium amagasakiense* and *Aspergillus niger* glucose oxidases as a basis for modeling substrate complexes. Acta Crystallogr D Biol Crystallogr 1999;55:969–77.

[4] Klinman JP. How do enzymes activate oxygen without inactivating themselves? Acc Chem Res 2007;40:325–33.

[5] Wongnate T, Sucharitakul J, Chaiyen P. Identification of a catalytic base for sugar oxidation in pyranose 2-oxidase reaction. ChemBioChem 2011;12:2577–86.

[6] Sucharitakul J, Wongnate T, Chaiyen P. Kinetic isotope effects on the noncovalent flavin mutant protein of pyranose 2-oxidase reveal insights into the flavin reduction mechanism. Biochemistry 2010;49:3753–65.

[7] Prongjit M, Sucharitakul J, Wongnate T, Haltrich D, Chaiyen P. Kinetic mechanism of pyranose 2-oxidase from *trametes multicolor*. Biochemistry 2009;48:4170–80.

[8] Pitsawong W, Sucharitakul J, Prongjit M, Tan TC, Spadiut O, Haltrich D, Divne C, Chaiyen P. A conserved active-site threonine is important for both sugar and flavin oxidations of pyranose 2-oxidase. J Biol Chem 2010;285:9697–705.

[9] Hallberg BM, Leitner C, Haltrich D, Divne C. Crystal structure of the 270 kDa homotetrameric lignin-degrading enzyme pyranose 2-oxidase. J Mol Biol 2004;341:781–96.

[10] Sucharitakul J, Prongjit M, Haltrich D, Chaiyen P. Detection of a C4a-hydroperoxyflavin intermediate in the reaction of a flavoprotein oxidase. Biochemistry 2008;47:8485–90.

[11] Ander P, Marzullo L. Sugar oxidoreductases and veratryl alcohol oxidase as related to lignin degradation. J Biotechnol 1997;53:115–31.

[12] Artolozaga MJ, Kubátová E, Volc J, Kalisz HM. Pyranose 2-oxidase from Phanerochaete chrysosporium-further biochemical characterization. Appl Microbiol Biotechnol 1997;47: 508–14.

[13] Schafer A, Bieg S, Huwig A, Kohring G, Giffhorn F. Purification by Immunoaffinity Chromatography, Characterization, and Structural Analysis of a Thermostable Pyranose Oxidase from the White Rot Fungus *Phlebiopsis gigantean*. Appl Environ Microbiol 1996;62:2586–92.

[14] Shin KS, Youn HD, Han YH, Kang SO, Hah YC. Purification and characterization of D-glucose oxidase from white-rot fungus *Pleurotus ostreatus*. Eur J Biochem 1993;215:747–52.

[15] Janssen FW, Ruelius HW. Pyranose oxidase from *Polyporus obtusus*. Methods Enzymol 1975;41:170–3.

[16] Baute L. Clinical trial with lorazepam in pre-operative anxiety. Acta Anaesthesiol Belg 1977;28:123–31.

[17] Olsson L, Mandenius CF, Kubatova E, Volc J. Immobilization of pyranose oxidase (*Phanerochaete chrysosporium*): characterization of the enzymic properties. Enzyme Microb Technol 1991;13:755–9.

[18] Giffhorn F. Fungal pyranose oxidases: occurrence, properties and biotechnical applications in carbohydrate chemistry. Appl Microbiol Biotechnol 2000;54:727–40.

[19] Spadiut O, Leitner C, Salaheddin C, Varga B, Vertessy BG, Tan TC, Divne C, Haltrich D. Improving thermostability and catalytic activity of pyranose 2-oxidase from Trametes multicolor by rational and semi-rational design. FEBS J 2009;276:776–92.

[20] Halada P, Leitner C, Sedmera P, Haltrich D, Volc J. Identification of the covalent flavin adenine dinucleotide-binding region in pyranose 2-oxidase from *Trametes multicolor*. Anal Biochem 2003;314:235–42.

[21] Kujawa M, Ebner H, Leitner C, Hallberg BM, Prongjit M, Sucharitakul J, Ludwig R, Rudsander U, Peterbauer C, Chaiyen P, Haltrich D, Divne C. Structural basis for substrate binding and regioselective oxidation of monosaccharides at C3 by pyranose 2-oxidase. J Biol Chem 2006;281:35104–15.

[22] Tan TC, Pitsawong W, Wongnate T, Spadiut O, Haltrich D, Chaiyen P, Divne C. H-bonding and positive charge at the N5/O4 locus are critical for covalent flavin attachment in trametes pyranose 2-oxidase. J Mol Biol 2010;402:578–94.

[23] Hecht HJ, Kalisz HM, Hendle J, Schmid RD, Schomburg D. Crystal structure of glucose oxidase from *Aspergillus niger* refined at 2.3 A° resolution. J Mol Biol 1993;229:153–72.

[24] Vrielink A, Ghisla S. Cholesterol oxidase: biochemistry and structural features. FEBS J 2009;276:6826–43.

[25] Quaye O, Lountos GT, Fan F, Orville AM, Gadda G. Role of Glu312 in binding and positioning of the substrate for the hydride transfer reaction in choline oxidase. Biochemistry 2008;47:243–56.

[26] Spadiut O, Tan TC, Pisanelli I, Haltrich D, Divne C. Importance of the gating segment in the substrate-recognition loop of pyranose 2-oxidase. FEBS J 2010;277:2892–909.

[27] Tan TC, Haltrich D, Divne C. Regioselective control of β-d-glucose oxidation by pyranose 2-oxidase is intimately coupled to conformational degeneracy. J Mol Biol 2011;409:588–600.

[28] Chosrowjan H, Taniguchi S, Wongnate T, Sucharitakul J, Chaiyen P, Tanaka F. Conformational heterogeneity in pyranose 2-oxidase from Trametes multicolor revealed by ultrafast fluorescence dynamics. J Photochem Photobiol A Chem 2012;234:44–8.

[29] Taniguchi S, Chosrowjan H, Wongnate T, Sucharitakul J, Chaiyen P, Tanaka F. Ultrafast fluorescence dynamics of flavin adenine dinucleotide in pyranose 2-oxidases variants and their complexes with acetate: conformational heterogeneity with different dielectric constants. J Photochem Photobiol A Chem 2012 (In revision).

[30] Sucharitakul J, Wongnate T, Chaiyen P. Hydrogen peroxide elimination from C4a-hydroperoxyflavin in a flavoprotein oxidase occurs through a single proton transfer from flavin N5 to a peroxide leaving group. J Biol Chem 2011;286:16900–9.

[31] Massey V. Activation of molecular oxygen by flavins and flavoproteins. J Biol Chem 1994;269:22459–62.

[32] Thotsaporn K, Chenprakhon P, Sucharitakul J, Mattevi A, Chaiyen P. Stabilization of C4a-hydroperoxyflavin in a two-component flavin-dependent monooxygenase is achieved through interactions at flavin N5 and C4a atoms. J Biol Chem 2011;286:28170–80.

[33] Bright HJ, Gibson QH. The oxidation of 1-deuterated glucose by glucose oxidase. J Biol Chem 1967;242:994–1003.

[34] Kohen A, Jonsson T, Klinman JP. Effects of protein glycosylation on catalysis: changes in hydrogen tunneling and enthalpy of activation in the glucose oxidase reaction. Biochemistry 1997;36:2603–11.

[35] Swoboda BE, Massey V. Purification and properties of the glucose oxidase from *Aspergillus niger*. J Biol Chem 1965;240:2209–15.

[36] Takegawa K, Fujiwara K, Iwahara S, Yamamoto K, Tochikura T. Effect of deglycosylation of N-linked sugar chains on glucose oxidase from *Aspergillus niger*. Biochem Cell Biol 1989;67:460–4.

[37] Kalisz HM, Hecht HJ, Schomburg D, Schmid RD. Effects of carbohydrate depletion on the structure, stability and activity of glucose oxidase from *Aspergillus niger*. Biochim Biophys Acta 1991;1080:138–42.

[38] Cavener DR. GMC oxidoreductases. A newly defined family of homologous proteins with diverse catalytic activities. J Mol Biol 1992;223:811–4.

[39] Lee J, Ahn H, Choi I, Boese M, Park MJ. Enhanced charge transport in enzyme-wired organometallic block copolymers for bioenergy and biosensors. Macromolecules 2012;45:3121–8.
[40] Niu X, Chen C, Zhao H, Chai Y, Lan M. Novel snowflake-like Pt-Pd bimetallic clusters on screen-printed gold nanofilm electrode for H_2O_2 and glucose sensing. Biosensors and Bioelectronics 2012 (In press).
[41] Alvarez-Icaza M, Kalisz HM, Hecht HJ, Aumann KD, Schomburg D, Schmid RD. The design of enzyme sensors based on the enzyme structure. Biosens Bioelectron 1995;10:735–42.
[42] Chan TW, Bruice TC. One and two electron transfer reactions of glucose oxidase. J Am Chem Soc 1977;99:2387–9.
[43] Wohlfahrt G, Trivi S, Zeremski J, Pericin D, Leskovac V. The chemical mechanism of action of glucose oxidase from *Aspergillus niger*. Mol Cell Biochem 2004;260:69–83.
[44] Bright HJ, Appleby M. The pH dependence of the individual steps in the glucose oxidase reaction. J Biol Chem 1969;244:3625–34.
[45] Massey V, Schopfer LM, Anderson RF. Structural determinants of the oxygen reactivity of different classes of flavoproteins. In Oxidases and Related Redox Systems; King TE, Mason HS, Morrison M. Eds.; Alan Liss: New York, 1988:147–66.
[46] Roth JP, Klinman JP. Catalysis of electron transfer during activation of O_2 by the flavoprotein glucose oxidase. Proc Natl Acad Sci USA 2003;100:62–7.

10 Toward understanding the mechanism of oxygen activation by flavoprotein oxidases

Marilyn Schuman Jorns

Abstract

Several common themes have emerged from our recent studies on the mechanism of oxygen activation by monomeric sarcosine oxidase (MSOX), N-methyltryptophan oxidase (MTOX), and heterotetrameric sarcosine oxidase (TSOX). MSOX and MTOX are members of a family of monomeric amino acid oxidases that contain 1 mol of covalently bound FAD. TSOX is a bifunctional, multimeric (αβγδ) enzyme that contains 1 mol each of noncovalently bound FAD and covalently bound FMN. FAD is bound to the β-subunit which exhibits structural and sequence homology with MSOX and MTOX; FMN is located at the interface between the TSOX α- and β-subunits. MSOX and MTOX contain separate sites for amino acid oxidation and oxygen reduction above opposite sides of the flavin ring (*re*- and *si*-face, respectively). The corresponding sites in TSOX are located at different flavins (FAD and FMN, respectively). A single basic residue is solely responsible for oxygen activation by MSOX and MTOX (Lys265 and Lys259, respectively), similar to that previously observed for His516 in glucose oxidase. Oxygen activation by TSOX is mediated by a pair of basic residues (βLys172, βLys278). An unobstructed or gated tunnel provides a postulated route for oxygen diffusion to the active site in TSOX and MSOX, respectively. Chloride acts as a putative oxygen surrogate that binds to the side chain of Lys265 in MSOX or His516 in glucose oxidase. The positively charged residues at the oxygen activation sites in MSOX, MTOX, TSOX, and glucose oxidase provide an apparent pre-organized binding site for superoxide anion that accelerates the critical initial 1-electron reduction of oxygen by lowering the reorganization energy of the reaction. Comparison with other enzymes indicates that, although electrostatic catalysis is frequently a contributing factor, flavo-oxidases exhibit considerable diversity in the mechanism of oxygen activation, as discussed.

10.1 Introduction

The ability to activate the reduction of molecular oxygen underpins all aerobic biology. Flavoprotein oxidases accelerate the reduction of oxygen to hydrogen peroxide, a highly reactive metabolite that is both a cytotoxin and a cell signaling molecule. Defining the principles that govern the oxygen reactivity of the reduced coenzyme and the nature of oxygen diffusion pathways is one of the last great frontiers in flavoprotein

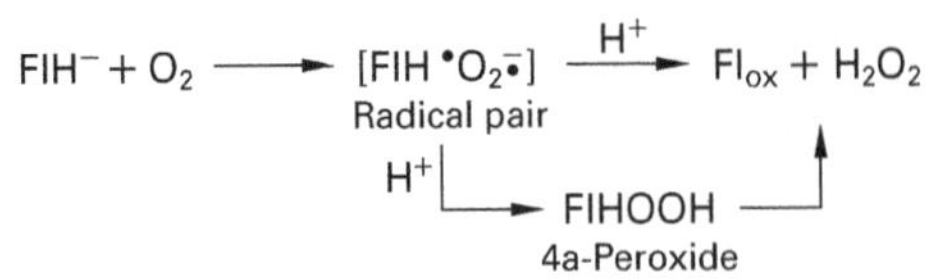

Fig. 10.1: Possible mechanisms for oxygen reduction by flavoprotein oxidases.

enzymology and an area of intense current interest. The 2-electron reduction of molecular oxygen by free reduced flavin is thermodynamically favorable but is spin-forbidden. Consequently, the reaction proceeds via an initial, rate-determining, 1-electron step that is spin-allowed but energetically unfavorable. Oxygen activation by flavoprotein oxidases depends on their ability to accelerate this step. The second electron transfer step is intimately tied to proton transfers and may (i) directly result in the formation of hydrogen peroxide, a path thought to occur in most flavoprotein oxidases [1]; or (ii) yield a peroxy intermediate that generates hydrogen peroxide in a subsequent step [2] (▸Fig. 10.1).

This chapter will first describe recent studies conducted in our laboratory on the mechanism of oxygen activation by flavo-oxidases. We will then attempt to integrate our findings into a more global view of the current state of knowledge, identifying common themes where possible but also addressing the apparent mechanistic diversity exhibited by this remarkable class of enzymes. Our studies have focused on three enzymes: monomeric sarcosine oxidase (MSOX), N-methyltryptophan oxidase (MTOX), and heterotetrameric sarcosine oxidase (TSOX). MSOX and MTOX are stable 44 kDa monofunctional enzymes of known structure that contain 1 mol of covalently bound FAD. The enzymes catalyze the oxidative demethylation of N-methyl-substituted amino acids in reactions that proceed via an imine intermediate (CH_2=NH-CH-CO_2^-) that is subsequently hydrolyzed to produce formaldehyde and the unsubstituted amino acid [3,4]. MSOX and MTOX are members of a family of prokaryotic and eukaryotic amino acid oxidases that contain covalently bound flavin. TSOX is a multimeric (αβγδ) enzyme that contains 1 mol each of noncovalently bound FAD and covalently bound FMN. TSOX is a bifunctional enzyme that catalyzes sarcosine oxidation and the transfer of the oxidized methyl group from sarcosine imine to tetrahydrofolate. FAD is bound to the β-subunit of TSOX; FMN is bound at the interface between the α- and β-subunits [5]. Despite considerable differences in structural and functional complexity, it is noteworthy that the monomeric enzymes, MSOX and MTOX, exhibit significant sequence and structural homology with the β-subunit in the heterotetrameric enzyme, TSOX.

10.2 Results and discussion

10.2.1 Lys265 is the oxygen activation site in MSOX

The single flavin in MSOX exhibits two functions: (i) sarcosine oxidation; (ii) oxygen reduction. Steady-state kinetic studies indicate that oxygen reacts with a reduced enzyme-imine complex [6]. Crystal structures of enzyme-inhibitor complexes show that the active site for sarcosine oxidation is located above the *re*-face of the flavin ring of FAD [3,7]. A clue regarding the possible location of the oxygen activation site was sought by comparing the MSOX flavin environment with that observed for FAD bound to the β-subunit of TSOX. FAD is the site of sarcosine oxidation in TSOX. However, reduced FAD bound to TSOX is totally unreactive towards oxygen. We found that MSOX and the β-subunit of TSOX contain highly similar active sites for sarcosine oxidation above the *re*-face of the flavin ring of FAD. However, striking differences are found above the *si*-face of the flavin ring, suggesting that this region might contain the oxygen activation site in MSOX. Two basic residues, Arg49 and Lys265, are found above the *si*-face in MSOX but are not present in TSOX (►Fig. 10.2).

Arg49 is essential for covalent flavin attachment in MSOX [8]. Mutation of Lys265 to a neutral residue (Ala, Met, Gln) results in a ~8000-fold decrease in the apparent turnover rate whereas a ~200-fold decrease is observed for the chemically conservative Lys265 to Arg mutation [9] (►Tab.10.1).

To determine whether mutation of Lys265 affects sarcosine binding or the rate of electron transfer to FAD, we used a stopped-flow spectrophotometer to monitor events occurring within ~1 millisecond after mixing Lys265Met, Lys265Arg, or wild-type

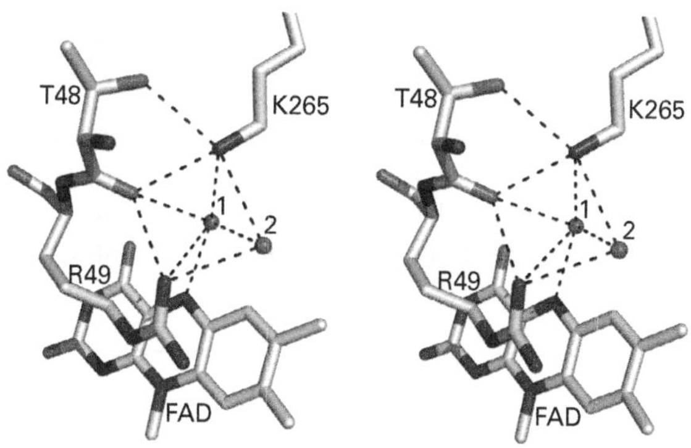

Fig. 10.2: Stereo-view of the region above the *si*-face of the flavin ring in MSOX.

Tab. 10.1: Effect of Lys265 mutations on MSOX turnover and the kinetics of reductive and oxidative half-reactions[a]

Preparation	$k_{cat\,app}$ (s^{-1})	K_d (mM)[b]	k_{red} (s^{-1})[b]	k_2 (s^{-1})[b]	k_{ox} ($M^{-1}\,s^{-1}$)[c]
WT	45.8	13.0	140	fast	2.83×10^5
K265M	0.0076	58	15.5	4.8	35.1
K265R	0.272	34	36.8	6.06	1140
K265A	0.0066				
K265Q	0.0049				
free flavin					250[d]

[a] Reactions were conducted in 100 mM potassium phosphate, pH 8.0, at 25 °C
[b] Reductive half-reaction parameters as specified in the scheme shown in ►Fig. 10.4
[c] Second-order rate constant for the reaction of reduced enzyme with oxygen
[d] Data previously reported [1]

enzyme with sarcosine under anaerobic conditions [9]. The reaction with wild-type or mutant MSOX proceeds via the formation of a spectrally detectable enzyme-substrate complex, intermediate I (►Figs. 10.3 and 10.4)).

In a second step, the mutant enzyme-substrate complexes are converted to an unusual reduced flavin species, intermediate II. The spectral properties of intermediate II are attributed to charge-transfer interaction between the electron-rich anionic reduced flavin and the electron-poor protonated imine product. Intermediate II is not detected with wild-type MSOX, a feature attributed to its rapid conversion to intermediate III. Intermediate III exhibits a typical reduced flavin spectrum that provides no evidence for charge-transfer interaction and is observed with both wild-type and mutant enzyme. We postulate that formation of intermediate III involves conversion of the protonated imine to the neutral imine, a species unlikely to act as a charge-transfer acceptor. Mutation of Lys265 to Met or Arg causes a less than 5-fold decrease in the stability of enzyme-sarcosine complex (K_d) and a less than 10-fold decrease in the rate of electron transfer from sarcosine to FAD (k_{red}) (►Tab. 10.1). The rate-determining step in the reductive half-reaction with the mutant enzymes, conversion of intermediate II to III (k_2), is 30-fold slower than the slow step with wild-type enzyme (k_{red}).

The relatively modest effects of the Lys265 substitutions on the reductive half-reaction cannot account for the much larger effects on the apparent turnover rate. To evaluate the effect of the mutations on oxygen reactivity, we mixed anaerobic solutions of the reduced enzyme with buffer containing different concentrations of oxygen. The reactions observed with wild-type and mutant enzyme exhibit a linear dependence on oxygen concentration [9]. A dramatic ~8000-fold decrease in oxygen reactivity is observed with Lys265Met whereas a ~250-fold decrease is seen with Lys265Arg. Significantly,

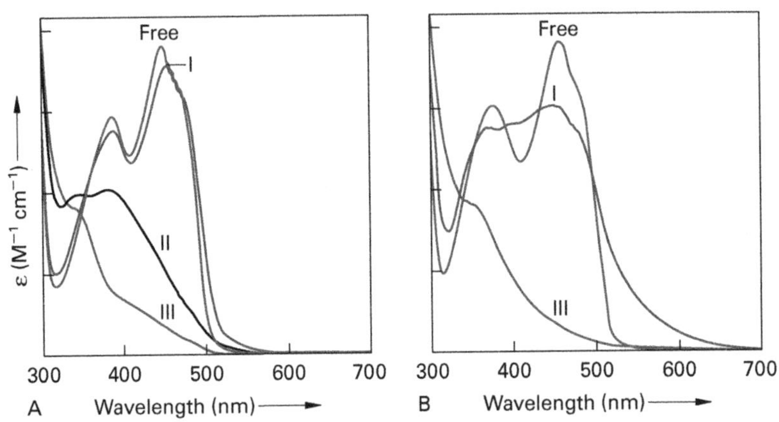

Fig. 10.3: Spectral properties of intermediates detected by monitoring the anaerobic reduction of Lys265Met (A) or wild-type MSOX (B) by sarcosine in a stopped-flow spectrophotometer.

$$\underset{}{\text{E-FAD}_{ox} + \text{S}} \underset{}{\overset{K_d}{\rightleftharpoons}} \underset{\text{I}}{\text{E-FAD}_{ox}\bullet\text{S}} \xrightarrow{k_{red}} \underset{\text{II}}{\text{E-FADH}^-\bullet\text{ImH}^+} \xrightarrow[-\text{H}^+]{k_2} \underset{\text{III}}{\text{E-FAD}^-\bullet\text{Ir}}$$

Fig. 10.4: Postulated mechanism for the reduction of MSOX or MTOX by sarcosine or N-methyltryptophan, respectively.

the reaction of reduced Lys265Met with oxygen is even slower (7-fold) than the sluggish reaction observed with free reduced flavin (▶Tab. 10.1).

The properties observed for mutants containing a neutral residue substitution at Lys265 strongly suggested that a basic residue at position 265 is essential for oxygen activation. This scenario cannot, however, explain the substantial decrease in oxygen reactivity observed for the chemically conservative Lys265Arg mutant. Crystal structures determined for Lys265Met and Lys265Arg show that the mutations do not affect the overall protein structure or the structure of the active site for sarcosine oxidation above the *re*-face of the flavin. On the *si*-face, the side chain of Met265 is nearly congruent with that observed for Lys265 in wild-type MSOX. In contrast, the side chain of Arg265 is dramatically shifted (~5 Å) compared with Lys265 and points in the opposite direction [10]. Apparently, the ~10% larger Arg side chain is not readily accommodated in the space occupied by Lys265 in wild-type MSOX. We conclude that the greatly decreased oxygen reactivity observed with Lys265Met is caused by the elimination of a positive charge whereas the reduced reactivity observed with Lys265Arg is attributed to an effective displacement of the positive charge away from the *si*-face of the flavin. Overall, the results provide compelling evidence that Lys265 is the site of oxygen activation in MSOX.

10.2.2 Lys259 is the oxygen activation site in MTOX

Lys259 is located above the *si*-face of the flavin ring in MTOX, similar to that observed for Lys265 in MSOX. The lower-resolution MTOX structure does not contain water molecules [4]. A water molecule has been modeled into the putative *si*-face active site in MTOX on the basis of an alignment with MSOX. The model suggests that Lys259 is hydrogen bonded to FAD(N5) via a bridging water molecule (▶Fig. 10.5).

Mutation of Lys259 to a neutral residue (Gln, Ala, Met) results in an ~2500-fold decrease in the apparent turnover rate [11] (▶Tab. 10.2). Reductive half-reaction studies with Lys259Gln and N-methyl-L-tryptophan show that the reaction proceeds with the formation of three intermediates (I, II, III) (▶Fig. 10.6A) that exhibit spectral properties similar to those observed with the analogous MSOX mutant, Lys265Met. Furthermore, only two intermediates (I, III) are detected with wild-type MTOX (▶Fig.10.6B), as observed

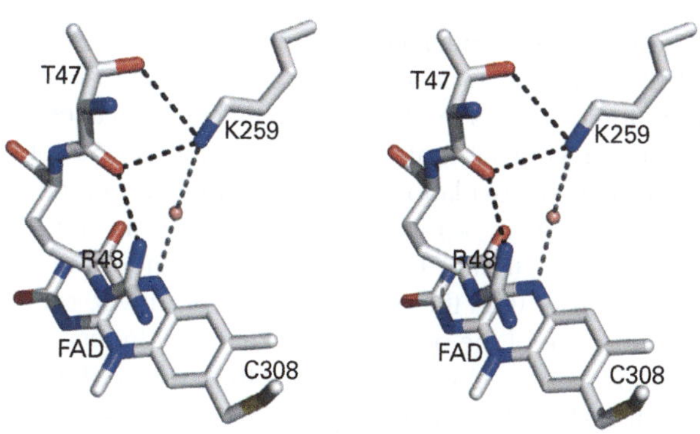

Fig. 10.5: Stereo-view of the region above the *si*-face of the flavin ring in MTOX.

Tab. 10.2: Effect of Lys259 mutations on MTOX turnover and the kinetics of reductive and oxidative half-reactions[a]

Preparation	$k_{cat\ app}$ (s^{-1})	K_d (mM)[b]	k_{red} (s^{-1})[b]	k_2 (s^{-1})[b]	k_{ox} ($M^{-1}\ s^{-1}$)[c]
WT	21.4	973	129	fast	2.13 x 10^5
K259Q	0.0166	569	2.13	0.244	87
K259A	0.0083				
K259M	0.0021				
free flavin					250[d]

[a] Reactions were conducted in 100 mM potassium phosphate, pH 8.0, at 25 °C
[b] Reductive half-reaction parameters as specified in the scheme shown in ▶Fig. 10.4
[c] Second-order rate constant for the reaction of reduced enzyme with oxygen
[d] Data previously reported [1]

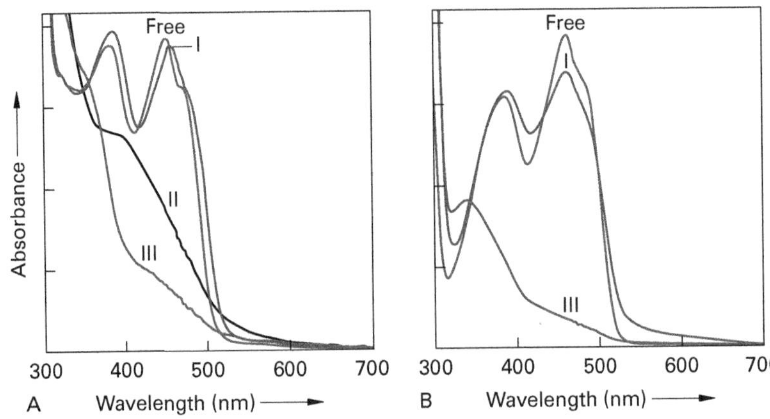

Fig. 10.6: Spectral properties of intermediates detected by monitoring the anaerobic reduction of Lys259Gln (A) or wild-type MTOX (B) by N-methyltryptophan in a stopped-flow spectrophotometer.

with wild-type MSOX. The strikingly similar kinetics and intermediates indicate that substrate reduction of MTOX is likely to proceed via the same mechanism proposed for MSOX (▶Fig. 10.4). Substitution of Lys259 in MTOX with Gln results in a less than 2-fold change in the stability of the enzyme-substrate complex (K_d) and a ~60-fold decrease in the rate of N-methyl-L-tryptophan oxidation (k_{red}). The rate-determining step in the reductive half-reaction with Lys259Gln (k_2) is ~500-fold slower than the slowest step observed with wild-type MTOX (k_{red}) (▶Tab. 10.2). Nevertheless, the value obtained for k_2 with the mutant is ~15-fold faster than the observed rate of turnover, suggesting that turnover may be limited by the rate of the oxidative half-reaction. Indeed, the reaction of reduced Lys259Gln with oxygen is ~2500-fold slower than with reduced wild-type MTOX and also slower (~3-fold) than the reaction with free reduced flavin (▶Tab. 10.2) [11]. The results show that Lys259 is the site of oxygen activation in MTOX.

10.2.3 A pair of lysines comprise the oxygen activation site in TSOX

As noted above, FAD is bound to the β-subunit of TSOX and the reduced coenzyme does not react with oxygen. Instead, electrons from reduced FAD are transferred to FMN, located at the interface between the α- and β-subunits, where oxygen is reduced to hydrogen peroxide [5,12]. Results obtained with MSOX and MTOX suggested that oxygen activation by TSOX might be mediated by a basic residue near the FMN site. In fact, we found that the N(5) position of FMN in TSOX is flanked by a pair of lysine residues, βLys172 and βLys278 (▶Fig. 10.7).

Substitution of either lysine with a neutral residue results in a ~10- to ~20-fold decrease in the apparent turnover rate whereas a ~100-fold decrease is observed when both lysines are replaced by neutral residues (βK172Q/βK278M). The anaerobic reduction of wild-type TSOX or the double mutant with sarcosine proceeds via the initial formation of an enzyme-sarcosine charge-transfer complex (E_{ox} + S ⇔ E_{ox}-S), characterized by the dissociation constant K_d, followed by electron transfer to FAD (E_{ox}-S ⇒ EH_2-P), characterized by the rate constant for reduction, k_{red}. The βK172Q/βK278M mutation does not affect the stability of the enzyme-substrate complex but does cause a small (2-fold) increase in the rate of electron transfer (▶Tab. 10.3). The rate of sarcosine oxidation by the double mutant is ~1000-fold faster than turnover, indicating that

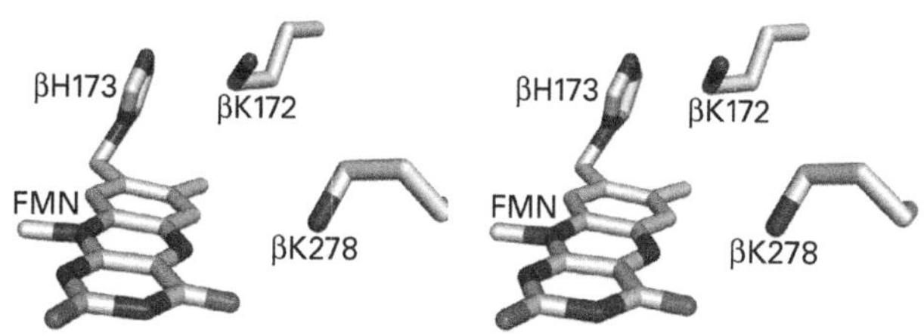

Fig. 10.7: Stereo-view of the FMN active site in TSOX.

Tab. 10.3: Effect of βLys172 and/or βLys278 mutations on TSOX turnover and the kinetics of the reductive and oxidative half-reactions[a]

Preparation	$k_{cat\ app}$ (s^{-1})	Reductive half-reaction[b]		Oxidative half-reaction[b]		
		K_d (mM)	k_{red} (s^{-1})	$K_{d\ app}$ (mM)	k_{lim} (s^{-1})	$k_{lim}/K_{d\ app}$ ($M^{-1}s^{-1}$)
WT	13.6	21	61.6	0.71	443	6.2×10^5
βK172Q/βK278M	0.123	26	115	1.3[c]	0.13[c]	100[c]
βK172Q	0.66					
βK278M	1.10					
free flavin						250[d]

[a] Except as noted, reactions were monitored in 100 mM sodium pyrophosphate buffer, pH 9.0, at 25 °C
[b] Reductive and oxidative half-reaction parameters as discussed in the text
[c] Data obtained at 6 °C where rates are ~5-fold slower than at 25 °C
[d] Data previously reported [1]

some other step must be rate-limiting during turnover. On the other hand, sarcosine oxidation by wild-type TSOX is only ~5-fold faster than turnover and may be partially rate-limiting.

Second-order kinetics are typically observed for the reaction of reduced flavoprotein oxidases with oxygen. The kinetics are consistent with a simple bimolecular reaction but do not rule out a mechanism involving the formation of a Michaelis complex that is unstable ($K_d \geq 10$ mM) and therefore difficult to detect within the accessible range of oxygen concentrations. Unlike most oxidases, the oxidation of 2-electron reduced wild-type TSOX or the βK172Q/βK278M mutant exhibits a hyperbolic dependence on the concentration of oxygen (►Fig. 10.8), a feature previously observed only with cholesterol oxidase [13].

The limiting rate of the oxidative half-reaction with wild-type TSOX (k_{lim}) is ~30-fold faster than turnover. The observed oxidative half-reaction parameters can be used to estimate an apparent second-order rate constant for the reaction of reduced wild-type TSOX with oxygen ($k_{lim}/K_{d\ app} = 6.2 \times 10^5\ M^{-1}\ s^{-1}$). It is worth noting that the calculated value is similar to rate constants obtained for the second-order reactions observed with wild-type MSOX or MTOX ($k_{ox} = 2.1$ or $2.8 \times 10^5\ M^{-1}\ s^{-1}$, respectively). The limiting rate of the oxidative half-reaction with the TSOX double mutant, βK172Q/βK278M, is about 3 orders of magnitude slower than observed with wild-type enzyme. The apparent second-order rate constant estimated for the reaction of the reduced mutant with oxygen is comparable to the rate observed with free reduced flavin (►Tab. 10.3). The results indicate that βLys172 and βLys278 play a crucial role in oxygen activation by TSOX.

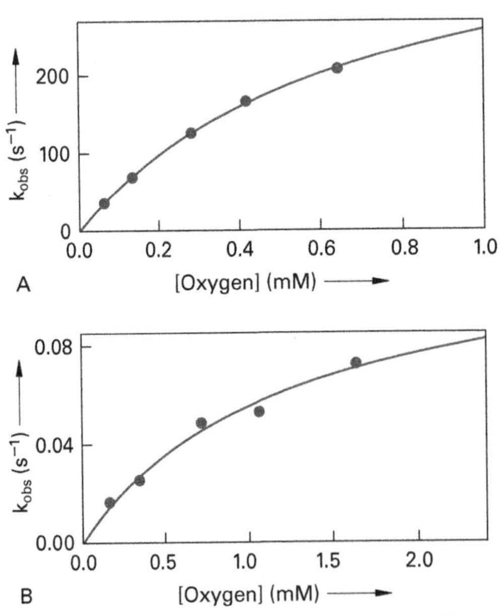

Fig. 10.8: Effect of oxygen concentration on the observed rate of oxidation of 2-electron reduced TSOX (EH_2). Results obtained wild-type enzyme and the βLys172Gln/βLys278Met double mutant are shown in panels A and B, respectively.

10.2.4 Probing the oxygen activation site in MSOX using chloride as an oxygen surrogate

The kinetics observed for the self-exchange reaction between oxygen and superoxide anion indicate that the reorganization energy required to change the configuration of the surrounding medium (λ_{out}) constitutes the major energy barrier in the 1-electron reduction of oxygen to superoxide anion [14]. The presence of a pre-organized binding site for superoxide anion could decrease λ_{out} and provide a mechanism by which an oxidase could accelerate the 1-electron reduction of oxygen. Lys265 in MSOX forms hydrogen bonds with a pair of active site waters (WAT1, WAT2), one of which is also hydrogen bonded to the N(5) position of FAD (►Fig. 10.2). The pocket occupied by these water molecules might provide a pre-organized binding site for superoxide anion. This scenario requires that oxygen displace at least one water molecule and occupy a polar reaction site. Nonpolar gases like xenon have traditionally been used to investigate oxygen-binding sites in nonpolar cavities. Recent studies with urate oxidase and other cofactor-less oxidases show that chloride ion is an alternate oxygen surrogate that can be used to probe polar oxygen-binding sites. Notably, chloride and oxygen bind at the same hydrophilic site in urate oxidase, displacing a tightly hydrogen-bonded water molecule, but not at a known xenon-binding site [15,16].

Flavoproteins exhibit absorption spectra that are highly sensitive to changes in the local environment, suggesting that chloride binding at the oxygen activation site in MSOX might be detectable by a perturbation of the flavin absorption spectrum. In fact, a spectrally detectable complex is formed upon titration of MSOX with chloride (K_d = 180 mM) in a reaction accompanied by a dramatic 20-nm bathochromic shift in the near-UV absorption band of the free enzyme (λ_{max} = 374 nm) (►Fig. 10.9A) [17].

Carboxylic acid derivatives that resemble sarcosine also perturb the absorption spectrum of MSOX but these anions bind at the substrate oxidation site above the *re*-face of the flavin ring [3,7]. Mutation of Lys265 to Met does not effect the binding of sarcosine analogs but does eliminate chloride binding [9,17], suggesting that chloride and sarcosine analogs bind at different sites. We reasoned that the putative presence of chloride at the oxygen activation site should not block the binding of a sarcosine analog on the opposite face of the flavin ring. To test this hypothesis, a titration was conducted in the presence of excess chloride (2.0 M) using methylthioacetate (MTA), a sarcosine analog that forms a charge-transfer complex with MSOX. The titration resulted in the apparent formation of an enzyme-chloride-MTA ternary complex that exhibits an intense charge-transfer band at 528 nm, as judged by the maximum in the difference spectrum calculated for 100% complex formation (►Fig. 10.9B). The charge-transfer band attributed to the ternary complex is hypsochromically shifted compared to that observed for the MSOX-MTA binary complex (λ_{max} = 532 nm) (►Fig. 10.9B). The dissociation constant estimated for the binding of MTA to the MSOX-chloride complex (K_d = 4.90 mM) exhibits a modest increase (~2-fold) compared with that observed for the binding of MTA to chloride-free MSOX (K_d = 2.43 mM). The results indicate that chloride and MTA occupy different binding sites.

To obtain definitive evidence for the postulated chloride binding site, crystals of free MSOX were soaked in sodium chloride. The crystal structure of the resulting MSOX-chloride binary complex shows that chloride binds above the *si*-face of the flavin ring at nearly the same position occupied by WAT2 in ligand-free enzyme (►Fig. 10.10). The bound chloride forms hydrogen bonds to Lys265:NZ, Arg49:NH1, WAT1, and WAT3. The arrangement of the four ligands exhibits approximate tetrahedral geometry. (WAT3

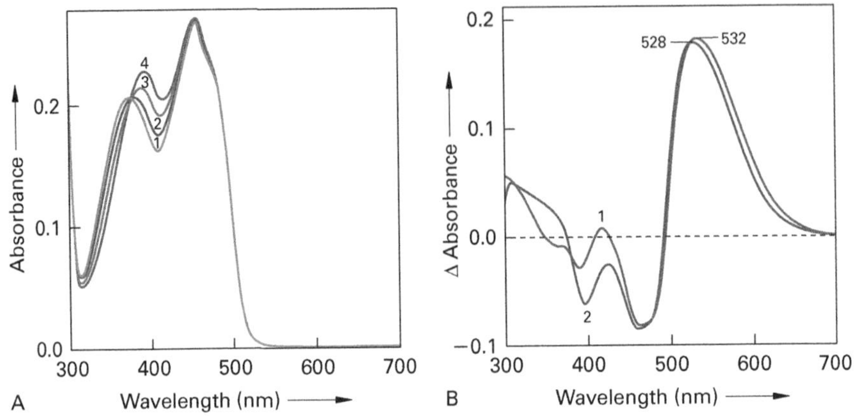

Fig. 10.9: Formation of MSOX-chloride binary and MSOX-chloride-MTA ternary complexes. Panel A shows a titration of MSOX with chloride. Curves 1–3 were recorded in the presence of 0, 72 or 316 mM potassium chloride, respectively. Curve 4 is the spectrum calculated for 100% complex formation. Panel B shows difference spectra calculated for 100% formation of the MSOX-MTA binary complex (curve 1, red line) and the MSOX-chloride-MTA ternary complex (curve 2, blue line).

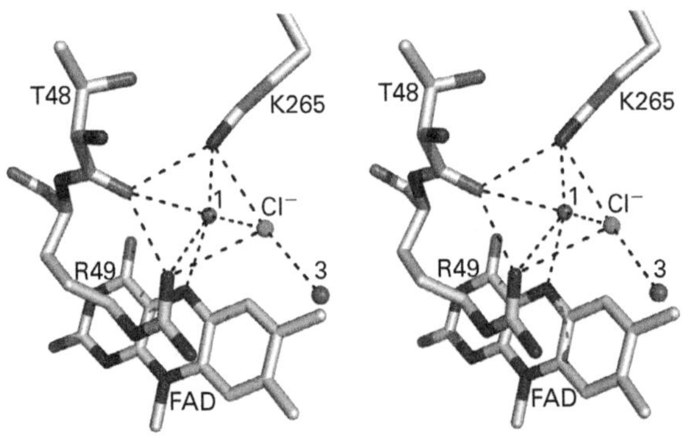

Fig. 10.10: Stereo-view of the region above the *si*-face of the flavin ring in the MSOX-chloride binary complex.

is not detected in ligand-free MSOX or the MSOX-MTA complex but is sometimes found in complexes with other analogs [17].) We also determined the structure of an enzyme-chloride-MTA ternary complex, produced by soaking crystals of free MSOX in buffer containing sodium chloride and MTA. The crystal structure of the ternary complex shows that chloride is bound above the *si*-face of the flavin ring (►Fig. 10.11C) at nearly the same position occupied by WAT2 in the MSOX-MTA binary complex (►Fig. 10.11A) or ligand-free enzyme (►Fig. 9.2). MTA is bound at the same position above the *re*-face of the flavin in the ternary complex (►Fig. 10.11D) and the binary MSOX-MTA complex (►Fig. 10.11C). The observed crystal structures of the binary and ternary chloride-containing complexes provide compelling evidence in support of deductions made based on results obtained in solutions studies [17].

We postulate that aerobic turnover of MSOX results in the transient formation of a ternary complex containing superoxide anion and sarcosine imine bound at opposite sides of the flavin ring, with FAD present as a neutral radical (►Fig. 10.12).

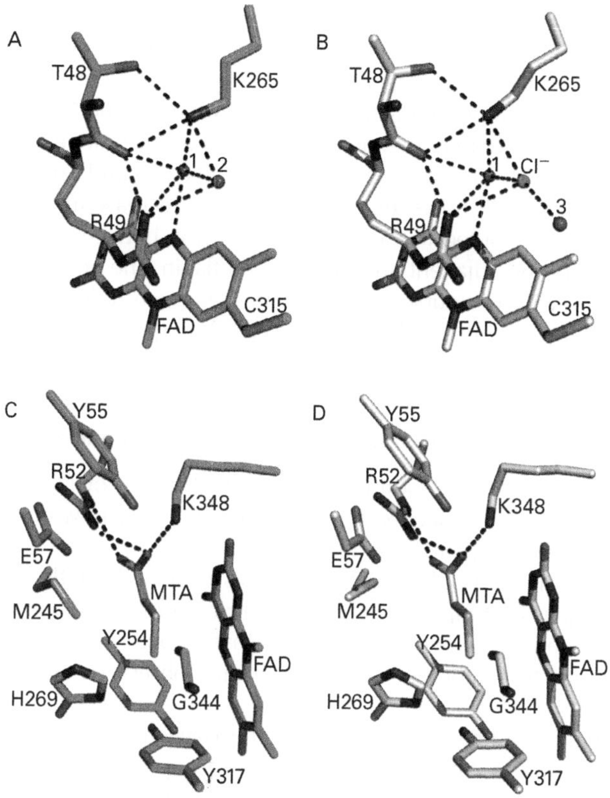

Fig. 10.11: Comparison of the regions above the *si*- and *re*-face of the flavin ring in the MSOX-chloride-MTA ternary complex with the MSOX-MTA binary complex. Panels A and C show the regions above the *si*- and *re*-face of the flavin in the MSOX-MTA binary complex. Panels B and D show the regions above the *si*- and *re*-face of the flavin in the MSOX-chloride-MTA ternary complex.

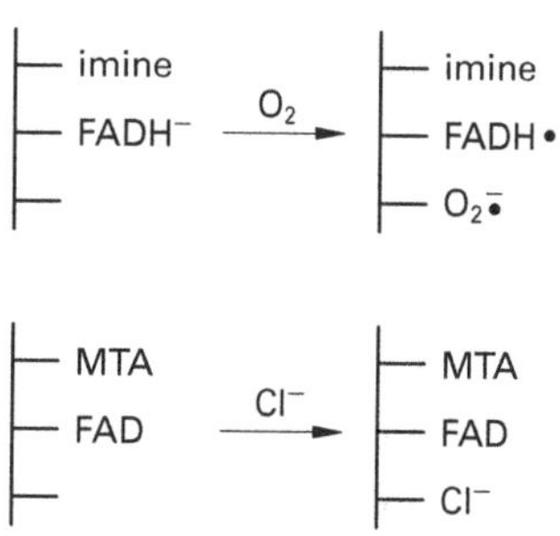

Fig. 10.12: The MSOX-chloride-MTA complex as a model for a postulated biradical intermediate formed upon reaction of oxygen with the reduced enzyme-imine complex.

The postulated biradical intermediate is consistent with steady-state kinetic studies which show that oxygen reacts with a reduced enzyme-imine complex [6] and the fact that the reaction must proceed via an initial, spin-allowed, 1-electron reduction of oxygen (►Fig. 10.1). The best available structural model for this reactive intermediate is provided by the crystal structure obtained for the MSOX-chloride-MTA ternary complex. The postulated biradical intermediate may be formed in a simple bimolecular reaction of oxygen with the reduced enzyme-imine complex or may involve formation of a Michaelis complex that contains molecular oxygen, imine and the 1,5-dihydroFAD anion (EFADH$^-$-O_2-imine). We hoped to generate a model for this reduced enzyme ternary complex by using chloride and MTA as surrogates for oxygen and sarcosine imine, respectively. Reduction of MSOX, however, results in a large decrease in chloride binding affinity [17], a feature attributed to electrostatic repulsion between the halide ion and the negative charge of the adjacent anionic reduced flavin. The negative charge of the reduced flavin anion will, of course, be lost upon 1-electron reduction of oxygen and, consequently, will not interfere with the proposed binding of superoxide anion in the same pocket occupied by chloride in complexes formed with oxidized MSOX.

10.2.5 Oxygen access to the proposed activation sites in TSOX and MSOX

A 36 Å long, unobstructed tunnel is found in the crystal structure of TSOX that leads from the protein surface to the FMN oxygen activation site and is wide enough to accommodate a molecule of oxygen [5] (►Fig. 10.13).

The results show that the unusual saturation kinetics observed for the reaction of reduced TSOX with oxygen (see ►Fig. 10.8) cannot be attributed to gated entry of oxygen via a tunnel that can alternate between open and closed conformations, a scenario suggested to give rise to the saturation kinetic behavior observed with cholesterol oxidase [13].

Unlike TSOX, a tunnel leading from solvent to the oxygen activation site above the *si*-face of the flavin ring in MSOX is not detected in any of the available crystal structures. As an alternative, we considered the possibility that oxygen might enter the protein via the sarcosine oxidation site above the *re*-face and then diffuse over to the opposite face of the flavin ring. The *re*-face of the flavin ring is accessible to solvent in ligand-free oxidized enzyme. However, a loop that controls access to the *re*-face

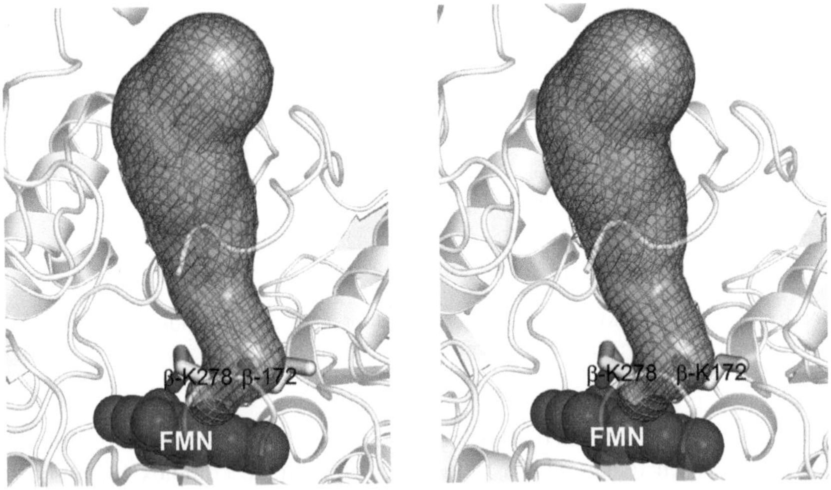

Fig. 10.13: Stereo-view of a putative oxygen tunnel in TSOX. FMN is in space-fill. βLys172 and βLys278 are shown as sticks.

adopts a closed conformation in ligand-free reduced MSOX and in reduced or oxidized enzyme-inhibitor complexes [3,7,17]. The crystallographic analysis suggested that oxygen access to the reduced flavin is probably mediated by dynamic motions of the protein. Temperature-accelerated molecular dynamic (TAMD) simulations permit large-scale conformational sampling of proteins in computationally accessible times [18]. The side chain of His45 lies about 8 Å above the *si*-face of the flavin ring. Spectacular motion of His45 is observed in TAMD studies with a reduced MSOX-inhibitor complex. During the simulation, the side chain of His45 moves nearly 9 Å away from the position observed in the crystal structure and points towards the protein surface rather than the interior of the protein, as observed in the crystal (▶Fig. 10.14). No global structural changes are detected during the simulation. The motion of His45, however, results in an unanticipated opening of a tunnel leading directly to the *si*-face of the flavin ring. A similar tunnel is found by the program MOLE [19] upon analysis of an MSOX mutant in which the side chain of His45 was eliminated by *in silico* mutagenesis (▶Fig. 10.15). The results suggest that His45 might act as a gatekeeper that controls gas diffusion from solution into the oxygen activation site.

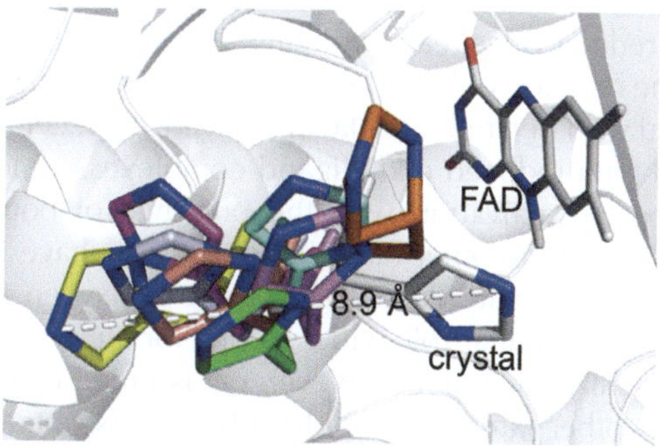

Fig. 10.14: The mobility observed for His45 in TAMD simulations is superimposed on the observed MSOX crystal structure, shows as a cartoon with FAD and His45 drawn as sticks.

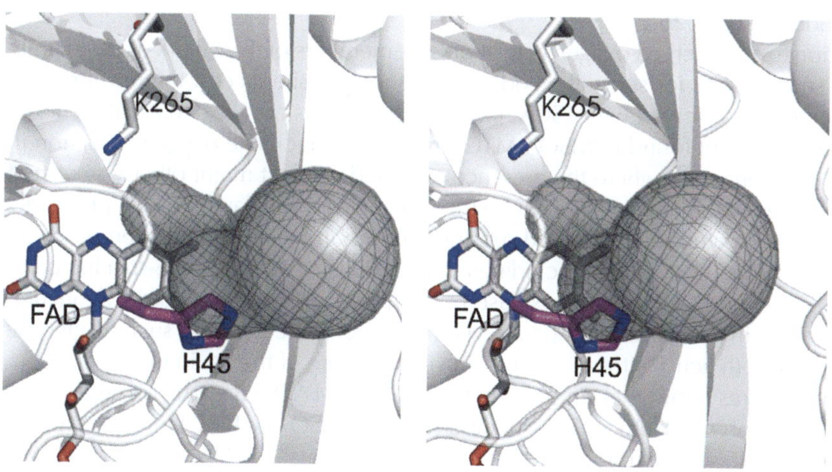

Fig. 10.15: Stereo-view of a potential oxygen tunnel found by analysis using the program MOLE of an MSOX mutant, His45Gly. The path is blocked in the observed crystal structure of wild-type enzyme by the side chain of His45 (drawn as sticks).

10.3 Common themes and mechanistic diversity

Our studies show that oxygen activation by MSOX and MTOX is mediated by a single lysine residue that is located above the *si*-face of the flavin ring and hydrogen bonded to the N(5) position of FAD via a bridging water molecule (Lys:NZ-Wat-FAD:N5 motif). The positively charged lysine side chain binds chloride, an apparent oxygen surrogate, and provides a postulated pre-organized binding site for superoxide anion that accelerates oxygen reduction by lowering the reorganization energy of the reaction. Studies by Klinman and Roth provide compelling evidence in support of a similar role for a positively charged residue, His516, in oxygen activation by glucose oxidase [20–22]. It is worth noting that His516 is hydrogen bonded to the N(5) position of FAD via a bridging water and its protonated form is required for chloride binding [17]. Our studies show that an analogous electrostatic role in oxygen activation by TSOX is mediated by a pair of lysine residues that flank the N(5) position of FMN.

The lysine in the Lys:NZ-Wat-FAD:N5 motif is conserved in nearly all members of the MSOX-MTOX family of oxidases (*e.g.*, fructosyl amino acid oxidase, pipecolate oxidase), with the notable exception of nikD. NikD catalyzes a remarkable aromatization reaction comprising two redox cycles, each utilizing a molecule of oxygen as electron acceptor. Satisfyingly, the reaction of ligand-free reduced nikD with oxygen is very slow, similar to that observed with free reduced flavin. Instead, an accelerated rate of oxygen reduction is observed with reduced nikD-ligand complexes, a mechanism that apparently synchronizes oxygen activation with the presence of bound intermediate or aromatic product [23]. On the other hand, recent studies have identified the lysine conserved in fructosyl amino acid oxidase as the site of oxygen activation [24]. The results strongly suggest that a similar role will be found for the corresponding lysine in other members of this family. It cannot, however, be assumed that a similar consistency will be observed for other families, as judged by results obtained for a highly conserved histidine residue in the glucose-methanol-choline oxidase family. This histidine is solely responsible for oxygen activation by glucose oxidase but exhibits a different function in choline oxidase where oxygen activation is mediated, in part, by bound product [20,25,26].

In addition to a conserved oxygen activation site, members of the MSOX-MTOX family contain a separate site for substrate oxidation on the opposite face of the flavin ring. A similar scenario is observed with TSOX, except that the sites are located at different flavins. The arrangement found in MSOX-MTOX family may facilitate oxygen reduction in these and other enzymes where the presence of bound product might otherwise restrict access of oxygen to the reactive N(5)-C(4a) locus of the reduced flavin. In fact, physically distinct sites for oxygen reduction and substrate oxidation have recently been observed or postulated for other oxidases, including alditol oxidase, histone lysine-specific demethylase (LSD1) and D-amino acid oxidase [27–29]. Of course a dichotomy of function is not necessary in enzymes that operate via a ping-pong mechanism, such as glucose oxidase where both active sites are found above the *si*-face of the flavin ring.

Crystal structures show that an apparent Lys:NZ-Wat-FAD:N5 motif is present in a number of flavoproteins other than members of the MSOX-MTOX family, including polyamine oxidase, monoamine oxidase (A and B isoforms), LSD1, L-amino acid oxidase, and 6-hydroxynicotine oxidase [30–35]. A role in oxygen activation has been postulated for the lysine in LDS1 [28] but the catalytic function of the lysine in these enzymes is largely unknown, with the notable exception of polyamine oxidase.

Surprisingly, the lysine in polyamine oxidase exhibits a relatively minor, nonelectrostatic role in the reaction of the reduced enzyme with oxygen [36,37]. It is conceivable that oxygen activation by polyamine oxidase might be mediated, at least in part, by the positively charged product, as proposed for choline oxidase, D-amino acid oxidase and monoamine oxidase [25,26,29,38,39]. The rate accelerations attributed to the presence of a positively charged product in choline oxidase, D-amino acid oxidase, or monoamine oxidase are, however, relatively modest (10- to 75-fold) in comparison to those observed for positively charged active site lysine or histidine residues. In these enzymes oxygen activation appears to require the participation of one or more factors in addition to bound product, as observed for a nonpolar residue in choline oxidase [40]. It is worth noting that even a modest role for electrostatic catalysis is apparently ruled out in some enzymes (*e.g.* alditol oxidase) by the absence of a nearby basic residue or a positively charged reaction product.

The global view of our current state of knowledge indicates that electrostatic catalysis by active site basic residues exerts a dominant role in enhancing the oxygen reactivity of the reduced coenzyme in some enzymes, such as members of the MSOX-MTOX family. However, oxygen reactivity in other enzymes is apparently modulated by an interplay of factors, such as bound product, active site polarity [40,41] and/or accessibility [42]. Furthermore, structural and molecular dynamic studies indicate that the pathway(s) employed by oxygen to reach the active site are also likely to exhibit considerable diversity and run the gamut from a single oxygen tunnel that is either freely accessible or gated, as in TSOX and MSOX, respectively, to multiple entry paths created by dynamic motions of the protein [5,28,29,43–46].

10.4 References

[1] Massey V. Activation of molecular oxygen by flavins and flavoproteins, J. Biol. Chem. 1994;269:22459–62.

[2] Sucharitakul J, Prongjit M, Haltrich D, Chaiyen P. Detection of a C4a-hydroperoxyflavin intermediate in the reaction of a flavoprotein oxidase, Biochemistry 2008;47:8485–90.

[3] Trickey P, Wagner MA, Jorns MS, Mathews FS. Monomeric sarcosine oxidase: Structure of a covalently-flavinylated secondary amine oxidizing enzyme, Structure 1999;7:331–45.

[4] Ilari A, Bonamore A, Franceschini S, Fiorillo A, Boffi A, Colotti G. The X-ray structure of N-methyltryptophan oxidase reveals the structural determinants of substrate specificity, Proteins 2008;71:2065–75.

[5] Chen Z, Hassan-Abdallah A, Zhao G, Jorns MS, Mathews FS. Heterotetrameric sarcosine oxidase: Structure of a diflavin metalloenzyme at 1.85 Å resolution, J. Mol. Biol. 2006;360:1000–18.

[6] Wagner MA, Jorns MS. Monomeric sarcosine oxidase: 2. Kinetic studies with sarcosine, alternate substrates and substrate analogs, Biochemistry 2000;39:8825–29.

[7] Wagner MA, Trickey P, Chen Z, Mathews FS, Jorns MS. Monomeric sarcosine oxidase: 1. Flavin reactivity and active site binding determinants, Biochemistry 2000;39:8813–24.

[8] Hassan-Abdallah A, Zhao G, Jorns MS. Covalent flavinylation of monomeric sarcosine oxidase: Identification of a residue essential for holoenzyme biosynthesis, Biochemistry 2008;47:1136–43.

[9] Zhao G, Bruckner RC, Jorns MS. Identification of the oxygen activation site in monomeric sarcosine oxidase: Role of Lys265 in catalysis, Biochemistry 2008;47:9124–35.

[10] Jorns MS, Chen Z, Mathews FS. Structural characterization of mutations at the oxygen activation site in monomeric sarcosine oxidase, Biochemistry 2010;49:3631–39.

[11] Bruckner RC, Winans J, Jorns MS. Pleiotropic impact of a single lysine mutation on biosynthesis of and catalysis by N-methyltryptophan oxidase, Biochemistry 2011;50: 4949–62.

[12] Jorns MS. Properties and catalytic function of the two nonequivalent flavins in sarcosine oxidase, Biochemistry 1985;24:3189–94.

[13] Piubelli L, Pedotti M, Molla G, Feindler-Boeckh S, Ghisla S, Pilone MS, Pollegioni L. On the oxygen reactivity of flavoprotein oxidases - An oxygen access tunnel and gate in *Brevibacterium sterolicum* cholesterol oxidase, J. Biol. Chem. 2008;283:24738–47.

[14] Lind J, Shen X, Merenyi G, Jonsson BO. Determination of the rate constant of self-exchange of the $O_2/O_2^{\cdot-}$ couple in water by $^{18}O/^{16}O$ isotope marking. J. Am. Chem. Soc. 1989;111:7654–55.

[15] Gabison L, Chiadmi M, El Hajji M, Castro B, Colloc'h N, Prange T. Near-atomic resolution structures of urate oxidase complexed with its substrate and analogues: the protonation state of the ligand, Acta Crystallogr. Sect. D 2010;66:714–24.

[16] Colloc'h N, Gabison L, Monard G, Altarsha M, Chiadmi M, Marassio G, Santos JSDO, El Hajji M, Castro B, Abraini JH, Prange T. Oxygen pressurized X-ray crystallography: Probing the dioxygen binding site in cofactorless urate oxidase and implications for its catalytic mechanism, Biophys. J. 2008;95:2415–22.

[17] Kommoju PR, Chen Z, Bruckner RC, Mathews FS, Jorns MS. Probing oxygen activation sites in two flavoprotein oxidases using chloride as an oxygen surrogate, Biochemistry 2011;50: 5521–34.

[18] Maragliano L, Vanden Eijnden E. A temperature accelerated method for sampling free energy and determining reaction pathways in rare events simulations, Chem. Phys. Lett. 2006;426:168–75.

[19] Petrek M, Kosinová P, Koca J, Otyepka M. MOLE: A Voronoi diagram-based explorer of molecular channels, pores, and tunnels, Structure 2007;15:1357–63.

[20] Klinman JP. How do enzymes activate oxygen without inactivating themselves?, Acc. Chem. Res. 2007;40:325–33.

[21] Roth JP, Klinman JP. Catalysis of electron transfer during activation of O_2 by the flavoprotein glucose oxidase., Proc. Natl. Acad. Sci. USA 2003;100:62–67.

[22] Roth JP, Wincek R, Nodet G, Edmondson DE, McIntire WS, Klinman JP. Oxygen isotope effects on electron transfer to O_2 probed using chemically modified flavins bound to glucose oxidase, J. Am. Chem. Soc. 2004;126:15120–31.

[23] Kommoju PR, Bruckner RC, Ferreira P, Carrell CJ, Mathews FS, Jorns MS. Factors that affect oxygen activation and coupling of the two redox cycles in the aromatization reaction catalyzed by nikD, an unusual amino acid oxidase, Biochemistry 2009;48:9542–55.

[24] McDonald CA, Fagan RL, Collard F, Monnier VM, Palfey BA. Oxygen reactivity in flavoenzymes: Context matters, J. Am. Chem. Soc. 2011;133:16809–11.

[25] Ghanem M, Gadda G. On the catalytic role of the conserved active site residue His466 of choline oxidase, Biochemistry 2005;44:893–904.

[26] Gadda G, Fan F, Hoang JV. On the contribution of the positively charged headgroup of choline to substrate binding and catalysis in the reaction catalyzed by choline oxidase, Arch. Biochem. Biophys. 2006;451:182–87.

[27] Forneris F, Heuts DPHM, Delvecchio M, Rovida S, Fraaije MW, Mattevi A. Structural analysis of the catalytic mechanism and stereo selectivity in *Streptomyces coelicolor* alditol oxidase, Biochemistry 2008;47:978–85.

[28] Baron R, Binda C, Tortorici M, McCammon JA, Mattevi A. Molecular mimicry and ligand recognition in binding and catalysis by the histone demethylase LSD1-CoREST complex, Structure 2011;19:212–20.

[29] Saam J, Rosini E, Molla G, Schulten K, Pollegioni L, Ghisla S. O_2 reactivity of flavoproteins: Dynamic access of dioxygen to the active site and role of a H^+ relay system in D-amino acid oxidase, J. Biol. Chem. 2010;285:24439–46.

[30] Binda C, Li M, Hubalek F, Restelli N, Edmondson DE, Mattevi A. Insights into the mode of inhibition of human mitochondrial monoamine oxidase B from high-resolution crystal structures, Proc. Natl. Acad. Sci. USA 2003;100:9750–55.

[31] Binda C, Coda A, Angelini R, Federico R, Ascenzi P, Mattevi A. A 30 angstrom long U-shaped catalytic tunnel in the crystal structure of polyamine oxidase, Structure 1999; 7:265–76.

[32] Son S, Ma J, Kondou Y, Yoshimura M, Yamashita E, Tsukihara T. Structure of human monoamine oxidase A at 2.2 Å resolution: The control of opening the entry for substrates/inhibitors, Proc. Natl. Acad. Sci. USA 2008;105:5739–44.

[33] Pawelek PD, Cheah J, Coulombe R, Macheroux P, Ghisla S, Vrielink A. The structure of L-amino acid oxidase reveals the substrate trajectory into an enantiomerically conserved active site, EMBO J. 2000;19:4204–15.

[34] Chen Y, Yang Y, Wang F, Wan K, Yamane K, Zhang Y, Lei M. Crystal structure of human histone lysine-specific demethylase 1 (LSD1), Proc. Natl. Acad. Sci. USA 2006;103:13956–61.

[35] Kachalova GS, Bourenkov GP, Mengesdorf T, Schenk S, Maun HR, Burghammer M, Riekel C, Decker K, Bartunik HD. Crystal structure analysis of free and substrate-bound 6-hydroxy-L-nicotine oxidase from *Arthrobacter nicotinovorans*, J. Mol. Biol. 2010;396:785–99.

[36] Fiorillo A, Federico R, Polticelli F, Boffi A, Mazzei F, Di Fusco M, Ilari A, Tavladoraki P. The structure of maize polyamine oxidase K300M mutant in complex with the natural substrates provides a snapshot of the catalytic mechanism of polyamine oxidation, FEBS J. 2011;278:809–21.

[37] Pozzi MH, Fitzpatrick PF. A lysine conserved in the monoamine oxidase family is involved in oxidation of the reduced flavin in mouse polyamine oxidase, Arch. Biochem. Biophys. 2010;498:83–88.

[38] Porter DJ, Voet JG, Bright HJ. Mechanistic features of the D-amino acid oxidase reaction studied by double stopped flow spectrophotometry, J. Biol. Chem. 1977;252:4464–73.

[39] Miller JR, Edmondson DE. Influence of flavin analogue structure on the catalytic activities and flavinylation reactions of recombinant human liver monoamine oxidases A and B, J. Biol. Chem. 1999;274:23515–25.

[40] Finnegan S, Agniswamy J, Weber IT, Gadda G. Role of valine 464 in the flavin oxidation reaction catalyzed by choline oxidase, Biochemistry 2010;49:2952–61.

[41] Hernandez-Ortega A, Lucas F, Ferreira P, Medina M, Guallar V, Martinez AT. Modulating O_2 reactivity in a fungal flavoenzyme: Involvement of aryl-alcohol oxidase Phe-501 contiguous to catalytic histidine, J. Biol. Chem. 2011;286:41105–14.

[42] Leferink NGH, Fraaije MW, Joosten HJ, Schaap PJ, Mattevi A, Van Berkel WJH. Identification of a gatekeeper residue that prevents dehydrogenases from acting as oxidases, J. Biol. Chem. 2009;284:4392–97.

[43] Coulombe R, Yue KQ, Ghisla S, Vrielink A. Oxygen access to the active site of cholesterol oxidase through a narrow channel is gated by an Arg-Glu pair, J. Biol. Chem. 2001;276: 30435–41.

[44] Lario PI, Sampson N, Vrielink A. Sub-atomic resolution crystal structure of cholesterol oxidase: What atomic resolution crystallography reveals about enzyme mechanism and the role of the FAD cofactor in redox activity, J. Mol. Biol. 2003;326:1635–50.

[45] Moustafa IM, Foster S, Lyubimov AY, Vrielink A. Crystal structure of LAAO from *Calloselasma rhodostoma* with an L-phenylalanine substrate: Insights into structure and mechanism, J. Mol. Biol. 2006;364:991–1002.

[46] Baron R, Riley C, Chenprakhon P, Thotsaporn K, Winter RT, Alfieri A, Forneris F, Van Berkel WJH, Chaiyen P, Fraaije MW, Mattevi A, McCammon JA. Multiple pathways guide oxygen diffusion into flavoenzyme active sites, Proc. Natl. Acad. Sci. USA 2009; 106:10603–08.

11 The acyl CoA dehydrogenases

Jung-Ja P. Kim, Niels Gregersen, Rikke Katrine Jentoft Olsen, and Sandro Ghisla

Abstract

ACADs constitute a family of flavoproteins that catalyze the first step, an α,β-dehydrogenation, in the β-oxidation of fatty acids conjugated to CoA. The three-dimensional structures of six members of this family have been solved; they have a high degree of similarity and are characterized by a typical "MCAD-fold". The soluble members of the family are homotetramers composed of dimers of dimers. The cofactor FAD and the substrates are bound at the active center in extended conformations wherein the α,β-position of the latter is positioned adjacent to the flavin N(5). A glutamate at the active center is positioned near the α-hydrogen of the substrate and serves in its abstraction as a proton, thereby initiating catalysis. During this process the β-hydrogen is concomitantly transferred as a hydride to the N(5) of the flavin thereby reducing it. In the next step, electrons are transferred from the reduced flavin to the respiratory chain via electron transfer(ring) flavoprotein (ETF) and ETF-ubiquinone oxidoreductase. Some salient points that are relevant in the context of the (bio)chemical mechanisms of catalysis are discussed. Genetic defects are common in ACADs. Because ACADs play a central role in energy production and amino acid catabolism, their deficiencies are manifested in a variety of pediatric disorders. These deficiencies are discussed in some details for the most important cases.

11.1 Introduction

The β-oxidation of fatty acids is a biochemical process of great importance that is found in essentially all organisms. It consists of four steps and is initiated by a desaturation/α,β-dehydrogenation of fatty acids conjugated to CoA. The importance of this dehydrogenation is evidenced by the occurrence of a quite large family of enzymes, the acyl-CoA dehydrogenases (ACADs), which carry out essentially the same chemistry. They differ, however, in their preference/selectivity for variants resulting from structural differences in the acyl chain. The first ACAD to be identified and isolated was short chain acyl-CoA dehydrogenase (SCAD) [1]. At the time it was named butyryl-CoA dehydrogenase and was reported by Mahler to be a copper enzyme [1]. This deduction likely originated in the bright green color of the purified protein. However, it was subsequently shown to be incorrect [2]. Shortly after the description of SCAD, Beinert's group reported on the discovery of "general" acyl-CoA dehydrogenase [3], the name reflecting its broad specificity with respect to chain length and structure of its substrate. This protein is nowadays named medium chain acyl-CoA dehydrogenase (MCAD) and is probably the best-studied member of the family [4,5]. In a pioneering study Beinert's group proposed a kinetic and biochemical mechanism for MCAD that has upheld excellently the test of time and can be applied to all ACADs [6]. Following this early work, additional members of the ACAD family were discovered, the latest newcomers being ACAD10 and ACAD11, which best accept substrates with chain lengths >20 carbons and thus can be included into the sub-group of the very-long chain ACADs (VLCAD) [7].

The function of ACADs is assumed to reside primarily in the catalysis of the aforementioned α,β-dehydrogenation step in the β-oxidation cycle of CoA-conjugates that are derived from various fatty acids and from amino acid catabolism. The resulting reducing equivalents are important contributors to the energetic needs of the cell. A further role is the control/regulation of metabolic processes related to β-oxidation [8]. On the other hand, the "size" of the ACAD family suggests that functions other than the mentioned ones are likely to exist. Thus, Vockley's group has recently reported that the expression of various ACADs (MCAD and ACAD9-11) in the human central nerve system is compartmentalized [7]. Based on this intriguing finding the authors suggest *"that β-oxidation in cerebellum participates in different functions other than generating energy, for example, the synthesis and/or degradation of unique cellular lipids and catabolism of aromatic amino acids, compounds that are vital to neuronal function"* [7]. If substantiated this insight would be of great relevance. A further important aspect that will be addressed below regards the medical implications of genetic defects affecting the various members of the ACADs. Further, ACAD9 has been shown recently to be involved in the assembly of mitochondrial Complex I, the deficiency of which is the most frequent cause of oxidative phosphorylation disorders [9].

11.2 Overall structure of soluble ACADs

The structure of pig MCAD was the first of any ACAD to be determined crystallographically [10], followed by many other soluble ACAD structures, including SCAD [11], isobutyryl-CoA dehydrogenase (iBD) [12], isovaleryl-CoA dehydrogenase (i3VD) [13], glutaryl-CoA dehydrogenase (GD) [14], and *Megasphaera elsdenii* butyryl-CoA dehydrogenase (bSCAD) [15]. All soluble ACADs are homotetramers, and are arranged as dimers of dimers, as exemplified by the structure of MCAD in a binary complex with C8-CoA (►Fig. 11.1A).

All of these structures are very similar, having the "MCAD-fold" and, thus all references to other ACAD structures will use the MCAD labeling convention (►Fig. 11.1B). The monomer structure is composed of an N-terminal α-helical domain, β-sheet domain, and a C-terminal α-helical domain. The most important residue involved in catalysis of all known ACADs is a glutamate that functions as a general base. In MCAD, this base is Glu376, located in the loop between helices J and K (►Fig. 11.1B). Its homologous residues are conserved in all members of the ACAD family with the exception of LCAD and i3VD. Discussed below is the structural basis for substrate specificity for three representative soluble ACADs, including MCAD, SCAD and GD. The structure of VLCAD is also presented as a prototype for the membrane-bound ACAD subfamily. In addition, the uniqueness of the location of the catalytic residue in LCAD and i3VD is discussed. For details of the structures of other members of the ACAD family (iBD, i3VD

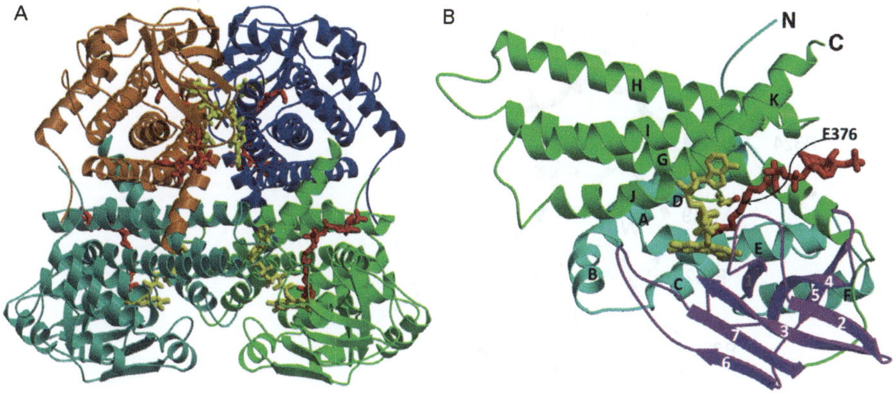

Fig. 11.1: Ribbon diagram of MCAD with bound C8-CoA. FAD and C8-CoA are shown in yellow and red sticks, respectively. (A) The tetramer is made of a dimer of dimers. Monomer-monomer interface in each dimer (blue and orange; and cyan and green) is extensive, while the dimer-dimer interactions (equatorial plane) are limited. Part of FAD is bound at the monomer-monomer interface. (B) A subunit of MCAD with bound C8-CoA. α-Helices are lettered and β-strands are numbered consecutively from the N-terminus. Both N- and C-termini are indicated. The catalytic base, Glu376, is located on the loop between helices J and K and shown near the Cα atom of the bound substrate. All figures are prepared using PyMol [159].

and i2VD) and acyl-CoA oxidases (the latter are closely related, peroxisomal enzymes), readers are referred to the earlier review by Kim and Miura [5].

11.2.1 Medium chain acyl-CoA dehydrogenase (MCAD)

As the prototypic ACAD, features common to all ACADs as well as those unique to MCAD will be discussed here. In all ACADs, the bound FAD is located in a crevice between the β-sheet domain and C-terminal helical domain of one monomer and the C-terminal helical domain of the other monomer of the dimeric unit of the tetramer (▶Fig. 11.1). The fatty acyl-CoA substrate is bound to the enzyme in an extended conformation: its acyl-chain is deeply buried inside the protein at the *re*-face of the FAD isoalloxazine ring, while the 3′AMP moiety of the CoA thioester is relatively exposed to the solvent. The Cα-Cβ moiety of the substrate is sandwiched between the isoalloxazine ring of FAD and the carboxylate group of Glu376[1]. As is detailed below, this arrangement is ideally suited for the carboxylate to abstract the *pro-R* hydrogen as a proton from the Cα atom and to transfer the Cβ *pro-R* hydrogen to the N-5 atom of the FAD ring as a hydride ion (▶Fig. 11.2, ▶Scheme 11.3) [5,16].

The thioester carbonyl oxygen is making hydrogen bonds with both 2′OH of the ribityl chain of FAD and the amide nitrogen of Glu376, ensuring the precise positioning and orientation of the substrate with respect to FAD and Glu376 for optimal catalysis. The substrate binding cavity of MCAD is deep enough to accommodate a fatty acyl chain length of up to 12 carbons (~12 Å from the substrate thioester carbonyl) (▶Fig. 11.3).

[1] Amino acids are numbered according to the sequence of the mature protein

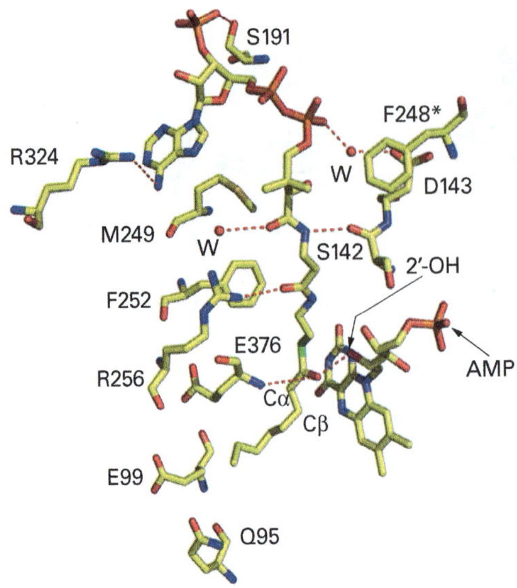

Fig. 11.2: Residues in the vicinity of C8-CoA binding cavity in MCAD. The bound ligand has an extended conformation. The fatty acyl chain of C8-CoA is deeply buried inside the enzyme molecule, while the 3-AMP-PPi portion is on the surface of the molecule. F284* is from the neighboring monomer of the dimer. Hydrogen bonds are shown as red dotted lines. The thioester sulfur atom is shown in green. Cα and Cβ atoms are labeled and are sandwiched between the flavin ring and the catalytic base, E376. The thioester carbonyl oxygen is hydrogen bonded to both 2′-OH of the ribityl chain of FAD and the main chain amide of E376.

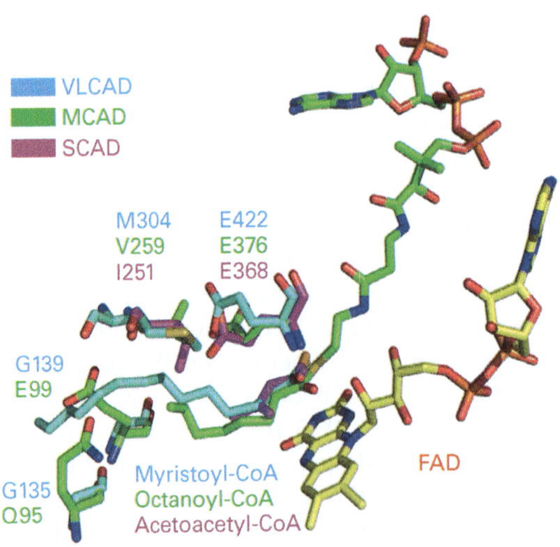

Fig. 11.3: Overlay of the residues involved in binding of the fatty acyl moiety to ACADs. SCAD (acetoacety-CoA, pink); MCAD (C8-CoA, green); and VLCAD (C14-CoA, light blue). For clarity, both FAD and the CoA moiety of C8-CoA from MCAD are shown. In MCAD, Glu99 and Gln95 form the "bottom" of the substrate binding cavity. In VLCAD, the corresponding residues are Gly139 and Gly135, respectively, making the cavity deeper and allowing for longer substrates to bind.

The base of the binding cavity is formed by Gln95 and Glu99 on helix E. In the absence of substrate, a string of well-ordered water molecules fill the binding cavity and no major changes in the tertiary or quaternary structure of MCAD have been observed upon binding of substrate. However, the side chains of a number of residues lining the substrate binding cavity adopt alternate conformations to accommodate substrate [10].

11.2.2 Short chain acyl-CoA dehydrogenase (SCAD)

As expected from the sequence alignment (see ▶Fig. 2 of reference [14], the overall structure of rat SCAD complexed with acetoacetyl-CoA is highly similar to MCAD (1.7 Å *r.m.s. deviation* between the main chain Cα atoms of the two structures) [17]. The structure also confirms results of mutagenesis studies that identified the catalytic residue being Glu368, the residue corresponding to Glu376 of MCAD [17]. SCAD lacks prolines corresponding to P257 and P258 on helix G in MCAD, both of which alter the trajectory of helix G and widen the substrate binding cavity. Instead, I251 on helix G in SCAD extends into the substrate binding cavity forming the bottom of the cavity and makes the binding cavity considerably shorter than seen in MCAD (▶Fig. 11.3). At a depth of ~7 Å measured from the substrate thioester carbonyl, the binding cavity limits substrate acyl-chain length of up to six carbons.

11.2.3 Glutaryl-CoA dehydrogenase (GD)

GD oxidatively decarboxylates glutaryl-CoA to crotonyl-CoA and CO2. Thus it possesses an additional decarboxylation activity that other ACADs do not have. The structures of GD with and without an alternate substrate, 4-nitrobutyryl-CoA have been determined [14]. In conjunction with kinetic studies [18], the structures reveal a plausible mechanism for the decarboxylation reaction [14]: the reaction is initiated by polarization of the 2-enoyl-CoA by the thioester carbonyl oxygen, and Arg94 stabilizes the transient crotonyl-CoA anion. Protonation of the crotonyl-CoA anion occurs by a 1,3-prototropic shift catalyzed by the conjugated acid of the general base, Glu370. A tight hydrogen-bonding network involving γ-carboxylate of the enzyme-bound glutaconyl-CoA, with Tyr369, Glu87, Arg94, Ser95, and Thr170, optimizes orientation of the γ-carboxylate for decarboxylation.

11.2.4 Very Long Chain Acyl-CoA Dehydrogenase (VLCAD)

Currently four ACADs have been identified that are specific for β-oxidation of long chain fatty acid thioesters, VLCAD, ACAD9, ACAD10 and ACAD11. Among them VLCAD is the best characterized both biochemically and structurally. There are two isoforms of VLCAD *in vivo*, the long form (615 residues) and the shorter form that lacks residues 7–28 of the long form due to an alternative splicing. The structure of the short form (hereafter simply referred to as VLCAD) has been determined [19]. Unlike the soluble ACADs, VLCAD is a homodimer with the first ~400 residues forming an MCAD-like domain and the unique C-terminal ~180 residues forming an α-helical bundle (α-domain 3), whose helical axis is almost perpendicular to the N-terminal helices (▶Fig. 11.4). A closer inspection of the two structures reveals that the interface between the N-terminal

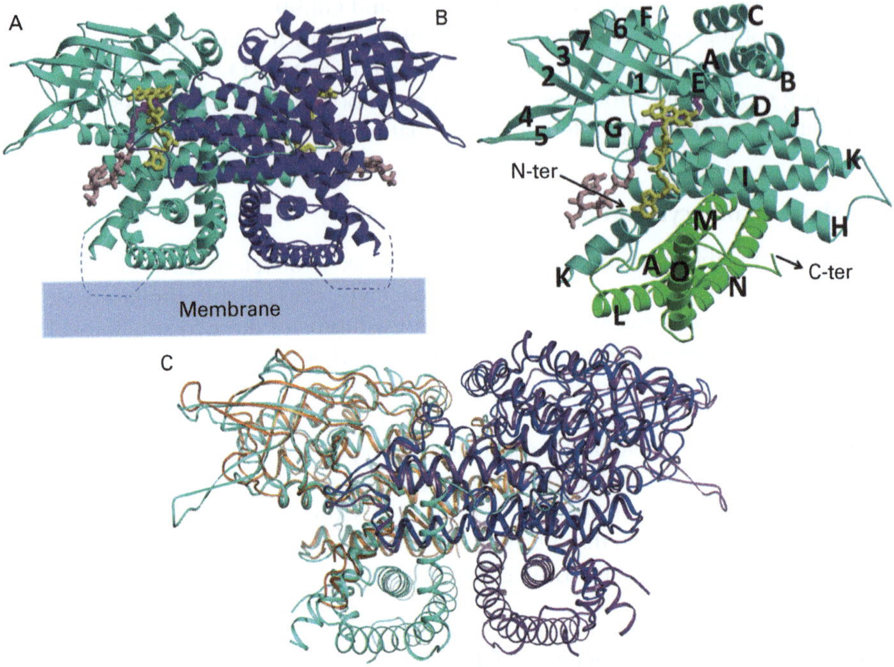

Fig. 11.4: Ribbon diagram of human VLCAD complexed with myristoyl-CoA. (A) A dimeric molecule is shown (cyan and purple monomers). FAD and C14-CoA are represented with sticks in yellow and pink/salmon, respectively. The 3-AMP-PPi portion of C14-CoA is disordered and rendered in salmon color. The disordered regions of the polypeptide (residues 446–478) is shown with dotted lines and proposed to be the membrane binding region. (B) Overall polypeptide folding of a monomer of VLCAD. Dotted line in cyan indicates the disordered region. Note the resemblance of the first 400 residues (shown in cyan) to the monomer structure of MCAD. The ~180 residue C-terminal domain forms an α-helical bundle (α-domain 3). (C) Overlay of the VLCAD dimer (cyan and purple) and a dimer of MCAD (orange and blue). The α-domain 3 of VLCAD is unique; however, the interface between α-domain 3 and the N-terminal 400 residue domain of VLCAD is very similar to that of the dimer-dimer interface in MCAD – see text.

and C-terminal domains in VLCAD is the same as the dimer-dimer interface of the tetrameric MCAD. Furthermore, the amino acid sequence of the C-terminal 180 residues has 14% sequence identity with those of the N-terminus, with which they align structurally, suggesting that the C-terminal α-domain 3 of VLCAD most likely originated from partial gene duplication. ▶Fig. 11.4C shows an MCAD dimer overlaid onto the VLCAD dimer, showing the similar folding between MCAD and the first 400 residues of VLCAD.

Although the crystals were obtained in the presence of myristoyl-CoA, the CoA moiety of the substrate is barely visible in the structure, presumably due to 1) mobility of the CoA moiety compared to the fatty acyl portion, and 2) partial hydrolysis of the thioester during the relatively long crystallization process (the CoA moiety shown in ▶Fig. 11.4 is modeled in). Glu422 (corresponding to Glu376 of MCAD) is situated at the *re*-face of the FAD isoalloxazine ring near the C2 and C3 atoms of the thioester, confirming its role as the catalytic base.

►Fig. 11.3 shows residues involved in the binding cavities for fatty acyl chains in SCAD, MCAD, and VLCAD. The depths of the binding cavities (measured from the carbonyl oxygen of the thioester) are approximately 8, 12 and 24 Å in SCAD, MCAD, and VLCAD, respectively. Thus, SCAD can accommodate short (C4-C6) acyl chains, MCAD can utilize C6-C12 acyl-CoAs, and VLCAD can accommodate acyl chain length as long as 24 carbons. However, VLCAD has very little activity for substrates with chain length less than 12 carbons [20], and the question is why. The depth of the cavity is part, but not all, of the story for chain length specificity. ►Fig. 11.4C shows an overlay of the structures of MCAD and SCAD in the vicinity of the CoA moiety-binding site. Although the adenosine-2′AMP moiety of the thioester is relatively exposed, in MCAD, there are several interactions between the adenosine-2′AMP-PPi portion of the CoA thioester and the polypeptide, including Ser191 with 2′phosphate, C6 amino group of adenine ring with Arg324 and Arg253, and some water molecule-mediated hydrogen bonds. However, in the VLCAD structure, the adenosine-2′phosphate-3′PPi is completely solvent-exposed, with no interactions with the polypeptide. This loss of binding energy is compensated by the additional hydrophobic interactions due to the longer fatty acyl chain in VLCAD. It has been shown that the alkyl-SCoA thioether binding energy for MCAD increases linearly with chain length (390 cal/CH_2 group) [21]. Therefore, in VLCAD, the substrate/product binding energy comes mainly from the acyl-moiety, while in other ACADs the binding energy arises from both fatty acyl group and the CoA moiety. The lack of an interaction between the AMP-PPi moiety of the substrate and the polypeptide is at least partially responsible for the lack of electron density attributable to substrate in the structure. VLCAD is a monotopic membrane binding protein. *In vitro* membrane binding studies with clinical mutations A450P and L462P have shown that, while these two mutants are active and stable, they have a greatly reduced ability to bind membrane, strongly suggesting that the region containing these residues must be involved in membrane binding [22,23]. However, residues 446–478, including A450 and L452, are disordered in the crystal structure. Based on an electrostatic surface analysis together with helical modeling of the disordered region, McAndrew *et al.* have proposed a model of VLCAD oriented on the mitochondrial membrane (►Fig. 11.4A). Further studies are needed to unequivocally confirm this hypothesis.

11.2.5 Position of the catalytic base in primary sequence

LCAD and i3VD are unique in that neither enzyme contains a glutamate at the position corresponding to Glu376 of MCAD, although significant sequence identity (35–45%) is otherwise conserved along the entire length of their polypeptide chains. Instead, both enzymes have either Gly (human LCAD) or Ala (human i3VD) at the corresponding position. Molecular modeling and site-directed mutagenesis studies of LCAD strongly suggest that Glu261 on helix G is the catalytic base [24]. The structure of human i3VD [13] reveals that Glu254 (corresponding to Glu261 of LCAD) is positioned near the C2-C3 bond of the bound CoA thioester, confirming it to be the catalytic base, in agreement with mutagenesis studies [25]. Although Glu367 (MCAD position) and Glu254 (i3VD position) are separated by more than 100 residues in the primary sequence, and thus positioned on different helices, they are topologically conserved in the three-dimensional structure of these proteins. i3VD and LCAD thus provide an excellent example of a catalytic residue that is not conserved in a family of closely related enzymes. While the mechanism by which such a migration of catalytic residue might occur in the ACAD family of enzymes is not clear, it is likely that it has evolved via a putative intermediate enzyme having Glu residues at both positions [26].

11.3 The basic biochemical mechanism of the α,β-dehydrogenation step

The overall mechanism catalyzed by ACADs consists of two half-reactions, oxidative and reductive, as is the case with most flavoproteins (►Scheme 11.1):

In the reductive half-reaction (a), an equivalent of H_2 is transferred from the substrate to the oxidized flavin cofactor (E~Fl_{ox}) thereby reducing it. This is followed by transfer of the reducing equivalents to the acceptor (Acc_{ox}) oxidized ETF in an oxidative half-reaction (b) wherein oxidized enzyme is reformed. Reduced ETF then goes on to transfer the equivalents to the respiratory chain via ETF-QO. The sequence of kinetic steps involved in catalysis by ACADs (►Scheme 11.2) has been best studied with MCAD [4], and the present, short discussion will refer to this case.

Catalysis is initiated by formation of an encounter complex E-Fl_{ox}~SH_2 that isomerizes and then undergoes dehydrogenation (species E-Fl_{ox}~SH_2 preceding k_3). This isomerization is rate-limiting with some substrates and is assumed to involve a change in ionization state [4]. k_3/k_{-3} represents the chemical step of dehydrogenation and involves formation of a complex of reduced enzyme with enoyl-CoA product (E-$Fl_{red}H^-$~S).

$$\text{E~Fl}_{ox} + \text{SH}_2 \xrightleftharpoons{(a)} \text{E~Fl}_{red}\text{H}^- + \text{S} + \text{H}^+$$

$$\text{E~Fl}_{red}\text{H}^- + \text{Acc}_{ox} \xrightleftharpoons{(b)} \text{E~Fl}_{ox} + \text{Acc}_{red}\text{H}^-$$

Scheme 11.1: Reactions catalyzed by ACADs. E~Fl and Acc stand for the redox forms of the enzyme and acceptor as denoted by the subscript (ox, red: oxidized or reduced), SH_2 and S are the substrate and product.

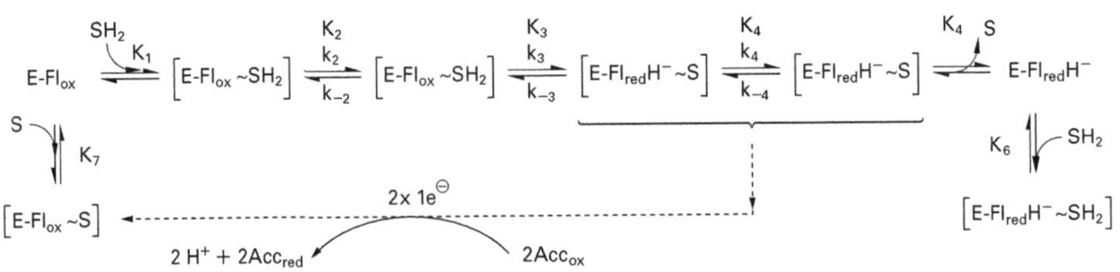

Scheme 11.2: Kinetic steps for catalysis by ACADs. The (minimal) scheme was developed for the case of MCAD, but its essentials are assumed to have general validity for ACADs. E: ACAD; Fl_{ox} and Fl_{red}: oxidized and reduced flavin cofactor; SH_2: acyl-CoA substrate; S: acyl-CoA enoyl product; k: single kinetic steps; K: equilibria. [Species] denote various complexes. K_1 to K_7 represent steps that are involved in the reductive half-reaction. In the presence of acceptors (Acc) such as ETF (native acceptor) or the ferricenium ion (artificial electron acceptor [162]) reduced enzyme species become reoxidized via the dotted lines (lower part of panel). Note that the H^+ balance has not been formalized. Adapted from [4].

This latter species is a charge transfer (CT) complex in which reduced flavin ($Fl_{red}H^-$) is the donor and enoyl-CoA (S) the acceptor [27]. This complex absorbs in the long-wavelength region, yielding the characteristic greenish-blue color that is observed upon addition of substrate to oxidized ACADs. A typical case is depicted in ▶Fig. 11.5 for MCAD. "k" terms in ▶Scheme 11.2 represent unidirectional steps while "K" represents equilibria. Interestingly, upon addition of substrate to MCAD in the absence of electron acceptors the system will attain an equilibrium encompassing all species that are not interconnected by the dashed arrows in ▶Scheme 11.2. The position of this equilibrium will depend on the reciprocal affinity of the species and ligands involved and on the values of the K's interconnecting these. In other words: the extent of apparent reduction

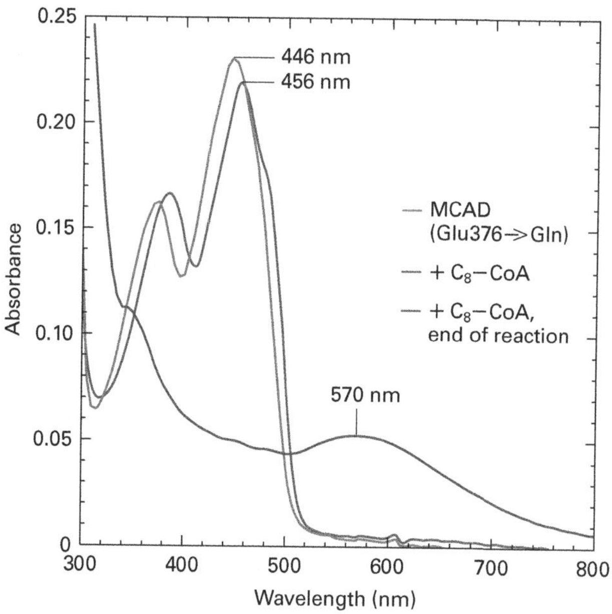

Fig. 11.5: Absorption spectra of MCAD. Green curve: purified Glu376Gln-MCAD [160], ≈16 μM in phosphate buffer pH 8 and at 5 °C. Immediately upon addition of a ≈25-fold molar excess octanoyl-CoA under anaerobic conditions the spectrum of the oxidized enzyme-C8-CoA complex was recorded (blue curve). Upon this a slow decrease of the absorption band centered at ≈450 nm ensues, which corresponds to reduction of enzyme bound oxidized flavin (see ▶Scheme 11.2). The final spectrum (red curve) was recorded after ≈15 h and represents the spectrum of (mainly) the complex of reduced enzyme with octenoyl-CoA. Further incubation over several days leads to a decrease/disappearance of the long-wavelength band centered at 570 nm, which corresponds to dissociation of the octenoyl-CoA from the complex (not shown). The Glu376Gln mutant was used since its reactions with substrate mimic those observed with wt-MCAD, but are much slower thus allowing a better characterization of intermediates; its spectral properties are closely similar to those of wt-MCAD. The spectrum of uncomplexed, reduced MCAD (or ACADs) is rather featureless down to 300 nm, it has a shoulder at ≈340 nm and essentially no absorption at $\lambda > 500$ nm; it is best obtained by chemical reduction of the oxidized enzyme, see *e.g.* ▶Fig. 1 in [47]. Note that the spectral differences between free oxidized enzyme and its complex with substrate can be used to monitor binding parameters of the latter. Adapted from [161].

of a given ACAD will depend on the values of the equilibrium constants K_{1-7} and on the concentrations of SH_2 and S.

The detailed chemical mechanism of the dehydrogenation step has been object of intensive studies and has been reviewed in [4]. Its essentials, as exemplified by MCAD are depicted in ▸Scheme 11.3:

The reaction is initiated by deprotonation of the acyl-CoA αH by the active center base Glu376 in its anionic form. Concomitantly the β-H is expulsed as a hydride which is accepted by the LUMO of the flavin at position N(5) this being equivalent to flavin reduction. Several mechanistically important details are addressed briefly in the following (for more in depth reviews we refer to [4,5]):

- While a glutamate is the catalytic base in the majority of ACADs, in long chain acyl-CoA dehydrogenase (LCAD) and isovaleryl-CoA dehydrogenase (i3VD) a corresponding Glu is placed at position 255 [26,28]; the overall topology at the active centers is conserved.
- From the three-dimensional structure (see above) one carboxylate oxygen of the H^+-abstracting base Glu376 in MCAD is positioned in the extension of the substrate αC-H bond, the oxygen being at ~4.5 Å from the αC [5,10].
- The substrate β-C-H function is positioned in the extension of the N(5) LUMO orbital the distance from the β-C to the N(5) center being ~3.3 Å [5,10].
- The stereochemistry of the reaction is *pro 2R,3R* [29,30].
- The breaking of the α- and β-H bonds is concerted in the case of MCAD and octanoyl CoA (single transition state) [4,31,32], although it may proceed via metastable carbanionic intermediates with other ACADs and substrates [33–35]. The profile of the catalyzed reaction (single transition state vs. intermediate) might differ depending on the enzyme and substrate and their interaction, *i.e.* on the extent of stabilization of the incipient carbanion in the given system.
- The pK_a of the αC-H group is lowered in the MCAD~SH_2 complex from >20 (free acyl-CoA) towards 7 [4,35–37].

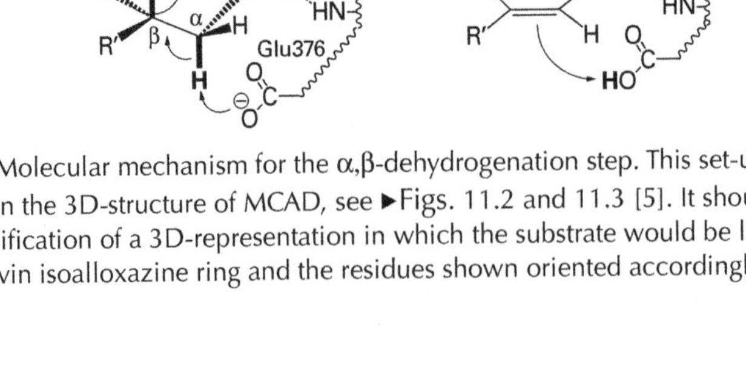

Scheme 11.3: Molecular mechanism for the α,β-dehydrogenation step. This set-up was derived mainly based on the 3D-structure of MCAD, see ▸Figs. 11.2 and 11.3 [5]. It should be noted that it is a 2D-simplification of a 3D-representation in which the substrate would be located on the plane of the flavin isoalloxazine ring and the residues shown oriented accordingly.

- The activation of the αC-H group (pK_a lowering) is brought about mainly by formation of two tight H-bonds with Glu376NH and FAD-2′OH [38]. This constitutes the equivalent of an "oxyanion hole", which serves to stabilize transient negative charges [39]. A further contribution to activation results from electronic effects of the (electron deficient) flavin [35,40–42] probably via interaction of the CT-type either in the ground state or involving the transition state, which will have anionic character.
- The pK_a of the active site base Glu376 is increased to ≥7, *i.e.* it has approximately the same value as that of the reaction partner as required for efficient H^+ transfer [38,43,44]. This effect is attributed mainly to desolvation at the active site induced by substrate binding [37].
- The activity of ACADs is strongly pH dependent (see below and ▶Fig. 11.6) reflecting the requirement of the active site base to be deprotonated at the outset of catalysis [4].
- The modulation (increase) of the reduction potential of the flavin in the E-Flox~SH_2 complex is an important aspect of catalysis. It is brought about by a combination of electrostatic effects in the active site. [45–49].

11.3.1 Chain length specificity and pH dependence

As pointed out in the introduction, single ACADs are not specific for a given structure of an acyl-CoA conjugate, but will accept similar substrates provided the structural differences are not too great. Thus each ACAD will exhibit a "proprietary" profile with respect to chain length structure as shown exemplarily in ▶Fig. 11.6A for 4 selected ACADs.

In addition the activity of ACADs is strongly dependent on pH, where in general – but not in every case – activity is low at low pH and increases with increasing pH. As illustrations, the cases of MCAD and LCAD are shown in ▶Fig. 11.6B. Whether the observed (apparent) pK's reflect ionizations of specific groups is unknown, since the underlying basis for the pH dependence is not understood in detail [4].

11.3.2 The oxidative half-reaction/interactions of ACADs with electron transfer flavoprotein (ETF)

The oxidative half-reaction is described by equation (b) in ▶Scheme 11.2 and consists in the transfer of two reducing equivalents from the reduced flavin of ACAD to the physiological acceptor electron transfer flavoprotein (ETF) and follows the sequence outlined in ▶Scheme 11.2. The transfer itself likely occurs in two one-electron steps [50]. ETF then transfers the reducing equivalents to the main mitochondrial respiratory chain via ETF-ubiquinone oxidoreductase (ETF-QO) [51]. ETF thus serves as the obligatory electron acceptor for at least 13 different flavoprotein dehydrogenases, including the 11 known ACADs as well as two enzymes involved in choline metabolism. ETF is a heterodimer (α subunit, 32 kDa; β-subunit, 28kDa) containing one FAD and one redox-inactive AMP per dimer [52,53]. The crystal structure of human ETF reveals that the protein folds into three distinct domains [54] (▶Fig. 11.7A). Domain I and the majority of domain II comprise the α-subunit, with a hinge that separates between them. Domain III is made entirely of the β-subunit. The FAD is located in a cleft between the two subunits, with

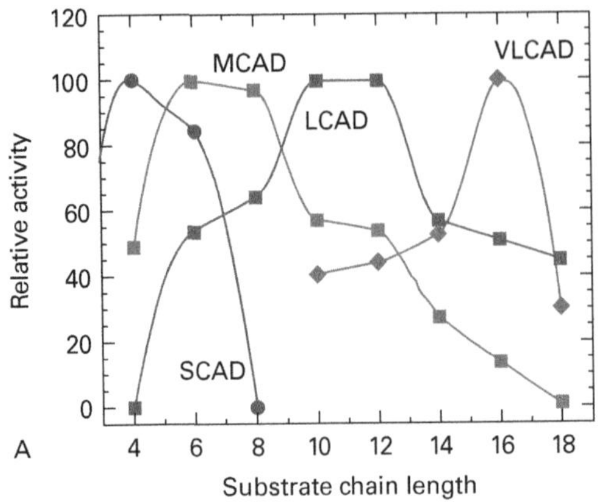

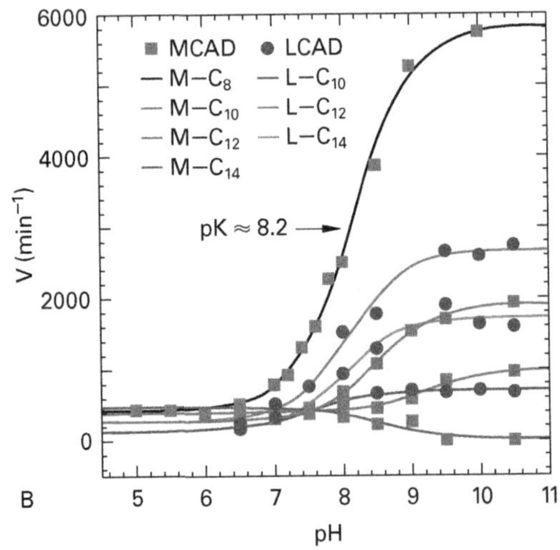

Fig. 11.6: Activity of ACADs as a function of substrate chain length and pH. (A) Dependence of the relative activity of ACADs from the substrate chain length. Activity values have been normalized. (B) pH dependence of the activity of MCAD and LCAD, and effect of the substrate chain length. Activities assessed with the ferricenium assay [162] in the presence of constant 250 mM KCl as electrolyte. "M" or "L" indicates either MCAD or LCAD, and the suffix the substrate chain length. Adapted from [4].

the majority residing in domain II. The AMP cofactor is buried deep within the β-subunit and plays a purely structural role (▶Fig. 11.7A).

The X-ray crystal structures of various ETFs (both human and *Paracoccus* ETF; wild-type and mutants) [54,55] and solution studies [56] indicate that the FAD-domain (domain II) is flexible. In addition, in the structure of MCAD complexed with ETF (MCAD:ETF), the electron density corresponding to the flavin binding domain (domain II) is barely visible [57]. Furthermore, the crystal structure of a mutant ETF (Eβ165A) in complex with MCAD (▶Fig. 11.7B) reveals that domain II is twisted about 30° with respect to

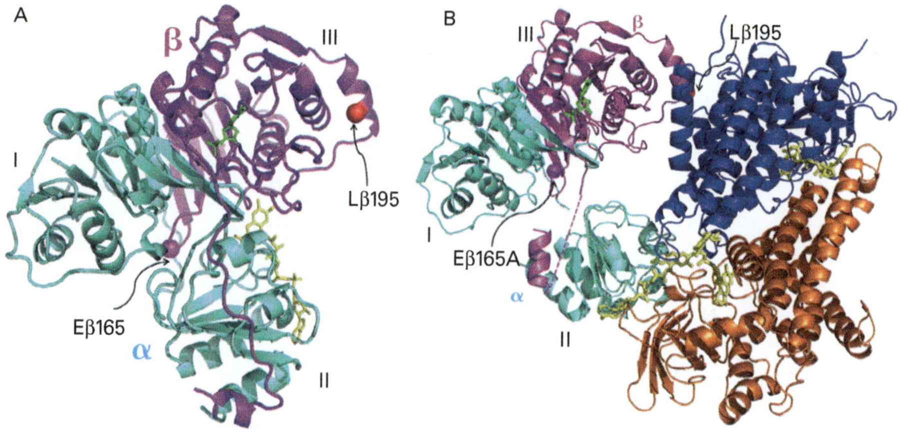

Fig. 11.7: Structures of ETF alone and in complex with MCAD. (A) The human ETF structure is comprised of three domains [54]. Domains I and II are contributed by the α-subunit (cyan), while domain III is made exclusively of the β-subunit (pink). FAD (yellow sticks) binds to domain II, while AMP (green sticks) binds inside of domain III. Locations of Lβ195 and Lβ165 are marked as red and pink balls, respectively. Lβ165 in the wild type structure makes hydrogen bond with Nα259 (not shown in the figure). (B) Structure of a mutant ETF (Eβ165A) in complex with MCAD. The color scheme for ETF is the same as in (A) and only a dimer of MCAD (blue and orange) is shown for clarity. The loop containing Lβ195 (in domain III, red ball) anchors to the hydrophobic area of the blue monomer of MCAD, and the highly mobile domain III (cyan) searches for an optimum position for fast electron transfer. Domain 3 in the complex structure is moved and about 30° twisted from the wild type structure (closed form). The dotted line indicates the disordered region of the ETF structure, presumably due to the high mobility of the domain II. Note that the flavins of ETF and the orange monomer of MCAD are in close proximity.

the rest of the ETF molecule so that the FAD domain is in a position optimal for fast electron transfer between the two flavins [58]. In the MCAD:ETF structure, Lβ195 of ETF is completely buried within a hydrophobic pocket lined by residues that are parts of α-helices A, C, and D of MCAD. It is also clear from the structure that ETF anchors to one monomer of MCAD while electron transfer occurs at the neighboring monomer. Taken together, Toogood *et al.* have proposed a dynamic multistate model in which ETF anchors to MCAD by a "recognition loop" located in domain III [50]. This loop acts as a static anchor and a highly mobile redox active FAD domain (Domain II) samples a range of conformations for an optimal electron transfer ("productive conformation" as opposed to the uncomplexed, "nonproductive conformation") (►Fig. 11.7B). It is likely that other ACADs would adopt the same mode for their interactions with ETF. However, it is not known whether ETF-QO (the electron acceptor of ETF) or non-ACAD-like dehydrogenases would use a similar strategy for their interaction with ETF.

11.3.3 The inhibition/inactivation of ACADs

Since binding of CoA conjugates to ACADs is in general tight, it is not surprising that there is a plethora of inhibitors and inactivators. One of the first examples is

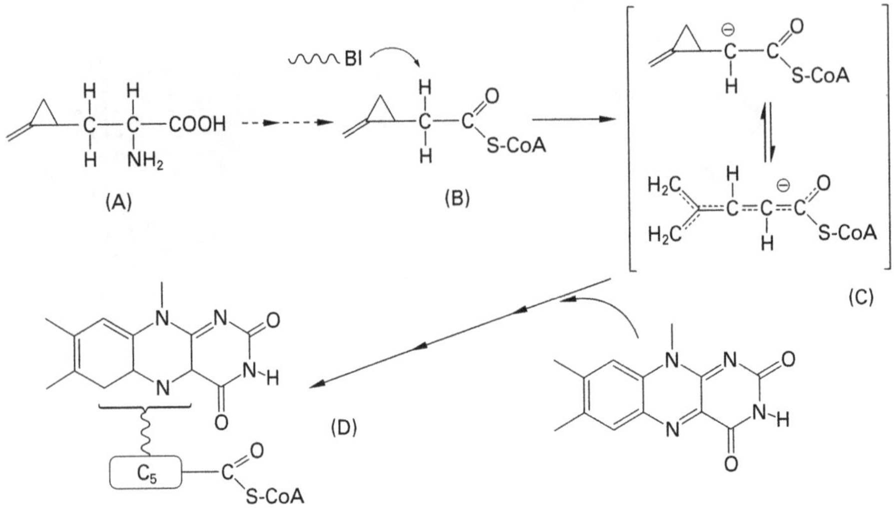

Scheme 11.4: Mode of inactivation of SCAD and MCAD by Hypoglycin A. The amino acid (A) is first degraded to methylenecyclopropylacetate, which is conjugated with CoA to yield methylenecyclopropylacetyl-CoA (B). This behaves as a "suicide inhibitor" and is probably deprotonated at its αC-H position to yield an activated intermediate (C). Its reaction with the active center flavin leads to formation of covalent adduct(s) that involve the flavin positions C(4a), N(5) and probably C(6). Adapted from [4,60].

CoA persulfide, a ligand that often copurifies with ACADs and gives the enzymes a characteristic green color that is attributed to a charge-transfer interaction [59]. Inhibitors in general exert their effects by tight, competitive binding to the active site, an example being the afore-mentioned CoA-persulfides. Inactivators on the other hand implement a modification of either the flavin cofactor or the protein. The mechanism by which this occurs varies from case to case and has been reviewed previously [4]. A historically relevant case is the inhibition by hypoglycin A [60]. This is an amino acid contained in the unripe ackee fruit, which, when ingested, induces Jamaican vomiting sickness [61,62]. Inhibition is not brought about by the hypoglycin itself, but by the CoA conjugate of methylenecyclopropylacetate, a metabolite of the former. As outlined in ▸Scheme 11.4 inactivation proceeds via a mechanism-based process, leading to covalent and irreversible modification of the flavin cofactor. The detailed mechanism of the modification is still unclear, although it might be initiated by abstraction of the methylenecyclopropyl-CoA αC-H to form a carbanionic species that subsequently attacks the oxidized flavin either nucleophilically or via a radical species.

11.3.4 Deficiencies of ACADs

Within the family of ACADs virtually every member is implicated in a recessively inherited single-gene defect or in secondary multiple ACAD deficiency, which is due to recessively inherited gene defects in electron-transferring flavoprotein (ETF) or ETF ubiquinone oxidoreductase (ETF-QO). Only the newly discovered ACADs 10 and 11 have yet to be associated with a specific disease [7,63,64]. ETF is the obligatory redox acceptor from functional ACADs and it transfers electrons to the respiratory chain via ETF-QO. Inherited defects in the genes coding for ETF and ETF-QO may illustrate the importance of the ACAD enzymes for cellular survival and death. Indeed, severe defects[2] due to mutations that abolish production of ETF or ETF-QO are lethal neonatally [70,71]. This is because ATP production from fatty acids and certain amino acids is dramatically decreased and because the functional block results in accumulation of acyl-CoA substrates. These, in turn, may be toxic, or be metabolized to toxic compounds, as illustrated for metabolites accumulated in MCAD deficiency [72], and more generally for a variety of metabolites in organic acidurias, including the other ACAD deficiencies [73]. Indeed, the accumulations are reflected in massive urinary excretion and accumulation in blood of metabolites originating from fatty acid oxidation, branched chain amino acid oxidation as well as glutaric acid metabolites from lysine, hydroxyl-lysine and tryptophan [74]. Although not so illustrative from the urine and blood metabolite profiles the same incompatibility with neonatal life holds also for long chain fatty acid oxidation (FAO) defects, such as VLCAD deficiency, where mRNA truncating defects result in neonatal death [75]. Actually, in these diseases there exists a genotype-phenotype relationship, which is not true for other ACAD disorders, such as MCAD, SCAD, i3VD and GD deficiencies [69,76–79]. Despite the fact that homozygosity or compounds heterozygosity for truncating gene defects are detected in the other ACAD disorders, except for SCAD deficiency, the total lack of enzyme protein does not in itself result in death, presumably because of overlapping/compensatory enzyme activities and lower toxicity of accumulated metabolites. SCAD deficiency is special, since no patients harboring truncating gene defects on both alleles have been detected [80]. Nor has any newborn with such defects been detected in the neonatal screening for metabolic disorders[3] [69]. The reason for this is not obvious, but it has been suggested that the total lack of SCAD gives rise to lethal amounts of butyric acid in the haploid cell state [80].

As discussed previously [70], FAO diseases caused by defects in ACAD genes comprise all types of mutations, although there is an overrepresentation of missense mutations. Especially after neonatal screening has been instituted in many countries, the identification of missense gene variation in biochemically affected newborns has accelerated, as will be discussed below for MCAD and ETF-QO deficiencies, and mentioned for VLCAD and ACAD9 deficiencies. SCAD deficiency is again special, and will be discussed separately. However, before we embark in these discussions the cellular management of FAO proteins with missense variations will be recapitulated.

---

[2] Mutations which abolish protein production may be big deletions, which delete the whole or part of the gene; small deletions or insertions, which are out of reading frame; splice defects, which delete whole or part of out of frame exons; stop codon alterations. All of these defects may produce truncated mRNA species, which are detected by the RNA surveillance systems and degraded by non-sense mediated decay [65,66] 'Mild' splice alteration as well as variations, which alter the function of splice enhancers or silencers, result in decreased amounts of mRNA and protein products [67]. Missense gene variations, which changes a single amino acid as well as deletions and insertions, which are in-frame, give rise to protein products in varying amounts and distorted conformation [68,69].

[3] Neonatal screening for the majority of ACAD diseases has been instituted in many countries. The screening is performed by determination of acylcarnitines by tandem mass spectrometry in blood collected on filter paper soon after birth.

11.4 Biogenesis of mitochondrial FAO proteins

An important reason for distinguishing between severe truncating gene defects, which do not give rise to any protein product, and defects that result in production of proteins, such as missense variant proteins, is that misfolded proteins may be detected by cellular protein quality control systems and thus affect the homeostasis of the cell in a variety of ways [68,81]. The effects on the cell are dependent on a number of factors: 1) the nature of the defect, which is a determining factor for the fate of the protein; 2) the efficiency of the quality control system, including possible effects of cofactors, such as FAD for ACAD and ETF/ETF-QO proteins, and 3) the presence of adverse cellular and/or environmental factors. Both 2) and 3) may modify the fate and therefore also the cellular consequences of misfolded proteins [68,81].

The possible fates of nuclear-encoded mitochondrial proteins, such as FAO enzyme proteins, are illustrated in ▸Fig. 11.8.

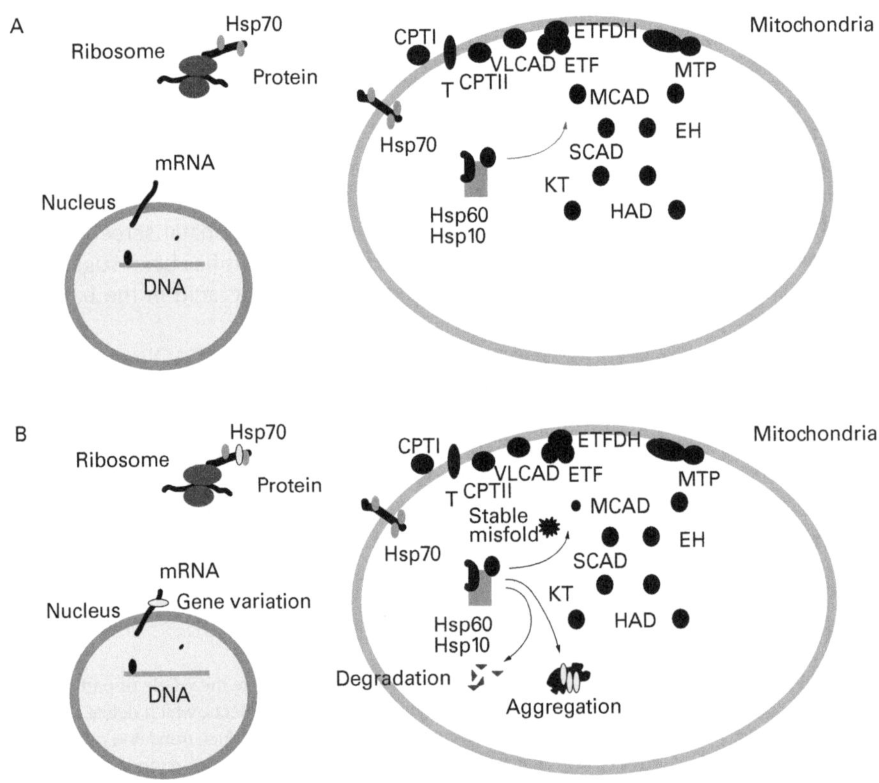

Fig. 11.8: (A) Biogenesis of fatty acid oxidation proteins. (B) Fate of fatty acid oxidation proteins carrying missense gene variations. For explanation, see text. The enzymes are: carnitine palmitoyl-CoA transferase I, CPTI; carnitine-acylcarnitine translocase, T; carnitine palmitoyl-CoA transferase II, CPTII; mitochondrial trifunctional protein, MTP; enoyl hydratases, EH; 3-hydroxy acyl-CoA dehydrogenases, HAD; 3-oxo acyl-CoA thiolases, KT; ETFDH = ETF-QO; for other proteins see list of abbreviations.

After transcription in the nucleus and translation in the cytosol, where the nascent polypeptide is shielded/protected by molecular chaperones such as cytosolic hsp70, and transported through the mitochondrial membranes (assisted by mitochondrial hsp70), some proteins may fold into their active conformation without further help, whereas others require further assistance from the hsp60/10 system to achieve the correct and functional conformation. Furthermore, many proteins whose function is dependent on cofactors need these for correct folding and optimal stability. This is illustrated by riboflavin deficient rats, which develop multiple acyl-CoA dehydrogenation defects and which show urinary excretion of metabolites similar to those excreted by human patients harboring ETF or ETF-QO defects [82]. In addition, such rats exhibit folding defective ETF/ETF-QO and ACADs in isolated mitochondria [83]. *In vitro* studies have shown that the temperature stability of wild-type human ETF-QO and SCAD is enhanced by FAD [84,85]. The effects of FAD on the fate of misfolding FAO enzymes have been discussed by Henriques *et al.*, and by Ames *et al.* [86,87]. The role of FAD as folding assistant (chaperone) will be discussed below, but first we describe the possible fates of misfolded FAO enzyme proteins, as illustrated in ▶Fig. 11.8B. If the missense variation is not located in splice regulatory elements or in the leader sequence, which is present in most FAO proteins, the transcription, translation and mitochondrial translocation are considered unaffected, as indicated for a large number of SCAD variant proteins [80]. However, once inside the mitochondrial matrix, there are a number of possible fates. The protein may be degraded immediately so that little or none of it reaches a stable or semi-stable conformation. Degradation is executed by mitochondrial proteases, such as Lon and ClpP, which together with the molecular chaperones constitute the mitochondrial protein quality control system, the main function of which is to eliminate misfolded and damaged proteins in order to avoid accumulation of protein aggregates [68]. Alternatively, the protein may not be fully degraded rapidly, and accumulate as aggregates. The extent to which this is the case for ACAD variants *in vivo* is not known, but *in vitro* experiments indicate that aggregates may well accumulate. Biogenesis studies of *in vitro* expressed ^{14}C-labelled variant SCAD proteins in isolated mouse mitochondria have shown that all of 27 tested misfolded proteins give rise to aggregates after prolonged association with hsp60 [80]. The association with the chaperone and aggregate is aggravated at elevated temperature, indicating that accumulation of misfolded ACAD proteins may contribute to clinical symptoms in affected patients, especially during fever episodes. Unfortunately, the experiments were not performed at deficiency or pharmacological doses of riboflavin, the precursor of FAD. This has been done, however, in a subsequent study where a number of ETF-QO missense variant proteins, have been investigated in HEK293 cells [84]. Under riboflavin depletion all variant proteins decrease in amount and activity compared to wild-type enzyme, whereas riboflavin supplementation rescues those variant proteins identified in patients who are responsive to riboflavin medication (see below). Moreover, it can be inferred from these *in vitro* experiments that FAD, whose intracellular levels are elevated as a result of riboflavin supplementation, facilitates protein folding early on in the process, probably in the nucleation phase. This result complements recent studies done with a missense variant of ETFβ, which together with ETFα constitutes the heterodimeric ETF protein (see ▶Fig. 11.7), where it is seen that FAD improves the conformational and proteolytic stability of the heterodimer [88]. Importantly, as for most of the ETF-QO variants examined [84], the studied missense mutation in the variant ETFβ protein is not located near the FAD binding site, supporting the notion that FAD acts in the general folding process, and not only to stabilize the FAD binding pocket. Further indications that FAD plays a role as a chaperone were reported by two groups. Many years ago Tanaka *et al.* investigated MCAD folding and tetramerization [89], while one of the present authors has shown that the prevalent MCAD variation, p.Lys329Glu (Lys304Glu)[4], which is located far from FAD binding, affects FAD binding [90]. A more detailed discussion of flavinylation of ACADs, ETF and ETF-QO is to be found in a recent review by Henriques [87]. Suffice to say here that in addition to the dependence on the molecular chaperones, which has been investigated intensively earlier [91,92], several variants of ACADs and ETF/ETF-QO are stabilized by FAD. As cellular FAD content is regulated by diet and different physiological stressors, this leads to fate 3) of misfolding. This could modulate certain 'mild' FAO protein variants that could exist in stable or semi-stable conformations, possessing more or less wild-type functionality. In summary, the residual function of certain missense and misfolding variants of ACAD, ETF and ETF-QO, is dependent on a variety of cellular and environmental factors, as measured both *in vitro* and *in vivo*. The question is thus to what extent these studies and ideas are relevant to the pathophysiology and treatment of patients with FAO and other ACAD disorders. As mentioned above, the discussion that follows will focus on MCAD, ETF-QO, VLCAD, ACAD9 and SCAD deficiencies, since these are the best understood cases.

11.5 MCAD deficiency

The reason for discussing MCAD deficiency in the present context is that it is the most thoroughly investigated ACAD disorder, both clinically and biochemically. This is due to the fact that it is the most frequently encountered disorder and as a result has been the driving force in the institution of neonatal screening for several ACAD disorders (see [93–95] and references therein). The first patient was described in 1976 [96], and the corresponding enzyme deficiency documented in 1982 [97]. In the early 1990s, when the major gene defect was identified as a missense variation, c.985A>G (p.Lys329Glu; Lys304Glu) [98–101], the total number collected world-wide with clinical and biochemical MCAD deficiency[5] was 172 [102]. 80% of these 172 patients were homozygous for c.985A>G and further 18% had this gene variation on one allele and another on the other. These other gene variations were subsequently identified in these 172 and in newly recognized patients and were found to comprise all types of gene variations, including big deletions, small deletions and insertions, stop codon and

---

[4] Amino acid numbering starts at start codon Met. Numbering without prefix p. are with numbering according to the mature protein

[5] Clinical and biochemical symptoms and signs in MCAD deficient patients are most often: Episodic attacks of hypoketotic hypoglycemia, vomiting, lethargy, which may develop into coma and death, as well as excretion of dicarboxylic acids and acylglycines in urine and accumulation of acylcarnitines in blood. The treatment during attacks consists of massive infusion of glucose/dextrose. In order to avoid attacks frequent feeding and avoidance of fasting, especially during feverish periods, should be instituted [95].

splice changes, as well as a number of different missense variations [70]. The expanding number of diagnosed patients, the relative high incidence in the order of 1 per 10,000[6], the improved diagnostic technologies, and the successful treatment were the driving force in the decision in many countries to perform neonatal screening for MCAD deficiency, and at the same time for other ACAD disorders [94,95,105]. Interestingly, the incidence as well as the number of 'new' missense variations in the MCAD gene, which has not been detected in symptomatic patients, has increased significantly after the institution of neonatal screening [69,94,95,105,106]. Indeed, from a prevalence of 80% homozygous patients carrying c.985A>G it has decreased to about 50% in neonatal screened babies, indicating the existence of mild gene variations, which give rise to variant proteins with mildly decreased enzyme activity and in biochemically detectable MCAD deficiency. Whether these milder forms of MCAD deficiency confer increased risk to developing symptomatic episodes has been debated; the conclusion from an expert meeting in 2009 is that the enhanced risk is not high, but that it cannot be neglected. It has been recommended that babies with these milder forms should be treated like those carrying c.985A>G and other gene variations identified in symptomatic patients [107]. Biochemically, a number of these mild variants has been investigated. Most intensively studied has been the c.199T→C variation, with a carrier incidence in the general population of about 1/500, which results in the MCAD variant p.Tyr67His (Tyr42His) [108]. This amino acid change probably perturbs the interaction between helices A, B and C and their link to the β-sheet domain. To judge the functional effect of this variant, an extensive investigation has been performed by O'Reily and coworkers [108]. Biogenesis in isolated mitochondria, structure analysis by circular dichroism, thermal stability and enzyme kinetic parameters, including FAD binding, have been compared *in vitro* to wild-type and to the disease-associated MCAD variant p.Lys329Glu. In the assay, which follows the biogenesis in isolated mitochondria, there is a tendency of the p.Tyr67His variant to form lower amounts of tetramer, and to degrade faster, especially at elevated temperature (41°C), compared to wild-type MCAD. The tendency to aggregate, on the other hand, was not higher than seen with wild-type protein, which contrasts low amounts of tetramer and elevated amounts of aggregated p.Lys329Glu-MCAD. The same authors [108] have analyzed tetramer and total protein amounts for variants by native and SDS based western blot analyses in cultured fibroblasts from individuals carrying the two variant and wild-type-MCADs in various combinations. These results corroborate the *in vitro* results. Circular dichroism studies have demonstrated that both the p.Tyr67His and p.Lys329Glu variants of MCAD have a somewhat perturbed secondary structure as compared to wild-type enzyme, with the greater effect seen with the more serious p.Lys329Glu variant, which also exhibited more compromised thermal stability. It is noteworthy in the present context that FAD binding showed the same trend. The spectral peak to peak ratio A280nm:450nm

[6] Since MCAD deficiency is autosomal recessively inherited and the prevalence of c.985A>G was 80%, the incidence was estimated from the carrier frequency of c.985A>G, which was determined in a number of countries, most comprehensively in a survey comprising European countries, the USA, Australia and Japan [103]. Especially in North-west Europe and the European colonised USA and Australia, the carrier incidence was high, 1: 100 and somewhat higher in some areas, indicating a strong founder effect from North-western Europe [104].

is 7.2 for p.Lys329Glu and 7.5 for p.Tyr67His, as compared to 7.5 for wild-type, again showing a slight perturbation in the mild variant in contrast to the significantly impaired binding seen in the p.Lys329Glu-MCAD variant. In conclusion, while the properties for MCAD p.Lys329Glu exhibit significant deviations from wild-type properties, those measured for MCAD p.Tyr67His show only slightly altered parameters, which are aggravated somewhat at higher temperature.

In another study Maier and co-workers have investigated ten purified MCAD variants, including p.Tyr67His and p.Lys329Glu with respect to tetramerization, protease sensitivity, thermal denaturation, thermal inactivation and kinetic stability, as well as kinetic parameters (but not FAD binding) [109]. The results are similar to those seen previously by O'Reily [108]. However, the variant proteins identified in neonatal screened babies and not in symptomatic patients disturbed all the structure and function of the variant MCAD proteins, some to a degree more pronounced than the p.Lys329Glu variant. These results indicate that there is a very poor correlation between the molecular defects, as measured *in vitro*, and the *in vivo* clinical symptoms. This is consistent with the observation that homozygosity for p.Lys329Glu is only predisposing, but not sufficient to develop appreciable clinical symptoms [94,110], and further substantiates the strategy that all babies with biochemical MCAD phenotype are at risk of developing symptomatic episodes [107].

With respect to riboflavin treatment, no positive reports have appeared to our knowledge, despite the fact that FAD may act as a chaperone [85,89,90] and that it has been shown *in vitro* that the chaperone Hsp60 can alleviate the structural dysfunction of a number of MCAD variant proteins, including the common disease associated p.Lys329Glu variant [92,109]. As will be discussed in relation to ETF-QO deficiency, there are a number of mild ETF-QO variants in patients, whose biochemical and clinical phenotypes improve with riboflavin treatment.

11.6 ETF-QO deficiency

Deficiency of ETF-QO affects all ACADs and recent results link the observed abnormality in patients with *in vitro* results in an illustrative way [84]. As mentioned above, ETF and ETF-QO deficiency give rise to multiple acyl-CoA dehydrogenation defect (MADD); a fatty acid oxidation disorder that exhibits a genotype-phenotype relationship [71]. Severe gene defects on both alleles leaving almost no mRNA expression and palmitic acid oxidation in patient fibroblasts (<5% of controls) are associated with a neonatal form with congenital abnormalities (Type I). Homozygosity or compound heterozygosity for splice or missense gene variation with severe structural and functional consequences for the encoded protein, and therefore possessing low but significant residual palmitic acid oxidation (8–12% of controls), are also associated with neonatal onset of the disease but without congenital abnormalities (Type II). A third type (Type III) of MADD is characterized with later onset of symptoms as seen in MCAD deficiency, sometimes limited to muscle symptoms (myopathy). These patients have at least one allele with a missense variation allowing considerable palmitic acid oxidation (28–52% of controls). Type I and II are considered severe MADD (S:MADD), while Type III represents a milder condition (M:MADD). The first cases of MADD were described in late 1970s and early 1980s. Although riboflavin is cofactor to both ETF and ETF-QO and to all the linked ACADs, riboflavin treatment proved to have little effect. However, two brothers from Denmark who were found to show the typical urinary organic acid profile responded well to riboflavin medication, both clinically and with decreased (but not normalized) excretion of metabolites [111]. Interestingly, long chain fatty acid oxidation using 1-[14C]-palmitic acid was not affected in cultured fibroblasts from the patients, but was affected when using U-[14C]-palmitic acid, indicating that medium and/or short chain fatty acid oxidation were decreased, compatible with the contemporary and later determinations of ACAD stabilities and activities in riboflavin deficient rats [83,112]. Later investigations, in which fibroblasts from one of the original Danish patients were cultured at low concentration of riboflavin, showed decreased intramitochondrial FAD concentration, indicating that defective maintenance of mitochondria FAD concentration rather than genetic defects in ETF/ETF-QO was responsible for the MADD in the patient, who showed remarkable response to riboflavin treatment [113,114]. Although a number of the enzymes and transporter proteins involved in riboflavin/FAD/FMN metabolism are known [87,115,116] and genetic defects in two of these have been identified [117,118], the majority of riboflavin responsive MADD (RR:MADD) patients were not genetically unraveled before the genes coding for ETFα and β as well as for ETF-QO were sequenced in a collection of eleven carefully selected patients (including the original Danish patients) with riboflavin-responsive myopathy [119]. Surprisingly, defects in the gene coding for ETF-QO were identified in all eleven. And interestingly, another five patients with the myopathic form of CoQ10 deficiency and diagnostic RR:MADD biochemical profiles also showed genetic defects in ETF-QO [120]. Each of these patients had at least one missense variation located in the FAD or ubiquinone binding domain of ETF-QO. Since the first reports of ETF-QO as the major disease-causing protein in RR:MADD, more than 80 additional RR:MADD patients have been described with similar gene defects. Due to a founder mutation (p.Ala84Thr) in the Han Chinese population, ETF-QO deficiency seems to be the most common cause of riboflavin-responsive myopathy in China with an expected incidence of 1:22,000 [121]. The molecular mechanisms for the riboflavin responsiveness in MADD were recently investigated by studying the influence of riboflavin on the steady-state level and activity of ETF-QO variants over-expressed in human HEK-293 cells [84]. ETF-QO variants (p.Gly472Arg, p.Leu334Pro and p.Leu138Arg) identified in patients with no or partial response to riboflavin treatment showed severe folding defects that could not be corrected by increasing cellular FAD content. On the other hand, ETF-QO variants (*e.g.*, p.Pro456Leu, p.Pro483Leu, p.Gly429Arg) identified in RR:MADD patients showed milder and riboflavin-responsive folding defects with increased hsp60 holding in the mitochondrial matrix. Interestingly, these variant proteins had decreased thermal stability even at supplemented concentrations of riboflavin, indicating that FAD does not completely correct the structural defects, and that this might allow access of oxygen to the partially reduced flavin or ubisemiquinones with increased production of reactive oxygen species (ROS) as documented for the p.Pro456Leu variant protein [84]. It remains to be clarified whether ROS production from variant ETF-QO proteins is a general finding in RR:MADD, and if so, how ROS impact CoQ10 deficiency and impair the function of some electron transport chain complexes as seen in pre-treatment muscle biopsies from RR:MADD patients.

11.7 VLCAD deficiency

Like MADD, VLCAD deficiency occurs in three forms, but riboflavin responsive forms have not been reported. Type I is a neonatal lethal cardiac and hepatic form with truncating or severely devastating missense variations on both alleles. Type II is the childhood form showing MCAD-like symptoms with hypoketotic hypoglycemia without cardiomyopathy and with at least one missense gene variation, which codes for a variant protein with some residual activity. Type III is the adolescent/adult form with myopathy and two missense variant proteins with residual activity [23,75,76]. Because the disorder may be lethal or give rise to life-threatening episodes, and because it is readily detected in neonatal screening programs by elevated C14:1-acylcarnitine in blood, it has been included in programs in many countries. The incidence in a number of countries has been estimated to about 1:40,000 to 1:170,000 [93,122,123]. Interestingly, although the incidence in the years before screening is not known, it is currently believed that the incidence has increased dramatically in the screened population, especially due to a prevalent mild variation, Val243Ala (p.Val283Ala)[7], which had previously been identified in patients with the myopathic form of VLCAD deficiency [75]. Indeed, among seven asymptomatic babies with biochemical indication of VLCAD deficiency and confirmed molecular genetic diagnosis, two patients were homozygous and four were compound heterozygous with Val243Ala coded from one allele and another missense variant protein from the other [123]. To judge the mechanism of VLCAD deficiency for this and five other disease associated missense variant VLCADs [75], Goetzman and co-workers have expressed them in a special bacterial system in which membrane association could be evaluated [23]. Two sets of interesting results have emerged from this study: first, three of the variants afforded either very little expressed protein and enzyme activity (Thr220Met, 3% of wild-type) or nearly zero (Arg429Trp, 0.2%; Arg573Trp, undetectable). The latter two variants have been shown to be associated with severe disease, corroborating that missense variations can be associated with the most severe lethal type of VLCAD deficiency, as mentioned above. Especially interesting, the (Val-243Ala) variant (which is associated to mild or no appreciable disease) shows only 22% of activity and less than 25% of protein compared to wild-type, indicating that minor amounts of VLCAD activity is sufficient to prevent toxic metabolite accumulation and serious cell damage. The correlation between amount of variant protein and enzyme activity further indicates that the Val243Ala change affects protein folding and results in its partial degradation, as discussed above. Whether the residual amount of functional protein has achieved the correct conformation or is slightly distorted is not known at the present time. It is possible that a loosening in structure may affect the channeling of electrons, as seen for some variant ETF-QO, which has been suggested to create sites

[7] The amino acid numbers is used differently in the cited literature. The mitochondrial leader sequence is 40 amino acids. Val243Ala designates mature VLCAD variant protein, and p.Val283Ala the variant with leader.

for ROS production [84]. This is an important question for future research, especially in light of the recent and ongoing studies concerning pharmacological enhancement of the amounts of missense VLCAD variants (see below). It is interesting to note that two of the investigated missense changes were located in the carboxy terminal in the extension that is not present in mitochondrial matrix localized ACADs, such as MCAD and SCAD. The carboxy terminal domain is believed to function in the dimeric VLCAD as an anchor to the mitochondrial inner membrane [19]. This membrane anchoring has been elegantly demonstrated in this study, where the amounts and activity of the variant protein is not decreased, but the association to the membrane is weakened [23], indicating the disease causing mechanism in these particular cases.

The traditional treatment of VLCAD deficiency is – like that for the other FAO defects – avoidance of fasting, sometimes combined with carbohydrate supplementation. Since the defect is located 'early' in FAO, supplementation with medium-chain triglycerides is often instituted [124]. Riboflavin treatment of patients with VLCAD deficiency has not been reported, nor have *in vitro* studies with riboflavin supplementation or depletion. However, beneficial effects of supplementation with riboflavin have been proposed from structural modeling and studies in cultured fibroblasts from patients with various severity of VLCAD deficiency [125]. Indeed, the authors of this work identified two missense amino acid variations, p.Arg366His (Arg326His) and p.Lys382Gln (Lys342-Gln), among 14 investigated disease associated missense changes in VLCAD, which may weaken the interaction with FAD but still possess significant residual amounts of expressed protein and enzyme activity. As seen for ETF-QO variants, it is such 'mild' amino acid alterations that compromise but do not destroy folding and stability that may be stabilized by the chaperone effect of FAD. Interestingly, the residual amount and activity may also be enhanced by inducers of the expression of fatty acid oxidation enzyme proteins [126]. Indeed, the fibroblasts carrying different combinations of missense variations have been treated with bezafibrate, a PPAR agonist. The amounts and activity of the variant VLCAD carrying mild missense amino acid changes, such as p.Arg366His (Arg326His) and p.Lys382Gln (Lys342Gln), are remarkably induced, whereas others predicted to affect the folding and stability severely, are not induced to appreciable amounts [125,126].

Another most interesting proof of concept treatment has been undertaken by the same group of researchers [127]. Resveratrol, the polyphenol found in red wine, has been given to cultured fibroblasts from patients carrying a variety of missense variant VL-CADs. The treatment increased the amount of VLCAD proteins and activity in control as well as patient cells. In short, the working mechanism was shown to be through sirtuin SIRT1and PGC-1α. The enhancement was in several cases more than 2-fold. What is not known at this time is whether an enhanced expression of slightly distorted variant VLCAD proteins after resveratrol or bezafibrate treatment will give rise to increased oxidative stress, as discussed for mildly misfolded ETF-QO proteins above. This must obviously be investigated before clinical trials are instituted.

11.8 ACAD 9 deficiency

ACAD9 is included in the present discussion about deficiencies in FAD containing ACADs and ETF-QO because patients with ACAD9 deficiency may be riboflavin responsive, and because ACAD9 evolutionarily is close to VLCAD (47% identical and 65% similar) [9,63]. Further, it has been postulated to be a new member of the ACAD enzyme family [128], and it has been suggested that defects in the regulatory region of the gene coding for ACAD9 are implicated in a FAO disorder [129]. However, and surprisingly, it turns out that homozygosity mapping [130], exome sequencing [131], or protein interaction investigations [9] all point to ACAD9 being an assembly factor (or chaperone) for Complex I of the mitochondrial respiratory chain. Like VLCAD, ACAD9 has a C-terminal tail and is functional as a dimer. While VLCAD is membrane-bound and involved in fatty acid oxidation, it is most likely that the C-terminal domain of ACAD9 has evolved to interact with at least two Complex 1 assembly factors, NDUFAF1 and Ecsit [9]. In cultured fibroblasts from two unrelated patients with isolated Complex I deficiency and homozygosity for a missense and compound heterozygosity for a missense and a truncating variation in the ACAD9 gene, respectively, reduced levels of NDUFAF1 and Ecsit were observed [9], strongly supporting the existence of the interaction and suggesting the disease mechanism. It was further shown by the same authors that ACAD9 does not function in long-chain FAO [9]. The simultaneous identification by three groups [9,130,131] of nine patients from five families with Complex I deficiency and carrying homozygous or compound heterozygous ACAD9 gene variations (eight missense and one truncating), support the notion that ACAD9 functions as a chaperone during assembly of Complex I. Interestingly, three of the patients originate from a large Dutch family, as originally described by Scholte in 1995 [132]. The patients exhibit exercise intolerance due to fatigue and pain, and a number of other symptoms compatible with a metabolic disorder, which has been documented to be an isolated Complex I deficiency. The patients respond well to large dosages of riboflavin, which nearly normalized Complex I activity. However, some metabolic dysfunction, as reflected in elevated blood lactate levels, remains [132]. This is a feature similar to that observed in riboflavin responsive MADD, as discussed above, where riboflavin treatment alleviated clinical symptoms and biochemical abnormalities, but did not normalize them [111]. As stated by the authors of the recent report [130], the folding and stability of the variant ACAD9 protein in the patients are probably compromised by the homozygous p.Arg532Trp missense variation, which is predicted to affect the FAD binding pocket in ACAD9. Thus, as discussed above, ACAD9 seems to be an example of a flavoprotein in which the FAD functions as a chaperone. How many patients with Complex I deficiency due to ACAD9 variations would benefit from riboflavin supplementation is not known at present, and the fact that riboflavin does not fully correct the metabolic dysfunction could indicate that the structure and function of the variant proteins cannot by fully restored by high cellular concentration of FAD, as discussed for ETF-QO and VLCAD variants above.

11.9 SCAD deficiency

It has been and still is enthusiastically debated whether SCAD deficiency is a disease entity in itself giving rise to a clinical state, or whether it is a susceptibility condition which requires additional genetic, cellular and/or environmental factors to become manifested clinically, or it is just a harmless biochemical/evolutionary condition [69,76,80,133–138].

Before trying to answer this question, it is necessary to discuss the current clinical, biochemical, molecular genetic, molecular pathogenetic and cellular pathological knowledge of SCAD deficiency.

11.9.1 Clinical aspects of SCAD deficiency

The clinical symptoms in the majority of children subsequently diagnosed with SCAD deficiency are quite different from the heart, liver and muscle pathologies seen in patients with other fatty acid oxidation defects. The predominant symptoms in patients with SCAD deficiency are neurological/neuromuscular, including failure to thrive, developmental delay, hypotonia, speech delay, and in some cases muscle weakness and seizures [80,137,138]. Although hypoglycemia is observed, it is not a major symptom as it is in MCAD deficiency. The reason for this is not obvious, and will be touched upon further in the relevant contexts below. Suffice is to say here that the penetrance to clinical disease for SCAD deficiency is extremely low, much lower than for MCAD deficiency, especially after the introduction of neonatal screening [137,139].

11.9.2 Biochemical aspects of SCAD deficiency

The first SCAD deficiency patients were diagnosed by determination of SCAD activity in cultured fibroblasts after detection of ethylmalonic acid in urine [140,141]. The fibroblasts investigated by Coates and co-workers from Tanaka's group showed initially 50% of the control activity, but after addition of anti MCAD antibodies, the activity was undetectable, confirming MCAD's ability to dehydrogenate butyryl-CoA, the substrate for SCAD. In hindsight, these overlapping activities may be an important reason for the low penetrance, especially since MCAD expression is dependent on the metabolic status of the cell, involving PPARs and PCG-1α, as discussed above in connection with VLCAD deficiency. This dependence on metabolic status is also reflected in the urinary biochemical markers of SCAD deficiency, urinary ethylmalonic acid and/or blood butyrylcarnitine, the amounts of which vary considerably from patient to patient [80]. This may also be a contributing reason that it is only the most vulnerable tissue, brain and muscle that are primarily affected.

In another investigation of three severely affected patients by Tanaka's group, it was shown that SCAD in fibroblasts from all three patients were produced in amounts comparable to control cells, although undetectable at steady-state by immunoblot analysis in two of the three cases, [142]. The authors concluded that the gene variations responsible for the SCAD deficiency in these cases must be missense mutations and not 'gross deletions, transcriptional mutants, or frame shift mutations leading to premature stop codons near the amino terminus' [142]. This finding is important in light of later molecular genetic results.

11.9.3 Molecular genetics of SCAD deficiency

The spectrum of gene variations identified in clinically affected patients as well as in babies identified in neonatal screening for SCAD deficiency is different from those in other fatty acid oxidation defects, such as MCAD and VLCAD deficiency [69]. First of all, these deficiencies comprise all types of gene variations, including out-of-frame deletions and insertions, stop codon and splice site changes, which do not produce any proteins, as well as in-frame deletions/insertions and missense variations, which give rise to expressed variants. In contrast, only a few truncating gene defects, resulting in no protein production have been identified in patients and neonate screened babies with SCAD deficiency, while more than 60 missense and a single in-frame deletion have been identified as of 2008 [69]. As discussed previously [138] and mentioned in the introduction to the section concerning ACAD deficiencies, the reason for this skewed spectrum is not obvious, but it has been suggested that total lack of SCAD may give rise to lethal levels of butyric acid in the haploid cell state [80]. Second, among the missense variations, two, c.511C>T (Arg147Trp; p.Arg171Trp) and c.625G>A (Gly185Ser; p.Gly209Ser) which are also present in the general population, are frequently encountered in patients and neonates with SCAD deficiency. Indeed, these two variations have been identified either in homozygous or compound heterozygous form in 4 and 14% in the general populations in USA and Denmark, respectively [143,144], and in 56% in a group of 293 symptomatic patients [69], indicating that they are 'mild' and may be considered as disease-associated susceptibility gene variations. In screened babies the prevalence is only 22% in a group of 45, in which at least 30 other variations are also found, suggesting that the screening procedure selects for the more severe forms of SCAD deficiency, without catching babies with susceptibility to clinical relevant SCAD deficiency. Nonetheless, the uncertainty about the clinical relevance of SCAD deficiency has caused SCAD deficiency to be withdrawn from the neonatal screening program in a number of countries, such as Germany, Australia and Denmark. This means that certain severe cases of SCAD deficiency will likely not be diagnosed early [145].

11.9.4 Molecular pathogenesis of SCAD deficiency

As mentioned above, a large number of missense gene variations have been identified in patients and screened newborn babies. A selection of these has been investigated with respect to their consequences for production, folding and stability of the coded protein variant. It seems that all investigated SCAD variants are misfolded. In one of the first identified patients, in whose fibroblasts unstable SCAD protein has been detected [142], compound heterozygosity for two gene variations, c.136C>T (Arg22Trp; p.Arg46Trp) and c.319C>T (Arg83Cys; p.Arg107Cys) have been identified [146]. These two gene variations have later been found in a number of patients [80,133,145] and investigated in model cells and isolated mitochondria [80,147–150]. The general conclusion concerning these and other 'severe' missense gene variations are that their encoded proteins are produced and translocated normally to mitochondria, where their folding is compromised. It is noteworthy that none of the amino acid changes directly affect binding to FAD or the substrate butyryl-CoA. As discussed above, the variants are processed by the mitochondrial protein quality control system, in which the chaperone part tries to rescue the misfolded proteins, and if this fails, the protease part tries to degrade them to avoid aggregation. Nevertheless, at least in isolated mitochondria the aggregation is pronounced for the majority of the 27 investigated missense variant SCAD proteins, among them Arg22Trp and Arg83Cys [80]. The Arg83Cys variant has also been shown to aggregate in a model astrocyte cell line [149]. As expected, the aggregation is aggravated at elevated temperature, which is relevant in patients during feverish episodes.

The processing of the two common variants Arg147Trp and Gly185Ser have also been examined extensively in model COS7 cells [144] and isolated mitochondria [80,148], and their physical-chemical properties of the purified proteins have also been studied [151]. Considering their high frequency in the general Caucasian population [143,144], it is surprising that the properties of these variants confer susceptibility to degradation and aggregation, however, to a lesser degree than the severe variant SCADs. They also produce significant amounts of tetrameric proteins in the isolated mitochondria, although this decreases at elevated temperature concomitant with an increase in aggregated SCAD [80]. The purified Arg147Trp and Gly185Ser variants also shows alterations compared to wild-type protein. While Arg147Trp retains normal catalytic efficiency, Gly185Ser SCAD exhibits reduced catalytic efficiency. The thermal and guanidine•HCl stability are decreased for both variants that, according to the authors, indicates decreased flexibility of the tertiary structure of the two variant SCAD proteins [151]. Unfortunately, FAD binding or cultivation of patient cells in riboflavin supplemented or depleted medium have not been performed, so the potential effectiveness of treatment with riboflavin cannot be assessed.

In conclusion, all these properties, together with the fact that MCAD possesses butyryl-CoA dehydrogenation activity (see ▶Fig. 11.6), support the notion that SCAD deficiency may be a severe susceptibility condition, which requires additional genetic, cellular and/or environmental factors to precipitate into a clinical condition. In this respect it is worth mentioning that riboflavin status and treatment have been evaluated clinically in 16 SCAD deficient patients [152], where it was found that blood FAD levels were within controls in all patients, but was lower in patients carrying the Gly185Ser variant in one or both alleles. Urinary ethylmalonic acid decreased and the clinical symptoms weakened in four of these patients after riboflavin treatment. Although the improvements persisted after ending the treatment period, this trial nevertheless indicates that some patients would benefit from riboflavin treatment. FAD deficiency may thus be one of the additional environmental factors, such as diet, needed for SCAD deficiency to precipitate clinical symptoms, as seen in ETF-QO deficiency [87].

11.9.5 Cellular pathological aspects of SCAD deficiency

In relation to the above discussion about the properties of SCAD variants, it is relevant to briefly discuss the cellular pathology due to SCAD deficiency. In connection with the studies on the misfolding of various SCAD variants, such as the severe Arg83Cys and the mild Gly185Ser SCAD, the effect on the mitochondrial proteome has been investigated [138,150,153]. The main results indicate that model and patient cells containing these SCAD variants are prone to oxidative stress, which may contribute to the pathology and development of clinical symptoms. Oxidative stress may be elicited directly by accumulated aggregates or – as discussed above for the other fatty acid oxidation defects – by inducing distorted protein conformations that result in leakage of electrons and/or access of oxygen and production of ROS. During metabolic stress, where accumulation of fatty acids increases, the effect may be to aggravate the oxidative stress, as showed for several organic acids, including ethylmalonic acid [154–158].

11.10 Acknowledgements

Studies performed in the Kim laboratory were supported by National Institute of Health Grant GM29076 (JJPK). Studies performed by the Danish research group were supported by grants from the Danish Medical Research Council and Lundbeck Foundation.

11.11 Abbreviations

ACAD	Acyl-CoA dehydrogenase
SCAD	Short chain acyl-CoA dehydrogenase
MCAD	Medium chain acyl-CoA dehydrogenase
LCAD	Long chain acyl-CoA dehydrogenase
VLCAD	Very long chain acyl-CoA dehydrogenase
iBD	Isobutyryl-CoA dehydrogenase
i3VD	Isovaleryl-CoA dehydrogenase
i2VD	"branched chain" acyl-CoA dehydrogenase
GD	Glutaryl-CoA dehydrogenase
ETF	electron transfer(ing) flavoprotein
ETF-QO	ETF-ubiquinone oxidoreductase
CT	charge transfer
MADD	multiple acyl-CoA dehydrogenation defect
LUMO	lowest unoccupied molecular orbital
FAO	fatty acid oxidation
ROS	reactive oxygen species

11.12 References

[1] Mahler HR. Studies on the fatty acid oxidizing system of animal tissues. IV. The prosthetic group of butyryl coenzyme A dehydrogenase. J Biol Chem 1954;206:13–26.
[2] Steyn-Parve EP, Beinert H. On the Mechanism of Dehydrogenation of Fatty Acyl Derivatives of Coenzyme A. The nature of the green color of butyryl dehydrogenase. JBiolChem 1958;233:853–61.
[3] Crane FL, Mii S, Hauge JG, Green DE, Beinert H. On the mechanism of dehydrogenation of fatty acyl derivatives of coenzyme A. J Biol Chem 1956;218:701–16.
[4] Ghisla S, Thorpe C. Acyl-CoA dehydrogenases. A mechanistic overview. Eur J Biochem 2004;271:494–508.
[5] Kim JJ, Miura R. Acyl-CoA dehydrogenases and acyl-CoA oxidases. Structural basis for mechanistic similarities and differences. Eur J Biochem 2004;271:483–93.
[6] Beinert H. Acyl Coenzyme A Dehydrogenases. In: Boyer PD, Lardy H, Myrback K, eds. Enzymes. 2 nd. ed. New York: Academic Press; 1963:447–66.
[7] He M, Pei Z, Mohsen A-W, et al. Identification and characterization of new long chain acyl-CoA dehydrogenases. Mol Genet Metab 2011;102:418–29.
[8] Bartlett K, Eaton S. Mitochondrial beta-oxidation. Eur J Biochem 2004;271:462–9.
[9] Nouws J, Nijtmans L, Houten SM, et al. Acyl-CoA dehydrogenase 9 is required for the biogenesis of oxidative phosphorylation complex I. Cell Metab 2010;12:283–94.
[10] Kim JJ, Wang M, Paschke R. Crystal structures of medium-chain acyl-CoA dehydrogenase from pig liver mitochondria with and without substrate. Proc Natl Acad Sci U S A 1993;90:7523–7.

[11] Battaile KP, Molin-Case J, Paschke R, et al. Crystal structure of rat short chain acyl-CoA dehydrogenase complexed with acetoacetyl-CoA: comparison with other acyl-CoA dehydrogenases. J Biol Chem 2002;277:12200–7.
[12] Battaile KP, Nguyen TV, Vockley J, Kim JJ. Structures of isobutyryl-CoA dehydrogenase and enzyme-product complex: comparison with isovaleryl- and short-chain acyl-CoA dehydrogenases. J Biol Chem 2004;279:16526–34.
[13] Tiffany KA, Roberts DL, Wang M, et al. Structure of human isovaleryl-CoA dehydrogenase at 2.6 A resolution: structural basis for substrate specificity. Biochemistry 1997;36:8455–64.
[14] Fu Z, Wang M, Paschke R, Rao KS, Frerman FE, Kim JJ. Crystal structures of human glutaryl-CoA dehydrogenase with and without an alternate substrate: structural bases of dehydrogenation and decarboxylation reactions. Biochemistry 2004;43:9674–84.
[15] Djordjevic S, Pace CP, Stankovich MT, Kim JJ. Three-dimensional structure of butyryl-CoA dehydrogenase from Megasphaera elsdenii. Biochemistry 1995;34:2163–71.
[16] Ghisla S, Thorpe C, Massey V. Mechanistic studies with general acyl-CoA dehydrogenase and butyryl-CoA dehydrogenase: evidence for the transfer of the beta-hydrogen to the flavin N(5)-position as a hydride. Biochemistry 1984;23:3154–61.
[17] Battaile KP, Mohsen AW, Vockley J. Functional role of the active site glutamate-368 in rat short chain acyl-CoA dehydrogenase. Biochemistry 1996;35:15356–63.
[18] Dwyer TM, Rao KS, Westover JB, Kim JJ, Frerman FE. The function of Arg-94 in the oxidation and decarboxylation of glutaryl-CoA by human glutaryl-CoA dehydrogenase. J Biol Chem 2001;276:133–8.
[19] McAndrew RP, Wang Y, Mohsen AW, He M, Vockley J, Kim JJ. Structural basis for substrate fatty acyl chain specificity: crystal structure of human very-long-chain acyl-CoA dehydrogenase. J BiolChem 2008;283:9435–43.
[20] Izai K, Uchida Y, Orii T, Yamamoto S, Hashimoto T. Novel fatty acid beta-oxidation enzymes in rat liver mitochondria. I. Purification and properties of very-long-chain acyl-coenzyme A dehydrogenase. J Biol Chem 1992;267:1027–33.
[21] Trievel RC, Wang R, Anderson VE, Thorpe C. Role of the carbonyl group in thioester chain length recognition by the medium chain acyl-CoA dehydrogenase. Biochemistry 1995;34:8597–605.
[22] Souri M, Aoyama T, Yamaguchi S, Hashimoto T. Relationship between structure and substrate-chain-length specificity of mitochondrial very-long-chain acyl-coenzyme A dehydrogenase. Eur J Biochem 1998;257:592–8.
[23] Goetzman ES, Wang Y, He M, Mohsen AW, Ninness BK, Vockley J. Expression and characterization of mutations in human very long-chain acyl-CoA dehydrogenase using a prokaryotic system. Mol Genet Metab 2007;91:138–47.
[24] Djordjevic S, Dong Y, Paschke R, Frerman FE, Strauss AW, Kim J-JP. Identification of the catalytic base in long chain acyl-CoA dehydrogenase. Biochemistry 1994;33:4258–64.
[25] Mohsen AW, Vockley J. Identification of the active site catalytic residue in human isovaleryl-CoA dehydrogenase. Biochemistry 1995;34:10146–52.
[26] Nandy A, Kieweg V, Kräutle FG, et al. Medium-long-chain chimeric human Acyl-CoA dehydrogenase: medium-chain enzyme with the active center base arrangement of long-chain Acyl-CoA dehydrogenase. Biochemistry 1996;35:12402–11.
[27] Massey V, Ghisla S. Role of charge-transfer interactions in flavoprotein catalysis. Ann N Y Acad Sci 1974;227:446–65.
[28] Lee HJ, Wang M, Paschke R, Nandy A, Ghisla S, Kim JJ. Crystal structures of the wild type and the Glu376Gly/Thr255Glu mutant of human medium-chain acyl-CoA dehydrogenase: influence of the location of the catalytic base on substrate specificity. Biochemistry 1996;35:12412–20.
[29] Biellmann JF, Hirth CG. Stereochemistry of the oxidation at the β–carbon of butyryl-SCoA. FEBS Letters 1970;9:55–6,335–6.
[30] La Roche HJ, Kellner M, Günther H, Simon H. Stereochemie der Butyryl-CoA-Dehydrogenase in *Clostridium Kluyveri*. Hoppe-Seyler's Z Physiol Chem 1971;352:399–402.

[31] Schopfer LM, Massey V, Ghisla S, Thorpe C. Oxidation-reduction of general acyl-CoA dehydrogenase by the butyryl-CoA/crotonyl-CoA couple. A new investigation of the rapid reaction kinetics. Biochemistry 1988;27:6599–611.

[32] Pohl B, Raichle T, Ghisla S. Studies on the reaction mechanism of general acyl-CoA dehydrogenase. Determination of selective isotope effects in the dehydrogenation of butyryl-CoA. Eur J Biochem 1986;160:109–15.

[33] Bhattacharyya S, Ma S, Stankovich MT, Truhlar DG, Gao J. Potential of Mean Force Calculation for the Proton and Hydride Transfer Reactions Catalyzed by Medium-Chain Acyl-CoA Dehydrogenase: Effect of Mutations on Enzyme Catalysis. Biochemistry 2005;44:16549–62.

[34] Poulsen T, Garcia-Viloca M, Gao J, Truhlar D. Free Energy Surface, Reaction Paths, and Kinetic Isotope Effect of Short-Chain Acyl-CoA Dehydrogenase. J Phys Chem B 2003;107:9567–78.

[35] Nishina Y, Sato K, Tamaoki H, et al. Molecular mechanism of the drop in the pKa of a substrate analog bound to medium-chain acyl-CoA dehydrogenase: implications for substrate activation. J Biochem (Tokyo) 2003;134:835–42.

[36] Vock P, Engst S, Eder M, Ghisla S. Substrate activation by acyl-CoA dehydrogenases: transition-state stabilization and pKs of involved functional groups. Biochemistry 1998;37:1848–60.

[37] Rudik I, Ghisla S, Thorpe C. Protonic equilibria in the reductive half-reaction of the medium-chain acyl-CoA dehydrogenase. Biochemistry 1998;37:8437–45.

[38] Engst S, Vock P, Wang M, Kim JJ, Ghisla S. Mechanism of activation of acyl-CoA substrates by medium chain acyl-CoA dehydrogenase: interaction of the thioester carbonyl with the flavin adenine dinucleotide ribityl side chain. Biochemistry 1999;38:257–67.

[39] Schowen KB, Limbach HH, Denisov GS, Schowen RL. Hydrogen bonds and proton transfer in general-catalytic transition-state stabilization in enzyme catalysis. Biochim Biophys Acta 2000;1458:43–62.

[40] Satoh A, Nakajima Y, Miyahara I, et al. Structure of the Transition State Analog of Medium-Chain Acyl-CoA Dehydrogenase. Crystallographic and Molecular Orbital Studies on the Charge-Transfer Complex of Medium-Chain Acyl-CoA Dehydrogenase with 3-Thiaoctanoyl-CoA. J Biochem (Tokyo) 2003;134:297–304.

[41] Rudik I, Thorpe C. Thioester enolate stabilization in the acyl-CoA dehydrogenases: the effect of 5-deaza-flavin substitution. Arch Biochem Biophys 2001;392:341–8.

[42] Dmitrenko O, Thorpe C, Bach RD. Effect of a Charge-Transfer Interaction on the Catalytic Activity of Acyl-CoA Dehydrogenase: A Theoretical Study of the Role of Oxidized Flavin. The journal of physical chemistry B 2003;107:13229–36.

[43] Thorpe C, Kim JJ. Structure and mechanism of action of the acyl-CoA dehydrogenases. Faseb J 1995;9:718–25.

[44] Stankovich MT, Sabaj KM, Tonge PJ. Structure/function of medium chain acyl-CoA dehydrogenase: The importance of substrate polarization [Review]. Archives of Biochemistry & Biophysics 1999;370:16–21.

[45] Johnson BD, Mancini-Samuelson GJ, Stankovich MT. Effect of Transition-State Analogues on the Redox Properties of Medium-Chain Acyl-CoA Dehydrogenase. Biochemistry 1995.

[46] Pellett JD, Sabaj KM, Stephens AW, et al. Medium-chain acyl-coenzyme A dehydrogenase bound to a product analogue, hexadienoyl-coenzyme A: effects on reduction potential, pK(a), and polarization. Biochemistry 2000;39:13982–92.

[47] Mancini-Samuelson GJ, Kieweg V, Sabaj KM, Ghisla S, Stankovich MT. Redox properties of human medium-chain acyl-CoA dehydrogenase, modulation by charged active-site amino acid residues. Biochemistry 1998;37:14605–12.

[48] Lamm TR, Kohls TD, Saenger AK, Stankovich MT. Comparison of ligand polarization and enzyme activation in medium- and short-chain acyl-coenzyme A dehydrogenase-novel analog complexes. Arch Biochem Biophys 2003;409:251–61.

[49] Lamm TR, Kohls TD, Stankovich MT. Activation of substrate/product couples by medium-chain acyl-CoA dehydrogenase. Arch Biochem Biophys 2002;404:136–46.

[50] Toogood HS, Leys D, Scrutton NS. Dynamics driving function: new insights from electron transferring flavoproteins and partner complexes. The FEBS journal 2007;274:5481–504.

[51] Ruzicka FJ, Beinert H. A new iron-sulfur flavoprotein of the respiratory chain. A component of the fatty acid beta oxidation pathway. J Biol Chem 1977;252:8440–5.
[52] Beinert H. Acyl-Coenzyme A Dehydrogenases. In: Boyer PD, Lardy H, K. M, eds. The Enzymes. New York: Academic Press; 1963:447–66.
[53] DuPlessis ER, Rohlfs RJ, Hille R, Thorpe C. Electron-transferring flavoprotein from pig and the methylotrophic bacterium W3A1 contains AMP as well as FAD. BiochemMolBiol Int 1994;32:195–9.
[54] Roberts DL, Frerman FE, Kim JJ. Three-dimensional structure of human electron transfer flavoprotein to 2.1-A resolution. ProcNatlAcadSciUSA 1996;93:14355–60.
[55] Roberts DL, Salazar D, Fulmer JP, Frerman FE, Kim JJ. Crystal structure of Paracoccus denitrificans electron transfer flavoprotein: structural and electrostatic analysis of a conserved flavin binding domain. Biochemistry 1999;38:1977–89.
[56] Chohan KK, Jones M, Grossmann JG, Frerman FE, Scrutton NS, Sutcliffe MJ. Protein dynamics enhance electronic coupling in electron transfer complexes. J Biol Chem 2001;276:34142–7.
[57] Toogood HS, van Thiel A, Basran J, Sutcliffe MJ, Scrutton NS, Leys D. Extensive domain motion and electron transfer in the human electron transferring flavoprotein.medium chain Acyl-CoA dehydrogenase complex. J Biol Chem 2004;279:32904–12.
[58] Toogood HS, van Thiel A, Scrutton NS, Leys D. Stabilization of non-productive conformations underpins rapid electron transfer to electron-transferring flavoprotein. J Biol Chem 2005;280:30361–6.
[59] Williamson G, Engel PC, Mizzer JP, Thorpe C, Massey V. Evidence That the Greening Ligand in Native Butyryl-CoA Dehydrogenase Is a CoA Persulfide. Journal of Biological Chemistry 1982;257:4314–20.
[60] Wenz A, Thorpe C, Ghisla S. Inactivation of general acyl-CoA dehydrogenase from pig kidney by a metabolite of hypoglycin A. J Biol Chem 1981;256:9809–12.
[61] Hassall CH, Reyle K. Hypoglycin A and B, two Biologically Active Polypeptides from Blighia sapida. Biochem J 1955;60:334–9.
[62] von Holt M, von Holt C, Böhm H. Metabolic Effects of Hypoglycin and Methylenecyclopropylacetic Acid. Biochim Biophys Acta 1966;125:11–21.
[63] Swigonova Z, Mohsen AW, Vockley J. Acyl-CoA dehydrogenases: Dynamic history of protein family evolution. J Mol Evol 2009;69:176–93.
[64] He M, Pei Z, Mohsen AW, et al. Identification and characterization of new long chain acyl-CoA dehydrogenases. Mol Genet Metab 2011;102:418–29.
[65] Frischmeyer PA, Dietz HC. Nonsense-mediated mRNA decay in health and disease. Hum MolGenet 1999;8:1893–900.
[66] Bhuvanagiri M, Schlitter AM, Hentze MW, Kulozik AE. NMD: RNA biology meets human genetic medicine. BiochemJ 2010;430:365–77.
[67] Nielsen KB, Sorensen S, Cartegni L, et al. Seemingly Neutral Polymorphic Variants May Confer Immunity to Splicing-Inactivating Mutations: A Synonymous SNP in Exon 5 of MCAD Protects from Deleterious Mutations in a Flanking Exonic Splicing Enhancer. Am J Hum Genet 2007;80:416–32.
[68] Gregersen N, Bross P, Vang S, Christensen JH. Protein misfolding and Human disease. Ann Rev Genom Human Genet 2006;7:103–24.
[69] Gregersen N, Andresen BS, Pedersen CB, Olsen RK, Corydon TJ, Bross P. Mitochondrial fatty acid oxidation defects--remaining challenges. J InheritMetab Dis 2008;31:643–57.
[70] Gregersen N, Bross P, Andresen BS. Genetic defects in fatty acid beta-oxidation and acyl-CoA dehydrogenases. EurJ Biochem 2004;271:470–82.
[71] Olsen RK, Andresen BS, Christensen E, Bross P, Skovby F, Gregersen N. Clear relationship between ETF/ETFDH genotype and phenotype in patients with multiple acyl-CoA dehydrogenation deficiency. Hum Mutat 2003;22:12–23.
[72] Schuck PF, Ferreira GC, Moura AP, et al. Medium-chain fatty acids accumulating in MCAD deficiency elicit lipid and protein oxidative damage and decrease non-enzymatic antioxidant defenses in rat brain. NeurochemInt 2009;54:519–25.

[73] Wajner M, Goodman SI. Disruption of mitochondrial homeostasis in organic acidurias: insights from human and animal studies. J BioenergBiomembr 2011;43:31–8.
[74] Frerman FE, Goodman SI. Defects of electron transfer flavoprotein and electron transfer flavoprotein-ubiquinone oxidoreductase: Glutaric aciduria type II. In: Scriver CR, Beaudet AL, Valle D, eds. The metabolic and molecular basis of inherited disease. 8th ed. New York 2001: 2357–65.
[75] Andresen BS, Olpin S, Poorthuis BJ, et al. Clear correlation of genotype with disease phenotype in very-long-chain acyl-CoA dehydrogenase deficiency. AmJ Hum Genet 1999; 64:479–94.
[76] Gregersen N, Andresen BS, Corydon MJ, et al. Mutation analysis in mitochondrial fatty acid oxidation defects: Exemplified by acyl-CoA dehydrogenase deficiencies, with special focus on genotype-phenotype relationship. Hum Mutat 2001;18:169–89.
[77] Arnold GL, Saavedra-Matiz CA, Galvin-Parton PA, et al. Lack of genotype-phenotype correlations and outcome in MCAD deficiency diagnosed by newborn screening in New York State. Mol Genet Metab 2010;99:263–8.
[78] Vockley J, Ensenauer R. Isovaleric acidemia: new aspects of genetic and phenotypic heterogeneity. AmJ MedGenet CSeminMedGenet 2006;142C:95–103.
[79] Christensen E, Ribes A, Merinero B, Zschocke J. Correlation of genotype and phenotype in glutaryl-CoA dehydrogenase deficiency. J InheritMetab Dis 2004;27:861–8.
[80] Pedersen CB, Kolvraa S, Kolvraa A, et al. The ACADS gene variation spectrum in 114 patients with short-chain acyl-CoA dehydrogenase (SCAD) deficiency is dominated by missense variations leading to protein misfolding at the cellular level. Hum Genet 2008;124:43–56.
[81] Gregersen N, Bross P. Protein misfolding and cellular stress: an overview. Methods Mol Biol 2010;648:3–23.
[82] Gregersen N. Riboflavin-responsive defects of beta-oxidation. JInherMetabDis 1985;8 Suppl 1:65–9.
[83] Nagao M, Tanaka K. FAD-Dependent Regulation of Transcription, Translation, Post- Translational Processing, and Post-Processing Stability of Various Mitochondrial Acyl-CoA Dehydrogenases and of Electron Transfer Flavoprotein and the Site of Holoenzyme Formation. JBiolChem 1992;267:17925–32.
[84] Cornelius N, Frerman FE, Corydon TJ, et al. Molecular mechanisms of riboflavin responsiveness in patients with ETF-QO variations and multiple acyl-CoA dehydrogenation deficiency. HumMol Genet 2012;21:3435–48.
[85] Lucas TG, Henriques BJ, Rodrigues JV, Bross P, Gregersen N, Gomes CM. Cofactors and metabolites as potential stabilizers of mitochondrial acyl-CoA dehydrogenases. BiochimBiophys-Acta 2011;1812:1658–63.
[86] Ames BN, Elson-Schwab I, Silver EA. High-dose vitamin therapy stimulates variant enzymes with decreased coenzyme binding affinity (increased K(m)): relevance to genetic disease and polymorphisms. AmJ ClinNutr 2002;75:616–58.
[87] Henriques BJ, Olsen RK, Bross P, Gomes CM. Emerging roles for riboflavin in functional rescue of mitochondrial beta-oxidation flavoenzymes. CurrMedChem 2010;17:3842–54.
[88] Henriques BJ, Rodrigues JV, Olsen RK, Bross P, Gomes CM. Role of flavinylation in a mild variant of multiple acyl-CoA dehydrogenation deficiency: a molecular rationale for the effects of riboflavin supplementation. J BiolChem 2009;284:4222–9.
[89] Saijo T, Kim JJ, Kuroda Y, Tanaka K. The roles of threonine-136 and glutamate-137 of human medium chain acyl-CoA dehydrogenase in FAD binding and peptide folding using site-directed mutagenesis: creation of an FAD-dependent mutant, T136D. ArchBiochemBiophys 1998;358:49–57.
[90] Kieweg V, Krautle FG, Nandy A, et al. Biochemical characterization of purified, human recombinant Lys304-- >Glu medium-chain acyl-CoA dehydrogenase containing the common disease-causing mutation and comparison with the normal enzyme. EurJBiochem 1997;246:548–56.

[91] Saijo T, Welch WJ, Tanaka K. Intramitochondrial folding and assembly of medium-chain acyl-CoA dehydrogenase (MCAD) - Demonstration of impaired transfer of K304E-variant MCAD from Its complex with Hsp60 to the native tetramer. JBiolChem 1994;269:4401–8.
[92] Bross P, Jespersen C, Jensen TG, et al. Effects of two mutations detected in medium chain acyl-CoA dehydrogenase (MCAD)-deficient patients on folding, oligomer assembly, and stability of MCAD enzyme. JBiolChem 1995;270:10284–90.
[93] Lund AM, Hougaard DM, Simonsen H, et al. Biochemical screening of 504,049 newborns in Denmark, the Faroe Islands and Greenland - Experience and development of a routine program for expanded newborn screening. Mol Genet Metab 2012.
[94] Andresen BS, Lund AM, Hougaard DM, et al. MCAD deficiency in Denmark. Mol Genet Metab 2012;106:175–88.
[95] Matern D, Rinaldo P. Medium-Chain Acyl-Coenzyme A Dehydrogenase Deficiency. In: Pagon RA, Bird TD, Dolan CR, Stephens K, Adam MP, eds. Gene Reviews. Seattle (WA)1993.
[96] Gregersen N, Lauritzen R, Rasmussen K. Suberylglycine excretion in the urine from a patient with dicarboxylic aciduria. ClinChimActa 1976;70:417–25.
[97] Kolvraa S, Gregersen N, Christensen E, Hobolth N. In vitro fibroblast studies in a patient with C6-C10-dicarboxylic aciduria: evidence for a defect in general acyl-CoA dehydrogenase. Clin Chim Acta 1982;126:53–67.
[98] Matsubara Y, Narisawa K, Miyabayashi S, et al. Identification of a common mutation in patients with medium-chain acyl-CoA dehydrogenase deficiency. BiochemBiophysResCommun 1990;171:498–505.
[99] Yokota I, Indo Y, Coates PM, Tanaka K. Molecular basis of medium chain acyl-coenzyme A dehydrogenase deficiency. An A to G transition at position 985 that causes a lysine-304 to glutamate substitution in the mature protein is the single prevalent mutation. JClinInvest 1990;86:1000–3.
[100] Kelly DP, Whelan AJ, Ogden ML, et al. Molecular characterization of inherited medium-chain acyl-CoA dehydrogenase deficiency. ProcNatlAcadSci(USA) 1990;87:9236–40.
[101] Gregersen N, Andresen BS, Bross P, et al. Molecular characterization of medium-chain acyl-CoA dehydrogenase (MCAD) deficiency: identification of a lys329 to glu mutation in the MCAD gene, and expression of inactive mutant enzyme protein in E. coli. Hum Genet 1991;86:545–51.
[102] Tanaka K, Yokota I, Coates PM, et al. Mutations in the medium chain acyl-CoA dehydrogenase (MCAD) gene. HumMutat 1992;1:271–9.
[103] Tanaka K, Gregersen N, Ribes A, et al. A survey of the newborn populations in Belgium, Germany, Poland, Czech Republic, Hungary, Bulgaria, Spain, Turkey, and Japan for the G985 variant allele with haplotype analysis at the medium chain Acyl-CoA dehydrogenase gene locus: clinical and evolutionary consideration. Pediatr Res 1997;41:201–9.
[104] Gregersen N, Winter V, Curtis D, et al. Medium-chain acyl-CoA dehydrogenase (MCAD) deficiency: the prevalent mutation G985 (K304E) is subject to a strong founder effect from northwestern Europe. Hum Hered 1993;43:342–50.
[105] Andresen BS, Dobrowolski SF, O'Reilly L, et al. Medium-Chain Acyl-CoA Dehydrogenase (MCAD) Mutations Identified by MS/MS-Based Prospective Screening of Newborns Differ from Those Observed in Patients with Clinical Symptoms: Identification and Characterization of a New, Prevalent Mutation That Results in Mild MCAD Deficiency. Am J Hum Genet 2001;68:1408–18.
[106] Smith EH, Thomas C, McHugh D, et al. Allelic diversity in MCAD deficiency: the biochemical classification of 54 variants identified during 5 years of ACADM sequencing. Mol Genet Metab 2010;100:241–50.
[107] Lindner M, Hoffmann GF, Matern D. Newborn screening for disorders of fatty-acid oxidation: experience and recommendations from an expert meeting. J InheritMetab Dis 2010;33:521–6.

[108] O'Reilly L, Bross P, Corydon TJ, et al. The Y42H mutation in medium-chain acyl-CoA dehydrogenase, which is prevalent in babies identified by MS/MS-based newborn screening, is temperature sensitive. EurJ Biochem 2004;271:4053–63.

[109] Maier EM, Gersting SW, Kemter KF, et al. Protein misfolding is the molecular mechanism underlying MCADD identified in newborn screening. Hum Mol Genet 2009;18: 1612–23.

[110] Grosse SD, Khoury MJ, Greene CL, Crider KS, Pollitt RJ. The epidemiology of medium chain acyl-CoA dehydrogenase deficiency: an update. Genet Med 2006;8:205–12.

[111] Gregersen N, Wintzensen H, Christensen SK, Christensen MF, Brandt NJ, Rasmussen K. C6-C10-dicarboxylic aciduria: investigations of a patient with riboflavin responsive multiple acyl-CoA dehydrogenation defects. PediatrRes 1982;16:861–8.

[112] Sakurai T, Miyazawa S, Furuta S, Hashimoto T. Riboflavin deficiency and beta-oxidation systems in rat liver. Lipids 1982;17:598–604.

[113] Gregersen N, Rhead W, Christensen E. Riboflavin responsive glutaric aciduria type II. In: Tanaka K, Coates PM, eds. Fatty acid oxidation: Clinical Biochemical and Molecular Aspects. New York: Alan R, Liss, INC; 1990:477–94.

[114] Rhead W, Roettger V, Marshall T, Amendt B. Multiple acyl-coenzyme A dehydrogenation disorder responsive to riboflavin: substrate oxidation, flavin metabolism, and flavoenzyme activities in fibroblasts. PediatrRes 1993;33:129–35.

[115] Henriques BJ, Bross P, Gomes CM. Mutational hotspots in electron transfer flavoprotein underlie defective folding and function in multiple acyl-CoA dehydrogenase deficiency. BiochimBiophysActa 2010;1802:1070–7.

[116] Torchetti EM, Brizio C, Colella M, et al. Mitochondrial localization of human FAD synthetase isoform 1. Mitochondrion 2010;10:263–73.

[117] Ho G, Yonezawa A, Masuda S, et al. Maternal riboflavin deficiency, resulting in transient neonatal-onset glutaric aciduria Type 2, is caused by a microdeletion in the riboflavin transporter gene GPR172B. HumMutat 2011;32:E1976–E84.

[118] Bosch AM, Abeling NG, Ijlst L, et al. Brown-Vialetto-Van Laere and Fazio Londe syndrome is associated with a riboflavin transporter defect mimicking mild MADD: a new inborn error of metabolism with potential treatment. J InheritMetab Dis 2011;34:159–64.

[119] Olsen RK, Olpin SE, Andresen BS, et al. ETFDH mutations as a major cause of riboflavin-responsive multiple acyl-CoA dehydrogenation deficiency. Brain 2007;130:2045–54.

[120] Gempel K, Topaloglu H, Talim B, et al. The myopathic form of coenzyme Q10 deficiency is caused by mutations in the electron-transferring-flavoprotein dehydrogenase (ETFDH) gene. Brain 2007;130:2037–44.

[121] Wang ZQ, Chen XJ, Murong SX, Wang N, Wu ZY. Molecular analysis of 51 unrelated pedigrees with late-onset multiple acyl-CoA dehydrogenation deficiency (MADD) in southern China confirmed the most common ETFDH mutation and high carrier frequency of c.250G>A. J Mol Med(Berl) 2011;89:569–76.

[122] Zytkovicz TH, Fitzgerald EF, Marsden D, et al. Tandem mass spectrometric analysis for amino, organic, and fatty acid disorders in newborn dried blood spots: a two-year summary from the New England Newborn Screening Program. ClinChem 2001;47:1945–55.

[123] Spiekerkoetter U, Sun B, Zytkovicz T, Wanders R, Strauss AW, Wendel U. MS/MS-based newborn and family screening detects asymptomatic patients with very-long-chain acyl-CoA dehydrogenase deficiency. J Pediatr 2003;143:335–42.

[124] Spiekerkoetter U, Lindner M, Santer R, et al. Management and outcome in 75 individuals with long-chain fatty acid oxidation defects: results from a workshop. J InheritMetab Dis 2009;32:488–97.

[125] Gobin-Limballe S, McAndrew RP, Djouadi F, Kim JJ, Bastin J. Compared effects of missense mutations in Very-Long-Chain Acyl-CoA Dehydrogenase deficiency: Combined analysis by structural, functional and pharmacological approaches. BiochimBiophysActa 2010;1802:478–84.

[126] Gobin-Limballe S, Djouadi F, Aubey F, et al. Genetic basis for correction of very-long-chain acyl-coenzyme A dehydrogenase deficiency by bezafibrate in patient fibroblasts: toward a genotype-based therapy. AmJHumGenet 2007;81:1133–43.
[127] Bastin J, Lopes-Costa A, Djouadi F. Exposure to resveratrol triggers pharmacological correction of fatty acid utilization in human fatty acid oxidation-deficient fibroblasts. HumMol Genet 2011;20:2048–57.
[128] Ensenauer R, He M, Willard JM, et al. Human acyl-CoA dehydrogenase-9 plays a novel role in the mitochondrial beta-oxidation of unsaturated fatty acids. J BiolChem 2005;280:32309–16.
[129] He M, Rutledge SL, Kelly DR, et al. A new genetic disorder in mitochondrial fatty acid beta-oxidation: ACAD9 deficiency. AmJ Hum Genet 2007;81:87–103.
[130] Gerards M, van den Bosch BJ, Danhauser K, et al. Riboflavin-responsive oxidative phosphorylation complex I deficiency caused by defective ACAD9: new function for an old gene. Brain 2011;134:210–9.
[131] Haack TB, Danhauser K, Haberberger B, et al. Exome sequencing identifies ACAD9 mutations as a cause of complex I deficiency. NatGenet 2010;42:1131–4.
[132] Scholte HR, Busch HF, Bakker HD, Bogaard JM, Luyt-Houwen IE, Kuyt LP. Riboflavin-responsive complex I deficiency. BiochimBiophysActa 1995;1271:75–83.
[133] Corydon MJ, Vockley J, Rinaldo P, et al. Role of common gene variations in the molecular pathogenesis of short-chain acyl-CoA dehydrogenase deficiency. Pediatr Res 2001;49:18–23.
[134] van Maldegem BT, Duran M, Wanders RJ, et al. Clinical, biochemical, and genetic heterogeneity in short-chain acyl-coenzyme A dehydrogenase deficiency. JAMA 2006;296:943–52.
[135] Waisbren SE, Levy HL, Noble M, et al. Short-chain acyl-CoA dehydrogenase (SCAD) deficiency: an examination of the medical and neurodevelopmental characteristics of 14 cases identified through newborn screening or clinical symptoms. Mol Genet Metab 2008;95:39–45.
[136] Jethva R, Bennett MJ, Vockley J. Short-chain acyl-coenzyme A dehydrogenase deficiency. Mol Genet Metab 2008;95:195–200.
[137] van Maldegem BT, Wanders RJ, Wijburg FA. Clinical aspects of short-chain acyl-CoA dehydrogenase deficiency. J InheritMetab Dis 2010;33:507–11.
[138] Pedersen CB, Zolkipli Z, Vang S, et al. Antioxidant dysfunction: potential risk for neurotoxicity in ethylmalonic aciduria. J InheritMetab Dis 2010;33:211–22.
[139] Rhead WJ, Allain D, van Calcar S, et al. Short-shain acyl-coenzyme a dehydrogenase (SCAD) and 3-methylcrotonyl-CoA carboxylase (MCC) deficiencied: tandem mass spectrometry newborn screening detect many clinically benign cases. J Inher Metab Dis 2002;25 (suppl 1):4.
[140] Amendt BA, Greene C, Sweetman L, et al. Short-chain acyl-coenzyme A dehydrogenase deficiency: clinical and biochemical studies in two patients. JClinInvest 1987;79:1303–9.
[141] Coates PM, Hale DE, Finocchiaro G, Tanaka K, Winter SC. Genetic deficiency of short-chain acyl-coenzyme A dehydrogenase in cultured fibroblasts from a patient with muscle carnitine deficiency and severe skeletal muscle weakness. JClinInvest 1988;81:171–5.
[142] Naito E, Indo Y, Tanaka K. Short chain acyl-coenzyme A dehydrogenase (SCAD) deficiency. Immunochemical demonstration of molecular heterogeneity due to variant SCAD with differing stability. JClinInvest 1989;84:1671–4.
[143] Nagan N, Kruckeberg KE, Tauscher AL, Snow BK, Rinaldo P, Matern D. The frequency of short-chain acyl-CoA dehydrogenase gene variants in the US population and correlation with the C(4)-acylcarnitine concentration in newborn blood spots. MolGenet Metab 2003;78:239–46.
[144] Gregersen N, Winter VS, Corydon MJ, et al. Identification of four new mutations in the short-chain acyl-CoA dehydrogenase (SCAD) gene in two patients: one of the variant alleles, 511C-->T, is present at an unexpectedly high frequency in the general population, as was the case for 625G-->A, together conferring susceptibility to ethylmalonic aciduria. Hum MolGenet 1998;7:619–27.

[145] Tein I, Elpeleg O, Ben-Zeev B, et al. Short-chain acyl-CoA dehydrogenase gene mutation (c.319C>T) presents with clinical heterogeneity and is candidate founder mutation in individuals of Ashkenazi Jewish origin. MolGenetMetab 2008;93:179–89.
[146] Naito E, Indo Y, Tanaka K. Identification of two variant short chain acyl-coenzyme A dehydrogenase alleles, each containing a different point mutation in a patient with short chain acyl-coenzyme A dehydrogenase deficiency. JClinInvest 1990;85:1575–82.
[147] Corydon TJ, Bross P, Jensen TG, et al. Rapid degradation of short-chain acyl-CoA dehydrogenase (SCAD) variants with temperature-sensitive folding defects occur after import into mitochondria. JBiolChem 1998;273:13065–71.
[148] Pedersen CB, Bross P, Winter VS, et al. Misfolding, degradation, and aggregation of variant proteins. The molecular pathogenesis of short chain acyl-CoA dehydrogenase (SCAD) deficiency. JBiolChem 2003;278:47449–58.
[149] Schmidt SP, Corydon TJ, Pedersen CB, Bross P, Gregersen N. Misfolding of short-chain acyl-CoA dehydrogenase leads to mitochondrial fission and oxidative stress. Mol Genet Metab 2010;100:155–62.
[150] Schmidt SP, Corydon TJ, Pedersen CB, et al. Toxic response caused by a misfolding variant of the mitochondrial protein short-chain acyl-CoA dehydrogenase. J InheritMetab Dis 2011;34:465–75.
[151] Nguyen TV, Riggs C, Babovic-Vuksanovic D, et al. Purification and Characterization of Two Polymorphic Variants of Short Chain Acyl-CoA Dehydrogenase Reveal Reduction of Catalytic Activity and Stability of the Gly185Ser Enzyme. Biochem 2002;41:11126–33.
[152] van Maldegem BT, Duran M, Wanders RJ, Waterham HR, Wijburg FA. Flavin adenine dinucleotide status and the effects of high-dose riboflavin treatment in short-chain acyl-CoA dehydrogenase deficiency. PediatrRes 2010;67:304–8.
[153] Gregersen N, Hansen J, Palmfeldt J. Mitochondrial proteomics-a tool for the study of metabolic disorders. J InheritMetab Dis 2012;35:715–26.
[154] Schuck PF, Leipnitz G, Ribeiro CA, et al. Inhibition of creatine kinase activity in vitro by ethylmalonic acid in cerebral cortex of young rats. NeurochemRes 2002;27:1633–9.
[155] Leipnitz G, Schuck PF, Ribeiro CA, et al. Ethylmalonic acid inhibits mitochondrial creatine kinase activity from cerebral cortex of young rats in vitro. NeurochemRes 2003;28:771–7.
[156] Barschak AG, Ferreira GC, Andre KR, et al. Inhibition of the electron transport chain and creatine kinase activity by ethylmalonic acid in human skeletal muscle. Metab Brain Dis 2006;21:11–9.
[157] Schuck PF, Busanello EN, Moura AP, et al. Promotion of lipid and protein oxidative damage in rat brain by ethylmalonic acid. NeurochemRes 2010;35:298–305.
[158] Schuck PF, Ferreira GC, Viegas CM, et al. Chronic early postnatal administration of ethylmalonic acid to rats causes behavioral deficit. BehavBrain Res 2009;197:364–70.
[159] DeLano WL. The PyMOL Users Manual: DeLano Scientific, Palo Alto, CA, USA; 2002.
[160] Bross P, Engst S, Strauss AW, Kelly DP, Rasched I, Ghisla S. Characterization of wild-type and an active site mutant of human medium chain acyl-CoA dehydrogenase after expression in Escherichia coli. J Biol Chem 1990;265:7116–9.
[161] Engst S, Bross P, Stiemke J, et al. Some Properties of Glu-376-Gln Active Site Mutant of Human Medium-Chain Acyl-CoA Dehydrogenase. In: Curti B, Ronchi, S, and Zanetti, G, ed. Flavins and Flavoproteins, Proceedings of the 10th Internat Symposium: W. DeGruyter; 1991:319–24.
[162] Lehman TC, Thorpe C. Alternate electron acceptors for medium-chain acyl-CoA dehydrogenase: use of ferricenium salts. Biochemistry 1990;29:10594–602.

12 Flavoproteins in oxidative protein folding

Colin Thorpe

Abstract

The introduction of structural disulfide bonds during oxidative protein folding involves the removal of two reducing equivalents for every disulfide generated. While multiple oxidants drive oxidative folding in eukaryotes, the flavin-linked sulfhydryl oxidases described in this Chapter play important roles in a variety of cellular locales. Members of the ERV family of sulfhydryl oxidases catalyze disulfide bond formation in the cytosol, mitochondrial intermembrane space (IMS), endoplasmic reticulum/Golgi, and extracellular matrix. The ERV domain is a compact bundle of helices that binds FAD tightly and presents a redox-active CxxC motif adjacent to the C4a position the flavin. A number of the smaller ERV-family sulfhydryl oxidases contain an additional shuttle disulfide, located within flexible N- or C-terminal extensions to the core FAD-binding domain to ferry reducing equivalents from client protein undergoing oxidation to the disulfide proximal to the flavin ring. While most of these sulfhydryl oxidases use molecular oxygen as terminal electron acceptor, the ERV enzymes occupying the mitochondrial IMS may also use cytochrome c, thereby avoiding the generation of hydrogen peroxide. This chapter also describes the discovery, structure and mechanism of the Quiescin-sulfhydryl oxidase (QSOX) enzymes. These more elaborate members of the ERV family are ancient fusions of one or two thioredoxin domains, a helix-rich region, and a single FAD-binding ERV domain. QSOXs directly insert disulfide bonds into unstructured proteins and can cooperate with protein disulfide isomerase to generate protein products with complex disulfide connectivity. Finally, the Ero1 sulfhydryl oxidases are briefly discussed. Ero1 enzymes also utilize bound FAD, together with proximal and shuttle disulfides, in a helix-rich domain that shows no obvious evolutionary kinship with the ERV-family of sulfhydryl oxidases.

12.1 Oxidative protein folding

Disulfide bonds constitute one of the most widely recognized forms of post-translational modification and their key roles in a range of cellular processes have been known for decades. Disulfide bonds stabilize a huge number of products of the secretory pathway in eukaryotes – from small peptides to mammoth extracellular structural proteins with hundreds of disulfide bridges. It is therefore surprising that the actual mechanisms for disulfide bond generation remain cryptic in all higher eukaryotes [1–5]. In the next paragraphs we introduce the process of oxidative folding and provide a historical perspective for the current uncertainties in the field.

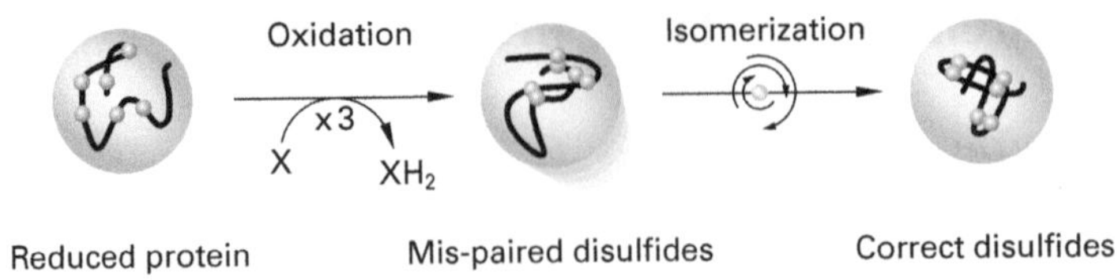

Fig. 12.1: Oxidative protein folding involves insertion and rearrangement of disulfide bonds. Typically, oxidation and isomerization steps operate in parallel.

Oxidative protein folding comprises two conceptual steps (▶Fig. 12.1). The first is the net generation of disulfide bonds via the removal of two reducing equivalents from each pair of cysteine residues. However, because no oxidant necessarily generates the correct pairings *de novo*, the second aspect of oxidative protein folding is the isomerization of mis-paired linkages by members of the protein disulfide isomerase family. It should be noted that ▶Fig. 12.1 is an oversimplification because oxidation and isomerization activities are comingled as protein folding progresses. Disulfide bonds can be generated in multiple locations in eukaryotic cells. The secretory system involves endoplasmic reticulum, Golgi and post-Golgi compartments of the cell. There is a distinct disulfide bond-forming system within the mitochondrial inter-membrane space that is crucial for the maturation of a number of proteins of the electron transport chain. While the cytosol of eukaryotes is generally too reducing to support the formation of structural disulfides, this locale is used for oxidative folding of coat proteins encoded by certain large double-stranded DNA viruses [6,7]. Finally, disulfide bonds formation can also occur extracellularly: at the cell surface and within the extracellular matrix (ECM) [1].

Many of the disulfide bonds generated in these diverse location are introduced by sulfhydryl oxidases catalyzing the reaction:

$$2\ PSH + O_2 \Rightarrow PS\text{-}SP + H_2O_2$$

All of the well-characterized sulfhydryl oxidases are flavin-linked, although a number of them were originally described as metalloenzymes. For example, the term sulfhydryl oxidase was coined by Rony *et al.* to describe a disulfide bond-generating activity in mammalian skin [8] that was later ascribed to a copper-dependent enzyme [9]. However subsequent cloning of the skin enzyme by Sato and colleagues [10], and biochemical studies from this laboratory [11] suggested that the metal content was adventitious, and that the true redox-active cofactor was flavin adenine dinucleotide (FAD). Another influential example was the iron-dependent enzyme from bovine milk [12]. A thorough reexamination of this enzyme showed that it, too, is flavin-dependent; the iron content of the older preparations likely derived from heavy contamination with the common milk protein lactoferrin – a protein with negligible sulfhydryl oxidase activity [13]. While the presence of metal-dependent sulfhydryl oxidases now seems questionable, there appears no mechanistic reason why they should not exist. Hence traces of copper and iron salts are potent catalysts of the oxidation of thiol compounds in aerobic buffers [14] and this chemical propensity could presumably be exploited to generate efficient disulfide bond-generating enzymes. It is ironic that the cryptic contamination of reagents and surfaces by redox-active transition metals has led to the widely-held notion that disulfide bond generation requires little, or nothing, in the way of catalysis. This, in turn, delayed the recognition that sulfhydryl oxidases play key roles in oxidative protein folding.

12.2 Convergent evolution of three classes of FAD-dependent sulfhydryl oxidases

Early work on the flavin-dependent sulfhydryl oxidases implicated a catalytic disulfide in redox communication with the flavin ring [15] and hence these studies could follow the precedents and approaches pioneered by Massey and Williams and their coworkers on the founding members of the **p**yridine **n**ucleotide **d**isulfide **o**xido**r**eductases (PN-DOR) superfamily [16–18]. In many of these enzymes a reduced pyridine nucleotide serves as the reductant for a substrate disulfide (via the mediation of FAD and one or more protein disulfide bonds that are in oxidation-reduction communication with the flavin). An example of the flow of reducing equivalents in eukaryotic thioredoxin reductases is shown [16–18]:

> NADPH (substrate) ⇒ FAD ⇒ -S-S- (proximal) ⇒ -S-S- (distal, or –S-Se-) ⇒ -S-S- (Thioredoxin substrate)

It therefore seemed reasonable that the sulfhydryl oxidases involved in oxidative protein folding might coopt the same protein scaffold by operating in reverse: oxidizing protein thiols, with eventual reduction of the proximal disulfide, the generation of reduced flavin, and the reduction of molecular oxygen (rather than NAD^+/$NADP^+$) [1,15]. Indeed, a number of fungal sulfhydryl oxidases are PNDOR superfamily members. However these flavoproteins typically form disulfide bonds with small substrates *e.g.* glutathione [19] and the epipolythiodiketopiperazine [20] and dithiolopyrrolone [21] antibiotics. In contrast, the flavoenzymes of oxidative protein folding that have been uncovered to date utilize two ancient flavin-binding scaffolds (ERV, Essential for Respiration and Vegetative Growth; and Ero1, ER oxidoreductin 1) which are entirely distinct from the architecture of PNDOR family members.

12.3 Two flavin-dependent pathways for protein disulfide bond generation in eukaryotes

The first flavin-dependent sulfhydryl oxidase to be discovered was noticed as a yellow band during chromatography of proteins isolated from rat seminal vesicles [22]. This monomeric ~66 kDa protein contained one molecule of FAD and oxidized a variety of monothiol and dithiol model substrates with the reduction of oxygen to hydrogen peroxide. Protein thiols derived from reduced RNase (see later) were found to be substrates of the enzyme [22]. Subsequently the evolutionary relationship between the seminal vesicle sulfhydryl oxidase and a human growth factor named Quiescin Q6 was recognized by Hoober *et al.* [23] and Benayoun *et al.* [24].

These Quiescin-sulfhydryl oxidase (QSOX) enzymes function as shown in ►Fig. 12.2A [1,25–28]. As will be detailed later, they can introduce disulfides bonds directly into

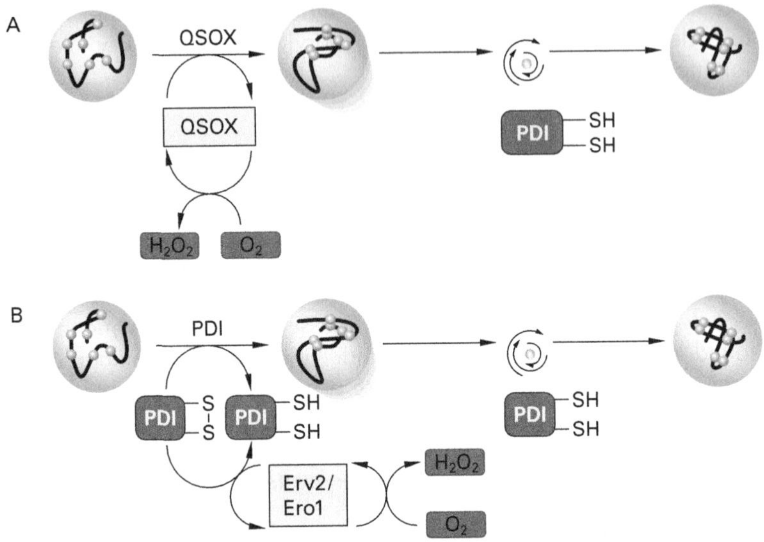

Fig. 12.2: Oxidative protein folding catalyzed by QSOX and Ero1. In panel A, QSOX inserts disulfides directly into unfolded reduced protein substrates while reduced PDI addresses mispairings. In panel B, oxidation of the reduced client protein is catalyzed by oxidized PDI and sustained indirectly by the reoxidation of reduced PDI by Erv2p or Ero1. PDI further serves as an isomerase as in panel A.

conformationally flexible protein or peptide substrates with the generation of hydrogen peroxide. A second enzyme, protein disulfide isomerase (PDI), corrects mispairings that may be introduced by the oxidase (▶Fig. 12.2A) [29,30]. Eukaryotic cells typically contain multiple PDI enzymes [30]. All share the presence of one or more thioredoxin domains with many of them carrying a redox-active disulfide at a loop-helix boundary [30]. Typically two amino acids separate the cysteine residues and these intervening residues can strongly modulate the redox potential of CxxC motifs [31–33]. The best understood member of the PDI family contains 4 thioredoxin domains arranged in an a-b-b′-a′ pattern as depicted in ▶Fig. 12.3.

The outermost thioredoxin domains of human PDI contain the rather oxidizing CxxC motifs (both here CGHC; with E'° values of about –180 mV [34]) compared to the strongly reducing prototypical thioredoxin from *Escherichia coli* (CGPC; ~ –270 mV [31–33]). The N-terminal (interchange) cysteine of each CxxC pair is solvent accessible and can form mixed disulfides with client protein and peptides. The C-terminal cysteine then serves a resolving role, remaining buried and inaccessible to direct interaction with substrates [30]. While it is agreed that PDI shuffles disulfides iteratively, in a process whose net

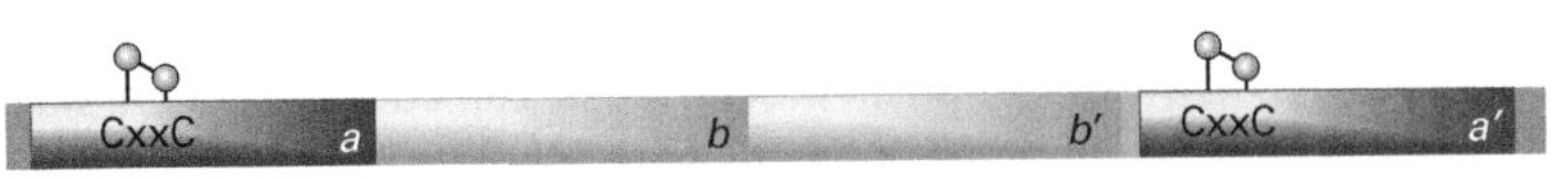

Fig. 12.3: Domain organization of a protein disulfide isomerase. Human PDI contains two redox-active thioredoxin domains (blue bars) carrying CxxC motifs. The inner two thioredoxin domains (grey bars) are redox-inactive.

directionality is dictated by the stability of the native fold, the precise mechanism of this isomerization is still debated. In the classic view, reduced PDI (using either a or a′ domains) reacts with a mis-paired disulfide forming a PDI-client protein mixed disulfide that can be resolved via an alternative pairing within the folding protein [29,30]. In this redox-neutral shuffling mode only reduced PDI and the mixed disulfide forms exist. However PDI can also function by cycles of reduction and oxidation involving both reduced and oxidized isomerase as distinct participants. Here isomerization starts with reduction of a mispaired disulfide by reduced PDI. The resulting oxidized PDI can then reoxidize the reduced client protein generating additional pairings that may contain correct linkages [29,30].

This ability of PDI to serve as an oxidant for reduced proteins is key to the "PDI-first" model of oxidative folding shown in ►Fig. 12.2B. Here, the net insertion of disulfides is sustained by reoxidation of reduced PDI by one of two dedicated flavin dependent oxidases abbreviated Ero1 [35,36] and Erv2 [37–39]. The two routes to the generation of protein disulfides shown in ►Fig. 12.2 are introduced here because they provide useful perspective for the QSOX enzymes to be described next. QSOXs are ancient fusions of PDI-like thioredoxin domains and Erv2-like flavoproteins that provide facile, self-contained, catalysts for protein disulfide bond generation.

12.3.1 Quiescin-sulfhydryl oxidases: structural aspects

Prior to the divergence of fungi from metazoans, a fusion of thioredoxin domains with a helix-rich flavin-binding module (hereafter abbreviated ERV) led to a facile and self-contained catalyst of oxidative protein folding [1,25]. These enzymes are found in most non-fungal eukaryotes; from the smallest free-living eukaryote (the marine algae, *Ostreococcus tauri*) to humans [27,28]. The enzymes from algae, plants and certain protists contain one complete PDI-like thioredoxin domain as well as a relatively oxidizing CxxC motif (►Fig. 12.4A).

In addition, metazoans have an adjacent second thioredoxin domain of unknown function (►Fig. 12.4B). This Trx2 domain lacks the CxxC motif as in the b-domains of PDI. Thereafter both types of QSOX are similar, with a helix-rich region (HRR) of about 100 amino acids followed by a domain that was originally found in a yeast growth factor named essential for respiration and vegetative growth (Erv1p [40]). The discovery that the sulfhydryl oxidase QSOX contained an ERV domain led to the finding that the yeast growth factors Erv1p and Erv2p were themselves small FAD-linked disulfide-generating enzymes and, reciprocally, identified the domain in QSOX responsible for flavin binding [1,25]. While crystal structures of full-length QSOXs have yet to appear, a structure of the HRR-ERV half of the human QSOX1 enzyme provided key insight into mechanism and intriguing clues as to the evolutionary origins of the HRR domain [41]. The HRR-ERV domain is a fusion of two similar 4-helix bundle domains that are arranged with the same topology as is found in the homodimeric Erv2p protein (see later). In addition to the FAD, each subunit of Erv2p contains a proximal CxxC motif that communicates reducing equivalents to the bound flavin [15,23,42–44]. During the evolution of QSOX the FAD binding site and proximal disulfide from the N-terminal domain of the HRR-ERV fusion were lost [41]. The resulting HRR-ERV pseudo-dimer then receives reducing equivalents from the thioredoxin domain as depicted in ►Fig. 12.4. As will be mentioned later, fusion of a redox-active thioredoxin domain to ERV domains

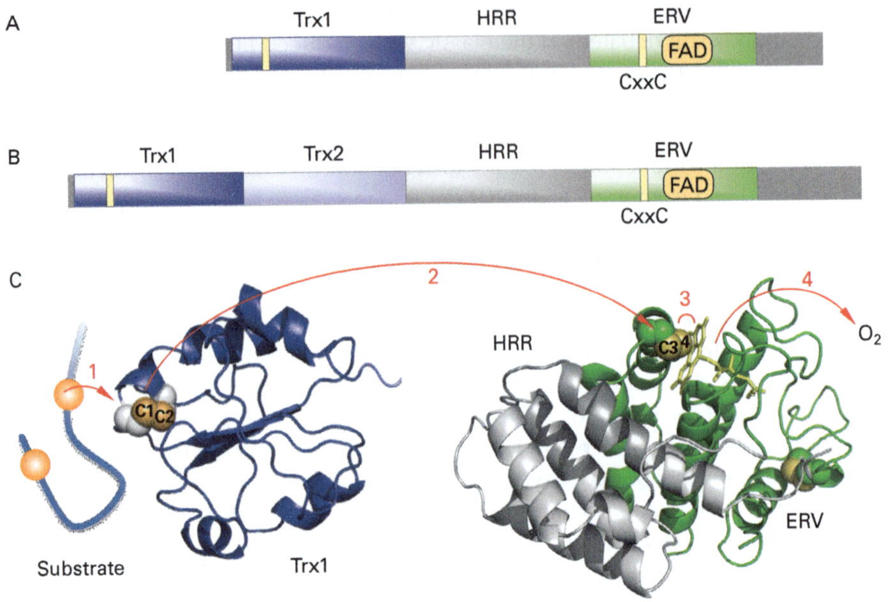

Fig. 12.4: Organization, structure and electron flow in Quiescin sulfhydryl oxidases. The QSOXs contain one or two thioredoxin (Trx) domains, followed by a helix-rich region (HRR) and the FAD binding ERV domain. Redox-active CxxC motifs are indicated by yellow bars in panels A and B. Panel A shows the simpler domain organization of QSOXs from plants, algae and protists. Metazoan QSOXs contain an additional redox-inactive thioredoxin domain (Trx2; panel B). Panel C shows crystal structures of a Trx1 domain paired with the HRR-Erv domain of human QSOX1. The flow of reducing equivalents during the oxidation of a dithiol substrate is depicted by red arrows. The spheres labeled C1-C2 represents the Trx CxxC motif, and C3-C4 the proximal disulfide next to the flavin ring (depicted in yellow sticks). The third CxxC motif (immediately above the ERV label) is of unknown function.

generates a facile catalyst capable of introducing disulfides into a huge range of small and large molecular weight thiol compounds.

12.3.2 Mechanistic studies of QSOX

Here we combine insights from the early studies on the avian egg white QSOX [15,44,45] with those from recombinant human [42] and protist [46] QSOXs. All QSOXs show substrate specificity profiles suggesting that reduced unfolded, or conformationally mobile, proteins are their preferred substrates [47,48]. Metazoan and protist QSOXs show very similar steady-state and pre-steady state kinetic behavior, although the latter lacks a second Trx domain [46]. All QSOXs sequences so far examined contain three conserved CxxC motifs. The first (C1-C2; ►Fig. 12.4C) is found in the redox-active Trx domain and is in oxidation-reduction communication with the second CxxC motif (C3-C4) placed adjacent to the flavin ring in the ERV domain. The third CxxC motif (C5-C6) has no known oxidation-reduction roles [41,42,46]; it is adjacent to a short 5th helix towards the C-terminus of the ERV domain (►Fig. 12.4C) with no obvious means of communication with other redox centers in the enzyme [41]. Mutation of one or both cysteines of this third CxxC motif showed little catalytic impact when assessed using model substrates of the enzyme *in vitro* [42,46]. In the cell, this completely conserved motif might participate in metal-binding, or function to secure QSOX to oxidation-reduction partners or to cellular structures.

Early partial proteolysis experiments with avian QSOX [44] and later site directed mutagenesis approaches showed [42,46,48] that thiol substrates primarily reduce the first CxxC motif located in the thioredoxin domain. Such reactions proceed via formation of a mixed disulfide between a substrate thiol and the solvent-exposed C1 sulfur atom (►Fig. 12.4C, step 1), followed by resolution from a second thiolate species in the substrate to generate the reduced thioredoxin domain. Although these thiol/disulfide exchange reactions are often regarded as chemically facile, they do have stringent steric requirements; the attacking thiolate species (S_A, ►Fig. 12.5) and the two sulfur atoms comprising the recipient disulfide (here formed by S_B-S_C) should be co-linear [48].

The mixed disulfide is then subject to attack from a second sulfur atom resolving the products of this second thiol/disulfide exchange step. QSOX family proteins do not seem to have significant highly specific binding sites for their thiol substrates and appear to function in a "hit-and-run" mode [48]. Hence monothiols, including glutathione, are relatively poor substrates because of the requirement that a second thiolate must resolve the mixed disulfide in a bimolecular reaction. In contrast dithiothreitol is a good substrate of QSOX family members, as are a seemingly limitless array of peptides and unfolded proteins that contain multiple cysteine residues [13,15,45–48]. QSOX enzymes generally cannot insert disulfide bonds into well-folded proteins that carry either buried or surface thiols presumably because of the requirements of attaining colinearity in both formation and resolution of the mixed disulfide required for net reduction of the enzyme [27,48]. Typical protein substrates of QSOX are prepared by reducing the disulfides of a secreted protein (using excess DTT in guanidine hydrochloride and then removing excess reagents by size-exclusion chromatography). The resulting reduced protein is conformationally very mobile because it lacks the disulfide bonds that normally prove greatly stabilizing. As one illustration, all eight thiols of reduced ribonuclease A react with mM DTNB in less than 100 ms [49] – much faster than a typical thiol found in a well-folded cytosolic protein. Given this flexibility, the size of the polypeptide chain undergoing oxidation by QSOX seems unimportant [47]; the enzyme can efficiently oxidize the B chain of reduced

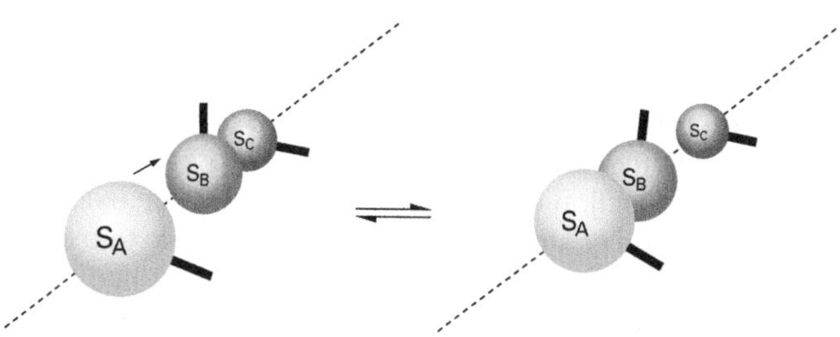

Fig. 12.5: Disulfide exchange reactions require colinearity between reacting sulfur atoms (S_A, S_B and S_C).

insulin and, conversely, rapidly generate disulfide bonds in reduced thyroglobulin (670 kDa). Similarly, there seems to be no strong dependence on the pI or the specific identity of a protein substrate [47]. A rate-limiting internal step in catalysis sets an upper limit of about 600–2000 disulfides formed per minute depending on the QSOX used [13,42,46,47]. Thus the rate-saturation observed with substrate thiols reflects the progressive expression of this kinetic limitation as the rate of the bimolecular step 1 increases (▶Fig. 12.4C, [46]). For this reason specificity toward thiol substrates is largely manifest by their K_m values [47]. While protein substrates with exceptionally low K_m values may be identified in future work, QSOX seems to be a versatile catalyst for the insertion of disulfide bonds into any protein, providing they are sufficiently conformationally mobile.

Step 2 in ▶Fig. 12.4 represents an inter-domain disulfide exchange reaction followed by reduction of the enzyme-bound FAD (step 3). While some uncertainty remains, it appears that the rate-limiting step in QSOX catalysis is the disengagement of this Trx-ERV inter-domain disulfide [46]. This species is signaled by a thiolate to flavin charge-transfer complex that decomposes with the appearance of reduced flavin when QSOX is mixed with thiol substrates under anaerobic conditions [45,46]. No evidence for the accumulation of a C4a thiol-flavin adduct in this phase of catalysis has been found for QSOX or any other ERV-family enzyme.

The reaction of the reduced flavin of QSOX with molecular oxygen shows the typical biomolecular dependence on oxygen, with a second-order rate constant of about $10^5\ M^{-1}s^{-1}$. Although QSOX enzymes appear to be typical flavoprotein oxidases, they do release a small amount of superoxide during turnover (amounting to about 0.5% of the total electron efflux) with the balance being hydrogen peroxide [50]. Surprisingly, several of the smaller ERV-fold sulfhydryl oxidases are comparatively prolific superoxide generators (see later).

Removal of the Trx CxxC disulfide (by mutation of either the C1 or the C2 cysteine residues) severely diminishes the catalytic efficiency of both mammalian and protist QSOXs because reducing equivalents can now only enter the oxidase via a direct reduction of the proximal (C3-C4) disulfide (▶Fig. 12.4C). While DTT is a significant substrate for the HRR-ERV fragment of QSOX, reduced RNase is now essentially a non-substrate [42,46]. For this reason, the fusion of a distal redox-active center in the Trx domain with the ERV domain provides a key catalytic advantage over the HRR-ERV fragment alone. Interestingly, a number of other ERV- and Ero1-family enzymes have distal disulfides, located in flexible elements of structure, that shuttle reducing equivalents to the proximal disulfide (see later).

12.3.3 QSOX can catalyze oxidative protein folding

In ▶Fig. 12.2A the direct substrates of QSOX are depicted as unfolded reduced proteins and the role of PDI is restricted to the isomerization of those mis-paired disulfides introduce during the net introduction of disulfide bonds. Unlike the oxidases Ero1 and Erv2p (▶Fig. 12.2B), *in vitro* studies showed that QSOX was unable to oxidize reduced PDI directly [51]. Hence the separate roles of oxidase and isomerase in oxidative protein folding could be reconstructed in model systems. Nanomolar levels of QSOX and micromolar concentrations of reduced PDI were able to catalyze the efficient oxidative refolding of reduced RNase in aerobic solutions without requiring a glutathione redox buffer [51]. However, this does not represent a particularly challenging application because RNase has a relatively simple disulfide connectivity (with four disulfide bonds and 105-disulfide isomers for the fully oxidized protein). More notably, comparable levels of QSOX and reduced PDI were able to efficiently refold riboflavin binding protein (with nine disulfide bonds and more than 34 million disulfide isomers). In this instance, the regain of binding activity could be followed continuously by the loss of the bright green fluorescence as riboflavin binds to the apo-binding protein. QSOX may find wider application in the heterologous expression of proteins and for *in vitro* folding of reduced proteins. A recent paper describes that coexpression of human QSOX with two different recombinant prion proteins allows their ready expression in *E. coli* [52]. Coexpression of QSOX in *E. coli* increased the yield of human tissue plasminogen activator, but not urokinase or chitinase [53].

12.3.4 Cellular roles of QSOX

Enzymological approaches have dominated the first decade of research into the QSOX enzyme family; consequently, many key aspects of their physiological roles remain to be clarified [27,28,38,39,54,55]. At least *in vitro,* the QSOX oxidases are clearly much more proficient at generating disulfide bonds than enzymes like Ero1p and Erv2p (see later). However such comparisons may be physiologically irrelevant, because they do not account for the segregation of activities to membrane surfaces, the formation of multienzyme complexes and the huge local concentrations of some of the PDI foldases in the secretory system of eukaryotes.

There are two QSOX isoforms in mammalian cells [1,25–28]. QSOX2 appears to be present at low concentrations in a wide range of mammalian tissues and remains poorly understood [27,28]. In contrast, QSOX1 is relatively abundant in tissues with a heavy secretory load, consistent with a role of this oxidase in disulfide bond generation. Intracellularly, QSOXs are observed in the ER, the Golgi, in secretory vesicles, and at the nuclear and plasma membrane surfaces, where they may be anchored by a single C-terminal transmembrane helix [25,54,56–58]. The shorter form of QSOX1, either generated as a splice variant lacking the last ~140 amino acids, or via removal of the transmembrane helix via proteolysis, is found secreted from a variety of cell types including lung fibroblasts [59], Chinese hamster ovary cells [24], and cultured prostate [60] and pancreatic [61] cancer cell lines. QSOXs are also found in the secretions from seminal vesicles [22], in milk [13], tears [43] and blood serum [62]. A number of possible roles for secreted forms of QSOX have been advanced. For example, QSOXs may participate in disulfide bond generation outside the cell and contribute to the formation of structures that are too large for completion intracellularly [54]. Secreted forms of QSOX may also play an antimicrobial role [22] or be involved in signaling mediated by hydrogen peroxide generation [56]. In addition, QSOX may contribute to extracellular oxidation-reduction cycles that modulate cellular adhesion and proliferation [26–28]. Although the specific physiological roles of QSOX1 remain cryptic, the enzyme is markedly overexpressed in a number of cancers including those of prostate [63] and pancreas [61,64]. In pancreatic cancer, this secretion leads to the accumulation of a peptide derived from the mobile C-terminal region of QSOX1 that is a promising blood serum marker for the disease. Further a partial knock-down of QSOX1 levels in pancreatic cancer cell lines drastically reduces their invasiveness [61]. Clearly the extracellular substrates of QSOX1 need to be identified.

12.4 Small ERV domain containing enzymes

The ERV flavin-binding domain in the QSOX enzymes is also found in a number of single-domain homodimeric sulfhydryl oxidases that are found in a variety of cellular locations in eukaryotes [39]. Selected members from this growing family of small sulfhydryl oxidases are discussed below. Most of these enzymes have an additional shuttle disulfide that communicates with the proximal disulfide housed in the FAD-binding domain. Representative examples of this varied architecture are shown in ▶Fig. 12.6, together with an indication of their intracellular locations.

12.4.1 Erv2p

Erv2p was first discovered as a protein sharing significant homology to the yeast growth factor Erv1p [37,65,66]. Erv2p is an exclusively fungal enzyme. When overexpressed in the ER, Erv2p can compensate for lack of the principal disulfide bond generating catalyst in *Saccharomyces cerevisae* (Ero1p [37], see later). Erv2p is believed to function by catalyzing the reoxidation of reduced PDI, as depicted in ▶Fig. 12.2B, although the modest *in vitro* activities observed suggest that other cellular partners of this oxidase may exist. The precise physiological roles of Erv2p remain uncertain [37–39].

The crystal structure of Erv2p was the first of a sulfhydryl oxidase to be published [67] and revealed a novel compact helix-rich FAD-binding fold (▶Fig. 12.7). The isoalloxazine ring is bound in the mouth of an irregular 4-helix bundle. The adenine moiety of FAD abuts a short fifth helix arranged orthogonally to the bundle axis [39,67]. The evolutionary

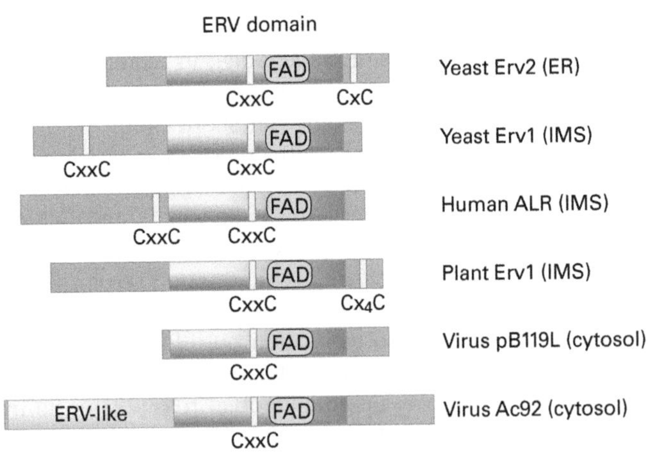

Fig. 12.6: Representative small sulfhydryl oxidases of the ERV-family. The core ERV domain is shown in green with the FAD prosthetic group and the CxxC motif depicted in yellow. The N- and C-terminal extension are shown in grey and the presence of an additional shuttle disulfide (Cx_nC) is indicated by a yellow bar. In all cases examined the wild-type proteins are homodimers, and communication between shuttle and proximal disulfides occurs across the subunit interface. The reported intracellular locations for these small sulfhydryl oxidases are noted in parentheses.

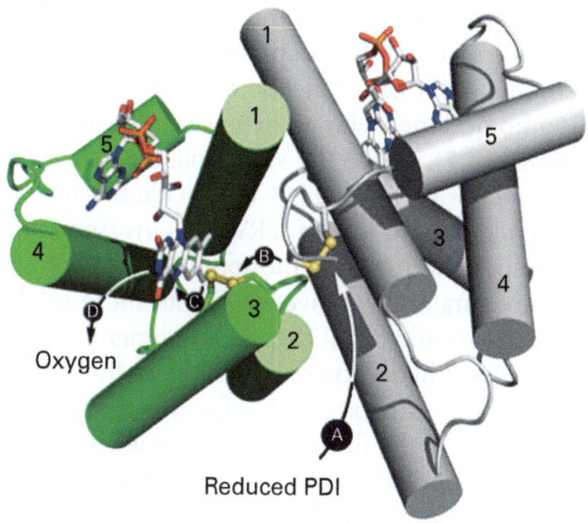

Fig. 12.7: The crystal structure of Erv2p. Helices 1 and 2 in the ERV fold participate in the dimer interface. The shuttle disulfide (at the C-terminus of the grey subunit) is reduced in step A and transfers reducing equivalents to the proximal disulfide of the other subunit (step B). Reduction of flavin and molecular oxygen follow (steps C and D).

origins of this unusual helix-rich FAD binding domain are unclear [39,67]. ERV-family proteins have not been encountered in prokaryotes or archaea, however the 4-helix bundle that comprises much of the FAD binding site is found in several prokaryotic/archaeal redox-active proteins [6]; for example those that carry binuclear iron centers (in the hemerythrin-like proteins [68]) or a heme group (as in cytochrome c′ [69]).

The adenine and isolloxazine rings are intercalated between a series of stacked aromatic residues. Starting from the top left of ▸Fig. 12.8: tyrosine, adenine, histidine, and tryptophan are layered over the *si* face of the flavin, with TYR covering parts of the re

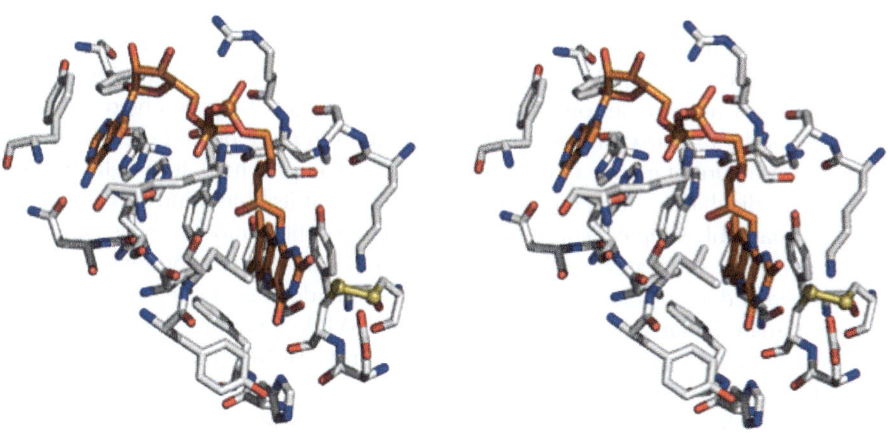

Fig. 12.8: A stereoview of the FAD binding site in yeast Erv2p. The isoalloxazine is intercalated within a stack of aromatic rings.

flavin face. The charge transfer thiol is also poised over the C4a position of the flavin *re* face [39,67]. A number of thiolate to oxidized flavin charge transfer interactions have been reported in ERV family members [45,46,70–72]; the corresponding C4a adducts have yet to be observed. Erv2p is a homodimer with an extensive intersubunit interface encompassing helices 1 and 2 (►Fig. 12.7). While the same interaction is used for the HRR-ERV pseudo-dimer of QSOX mentioned earlier, it is remarkable that two other modes of dimerization have been encountered in viral ERV proteins (see later). Hence the domains appear rather autonomous, and indeed monomeric forms of Erv2p have been generated and shown to be enzymatically active with both the model substrate DTT and reduced PDI [73]. Clearly the shuttle disulfide and the dimeric status of Erv2p are not essential to support the modest sulfhydryl oxidase activity of these enzymes.

Fass and coworkers have considered the structures of Erv family members and identified cavities of various sizes on the *si* face of the flavin that are candidates for the reduction of molecular oxygen [39,74]. Conventional measurements of the products of these sulfhydryl oxidases show formation of one molecule of hydrogen peroxide for every disulfide formed. While this is consistent with the expected behavior of simple flavoprotein oxidases, recent work shows that approximately 20% of the oxygen undergoing reduction by Erv2p escapes from the enzyme before a second reducing equivalent can convert this superoxide to hydrogen peroxide [50]. Thus about 10% of the electrons released by Erv2p emerge as the superoxide ion and this percentage can be increased about 3-fold using mutations which cause monomerization of the enzyme [50]. While this phenomenon is found in several other small sulfhydryl oxidase members, it is almost absent in the larger QSOX family members. The mechanism and significance of superoxide generation remain unclear [50].

12.4.2 Disulfide bond formation in the mitochondrial intermembrane space

The observation that a number of components of the mitochondrial respiratory chain, and proteins of the mitochondrial IMS, contain disulfide bridges led eventually to the model for disulfide bond generation depicted in ►Fig. 12.9A [75–78]. Certain proteins cross the outer mitochondrial membrane without a pre-sequence and are retained in the IMS by subsequent disulfide bond formation. MIA40 serves as a mediator between the unfolded IMS client proteins and yeast Erv1p, or its mammalian ortholog augmenter of liver regeneration (ALR) [75–78]. While many aspects of this pathway were established using yeast as a model organism, several key insights have emerged from the enzymology of the mammalian system to be described in more detail here. The three-dimensional structure of mammalian MIA40 reveals a compact helix-turn-helix hairpin stapled by two disulfides with a redox-active CPC motif at its N-terminus (►Fig. 12.9B, [79]). This remarkably simple disulfide-bridged helix-loop-helix module is also found in a number of MIA40 substrates [75–78]. Upon oxidation of IMS client proteins, reduced MIA40 delivers reducing equivalents to the shuttle disulfide of either yeast Erv1p or the long form of mammalian ALR (►Fig. 12.6, [75–81]). While a crystal structure of Erv1p has not yet appeared [77], structures of the short form of rat [82] and human [83] ALR show the typical helix-rich ERV-module mentioned earlier (►Fig. 12.9C).

Mammalian ALR was originally isolated from blood as a factor that accelerated the regeneration of rat liver tissue and was found to be homologous to the yeast growth factor Erv1p [84]. ALR has also been named heptatopoietin and Growth Factor ERV-like

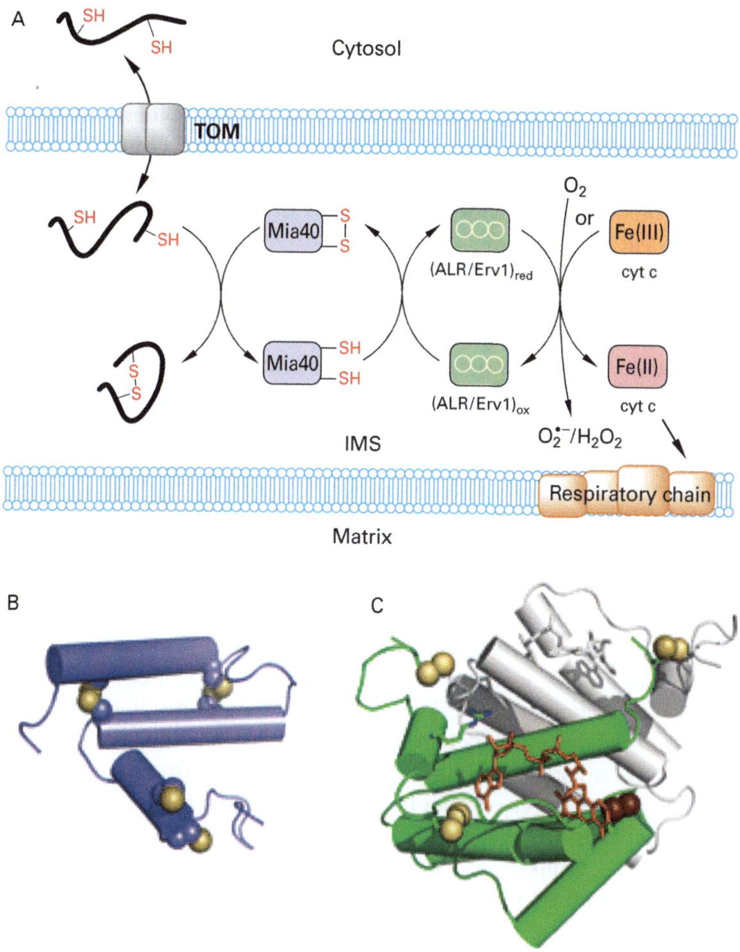

Fig. 12.9: Oxidative folding in the mitochondrial intermembrane space. Panel A shows a model for oxidative protein folding within the mitochondrial IMS. Either molecular oxygen or cytochrome c reoxidize reduced ALR/Erv1p (see the text). Panel B shows the 3D structure of human MIA40 with the CxC motif in its reduced form. The homodimeric flavoprotein human short-form ALR is shown in panel C. The proximal disulfide in the green subunit is depicted by orange spheres and the associated FAD as brown sticks. The intersubunit disulfides between green and grey subunit are shown as yellow spheres at the top left and top right of the Figure. ARG194 (to left; green and blue sticks) forms a H-bond with the ribityl ring of FAD at the convergence of N- and C-terminii.

(GFER) protein. These cytokine-like functions of ALR represent the short form of the mammalian protein (sfALR) that lacks the first 80 amino acids of the N-terminus (including the shuttle disulfide depicted in ▶Fig. 12.6). Additionally, sfALR has a variety of intracellular roles (see for example [83]). The longer splice variant, lfALR, plays a key role in disulfide generation in the IMS (▶Fig. 12.9A, [75–78]) and also functions in the cytosol, where it modulates the dynamics of mitochondrial fission and fusion [85] and serves to regulate the proliferation of hematopoietic stem cells [86]. A human mutation of lfALR has been described in which an R194H mutation led to delayed development, muscle myopathy, cataract and hearing loss [87]. R194 participates in a network of H-bonds between the FAD ribityl chain and residues at the N-terminus of one subunit and the C-terminus of the other (►Fig. 12.9C). While the R194H mutant protein remains fully active, it is less stable than the wild-type protein and shows greatly enhanced conformational flexibility, as evident by the loss of about 40% of the resonances in TROSY-HSQC NMR spectra [83].

The initial enzymological characterization of sfALR showed it to be a modest sulfhydryl oxidase with the model substrate DTT [88]. Strikingly, the K_m for oxygen found in these steady state experiments was 240 µM – a value much higher than expected for a normal flavoproteins oxidase. Indeed cytochrome c was found to be a much better electron acceptor than oxygen for both sfALR and lfALR [72,88]. This cytochrome c pathway could then deliver the reducing equivalents generated during oxidative protein folding in the IMS directly to the respiratory chain without the obligatory generation of reactive oxygen species. While this model has been widely adopted [75–78], cytochrome c peroxidase may also consume the hydrogen peroxide liberated during the oxidative folding in the IMS [89]. Further studies on the oxygen reactivity of these ERV-family members led to the surprising finding that ALR and Erv2p generate significant levels of superoxide during aerobic turnover [50]. In these experiments, superoxide was initially detected by its ability to stimulate the aerobic oxidation of several water-soluble phosphine substrates [90], but superoxide is also released during the turnover of thiol substrates [50]. It remains to be seen whether the ability of short and long forms of ALR to release superoxide has physiological significance in the context of cellular signaling events [50].

Finally, Fass and coworkers have shown that the crystal structure of the *Arabidopsis* mitochondrial Erv1 [91] is very similar to that for other members of the ERV-family [74]. In *At*Erv1 the shuttle disulfide is a Cx4C motif located within a disordered C-terminal extension (►Fig. 12.6). Unlike the yeast protein Erv2p, *At*Erv1 shows appreciable oxidase activity towards reduced thioredoxin [74].

12.4.3 Viral ALR proteins

Many viruses that infect eukaryotes utilize the comparatively oxidizing environment of the host secretory network to facilitate the insertion of disulfides into their envelope proteins prior to emergence from the cell. In contrast, a number of large double-stranded DNA viruses, including pox- and mimi-viruses, subvert the local oxidation-reduction poise of a part of the cytosol (the viral factory) for this same purpose. Several of these ERV-fold viral sulfhydryl oxidases have been studied [6,92–95]. While they show comparable placement of helices, proximal disulfide, and FAD prosthetic groups as their eukaryotic counterparts, there is a marked variation in the interface used for dimerization. The sulfhydryl oxidase pB119L from African Swine Fever virus dimerizes using helices 2 and 3, whereas a mimiviral oxidase employs helices 1 and 2 in common with yeast Erv2p and all other known eukaryotic oxidases [6,94]. A further variation is found in the product of the Ac92 gene of the *Autographa californica* nuclear polyhedrosis virus [96]; this baculovirus sulfhydryl oxidase employs helices 3 and 4 to provide the dimerization interface [39,95]. Remarkably, each subunit of the baculoviral enzyme contains two domains, in contrast to the single domain expected for these small

ERV-family oxidases. The first helix-rich domain of the baculovirus oxidase lacks FAD and the proximal disulfide and resembles the HRR domain previously believed to be unique to QSOX [41]. The C-terminal region of the Ac92 gene product represents a typical ERV domain with the expected complement of redox centers. The finding of a structural counterpart to the QSOX HRR-ERV domain fusion in a baculovirus is intriguing from an evolutionary perspective. Further examination of the nucleocytoplasmic large DNA virus families may provide additional ERV variants that may have been present early in the evolution of the eukaryotes.

Unlike the eukaryotic enzymes, the viral oxidases studied to date lack a shuttle disulfide (▶Fig. 12.6). Transmission of reducing equivalent from the viral client proteins undergoing folding in the viral factory is facilitated by one or more proteins that themselves carry redox-active disulfides [6]. A number of these protein mediators contain CxxC motifs but bear no obvious relationship to the thioredoxins [6]. For example pA151R from African swine fever virus is a 151-residue protein with a C-terminal CxxC motif that appears to transmit reducing equivalents to the cognate oxidase as follows:

viral client reduced proteins (*e.g.* pE248R) ⇒ pA151R ⇒ (proximal disulfide ⇒ FAD) ⇒ oxygen

12.5 Ero1

In addition to the ERV-fold oxidases which have formed the focus of this chapter, there is an another family of FAD-dependent sulfhydryl oxidases that are believed to play a significant role in disulfide bond formation in the ER of eukaryotes from yeast to mammals. Ero1 was first identified in *S. cerevisae* [35,36] and later two paralogs (Ero1α and Ero1β) were found in mammals [97,98]. These oxidases are believed to function by direct oxidation of one or more of the PDI proteins in the ER (▶Fig. 12.2B). While Ero1 is essential in *S. cerevisiae* and *Caenorhabditis elegans*, mice strains that are almost completely devoid of both Ero1α and Ero1β showed only a modest phenotype [99]. This surprising observation has led to a search for additional cellular catalysts for disulfide bond formation, and to the realization that mammals have multiple pathways for the insertion of disulfide bonds during oxidative folding. A number of recent reviews of the physiological and regulatory aspects of the Ero1 proteins have appeared [100–104]. The crystal structure of yeast Ero1p (▶Fig. 12.10, [105]) and human Ero1α [106] have provided key insights into the mechanism of these enzymes. Ero1 proteins are monomeric and bind FAD in a single helix-rich domain [105,106]. A proximal disulfide is located at a turn-helix boundary as is observed with ERV-family proteins. Although both the Ero1 and ERV families bind FAD comparably within 4-helix bundles, the connectivity of helices differs in the two proteins classes and there is no evident sequence similarity between them [105]. In both the yeast and mammalian enzymes a mobile loop (colored red in the yeast structure in ▶Fig. 12.10) carries the shuttle disulfide that communicates with PDI analogues. The mobility of this loop is constrained by regulatory disulfides whose reduction activates both the yeast and mammalian oxidases [105–107]. Mammalian and yeast enzymes achieve this activation with different regulatory disulfide pairings. Both yeast and mammalian Ero1 proteins prove modest catalysts of PDI analogs in terms of turnover number determined from *in vitro* experiments [108–110]. The K_m values for reduced PDI analogues has not been reported.

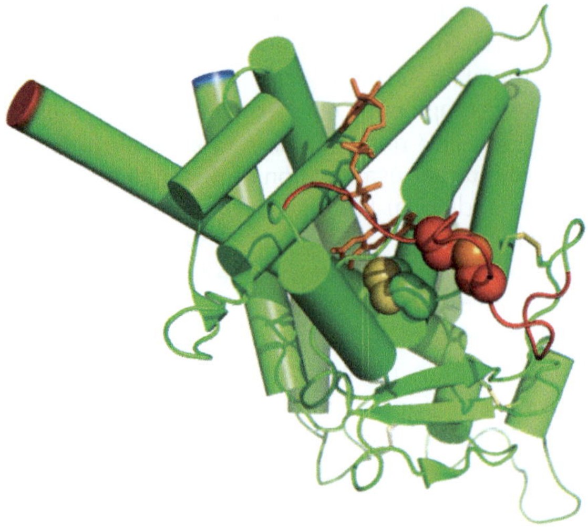

Fig. 12.10: The crystal structure of yeast Ero1p. The proximal disulfide (gold spheres) is adjacent to the C4a locus of the isoalloxazine ring (brown sticks). The shuttle Cx_4C disulfide is located on a mobile loop (highlighted in red) where it can receive reducing equivalents from reduced PDI. Additional disulfide bonds shown in gold sticks play a regulatory function; their reduction stimulates the oxidation of reduced PDI by Ero1p. N- and C-terminii are depicted in blue and red respectively.

12.6 Conclusions

This chapter has described the structure and mechanism of a number of flavin-dependent sulfhydryl oxidases. Of these, the ERV-family oxidases are the most structurally and catalytically diverse, ranging from the minimalist disulfide bond-generating enzymes of poxviruses, to the multi-domain QSOX enzymes that are found in most non-fungal eukaryotes. In contrast, the Ero enzymes are a structurally more conserved group of flavoproteins, with a restricted set of immediate protein substrates. A major challenge for the future will be to understand the biological role of each of these sulfhydryl oxidases and, in particular, to apportion their contribution to the total disulfide output of higher eukaryotes.

12.7 Acknowledgments

Work from the author's laboratory has been supported by NIH grant GM26643.

12.8 References

[1] Thorpe C, Coppock DL. Generating disulfides in multicellular organisms: Emerging roles for a new flavoprotein family. J Biol Chem 2007;282:13929–33.
[2] Sevier CS. New insights into oxidative folding. J Cell Biol 2010;188:757–8.
[3] Schulman S, Wang B, Li W, Rapoport TA. Vitamin K epoxide reductase prefers ER membrane-anchored thioredoxin-like redox partners. Proc Natl Acad Sci U S A;107:15027–32.

[4] Zito E, Melo EP, Yang Y, Wahlander A, Neubert TA, Ron D. Oxidative protein folding by an endoplasmic reticulum-localized peroxiredoxin. Molecular Cell 2010;40:787–97.
[5] Tavender TJ, Springate JJ, Bulleid NJ. Recycling of peroxiredoxin IV provides a novel pathway for disulphide formation in the endoplasmic reticulum. Embo J 2011;15:4185–97.
[6] Hakim M, Fass D. Cytosolic disulfide bond formation in cells infected with large nucleocytoplasmic DNA viruses. Antioxid Redox Sign 2010;13:1261–71.
[7] Senkevich TG, White CL, Koonin EV, Moss B. A viral member of the ERV1/ALR protein family participates in a cytoplasmic pathway of disulfide bond formation. Proc Natl Acad Sci U S A 2000;97:12068–73.
[8] Rony H, Schieff G, DM C, Rennagel W. Sulfhydryl oxidase in skin homogenates. J Investigative Dermatology 1958;30:43–50.
[9] Yamada H. [Localization in skin, activation and reaction mechanisms of skin sulfhydryl oxidase]. Nippon Hifuka Gakkai Zasshi 1989;99:861–9.
[10] Matsuba S, Suga Y, Ishidoh K, et al. Sulfhydryl oxidase (SOx) from mouse epidermis: molecular cloning, nucleotide sequence, and expression of recombinant protein in the cultured cells. J Dermatol Sci 2002;30:50–62.
[11] Brohawn SG, Rudik I, Thorpe C. Avian sulfhydryl oxidase is not a metalloenzyme: adventitious binding of divalent metal ions to the enzyme. Biochemistry 2003;42:11074–82.
[12] Janolino VG, Swaisgood HE. Isolation and characterization of sulfhydryl oxidase from bovine milk. J Biol Chem 1975;250:2532–8.
[13] Jaje J, Wolcott HN, Fadugba O, et al. A flavin-dependent sulfhydryl oxidase in bovine milk. Biochemistry 2007;46:13031–40.
[14] Munday R, Munday CM, Winterbourn CC. Inhibition of copper-catalyzed cysteine oxidation by nanomolar concentrations of iron salts. Free Radic Biol Med 2004;36:757–64.
[15] Hoober KL, Joneja B, White HB, III, Thorpe C. A Sulfhydryl Oxidase from Chicken Egg White. J Biol Chem 1996;271:30510–6.
[16] Argyrou A, Blanchard JS. Flavoprotein disulfide reductases: advances in chemistry and function. Prog Nucleic Acid Res Mol Biol 2004;78:89–142.
[17] Williams CH, Jr. Lipoamide dehydrogenase, glutathione reductase, thioredoxin reductase, and mercuric ion reductase-A family of flavoenzyme transhydrogenases. In: Müller F, ed. Chemistry and Biochemistry of Flavoenzymes. Chemistry and Biochemistry of Flavoenzymes: CRC Press; 1992:121–211.
[18] Williams CH, Jr. Mechanism and structure of thioredoxin reductase from Escherichia coli. Faseb J 1995;9:1267–76.
[19] de la Motte RS, Wagner FW. *Aspergillus niger* sulfhydryl oxidase. Biochemistry 1987;26:7363–71.
[20] Scharf DH, Remme N, Heinekamp T, Hortschansky P, Brakhage AA, Hertweck C. Transannular disulfide formation in gliotoxin biosynthesis and its role in self-resistance of the human pathogen Aspergillus fumigatus. J Am Chem Soc 2010;132:10136–41.
[21] Li B, Walsh CT. Streptomyces clavuligerus HlmI is an intramolecular disulfide-forming dithiol oxidase in holomycin biosynthesis. Biochemistry 2011;50:4615–22.
[22] Ostrowski MC, Kistler WS. Properties of a flavoprotein sulfhydryl oxidase from rat seminal vesicle secretion. Biochemistry 1980;19:2639–45.
[23] Hoober KL, Glynn NM, Burnside J, Coppock DL, Thorpe C. Homology between egg white sulfhydryl oxidase and quiescin Q6 defines a new class of flavin-linked sulfhydryl oxidases. J Biol Chem 1999;274:31759–62.
[24] Benayoun B, Esnard-Feve A, Castella S, Courty Y, Esnard F. Rat seminal vesicle FAD-dependent sulfhydryl oxidase. Biochemical characterization and molecular cloning of a member of the new sulfhydryl oxidase/quiescin Q6 gene family. J Biol Chem 2001;276:13830–7.
[25] Thorpe C, Hoober K, Raje S, et al. Sulfhydryl oxidases: emerging catalysts of protein disulfide bond formation in eukaryotes. Arch Biochem Biophys 2002;405:1–12.
[26] Coppock DL, Thorpe C. Multidomain flavin-dependent sulfhydryl oxidases. Antioxid Redox Signal 2006;8:300–11.

[27] Heckler EJ, Rancy PC, Kodali VK, Thorpe C. Generating disulfides with the Quiescin-sulfhydryl oxidases. Biochim Biophys Acta 2008;1783:567–77.
[28] Kodali VK, Thorpe C. Oxidative protein folding and the Quiescin-sulfhydryl oxidase family of flavoproteins. Antioxid Redox Signal 2010;13:1217–30.
[29] Wilkinson B, Gilbert HF. Protein disulfide isomerase. BBA-Proteins Proteom 2004;1699:35–44.
[30] Hatahet F, Ruddock LW. Protein Disulfide Isomerase: A Critical Evaluation of Its Function in Disulfide Bond Formation. Antioxid Redox Signal 2009;11:2807–50.
[31] Huber-Wunderlich M, Glockshuber R. A single dipeptide sequence modulates the redox properties of a whole enzyme family. Fold Design 1998;3:161–71.
[32] Krause G, Lundstrom J, Barea JL, Pueyo de la Cuesta C, Holmgren A. Mimicking the active site of protein disulfide-isomerase by substitution of proline 34 in Escherichia coli thioredoxin. J Biol Chem 1991;266:9494–500.
[33] Chivers PT, Prehoda KE, Raines RT. The CXXC motif: a rheostat in the active site. Biochemistry 1997;36:4061–6.
[34] Lundstrom J, Holmgren A. Determination of the reduction-oxidation potential of the thioredoxin-like domains of protein disulfide-isomerase from the equilibrium with glutathione and thioredoxin. Biochemistry 1993;32:6649–55.
[35] Pollard MG, Travers KJ, Weissman JS. Ero1p: a novel and ubiquitous protein with an essential role in oxidative protein folding in the endoplasmic reticulum. Mol Cell 1998;1:171–82.
[36] Frand AR, Kaiser CA. The ERO1 gene of yeast is required for oxidation of protein dithiols in the endoplasmic reticulum. Mol Cell 1998;1:161–70.
[37] Sevier CS, Cuozzo JW, Vala A, Aslund F, Kaiser CA. A flavoprotein oxidase defines a new endoplasmic reticulum pathway for biosynthetic disulphide bond formation. Nat Cell Biol 2001;3:874–82.
[38] Sevier C. Erv2 and Quiescin Sulfhydryl Oxidases: Erv-Domain Enzymes Associated with the Secretory Pathway. Antioxid Redox Sign 2012;16:800–8.
[39] Fass D. The Erv family of sulfhydryl oxidases. Biochim Biophys Acta 2008;1783:557–66.
[40] Lisowsky T, Stein G. Functional comparison of the yeast scERV1 and scERV2 genes. Yeast 1998;14:171–80.
[41] Alon A, Heckler E, Thorpe C, Fass D. QSOX contains a pseudo-dimer of functional and degenerate sulfhydryl oxidase domains. FEBS Lett 2010;584:1521–5.
[42] Heckler EJ, Alon A, Fass D, Thorpe C. Human quiescin-sulfhydryl oxidase, QSOX1: probing internal redox steps by mutagenesis. Biochemistry 2008;47:4955–63.
[43] Raje S, Glynn N, Thorpe C. A continuous fluorescence assay for sulfhydryl oxidase. Anal Biochem 2002;307:266–72.
[44] Raje S, Thorpe C. Inter-domain redox communication in flavoenzymes of the quiescin/sulfhydryl oxidase family: role of a thioredoxin domain in disulfide bond formation. Biochemistry 2003;42:4560–8.
[45] Hoober KL, Thorpe C. Egg white sulfhydryl oxidase: Kinetic mechanism of the catalysis of disulfide bond formation. Biochemistry 1999;38:3211–7.
[46] Kodali VK, Thorpe C. Quiescin sulfhydryl oxidase from Trypanosoma brucei: catalytic activity and mechanism of a QSOX family member with a single thioredoxin domain. Biochemistry 2010;49:2075–85.
[47] Hoober KL, Sheasley SS, Gilbert HF, Thorpe C. Sulfhydryl oxidase from egg white: a facile catalyst for disulfide bond formation in proteins and peptides. J Biol Chem 1999;274:22147–50.
[48] Codding JA, Israel BA, Thorpe C. Protein Substrate Discrimination in the Quiescin Sulfhydryl Oxidase (QSOX) Family. Biochemistry 2012;51:4226–35.
[49] Ramadan D, Rancy PC, Nagarkar RP, Schneider JP, Thorpe C. Arsenic(III) species inhibit oxidative protein folding in vitro. Biochemistry 2009;48:424–32.

[50] Daithankar VN, Wang W, Trujillo JR, Thorpe C. Flavin-linked Erv-family sulfhydryl oxidases release superoxide anion during catalytic turnover. Biochemistry 2012;51:265–72.
[51] Rancy PC, Thorpe C. Oxidative Protein Folding in vitro: a Study of the Cooperation between Quiescin-sulfhydryl Oxidase and Protein Disulfide Isomerase. Biochemistry 2008;47:12047–56.
[52] Abskharon RN, Ramboarina S, El Hassan H, et al. A novel expression system for production of soluble prion proteins in E. coli. Microb Cell Fact 2012;11:6.
[53] Lobstein J, Emrich CA, Jeans C, Faulkner M, Riggs P, Berkmen M. SHuffle, a novel Escherichia coli protein expression strain capable of correctly folding disulfide bonded proteins in its cytoplasm. Microb Cell Fact 2012;11:56.
[54] Chakravarthi S, Jessop CE, Willer M, Stirling CJ, Bulleid NJ. Intracellular catalysis of disulphide bond formation by the human sulphydryl oxidase, QSOX1. Biochem J 2007;404:403–11.
[55] Rutkevich LA, Williams DB. Vitamin K epoxide reductase contributes to protein disulfide formation and redox homeostasis within the endoplasmic reticulum. Molecular Biol Cell 2012.
[56] Mairet-Coello G, Tury A, Esnard-Feve A, Fellmann D, Risold PY, Griffond B. FAD-linked sulfhydryl oxidase QSOX: topographic, cellular, and subcellular immunolocalization in adult rat central nervous system. J Comp Neurol 2004;473:334–63.
[57] Tury A, Mairet-Coello G, Poncet F, et al. QSOX sulfhydryl oxidase in rat adenohypophysis: localization and regulation by estrogens. J Endocrinol 2004;183:353–63.
[58] Wittke I, Wiedemeyer R, Pillmann A, Savelyeva L, Westermann F, Schwab M. Neuroblastoma-derived sulfhydryl oxidase, a new member of the sulfhydryl oxidase/Quiescin6 family, regulates sensitization to interferon gamma-induced cell death in human neuroblastoma cells. Cancer Res 2003;63:7742–52.
[59] Coppock DL, Cina-Poppe D, Gilleran S. The Quiescin Q6 gene (QSCN6) is a fusion of two ancient gene families: thioredoxin and ERV1. Genomics 1998;54:460–8.
[60] Martin DB, Gifford DR, Wright ME, et al. Quantitative proteomic analysis of proteins released by neoplastic prostate epithelium. Cancer Res 2004;64:347–55.
[61] Katchman BA, Antwi K, Hostetter GH, et al. Quiescin Sulfhydryl Oxidase 1 Promotes Invasion of Pancreatic Tumor cells Mediated by Matrix Metalloproteinases. Mol Cancer Res 2011.
[62] Zanata SM, Luvizon AC, Batista DF, et al. High levels of active quiescin Q6 sulfhydryl oxidase (QSOX) are selectively present in fetal serum. Redox Rep 2005;10:319–23.
[63] Ouyang X, DeWeese TL, Nelson WG, Abate-Shen C. Loss-of-function of Nkx3.1 promotes increased oxidative damage in prostate carcinogenesis. Cancer Res 2005;65:6773–9.
[64] Antwi K, Hostetter G, Demeure MJ, et al. Analysis of the plasma peptidome from pancreas cancer patients connects a peptide in plasma to overexpression of the parent protein in tumors. J Proteome Res 2009;8:4722–31.
[65] Lee J, Hofhaus G, Lisowsky T. Erv1p from Saccharomyces cerevisiae is a FAD-linked sulfhydryl oxidase. FEBS Letters 2000;477 (1–2):62–6.
[66] Gerber J, Muhlenhoff U, Hofhaus G, Lill R, Lisowsky T. Yeast ERV2p is the first microsomal FAD-linked sulfhydryl oxidase of the Erv1p/Alrp protein family. J Biol Chem 2001;276:23486–91.
[67] Gross E, Sevier CS, Vala A, Kaiser CA, Fass D. A new FAD-binding fold and intersubunit disulfide shuttle in the thiol oxidase Erv2p. Nat Struct Biol 2002;9:61–7.
[68] French CE, Bell JM, Ward FB. Diversity and distribution of hemerythrin-like proteins in prokaryotes. FEMS Microbiol Lett 2008;279:131–45.
[69] Weber PC, Salemme FR, Mathews FS, Bethge PH. On the evolutionary relationship of the 4-alpha-helical heme proteins. The comparison of cytochrome b562 and cytochrome c′. J Biol Chem 1981;256:7702–4.
[70] Hofhaus G, Lee JE, Tews I, Rosenberg B, Lisowsky T. The N-terminal cysteine pair of yeast sulfhydryl oxidase Erv1p is essential for in vivo activity and interacts with the primary redox centre. Eur J Biochem 2003;270:1528–35.

[71] Wang W, Winther JR, Thorpe C. Erv2p: characterization of the redox behavior of a yeast sulfhydryl oxidase. Biochemistry 2007;46:3246–54.
[72] Daithankar VN, Farrell SR, Thorpe C. Augmenter of liver regeneration: substrate specificity of a flavin-dependent oxidoreductase from the mitochondrial intermembrane space. Biochemistry 2009;48:4828–37.
[73] Vala A, Sevier CS, Kaiser CA. Structural determinants of substrate access to the disulfide oxidase Erv2p. J Mol Biol 2005;354:952–66.
[74] Vitu E, Bentzur M, Lisowsky T, Kaiser CA, Fass D. Gain of function in an ERV/ALR sulfhydryl oxidase by molecular engineering of the shuttle disulfide. J Mol Biol 2006;362:89–101.
[75] Chacinska A, Koehler CM, Milenkovic D, Lithgow T, Pfanner N. Importing mitochondrial proteins: machineries and mechanisms. Cell 2009;138:628–44.
[76] Herrmann JM, Riemer J. Mitochondrial Disulfide Relay: Redox-regulated Protein Import into the Intermembrane Space. J Biol Chem 2012;287:4426–33.
[77] Endo T, Yamano K, Kawano S. Structural Basis for the Disulfide Relay System in the Mitochondrial Intermembrane Space. Antioxid Redox Signal 2010;13:1359–73.
[78] Deponte M, Hell K. Disulfide bond formation in the intermembrane space of mitochondria. J Biochem 2009;146:599–608.
[79] Banci L, Bertini I, Cefaro C, et al. MIA40 is an oxidoreductase that catalyzes oxidative protein folding in mitochondria. Nat Struct Mol Biol 2009;16:198–206.
[80] Banci L, Bertini I, Calderone V, et al. An Electron-Transfer Path through an Extended Disulfide Relay System: The Case of the Redox Protein ALR. J Am Chem Soc 2012.
[81] Banci L, Bertini I, Calderone V, et al. Molecular recognition and substrate mimicry drive the electron-transfer process between MIA40 and ALR. Proc Natl Acad Sci USA 2011;108:4811–6.
[82] Wu CK, Dailey TA, Dailey HA, Wang BC, Rose JP. The crystal structure of augmenter of liver regeneration: A mammalian FAD-dependent sulfhydryl oxidase. Protein Sci 2003;12:1109–18.
[83] Daithankar VN, Schaefer SA, Dong M, Bahnson BJ, Thorpe C. Structure of the human sulfhydryl oxidase augmenter of liver regeneration and characterization of a human mutation causing an autosomal recessive myopathy. Biochemistry 2010;49:6737–45.
[84] Hagiya M, Francavilla A, Polimeno L, et al. Cloning and sequence analysis of the rat augmenter of liver regeneration (ALR) gene: expression of biologically active recombinant ALR and demonstration of tissue distribution. Proc Natl Acad Sci USA 1994;91:8142–6.
[85] Todd LR, Damin MN, Gomathinayagam R, Horn SR, Means AR, Sankar U. Growth Factor erv1-like Modulates Drp1 to Preserve Mitochondrial Dynamics and Function in Mouse Embryonic Stem Cells. Mol Biol Cell 2010;21:1226–36.
[86] Sankar U, Means AR. Gfer is a critical regulator of HSC proliferation. Cell Cycle 2011;10:2263–8.
[87] Di Fonzo A, Ronchi D, Lodi T, et al. The mitochondrial disulfide relay system protein GFER is mutated in autosomal-recessive myopathy with cataract and combined respiratory-chain deficiency. Am J Hum Genet 2009;84:594–604.
[88] Farrell SR, and Thorpe, C. Augmenter of liver regeneration: a flavin dependent sulfhydryl oxidase with cytochrome C reductase activity. Biochemistry 2005;44:1532–41.
[89] Dabir DV, Leverich EP, Kim SK, et al. A role for cytochrome c and cytochrome c peroxidase in electron shuttling from Erv1. Embo J 2007;26:4801–11.
[90] Cline DJ, Redding SE, Brohawn SG, Psathas JN, Schneider JP, Thorpe C. New water-soluble phosphines as reductants of peptide and protein disulfide bonds: reactivity and membrane permeability. Biochemistry 2004;43:15195–203.
[91] Levitan A, Danon A, Lisowsky T. Unique features of plant mitochondrial sulfhydryl oxidase. J Biol Chem 2004;279:20002–8.
[92] Senkevich TG, White CL, Koonin EV, Moss B. Complete pathway for protein disulfide bond formation encoded by poxviruses. Proc Natl Acad Sci U S A 2002;99:6667–72.

[93] Rodriguez I, Redrejo-Rodriguez M, Rodriguez JM, Alejo A, Salas J, Salas ML. African swine fever virus pB119L protein is a flavin adenine dinucleotide-linked sulfhydryl oxidase. J Virol 2006;80:3157–66.
[94] Hakim M, Fass D. Dimer interface migration in a viral sulfhydryl oxidase. J Mol Biol 2009;391:758–68.
[95] Hakim M, Mandelbaum A, Fass D. Structure of a baculovirus sulfhydryl oxidase, a highly divergent member of the erv flavoenzyme family. J Virol 2011;85:9406–13.
[96] Long CM, Rohrmann GF, Merrill GF. The conserved baculovirus protein p33 (Ac92) is a flavin adenine dinucleotide-linked sulfhydryl oxidase. Virology 2009;388:231–5.
[97] Cabibbo A, Pagani M, Fabbri M, et al. ERO1-L, a human protein that favors disulfide bond formation in the endoplasmic reticulum. J Biol Chem 2000;275:4827–33.
[98] Pagani M, Fabbri M, Benedetti C, et al. Endoplasmic reticulum oxidoreductin 1-lbeta (ERO1-Lbeta), a human gene induced in the course of the unfolded protein response. J Biol Chem 2000;275:23685–92.
[99] Zito E, Chin KT, Blais J, Harding HP, Ron D. ERO1-beta, a pancreas-specific disulfide oxidase, promotes insulin biogenesis and glucose homeostasis. Journal Cell Biol 2010;188:821–32.
[100] Araki K, Inaba K. Structure, mechanism, and evolution of Ero1 family enzymes. Antioxid Redox Sign 2012;16:790–9.
[101] Sevier CS, Kaiser CA. Ero1 and redox homeostasis in the endoplasmic reticulum. Biochim Biophys Acta 2008;1783:549–56.
[102] Ramming T, Appenzeller-Herzog C. The physiological functions of mammalian endoplasmic oxidoreductin 1: on disulfides and more. Antioxid Redox Sign 2012;16:1109–18.
[103] Bulleid NJ, Ellgaard L. Multiple ways to make disulfides. Trends in Biochem Sci 2011; 36:485–92.
[104] Tavender TJ, Bulleid NJ. Molecular mechanisms regulating oxidative activity of the Ero1 family in the endoplasmic reticulum. Antioxid Redox Signal 2010;13:1177–87.
[105] Gross E, Kastner DB, Kaiser CA, Fass D. Structure of Ero1p, source of disulfide bonds for oxidative protein folding in the cell. Cell 2004;117:601–10.
[106] Inaba K, Masui S, Iida H, Vavassori S, Sitia R, Suzuki M. Crystal structures of human Ero1alpha reveal the mechanisms of regulated and targeted oxidation of PDI. Embo J 2010; 29:3330–43.
[107] Gross E, Sevier CS, Heldman N, et al. Generating disulfides enzymatically: reaction products and electron acceptors of the endoplasmic reticulum thiol oxidase Ero1p. Proc Natl Acad Sci U S A 2006;103:299–304.
[108] Chambers JE, Tavender TJ, Oka OB, Warwood S, Knight D, Bulleid NJ. The reduction potential of the active site disulfides of human protein disulfide isomerase limits oxidation of the enzyme by Ero1alpha. J Biol Chem 2010;285:29200–7.
[109] Baker KM, Chakravarthi S, Langton KP, Sheppard AM, Lu H, Bulleid NJ. Low reduction potential of Ero1alpha regulatory disulphides ensures tight control of substrate oxidation. Embo J 2008;27:2988–97.
[110] Vitu E, Kim S, Sevier CS, et al. Oxidative activity of yeast Ero1p on protein disulfide isomerase and related oxidoreductases of the endoplasmic reticulum. J Biol Chem 2010; 285:18155–65.

13 Glutamate synthase

Maria Antonietta Vanoni

Abstract

Glutamate synthases (GltS) form a family of complex iron-sulfur flavoproteins catalyzing the reductive synthesis of L-glutamate from 2-oxoglutarate and L-glutamine, which comprises NAD(P)H and Fd-dependent forms. GltS control and coordinate the glutaminase and synthase activities taking place at distinct catalytic subsites through mechanisms that have been deduced from a combination of kinetics, equilibrium titrations, structural and modeling studies, which will be here discussed also in the light of the properties of related enzymes.

13.1 Introduction

Glutamate synthase (GltS) is a complex iron-sulfur flavoprotein that catalyzes the reductive transfer of the amide group of L-glutamine (L-Gln) to the C(2) carbon of 2-oxoglutarate (2-OG) using reduced pyridine nucleotides, NADPH or NADH, or reduced ferredoxin (Fd) as the electron donor, depending on the GltS form (Eq. 13.1).

GltS is responsible with glutamine synthetase (GS, Eq. 13.2) of ammonia assimilation forming the so-called GS/GOGAT pathway (with GOGAT being the initial acronym assigned to GltS from L-glutamine:2-oxoglutarate amidotransferase) in bacteria, yeast and plants. The overall reaction (Eq. 13.3) is similar to that of glutamate dehydrogenase except for the additional consumption of one ATP molecule, which is the price to pay in order to assimilate ammonia when its concentration is low [1–5].

$$\text{L-glutamine} + \text{2-oxoglutarate} + 2e^- + 2H^+ \rightarrow 2\ \text{L-glutamate} \qquad \text{(Eq. 13.1)}$$

$$\text{L-glutamate} + NH_4^+ + \text{ATP} \rightarrow \text{L-glutamine} + \text{ADP} + P_i \qquad \text{(Eq. 13.2)}$$

$$NH_4^+ + \text{2-oxoglutarate} + 2e^- + 2H^+ + \text{ATP} \rightarrow \text{L-glutamate} + \text{ADP} + P_i \qquad \text{(Eq. 13.3)}$$

GltS belongs to the class of L-Gln-dependent amidotransferases, which make available the ammonia unit carried by L-Gln [4,5]. The demonstrated essentiality of GltS in model organisms and pathogens makes it an attractive target for novel drug design and/or metabolic engineering ([6], and references therein).

GltS exhibit a modular architecture resulting from recruitment of functional domains in various combinations, and fall in four classes as discussed in detail in [7] (▶Fig. 13.1).

13.1.1 NADPH-GltS

Bacteria contain an NADPH-dependent GltS form (NADPH-GltS, [1,2,5]). The *Azospirillum brasilense* enzyme (Ab-GltS) eventually became the best studied GltS form, and will be here taken as the model of bacterial NADPH-GltS [2,5–11], also in the light of the conservation of the sequences of its large or α subunit (αGltS, 162 kDa for the *A. brasilense* enzyme) and small or β-subunit (βGltS, 52.3 kDa, ▶Fig. 13.1), which form a stable and catalytically active αβ-protomer. The characterization of the Ab-GltS αβ holoenzyme, and of the αGltS and βGltS forms expressed separately in *E. coli* cells led to the unambiguous determination of the identity and the distribution of the GltS cofactors and of the catalytic subsites (▶Fig. 13.2).

13.1.2 Fd-GltS

Cyanobacteria, algae and plastids of higher plants contain a ferredoxin (Fd)-dependent form of GltS (FdGltS), which uses reduced Fd (Fd_{red}) as the electron donor for L-Glu synthesis from L-Gln and 2-OG [3,12–16]. FdGltS consists of a single polypeptide chain functionally and structurally similar to αGltS.

13.1.3 NADH-GltS

A GltS form has been isolated from *Saccharomyces cerevisiae*, lupin, *Medicago sativa* and *Bombyx mori* [7]. It is a single polypeptide chain of ≈200 kDa corresponding to the fusion of the bacterial α and βGltS, which uses NADH instead of NADPH as the reductant and is, therefore, defined as the NADH-GltS form. From genome and cDNA sequencing data, and measurements of GltS activity in cell extracts or partially purified enzyme preparations, NADH-GltS is present in yeast, plants, and lower animals, but no GltS is found in higher animals [6,7,10]. NADH-GltS has been poorly characterized biochemically, but the strong sequence similarity with NADPH-GltS allows us to infer most of its properties from those of the bacterial enzyme.

13.1.4 Archeal GltS

A fourth class of GltS includes the putative archeal enzyme, which is also found in some hyperthermophilic bacteria as the result of lateral gene transfer [7,17]. In *Methanococcus jannaschii* and *Archeoglobus fulgidus* the open reading frames (ORF) assigned to the putative archeal GltS consists of approximately 500 residues and appear to be formed by an N-terminal domain containing the cysteine signature for the formation of two [4Fe-4S] clusters followed by a domain similar to the αGltS and FdGltS synthase domain (▶Figs. 13.1 and 13.2; [7]). ORF likely to correspond to the additional domains and/or subunits of NADPH- and FdGltS are also found [17], but this enzyme form has not been characterized yet, so that its properties can only be deduced from sequence analyses and the properties of NADPH- and Fd-GltS.

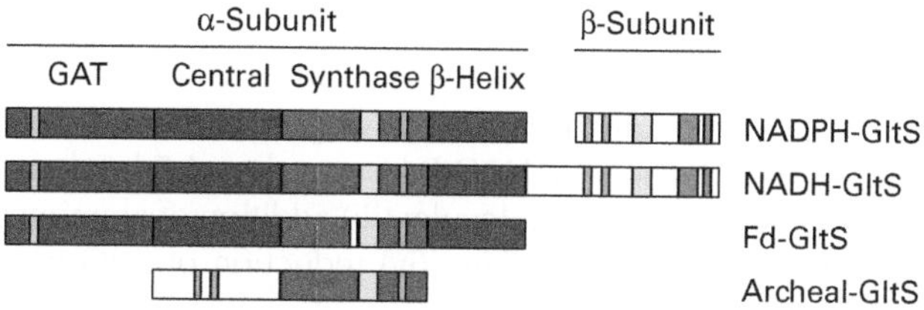

Fig. 13.1: Scheme of the domain structure of GltS. The numbering of the *A. brasilense* NADPH-GltS α and β-subunits is indicated. From left: Type II (PurF-type) L-glutamine-dependent amidotransferase (GAT) domain, residues 1–423 [51], blue. Residues 50–175 corresponding to an insert of the GAT domain typical of GltS [7,51], light blue. Red: central domain, residues 423–779. Green: Synthase domain (residues 780–1203) containing FMN and the [3Fe-4S] cluster with the conserved residues identifying the FMN binding site in comparison to enzymes of the flavocytochrome b_2 family [7,37] (yellow) and the location of Cys1192, 1108 and 1113, the ligands of the [3Fe-4S] cluster (orange). Purple: β-helix, residues 1203–1472. Orange: Cys-rich regions for the formation of the [4Fe-4S] clusters of βGltS and NADH-GltS. Yellow: residues 149–177 matching the consensus sequence for the formation of the adenylate binding site of FAD. Light green: residues 291–339 marking the NADPH-binding site. Cyan, residues 432–442 matching the second FAD consensus sequence. In FdGltS, the white bar indicates residues 907–933 forming the Fd-loop. The *Methanococcus jannaschii* putative GltS [17] is taken as a model of the archeal GltS. From the N-terminus, the orange bars indicate the Cys-rich regions with spacing suggesting the formation of three [4Fe-4S] clusters, one of which replaces the [3Fe-4S] center of other GltS forms [7].

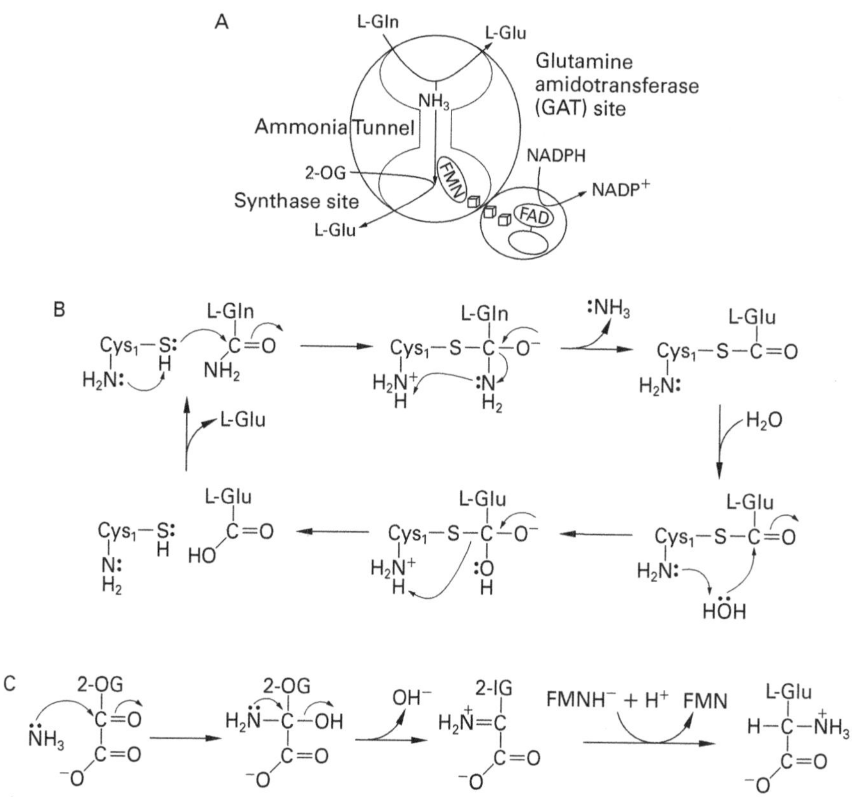

Fig. 13.2: Model of NADPH-GltS (A) and proposed mechanism of the glutaminase (B) and synthase (C) reactions.

13.2 The GltS-catalyzed reactions

The GltS reaction (Eq. 13.1) can formally be broken down into a minimum of four steps corresponding to: (i) enzyme reduction by NAD(P)H or reduced Fd (Fd_{red}); (ii) L-Gln hydrolysis to release the nascent ammonia molecule; (iii) addition of ammonia to 2-OG to yield the 2-iminoglutarate (2-IG) intermediate; (iv) reduction of 2-IG to L-Glu and regeneration of the oxidized enzyme (►Fig. 13.2).

At the pH optimum (pH 7.5–8.0), the K_m values for the reduced pyridine nucleotide substrate are in the low μM range (*e.g.* 2–5 μM for the Ab-GltS, [18–20]); those for 2-OG and L-Gln have been reported to be 10–300 μM and 10–1000 μM, respectively ($K_{m,2\text{-}OG}$, ~30 μM; $K_{m,\ L\text{-}Gln}$ ~300 μM for Ab-GltS). Turnover numbers (k_{cat}) of 50–60 s^{-1} have been calculated for Ab-GltS at 25 °C [2,19,20]. For the best studied NADPH-GltS forms, the overall reaction is well described by a two-site uni-uni bi-bi ping pong steady-state kinetic mechanism with 2-OG binding before L-Gln in the bi-bi segment [2,20] leading to a scheme for NAD(P)H-GltS architecture (►Fig. 13.2), consisting of:

- Site 1: the NADPH oxidizing site where NADPH binds and is oxidized with transfer of the NADPH 4-*proS* hydrogen, as a hydride anion, to the enzyme flavin located at this site (later identified as FAD, [21–23]);
- Site 2: the synthase site, where 2-OG is converted into the postulated 2-IG intermediate, upon addition of ammonia deriving from L-Gln hydrolysis, with final reduction of 2-IG to yield L-Glu by hydride transfer from the FMN coenzyme located at this site [21,24].

The presence of a distinct site or subsite for glutamine hydrolysis (glutaminase site or GAT site, from glutamine amidotransferase) was proposed on the basis of known properties of L-Gln-dependent amidotransferases in which the glutaminase and synthase sites may be even located on different subunits [5]. The sequential ordered mechanism of the bi-bi segment of the reaction, suggests a mechanism of communication between the sites, which is now known to be a common feature of amidotransferases [9,25–28]. Connection between the NADPH oxidizing and the synthase sites was proposed to be established by the GltS [Fe-S] clusters forming an intramolecular electron transfer chain [2].

FdGltS from cyanobacteria and plants are specific for plant-type [2Fe2S]-containing Fd [12,14,16]. For *in vitro* work, assays need to be carried out under anaerobiosis limiting kinetic studies. Fd_{red} is generated by including dithionite in the reaction mixtures. Dithionite itself is a poor electron donor, while some activity can be obtained by substituting methylviologen (MV) for Fd. K_m values for L-Gln are similar to those reported for NAD(P)H-GltS; K_m values for 2-OG are as high as 1 mM and those for Fd are in the 1-20 μM range [10–12,14]. For the *Synechocystis* PCC6803 FdGltS we measured K_m values for 2-OG of ~100 μM (at 2.5 mM L-Gln) and of 2.4 mM for L-Gln (at 2.3 mM 2-OG) in the presence of saturating Fd (21 μM). With saturating 2-OG and Fd, the apparent k_{cat}, extrapolated at infinite L-Gln concentration, was 125 s^{-1} at pH 7.5 and 25 °C [29,30].

GltS can convert 2-OG into L-Glu when ammonium chloride is substituted for L-Gln (Eq. 13.4).

$$NH_4^+ + \text{2-oxoglutarate} + 2e^- + 2H^+ \rightarrow \text{L-glutamate} + H_2O \qquad \text{(Eq. 13.4)}$$

Kinetic studies with Ab-GltS were consistent with a two-site uni-uni bi-bi ping-pong kinetic steady-state kinetic mechanism and the 2-site model proposed above [20]. At pH 7.5–8.0 the ammonia-dependent reaction proceeds with an apparent k_{cat} that is less than 10% that of the L-Gln-dependent activity. At the pH optimum (9.3–9.5) of the ammonia-dependent reaction the k_{cat} is similar to that of the L-Gln-dependent reaction at this pH, which in turn is half that measured at pH 7.5–8 [31]. However, the K_m for ammonia, which is essentially pH independent when the ammonia concentration is taken into account, is high (≈0.25 M, [20,31]) suggesting that the reaction may not be physiologically relevant. Also FdGltS catalyzes an ammonia-dependent reaction [29,30], but it has not been studied in detail.

A. brasilense NADPH-GltS catalyzes the oxidation of NADPH in the presence of synthetic electron acceptors. Iodonitrotetrazolium (INT) and ferricyanide appeared to accept electrons from the NADPH oxidizing site, while dichlorophenolindophenol could accept electrons from the synthase site. The calculated k_{cat} with these acceptors were 2–3-fold higher than that measured for the L-Gln-dependent reaction indicating that steps other than NADPH oxidation and electron transfer between the enzyme flavins are determining the turnover of the physiological reaction [32].

A L-Glu:INT oxidoreductase activity yielding 2-OG, ammonia and reduced INT was demonstrated for NADPH-GltS, αGltS and FdGltS providing a handy way to assay the last two enzymes during purification and for routine controls of preparations [19,24,29,30]. The maximum reaction velocity increased 10-fold from 7.5 to 9.5, with a sharp drop at higher pH, mainly due to protein instability. With NADPH-GltS, the k_{cat} at pH 9.5 was 10 s^{-1}, *i.e.* 20% that of the physiological reaction at pH 7.5 with a K_m for L-Glu of approximately 0.7 mM. With FdGltS, k_{cat} was ~3.4 s^{-1}, but $k_{cat}/K_{m,L\text{-}Glu}$ was 6-fold higher than that measured for NADPH-GltS due to a significantly lower value of $K_{m,LGlu}$ (~50 μM). At pH 7.5, k_{cat} values of ~0.3 s^{-1} and 0.6 s^{-1} were measured for NADPH- and Fd-GltS, respectively, with K_m values for L-Glu between 0.2 and 0.3 mM. The physiological relevance of this reaction is not known, but it is unlikely in the light of the requirement of an efficient electron acceptor. In this respect, early reports of a weak L-Glu:$NADP^+$ oxidoreductase activity at high pH values never found independent confirmation, and GltS exhibits negligible oxidase activity.

Whether GltS can hydrolyze L-Gln (Eq. 13.5) in the absence of 2-OG and/or electrons loaded on its redox centers has been studied in several ways also in relation to the properties of other amidotransferases [4,5,33].

$$\text{L-glutamine} + H_2O \rightarrow \text{L-glutamate} + NH_4^+ \qquad \text{(Eq. 13.5)}$$

The differences in chemical shift of L-[^{15}N-amido]-Gln, L-[^{15}N-amino]-Glu and ^{15}N-ammonium ions were exploited to monitor the Ab-GltS reaction by NMR spectroscopy under various conditions [21]. 2-OG and NADPH are required to observe L-Glu synthesis, with no loss or exchange of the transferred ammonia molecule into the solvent. The degree of coupling of the glutaminase and synthase reactions was also studied by measuring the kinetics and extent of production of L-[^{14}C-U]-Glu from L-[^{14}C-U]-Gln or 2-[^{14}C-U]-OG in the presence of the unlabeled form of the other substrates in discontinuous assays [24,29,30]. This approach demonstrated that also in *Synechocystis* PCC6804 FdGltS, L-Gln hydrolysis and L-Glu synthesis from 2-OG are tightly coupled. Thus, GltS tightly controls its glutaminase activity so that L-Gln is hydrolyzed only when the enzyme is in the correct reduced state and 2-OG is bound at the synthase site in order to avoid wasteful consumption of L-Gln and to favor the conversion of the (unstable) 2-IG intermediate into the stable L-Glu product (▶Fig. 13.2). As in the case of other amidotransferases, how the enzyme prevents the loss of the transferred ammonia into the solvent became evident when the three-dimensional structures of various amidotransferases, including αGltS and FdGltS, became available.

13.3 Flavins and iron-sulfur centers of GltS

The NADPH-GltS contains one FAD and one FMN coenzyme, which could be distinguished on the basis of their differential reactivity with sulfite. Only the flavin located at the synthase site of NADPH-GltS reacted with sulfite yielding a flavin N(5)-sulfite adduct (K_d, 1.3 mM, [22]), which was competitively displaced from the flavin by 2-OG (calculated K_d for the enzyme-2-OG complex, 6 μM) providing a means to distinguish between the GltS flavins. Accordingly, also αGltS and FdGltS FMN coenzyme were later found to react with sulfite (K_d, 2.3 mM for αGltS [24] and 11 mM for *Synechocystis* FdGltS), which was displaced by 2-OG (K_d for the enzyme-2-OG complex, 34 μM for αGltS and 17 μM for FdGltS, [29,30]). Other properties of the GltS will be discussed below, where appropriate.

Low-temperature EPR spectroscopy studies of "as isolated" (oxidized) NADPH-GltS and FdGltS showed spectra consistent with the presence of one $[3Fe\text{-}4S]^{+1}$ cluster [22,34,35]. Reduction of *A. brasilense* NADPH-GltS with NADPH (in the presence of glucose 6-phosphate and glucose 6-phosphate dehydrogenase as an NADPH-regenerating system to abolish product inhibition) led to reduction of one of the enzyme flavins (as determined by absorbance spectroscopy) and loss of the $[3Fe\text{-}4S]^{+1}$ cluster signal, which was substituted by an EPR spectrum dominated by a $g = 1.96$ signal with properties consistent with those of one $[4Fe\text{-}4S]^{+1}$ cluster/αβ-protomer [22]. The observed $g = 12$ signal attributable to the reduced $[3Fe\text{-}4S]^0$ center, further supporting the concept that this cluster is a component of the native and fully functional GltS. Anaerobic dithionite addition caused reduction of the enzyme flavins and of the $[3Fe\text{-}4S]^{+1}$ cluster, but only partial reduction of the [4Fe-4S] center, indicating that its potential is low. Photoreduction led to reduction of both enzyme flavins, of the $[3Fe\text{-}4S]^{+1}$ cluster to the $[3Fe\text{-}4S]^0$ state, and to that of two different $[4Fe\text{-}4S]^{+1}$ clusters, one corresponding to that reduced with NADPH and the other a novel, lower potential center. The observed complex signal may arise from two interacting spin systems due to two distinct $[4Fe\text{-}4S]^{+1,+2}$ clusters. Interestingly, also the $g = 12$ signal due to the reduced $[3Fe\text{-}4S]^0$ cluster was affected by reduction of the two $[4Fe\text{-}4S]^{+1,+2}$ clusters to the +1 state, as discussed in [22].

13.4 Localization of catalytic subsites and coenzymes

The cloning of the genes encoding *Escherichia coli* [36] and *A. brasilense* GltS subunits [37], and of the maize [38] and *Synechocystis* PCC6803 FdGltS [35,39] opened the way to sequence analyses, overexpression and site-directed mutagenesis studies, which were done with the *A. brasilense* enzyme for the NADPH-GltS form and, mainly, with the *Synechocystis* species for FdGltS.

βGltS is formed by a region with overall similarity to the adrenodoxin reductase class of FAD-dependent NADPH oxidoreductase preceded (▶Fig. 13.1) by an N-terminal extension containing two Cys-rich regions with spacing different from that of well-characterized [Fe-S] clusters [7,37]. In the 291–339 region, a GXGXXG motif is found in NADH-GltS, but a GXGXXA is present in NADPH-GltS suggesting that this marks the NADPH-binding site, while the 149–177 region maps the FAD binding site [7,15,37], as confirmed by site-directed mutagenesis [40]. The *A. brasilense* βGltS produced in *E. coli* cells contained one FAD coenzyme, and harbored the fully functional NADPH oxidizing site of NADPH-GltS [23,40] with a rate of reduction of the bound FAD by NADPH of ~900 s^{-1} (25 °C), similar to that of reduction of one of NADPH-GltS flavins, which is 20-fold higher than the k_{cat} of NADPH-GltS measured with the physiological substrates and 4–5-fold higher than the k_{cat} measured in the diaphorase activities with INT or $Fe(CN)_6^{-3}$. Only a small amount of flavin semiquinone was observed during reduction of βGltS, but a stable charge-transfer complex between reduced enzyme and $NADP^+$ was formed [23,40].

No iron was found associated with βGltS preparations so that the role of the N-terminal (conserved) Cys residues could be deduced from the similarity of βGltS with a class of β-like or GltD-like proteins or protein domains [7], the crystal structure of dihydropyrimidine dehydrogenase (DPD) [41] whose N-terminal domain is β-like [7,42], and site-directed mutagenesis of βGltS [43]. The class of β-like proteins now includes, beside the 149-652 region of *E. coli* AegA, *Rhodobacter capsulatus* GltX and the α subunit of *Pyrococcus furiosus* sulfide dehydrogenase, SudA [7,43], PH0876 and PH1873 from *Pyrococcus horikoshii* [44], the N-terminal region of the *Allochromatium vinosum dsrL* gene product [45], part of the *Thermococcus litoralis* NsoC, and *Entamoeba histolytica* EhNO1 and EhNO2 [46]. Overall, β-like proteins (or domains) appear to be a class of FAD-dependent NAD(P)H oxidoreductases that serve to make available reducing equivalents from NAD(P)H to a second subunit (or domain) through low to very low potential $[4Fe\text{-}4S]^{+1,+2}$ clusters encoded by the N-terminal Cys-rich extension.

The DPD structure [7,41–43] confirms the similarity of its N-terminal β-like region (residues 21–520) with the adrenodoxin reductase class of NAD(P)H oxidoreductases, and clearly shows that one [4Fe-4S] cluster (nFeS1 in DPD) is formed by DPD Cys 79, 82, 87 and 140 corresponding to βGltS Cys 47, 50, 55 and 108. The second DPD [4Fe-4S] cluster (nFeS2) is formed by Cys 91, 130 and 136 (corresponding to βGltS Cys 59, 98, 104) and Gln156 (corresponding to βGltS Glu124). The DPD nFeS2 and nFeS1 (in this order) form, with the two [4Fe-4S] clusters encoded by the C-terminal 8-Fe Fd-like domain, a linear electron-transfer chain connecting FAD (on the β-like domain) and FMN (on the dihydroorotate dehydrogenase-like domain, [41]). At variance with NADPH-GltS, DPD [4Fe-4S] clusters exhibit a very low potential so that they could not be reduced even under harsh conditions [42,47], while in NADPH-GltS holoenzyme, one could be reduced quantitatively with NADPH and both by photoreduction [22,29,48]. Substitution of Cys 47, 50, 55 and 59 with Ala residues in the *Azospirillum* βGltS had no effect on the properties of the isolated βGltS, but prevented formation of the αβ-protomer when co-expressed with the wild-type αGltS [43]. Thus, the NADPH-GltS α and β-subunits cooperate for the formation of the catalytically active αβ-protomer: αGltS is required for the formation of the [4Fe-4S] clusters encoded by the N-terminal region of βGltS, and the clusters are essential to structure such an N-terminal region and to allow its interaction with αGltS [43].

αGltS is very conserved in NADPH-GltS, and is very similar to the corresponding N-terminal 3/4 of the eukaryotic NADH-GltS and to FdGltS (▶Fig. 13.1, [7,37]). The mature protein is formed by an N-terminal domain similar to the Ntn-(Type 2 or PurF-type) glutamine amidotransferase domain with the catalytically essential Cys1 [5]. Conserved regions follow and include the FMN binding region, that forming the $[3Fe\text{-}4S]^{+1,0}$ cluster and a C-terminal domain [7], which will be discussed with the three-dimensional structure of the protein.

The recombinant αGltS contains the FMN coenzyme and the $[3Fe\text{-}4S]^{0,+1}$ of GltS [24]. A similar cofactor content was established for FdGltS [14,35]. L-Glu reduced αGltS and Fd-GltS at a rate slow enough to determine that FMN is the entry point of the reducing equivalents from L-Glu in the reverse of the synthase reaction (▶Fig. 13.2; [29]). αGltS and FdGltS catalyze the formation of L-Glu from L-Gln and 2-OG provided a suitable reducing system is present. In both cases, dithionite readily reduces the enzyme cofactors, as determined by anaerobic titrations, while photoreduction seems to damage the [3Fe-4S] cluster [24,29,48]. With both enzymes, the spectral changes were consistent with reduction of the [3Fe-4S] cluster preceding that of FMN, with no formation of flavin semiquinone forms. Addition of excess L-Gln (or ammonia) and 2-OG to fully reduced αGltS causes the rapid reoxidation of FMN only [24]. This is at variance with what was observed with a similar experiment carried out with fully (photo)reduced NADPH-GltS in which recovery of the starting oxidized spectrum was readily observed. With FdGltS, anaerobic addition of L-Gln and 2-OG to fully reduced enzyme also caused only the oxidation of FMN [29], but the reaction took 2 h to reach completion (▶Fig. 13.3). When the experiment was carried out in the presence of Fd that was also reduced with dithionite (to obtain a 1:1 $FdGltS_{red}$: Fd_{red} stoichiometry), addition of L-Gln and 2-OG led to biphasic absorbance changes: the spectrum recovered immediately after substrate addition (30 s) was consistent with the presence of 2-electron reduced FMN, oxidized [3Fe-4S] cluster and oxidized Fd (▶Fig. 13.3). The spectrum of the fully oxidized FdGltS was then slowly obtained. Thus, Fd_{red} needs to be bound to reduced FdGltS in order to activate its glutamate synthase reaction. The faster reaction observed when ammonia was substituted for L-Gln in the absence of Fd (▶Fig. 13.3) suggested that binding of Fd_{red} has an activatory effect on L-Gln hydrolysis and/or transfer of the released ammonia molecule to 2-OG (see also ▶Fig. 13.3C). Accordingly, catalytic amounts of αGltS could produce L-Glu from L-Gln and 2-OG in anaerobic assays in the presence of dithionite and MV [24,29] but the corresponding reaction with FdGltS was greatly enhanced when Fd was substituted for MV [29,30].

L-Gln hydrolysis was observed using αGltS in the absence of 2-OG and a reducing system [24], but not with FdGltS or NADPH-GltS αβ holoenzyme [29,30]. Thus, βGltS is required also to ensure coupling between glutamine hydrolysis and glutamate synthesis within αGltS, while the isolated (monomeric) FdGltS is sufficient to stabilize a conformation that ensures coupling.

13.5 Mid-point potential values of the GltS cofactors and electron transfer pathway between the GltS flavins

Estimates of the mid-point potential values (E_m) of FAD, FMN and of the [3Fe-4S] cluster were obtained, at pH 7.5 and 20 °C, from absorbance-monitored redox titrations

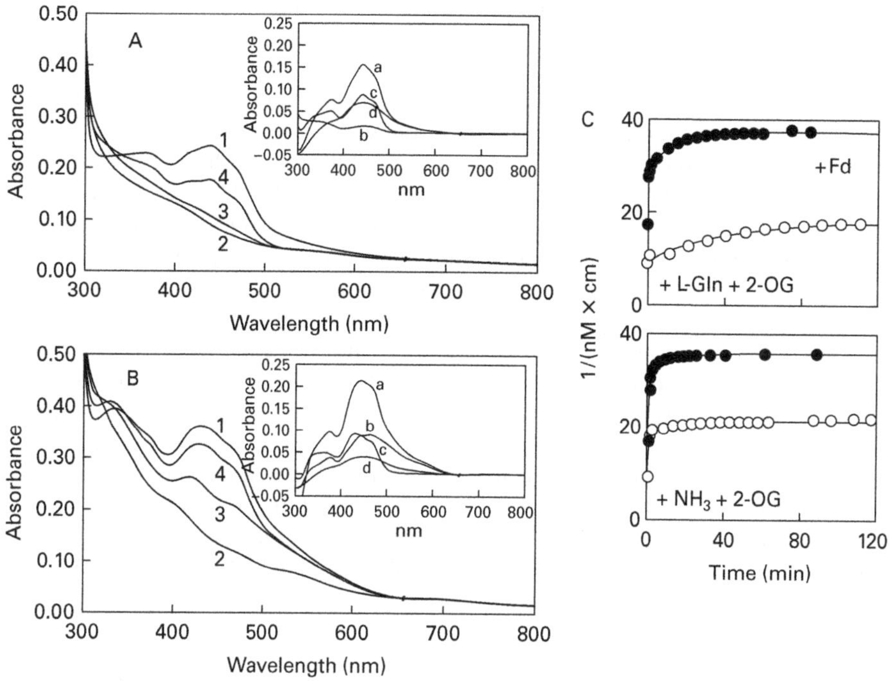

Fig. 13.3: Activatory effect of reduced Fd on the glutamate synthase reaction of FdGltS. (A) *Synechocystis* PCC6803 FdGltS (9.8 μM, 1) was reduced with 2.5-fold molar excess dithionite (2). L-Gln and 2-OG (~200-fold molar excess) were added. The spectra recorded immediately (1 min, 3) after the addition and at the end of the reaction (193 min, 4) are shown. In the inset the following difference spectra are shown: spectrum a, spectrum 1 – spectrum 2; spectrum b, spectrum 3-spectrum 2; spectrum c, spectrum 4 – spectrum 2; spectrum d, spectrum 1 – spectrum 4. (B) An experiment similar to that of Panel (A) was carried out including 13 μM Fd, which was also reduced with dithionite prior to L-Gln and 2-OG addition (3.3 mol dithionite/mol FdGltS). Spectra numbering as in Panel (A) except for spectrum 3 and spectrum 4 that were recorded 0.5 min and 90 min after substrates addition. (C) Time-course of the oxidation of Fd-GltS upon addition of L-Gln and 2-OG (upper panel, from the experiment in Panels (A) and (B)) or ammonia (750 nmol/nmol FdGltS) + 2-OG (125 nmol/nmol FdGltS, lower panel). The values of absorbance at 438 nm were converted to apparent extinction coefficients to normalize for the difference (~10%) in FdGltS concentration. Open symbols: no Fd; closed symbols: with Fd. The figures are reprinted with permission from [29]. Copyright 2002. American Chemical Society.

of the recombinant *Azospirillum* NADPH-GltS αβ holoenzyme and of its isolated subunits [48]. The E_m value of the βGltS-bound FAD was ~−340 mV, but it increased to ~−310 mV in the presence of 3-aminopyridine adenosine dinucleotide phosphate (AADP), a non-reducible $NADP^+$ analog that binds to βGltS (K_i ~1 μM). The latter formed a stable-charge transfer complex with the reduced protein, analogous to the reduced enzyme/$NADP^+$ complex observed during reductive titrations of βGltS with NADPH or photoreductions in the presence of $NADP^+$ [23,48]. With αGltS the E_m values of the FMN_{ox}/FMN_{hq} and of the $[3Fe\text{-}4S]^{+1}/[3Fe\text{-}4S]^0$ couples were determined to be ~−240 mV and –260 mV, respectively. The glutamine analog L-methionine sulfone (MetS) had no effect, while 2-OG lowered the E_m value of the [3Fe-4S] cluster to approximately –280 mV. For the NADPH-GltS αβ holoenzyme, the E_m value of the FAD_{ox}/FAD_{hq} couple was ~–300 mV and insensitive to the presence of AADP, a fact that correlates with the lack of formation of charge-transfer complexes between reduced enzyme and AADP or $NADP^+$. The E_m value of the FMN_{ox}/FMN_{hq} was ~–240 mV and that of the [3Fe-4S] cluster was –280 mV. 2-OG caused a ~30 mV decrease of the FMN E_m value and a dramatic 90 mV increase of the [3Fe-4S] cluster potential. AADP and MetS had no effect on the observed redox behavior of the GltS cofactors. No flavin semiquinone forms were detected, indicating that, in all cases, the oxidized (ox)/semiquinone (sq) couple is less negative than the semiquinone/hydroquinone (hq) couple by at least 120 mV. With this assumption, estimates of the E_m values of each $flavin_{ox}/flavin_{sq}$ and $flavin_{sq}/flavin_{hq}$ couple were obtained (►Fig. 13.4). A bifurcated scheme for the electron transfer pathway from FAD to FMN along the [Fe-S] centers, minimizing the number of thermodynamically uphill electron transfer steps, could be drawn [48]. Assuming a linear geometry of the chain formed by the flavins and the [Fe-S] clusters, as emerging from modeling studies of the NADPH-GltS αβ-protomer [49], an alternative scheme was proposed (►Fig. 13.4). It is characterized by the fact that the two electrons follow the same path with several thermodynamically uphill steps, which are possible within an overall thermodynamically favored process.

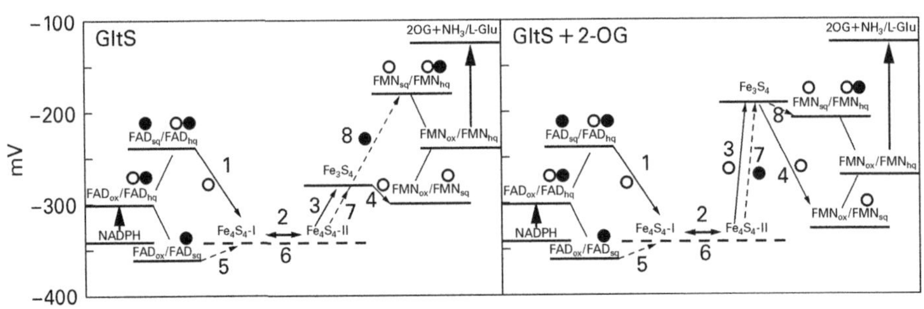

Fig. 13.4: Proposed electron transfer pathway in NADPH-GltS. On the basis of measurements of the E_m values of the flavin coenzymes and of the [3Fe-4S] centers of NADPH-GltS [48], and the structural model of the NADPH-GltS αβ-protomer [49] (see ►Fig. 13.9), the electron transfer from FAD (on βGltS) to FMN (on αGltS) through the two [4Fe-4S] centers (on βGltS) and the [3Fe-4S] cluster (on the αGltS) may take place as depicted in the schemes, which take into account the differences of the E_m values measured in the absence (left) or presence (right) of 2-OG. The following E_m values (20 °C, pH 7.5) have been taken into account: $NADP^+/NADPH$, –340 mV; FAD_{ox}/FAD_{hq}, –300 mV; (NH_3 +2-OG)/L-Glu, –126 mV; FAD_{ox}/FAD_{hq}, –240 mV (no 2-OG) and –270 mV (+1 mM 2-OG); $[3Fe\text{-}4S]^{+1}/[3Fe\text{-}4S]^0$, –260 mV (no 2-OG) and –190 mV (+1 mM 2-OG). The values of the oxidized/semiquinone and semiquinone/hydroquinone couples for FAD and FMN were calculated assuming a separation of at least 120 mV due to no detection of semiquinone species: FAD_{ox}/FAD_{sq}, –360 mV; FAD_{sq}/FAD_{hq}, –240 mV; FMN_{ox}/FAD_{sq}, –300 mV (+2-OG, -330 mV); FMN_{sq}/FMN_{hq}, –180 mV (+2-OG, –210 mV). The E_m of two $[4Fe\text{-}4S]^{+1,+2}$ centers of NADPH-GltS are arbitrarily set as equal to each other and in the range of that of NADPH. The numbers and arrows indicate the path of the first (open circles) and the second (closed circle, broken arrows) electron from reduced FAD.

Redox titrations of spinach FdGltS indicated that the FMN_{ox}/FMN_{hq} couples are equipotential or separated by less than 30 mV with E_m values between –170 and –225 mV [34,50]. A value of –200 mV was reported for the FMN_{ox}/FMN_{hq} couple with the *Synechocystis* FdGltS [35] and the E_m of the [3Fe-4S] cluster was estimated to be ~–160 mV still consistent with electron transfer from reduced Fd to FMN through the [3Fe-4S] center [11,29,48]. In the light of the 1:1 stoichiometry of the Fd/FdGltS complex determined by small-angle X-ray scattering (SAXS, see below) and the need of reduced Fd for activation of glutamate synthesis, it was also proposed [10] that FdGltS undergoes priming steps in which FMN and the [3Fe-4S] cluster are reduced at the expenses of three Fd_{red} molecules. Then, the catalytically competent (4-electron reduced) $Fd_{red}/FdGltS_{red}$ complex would be formed. The complex, would catalyze L-Glu synthesis from L-Gln and 2-OG by shuttling between 2- and 4-electron-reduced forms and consuming 2 Fd_{red} molecules for each catalytic cycle (see [10] for details and ►Fig. 13.8C).

13.6 Structure of αGltS and FdGltS and the mechanism of control and coordination of the partial activities

Crystallization experiments have been carried out with *A. brasilense* NADPH-GltS, the recombinant αGltS subunit and, later, *Synechococcus* PCC6803 FdGltS in the absence/presence of their substrates and ligands in various combinations. Using NADPH-GltS in the presence of MetS, as a L-Gln analog, and 2-OG, well-diffracting crystals were obtained, but they contained αGltS molecules only, with the β-subunit found as denatured protein in the crystallization droplets. No crystals were obtained starting from the isolated αGltS. With FdGltS suitable crystals were obtained with the unliganded enzyme, in the presence of MetS and 2-OG, and after reaction with 6-diaza-5-oxo-L-norleucine (DON).

The structure of αGltS (PDB ID, 1EA0) was solved at 3 Å resolution [51], with few disordered segments corresponding to surface loops (1172–1179 and 1194–1202), which had been found to be very sensitive to tryptic and chymotryptic attack in limited proteolysis experiments [32], and a C-terminal heptapeptide. An α_2 dimer was found in the asymmetric unit, which was, however, consistent with the fact that the isolated αGltS is the catalytically active unit. αGltS is formed by four domains of approximately similar sizes (►Figs. 13.1 and 13.5). The N-terminal domain (residues 1–422) is characterized by a four-layer α/β/β/α-topology similar to the fold of the NH_2–terminal nucleophile (Ntn) class of L-Gln amidotransferases (also known Type II or PurF-type), which includes as well-characterized members L-glutamine phosphoribosyl pyrophosphate amidotransferase (PRPP-AT), glucosamine 6-phosphate synthase (GlmS) and asparagine synthetase (AS) (see [9,26,28,33] for recent reviews). GltS is an exception because of the presence of an insert (residues 50–175, [7,37,51]), which forms an additional subdomain consisting of a four-stranded antiparallel β-sheet and two α-helices. MetS, in this domain, maps the L-Gln binding site, but, acting as an analog of the Cys1 γ-glutamylthioester intermediate, induces a conformational change that pushes away the Cys1 thiol from the reaction center to a conformation in which it is unable to initiate catalysis (►Figs. 13.2 and 13.5). The glutamine amidotransferase (GAT) domain is followed by a so-called "central domain" (residues 423–779), which is formed by highly conserved regions in GltS [7]. It has no sequence or structural similarity with other

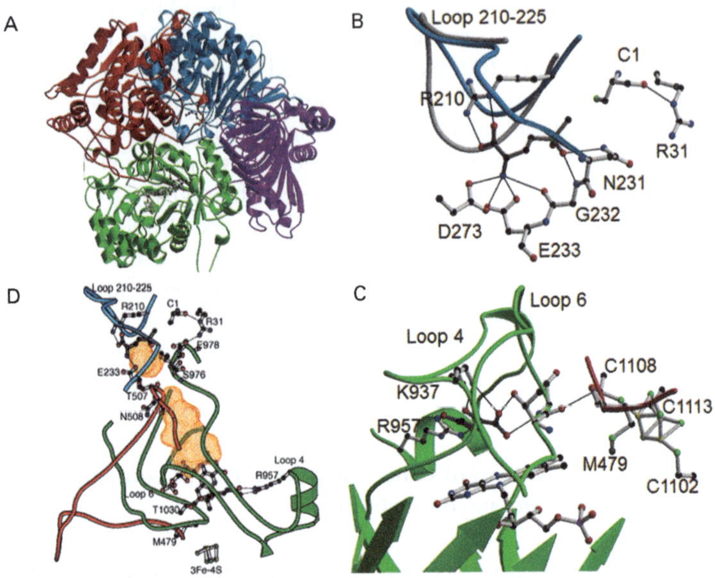

Fig. 13.5: Structure of the α subunit of glutamate synthase. (A) Ribbon diagram of the three-dimensional structure of αGltS (PDB ID 1EA0, [51]) with MetS in the GAT (blue) domain and with FMN, 2-OG and the [3Fe-4S] center in the synthase (green) domain as sticks. Color code as in ►Fig. 13.2. (B) Detail of the glutaminase site with bound MetS. Loop 210–225 (blue), the Q loop of amidotransferase is in an open conformation as opposed to the closed conformation found in the structure of other amidotransferases in complex with glutamine analogs, as exemplified by the 73–90 loop of PRPP-AT (light grey, 1ECC, [60]). (C) Detail of the synthase domain with 2-OG, FMN and the [3Fe-4S] cluster in sticks. (D) Detail of the core of αGltS highlighting the ammonia tunnel and the elements that might be involved in activation of the glutaminase reaction and tunnel opening by sensing the ligation and redox state of the synthase domain. Color code as in ►Fig. 13.1. Reprinted from [51]; Copyright 2000 with permission from Elsevier.

known proteins and has a α/β-topology, but shows structural similarity with proteins having $(\beta/\alpha)_8$-topology, and, especially, the FMN binding domain of GltS. This central domain can be described as an incomplete $(\beta/\alpha)_8$-barrel consisting of β-strand 1, β/α units 5, 6, 7 and 8. The two stands topologically equivalent to strands 1 and 5 of a regular $(\beta/\alpha)_8$-barrel are linked by a helix (residues 552–564).

The FMN binding domain (residues 780–1203) follows. It contains the FMN coenzyme, the [3Fe-4S] cluster and the 2-OG binding site. It is a $(\beta/\alpha)_8$-barrel typical of the FMN binding domain of the enzymes of the flavocytochrome b_2 family, with the highest degree of similarity in the FMN binding region, as predicted by sequence analyses [7,37] so that contacts between FMN and the protein resemble those observed in this class of enzymes. Accordingly, Lys999 is in contact (3.3 Å) with the N(1)-O(2) locus of the isoalloxazine. The [3Fe-4S] cluster is bound to a loop (residues 1101–1118) that provides the Cys1102, 1108 and 1113 ligands of the cluster and exhibits the geometry and stereochemistry found in the [3Fe-4S] cluster of fumarate reductase. The shortest interatomic distance between the cluster and FMN is 6.9 Å (between atoms C7M of

FMN and Fe4 of the cluster) with Met479 (on a 470–520 elongated loop emerging from the central domain) being in van der Waals contact with both of these atoms with its S and C side-chain atoms. This residue is particularly interesting in the light of the fact that it is the only GltS residue falling in an energetically disallowed region of the Ramachandran plot, which may be triggered for (regulatory) conformational changes upon cofactors reduction and 2-OG binding to the FMN/synthase site of GltS.

The C-terminal domain (residues 1203–1472) is a right-handed β-helix formed by seven turns, a fold so far unique to GltS. The domain plays an important structural role in GltS as emerging from the analysis of the structure of αGltS and of the NADPH-GltS oligomer (see below).

The structure of αGltS revealed the presence of an intramolecular ammonia tunnel for the transfer of ammonia released from L-Gln in the glutaminase site of the GAT domain to 2-OG in the synthase site across a distance of ~30 Å (▶Fig. 13.5). The core of αGltS contains two cavities separated by an obstruction mainly formed by backbone atoms of residues Thr507 and Asn508 of the 470–520 loop of the central domain and of residues Ser976 and Ile977 of loop 4 of the synthase domain. The first cavity is roughly spherical in shape and is located near the L-Gln binding site. It is lined by residues belonging to the GAT domain, to the 470–520 loop of the central domain and to loop 4 of the FMN binding domain. The cavity is not shielded from solvent due to the open conformation of the 210–225 loop. This loop is conserved in Ntn-amidotransferases, where it is defined as the Q-loop and acts as a lid of the glutaminase site shifting from an open to a closed conformation on glutamine binding to promote glutamine hydrolysis and avoid loss of the released ammonia into the solvent (▶Fig. 13.5). The second cavity is a cone extending towards the 2-OG binding site and is fully shielded from solvent. It is at the core of the protein with its wall mainly formed by residues of loop 4 and loop 6 of the FMN binding domain and of the 470–520 loop of the central domain. The cavity is decorated by backbone and side-chain carbonyl oxygen atoms and aliphatic groups, which confer an overall hydrophilic character. Among them are Pro509 and Pro510 (of the central domain) and Pro938, Pro968, Pro969 and Pro970 of loop 4 of the FMN binding domain, which are invariant residues in highly conserved αGltS and FdGltS segments [7]. The cavity is the intramolecular tunnel, allowing the transfer of ammonia released from L-Gln to 2-OG bound in front of the FMN coenzyme in the synthase domain, a feature common to all amidotransferases. Interestingly, such an ammonia tunnel in these enzymes is a case of convergent evolution in that it is formed by the unrelated synthase domains rather than by the conserved GAT domain, a fact that might explain differences in the tunnel structures and in the mechanisms of cross-activation of the glutaminase and synthase activities observed in these enzymes [9,27,33]. In GltS, for ammonia to be transferred across the tunnel, the obstruction formed by residues 507–508 and 976–977 should be removed through breathing motions or, more likely, through conformational changes triggered by the enzyme ligation and redox state.

The *Synechocystis* PCC6803 FdGltS was crystallized in the presence of MetS, DON, and 2-OG in various combinations [9,52,53]. Crystals were obtained for FdGltS in the free state (1LM1, 2.7 Å resolution), in complex with 2-OG (1LLW, 1OFD, 2.0 Å resolution) and in complex with 2-OG after reaction of Cys1 with DON to yield to covalent complex with 5-oxo-L-norleucine (ONL; 1OFE, 2.45 Å resolution). Soaking of the crystals with L-Glu led to determining the structure of reduced FdGltS (1LLZ). The

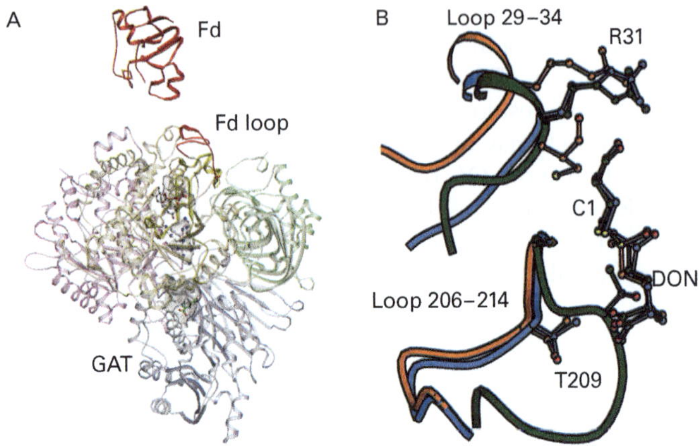

Fig. 13.6: Structure of FdGltS. (A) Ribbon diagrams of *Synechocystis* FdGltS and Fd. FdGltS domains are in light blue (GAT), magenta (central), yellow (synthase) and green (C-terminal β-helix). The Fd loop is in red. Loop 4 and loop 6, which are the elements of the synthase domain forming part of the tunnel wall and connecting the synthase and GAT domains are in a darker shade of yellow. The position of Fd (red, 1OFF, [53]) is not based on experimental evidence. Water molecules in the ammonia tunnel are as blue spheres; 2-OG is in red sticks. The ONL covalently bound to Cys-1 is in green; FMN is in sticks and CPK colors. (B) Detail of the glutaminase site of FdGltS that had formed a covalent Cys1-ONL adduct (1OFE , blue, [53]) superposed with the corresponding regions of αGltS in complex with MetS (1EA0, orange, [51]) and PRPP-AT that had reacted with L-DON to give the covalent Cys1/ONL adduct (1ECC, green, [60]). Loop 29–34 of FdGltS corresponds to loop 29–24 of αGltS and loop 24–29 of PRPP-AT (1ECC) Loop 206–214 of FdGltS corresponds to loop 210–218 of αGltS (see ▸Fig. 13.5) and loop 73–81 of PRPP-AT (1ECC). Reprinted with permission from [53]; Copyright 2003. Elsevier.

overall structures of the FdGltS forms are similar to each other and to that of αGltS with the notable exception of the presence of a loop (residues 907–933) at the surface of the protein and in proximity (14 Å) of the [3Fe-4S] cluster that corresponds to the ~17 residues insert at position 897–898 of αGltS found in plant-type FdGltS (▸Figs. 13.2 and 13.6). This "Fd-loop" may be implicated in docking or sensing bound Fd_{red}, which is required for Fd-GltS reduction and must be bound to activate the FdGltS reaction (▸Fig. 13.3, [29]). Although it was possible to crystallize *Synechocystis* [2Fe-2S] Fd (1OFF), which was similar to other plant-type Fd, co-crystallization of the *Synechocystis* Fd/Fd-GltS complex was not successful [53]. Thus, attempts to obtain a model of the complex by SAXS were done [53]. This approach showed that FdGltS and Fd are monomeric in solution and that they likely form a 1:1 Fd:FdGltS complex. The experimental scattering data could be well reproduced assuming binding of Fd near the Fd-loop of Fd-GltS and at a distance compatible with electron transfer from the Fd [2Fe-2S] cluster to the FdGltS [3Fe-4S] center. A report of the successful co-crystallization of *Leptolyngbya boryana* Fd/FdGltS complex has recently appeared [54] so that the high-resolution structure of this interesting complex may soon be available.

Overall, the αGltS and FdGltS structures available, in comparison to the known functional properties of GltS and of the related Ntn-amidotransferases, are consistent with the mechanistic scheme of the glutaminase and synthase reactions depicted in

►Fig. 13.2, and indicate which conformational changes might place during the catalytic cycle to control and coordinate the activities across the ammonia tunnel.

The glutaminase reaction is assisted by the free protein amino group that deprotonates Cys1 side chain, protonates the leaving ammonia and perhaps also deprotonates the incoming water molecule to complete glutamine hydrolysis. The two transient tetrahedral intermediates leading to and from the covalent Cys1 γ-glutamylthioester intermediate are stabilized by the oxyanion hole that in FdGltS is formed by the side chain of Asn227 (corresponding to Asn231 of αGltS) and the backbone nitrogen atom of Gly228 (Gly232 in αGltS). The Q loop (210–225 in αGltS and 206–214 in FdGltS) must close onto the glutaminase site when glutamine binds to provide a solvent-shielded environment so that the released ammonia molecule is not lost into the solvent. Furthermore, closure of this loop would allow the formation of a hydrogen bond between Thr209 (FdGltS numbering) and L-Gln/L-Glu C(5) carbonyl oxygen as found in the ligand-bound (Q loop closed) conformation of PRPP-AT [55] and GlmS [56] (►Figs. 13.6 and 13.2). In the FMN/synthase domain, 2-OG is held in place by a complex network of interactions. Interestingly, the C(2) carbon of 2-OG is 4 Å far from FMN-N(5) position and the 2-OG C(1)-C(2)O-C3 plane is not parallel to the FMN isoalloxazine ring (►Fig. 13.5C). In this geometry reduction of 2-OG is not possible, preventing a 2-OG reductase activity. On the contrary, 2-OG is at a position suitable for addition of ammonia from the tunnel to its C(2) carbon yielding the 2-IG intermediate. Lys972 (Lys937 in αGltS) is at hydrogen bonding distance from the carbonyl oxygen atom and may promote 2-IG formation by polarizing the carbonyl group and stabilizing the developing negative charge of the predicted tetrahedral intermediate. The 2-IG intermediate should then reposition in the active site to bring its C(2) at a distance (~3.5 Å) and geometry (C1-C2(N)-C3 plane parallel to the flavin) suitable for reduction of 2-IG to yield L-Glu upon hydride transfer from the reduced FMN N5 position. 2-IG reduction may be favored by Glu903 (Glu886 in αGltS) [9,53].

In none of the crystallographically determined structures, the tunnel obstruction at the entry point towards the glutaminase site was removed. However, the αGltS and FdGltS structures suggest how the presence of 2-OG, reduced cofactors and, for FdGltS, reduced Fd, may be signaled to the glutaminase site to activate it, and may cause the coordinated opening of the tunnel entry point (►Fig. 13.5D). The N-terminal part of loop 4 (968–1013 in FdGltS; 933–978 in αGltS) of the FMN domain interacts with the 907–933 Fd-loop (in FdGltS), 2-OG and is in proximity of the loop carrying the [3Fe-4S] cluster (in all GltS). Its C-terminal part forms part of the tunnel wall ending with residues Ser1011, Ile1012 (obstructing the tunnel entrance; residues Ser976 and Ile977 in αGltS) and Glu1013 (in contact with Cys1 free amino group; Glu978 in αGltS). Thus, small conformational changes triggered by 2-OG binding, [3Fe-4S] cluster reduction (and binding of Fd_{red} to FdGltS) may be efficiently transmitted through loop 4 across the tunnel. On the other side, the 423–786 loop of the central domain (470–520 loop of αGltS) may also participate in communication between the catalytic sites with Met475 of this loop (Met479 of αGltS) possibly sensing the cofactors redox state and the 2-OG presence. The loop ends with Thr503 and Asn504 (Thr507 and Asn508 in αGltS) that form the tunnel obstruction facing residues 1011 and 1012 (976 and 977 in αGltS) of loop 4. Thr503 also interacts with Glu229 (Glu233 in αGltS) which in turn is in contact with the L-Gln substrate amino group and is adjacent to N227 and G228, the residues forming the oxyanion hole in the glutaminase site (G232 and N231in αGltS). Overall, also small conformational changes of this loop, triggered by the ligation (and redox) state of the synthase site may be transmitted to the tunnel entrance and to the glutaminase site.

The proposed model was supported by two 4 ns molecular dynamics simulations in explicit solvent of αGltS in the oxidized and unliganded state (αGltS$_{free}$) and of reduced αGltS in complex with L-Gln and 2-OG (αGltS $_{bound}$) starting from the coordinates of αGltS in complex with MetS and 2-OG (αGltS $_{crystal}$, [57]). Removing MetS and 2-OG to yield the αGltS$_{free}$ form led to narrowing of both the L-Gln and 2-OG binding pockets and further constriction of the tunnel entry point (►Fig. 13.7). In the glutaminase site, Cys1 is further rotated away from the active site to an inactive conformation. When L-Gln is substituted for MetS in the αGltS crystallographic structure, Cys1 reorients from the inactive geometry observed in the crystallographic αGltS-MetS structure (and in the simulated αGltS$_{free}$, ►Figs. 13.7, 13.5A and 13.6B) to reach a position similar to that of Cys1 in the crystallographic structures of free FdGltS or of the covalent complex with

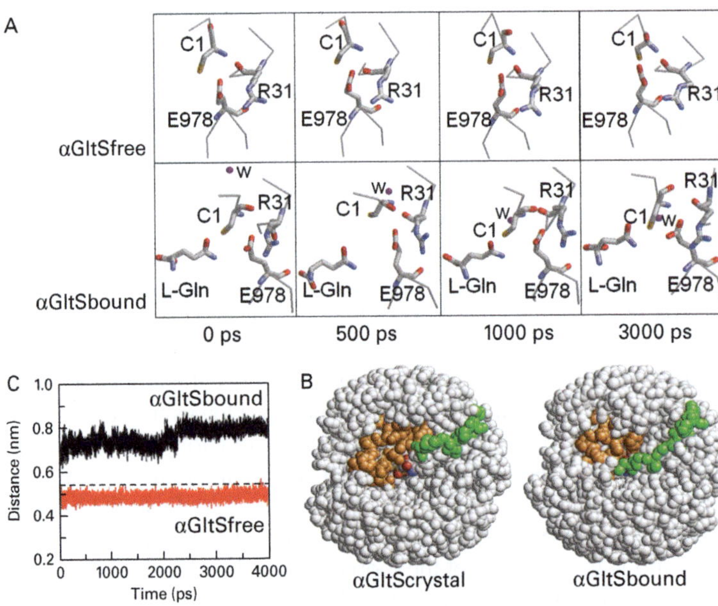

Fig. 13.7: Molecular dynamics simulations of αGltS. (A) Snapshots of the glutaminase site during the simulations of free (unliganded) αGltS and of the reduced protein in complex with 2-OG and L-Gln (αGltS$_{bound}$) starting from the crystallographic coordinates of αGltS in complex with MetS and 2-OG (1EA0, [51]). In αGltS$_{bound}$, Cys-1 reorients from the catalytically inactive position observed in the αGltS crystal to a position that allows nucleophilic attack of Cys-1 thiolate on L-Gln C(5) atom in the bound form and similar to that found in the active conformation of FdGltS [9,51,52]. (B) Comparison of the crystallographic structure of αGltS in comlex with MetS and 2-OG (1EA0, [51]) with that obtained at the end of the simulation of αGltS in complex with L-Gln and 2-OG. The Q loop (residues 210–225) is in orange, loop 263–271 in green, MetS or L-Gln in CPK colors. (C) Width of the ammonia tunnel entrance during the simulation of αGltS$_{free}$ (red) and αGltS$_{bound}$ (black) as calculated from the distance between the center of mass of residues Thr507–Asn508 and Ser976–Ile977 The dashed line represents the corresponding value observed in αGltS crystal structure. Reprinted with modifications from [57]; Copyright 2008. John Wiley and Sons.

ONL (►Fig. 13.6), which reflect an active conformation [9,51,53]. In this geometry, the Cys1 amino group is hydrogen bonding with the main chain carbonyl oxygen atom of Arg31, the side-chain of Glu978 and, importantly, the Sγ atom approaches L-Gln C(5) reaching a position suitable for nucleophilic attack. The Cys1 reorientation is accompanied by rigid body motions involving residues 6–32 and 73–99, which reflect the coordinated movements of Cys1, residues 15–28 and 29–34, postulated to take place to convert the inactive conformation of the glutaminase site observed in αGltS in complex with MetS into the active conformation observed with all FdGltS structures (►Fig. 13.6B, [9]). During the simulations the Q loop remained in an open conformation. However, in the $\alpha GltS_{bound}$ form another loop (residues 263–271) closed onto the glutaminase site and partially shielded it from solvent (►Fig. 13.7). Thus, this loop may functionally replace the Q loop in GltS. The simulation of $\alpha GltS_{bound}$ showed that 2-OG bound to the synthase site, reorients through rotations mainly around C2-C3 and C3-C4 bonds so that C2 moves further away from the flavin towards the tunnel to favor ammonia addition. In the simulations Lys931(corresponding to Lys966 of FdGltS) rather than Lys937 (corresponding to Lys972 of FdGltS, [9,53]) interacts with 2-OG O(2) promoting ammonia addition so that two different Lys residues may participate in 2-IG formation and reduction. Met479 also undergoes conformational changes during the simulations, while remaining in contact with FMN and the [3Fe-4S] cluster, supporting the concept that this residue may participate in the mechanisms that control and coordinate catalysis in GltS by sensing the enzyme ligation and redox state. Finally, in the $\alpha GltS_{bound}$ simulation, the tunnel entry point broadens from 4.8 Å or 5.4 Å (of the $\alpha GltS_{free}$ simulation and αGltS or FdGltS structures) to 6.7–8.7 Å (►Fig. 13.7). Interestingly, tunnel opening led to the entrance of several water molecules that interacted with groups lining the tunnel in a way resembling that observed in the crystal structures of FdGltS, which might mark the positions of ammonia during its journey towards 2-OG [53].

The experimental and computed structures of GltS highlighted the critical position of FdGltS Glu1013 (Glu978 in αGltS, ►Figs. 13.5D and 13.8), which is next to the residues 1011–1012 that obstruct the tunnel entrance (residues 976–977 in αGltS), where its δ-carboxylate group interacts with several key residues, namely: Cys1 α-amino group, the N-terminal nucleophile; Arg31 NH1 nitrogen, which in turn seems important for the correct positioning of Cys1 through an interaction with its main chain carboxylic oxygen; the Asn227 side chain amide nitrogen (part of the oxyanion hole, ►Figs. 13.5B and 13.8). The Glu1013 Asp, Asn and Ala substitutions in FdGltS led to progressive loss of activity supporting the importance of the negative charge of the residue at position 1013 by, presumably, affecting the precise geometry of the oxyanion hole, and slowing down formation of the γ-glutamyl thioester intermediate. The Glu1013Asn substitution caused uncoupling of the glutaminase and synthase reaction of FdGltS indicating that in this enzyme species removal of the hydrogen bonding interaction between Glu1013 carboxylate and Ser1011 side chain hydroxyl group may prevent the coordinated tunnel opening triggered by 2-OG (and electrons) binding to the synthase site. The Glu1013Asp substitution instead revealed a sigmoid dependence of the initial reaction velocity on L-Gln concentration that could be interpreted to indicate a two-step activation mechanism of the enzyme, especially in light of the fact that FdGltS is a monomer in solution; this may also apply to NADPH-GltS (►Fig. 13.8C). According to this hypothesis, binding of 2-OG and reduction of the cofactors at the synthase site partially activate the glutaminase site so that L-Gln may bind and complete the conformational change to yield the fully active

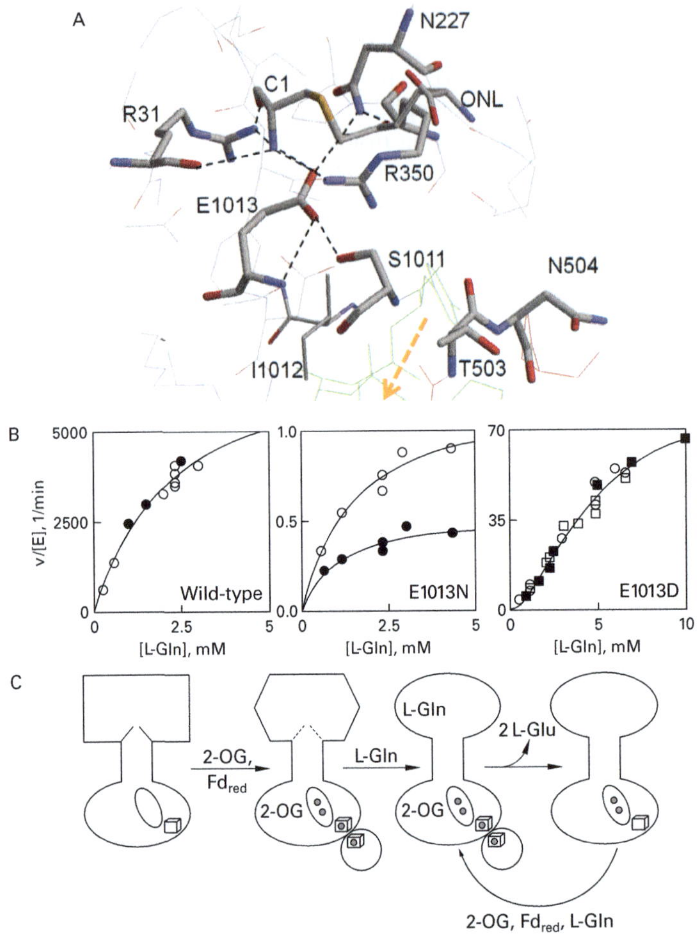

Fig. 13.8: Role of Glu1013 of FdGltS in the activation and coupling of the glutaminase and synthase reactions of GltS. (A) Details of the glutaminase site and tunnel entry point in the covalent adduct of FdGltS Cys1 with ONL (1OFE, [53]) highlighting the interactions of Glu1013. The gold arrow indicates the direction of the tunnel. The dashed lines indicate possible interactions among residues (minimum interatomic distances 3.5 Å). (B) Steady-state kinetic analyses of the FdGltS variants carried out in 50 mM Hepes/KOH buffer, pH 7.5, in the presence of 21 μM Fd, 4 mM sodium dithionite, FdGltS (19 nM, wild-type; 4 μM, E1013N; 2 μM, E1013D) and 2.5 mM 2-OG. The initial reaction velocity was determined by monitoring the time-course of L-Glu production from L-[U-^{14}C]-Gln (open symbols) or 2-[U-^{14}C]-OG (closed symbols) in discontinuous assays. (C) Proposed model for the two-step activation of glutamine hydrolysis and ammonia transfer. The initial oxidized enzyme (very low to no affinity for L-Gln; tunnel closed) is reduced at the expenses of three reduced Fd molecules, and forms a stable complex with reduced Fd (►Fig. 13.3 and [29]) stabilizing a partially active conformation of the glutaminase site. Binding of L-Gln to this species fully activates the glutaminase site and most likely induced tunnel opening so that the catalytic cycle can be completed. Since the return of the enzyme to the initial inactive conformation is slow (perhaps because a 2-electron reduced species is formed), the concentration of the active enzyme is set by L-Gln concentration during turnover. Symbols: FMN, oval; [Fe-S] clusters, cubes; electrons, circles. Panels (A) and (B) reprinted with permission from [30]; Copyright 2007. American Chemical Society.

glutaminase site. If, in the mutant, the equilibrium between the inactive and active glutaminase site lies towards the inactive conformation, and if the return of such a fully active conformation to the inactive one, after the first turnover, is relatively slow, the steady-state level of active enzyme is set by L-Gln concentration leading to the observed sigmoid kinetics. The slow return to the initial inactive conformation would be explained by taking into account that the enzyme may shuttle between a 2-electron and a 4-electron reduced state. This process would go undetected in the wild-type enzyme where the conformational changes may be fast and favored. A similar two-step activation process has been proposed also for other amidotransferases [26,58] reinforcing the concept that the convergent evolution that led to the formation of ammonia tunnels by the unrelated synthase sites of amidotransferases extends to the communication mechanisms between the glutaminase and synthase sites.

13.7 Structure of the NADPH-GltS αβ-protomer

No crystal forms of the NADPH-GltS αβ-protomer have been obtained. In order to gain structural information on this enzyme form we turned to small-angle X-ray scattering (SAXS), which allows structural data to be obtained in solution, and to low-temperature electron microscopy (cryoEM), which yields structural information from analysis of single particles. The combination of the two methods proved to be a powerful tool to gain insight into the structure of NADPH-GltS [49], which was found to mainly form 1.2 MDa $(\alpha\beta)_6$-hexamers. The experiments clarified that the previously determined tetrameric oligomeric state of NADPH-GltS likely depended on the underestimation of the mass during gel filtration experiments for the lack of suitable standards and chromatographic resins [8,19], as well as to the fact that lower mass particles (mainly αβ-monomers) are always present in NADPH-GltS preparations accounting for 10–20% of the protein present [49,59]. Shape reconstruction from the cryoEM particles images led to modeling of NADPH-GltS at 9.5 Å resolution, which allows secondary structure elements to be observed. The NADPH-GltS $(\alpha\beta)_6$ hexamer is well-described as being formed by three pillars, corresponding to the crystallographic α_2 dimers, with one copy of the β-subunit per each αGltS, located at the periphery (▶Figs.13.9A and B). Fitting the crystallographic structure of the α_2 dimer and a homology model of βGltS, based on the crystallographic structure of the corresponding domain of DPD, into the 9.5 Å-resolution cryoEM model provided insight into its structural features (▶Fig. 13.9C and D). The interprotomeric contacts involve α/α- and α/β-interactions of adjacent subunits, which are weaker than the α/α contacts of the crystallographic α_2 dimer. In both of these interprotomeric contacts, the β-helical domain of αGltS is involved, demonstrating the multiple structural role of the αGltS C-terminal region. In the isolated αGltS, the β-helix and the central domain maintain the GAT and synthase domains in place, contributing to stabilize the structure of αGltS and forming the ammonia tunnel [51]. With residues extending from the β-helical turns one αGltS establishes contacts with the FMN domain of the αGltS of the adjacent pillar. The α/β-interprotomeric contacts are between residues 370–385 of one βGltS and residues 1438–1449 of the αGltS belonging to the αβ-protomer of the neighboring pillar. The latter residues are at the C-terminus of the β-helical domain and belong to a α-helix resting on the β-helix side.

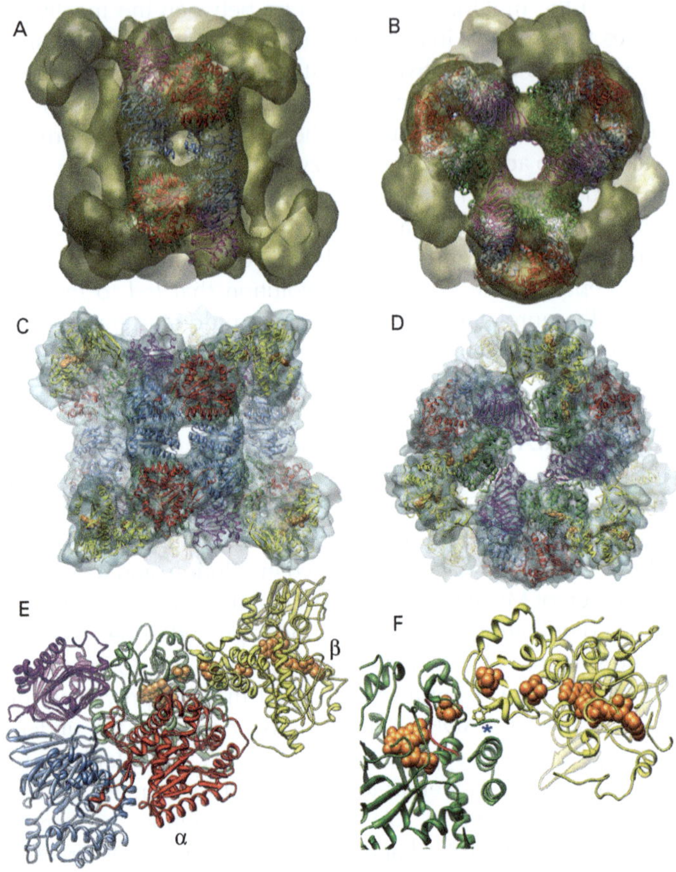

Fig. 13.9: Structure of the 1.2 MDa NADPH-GltS hexamer as deduced from cryo-electron microscopy, small-angle X-ray scattering and molecular modeling. (A) and (B) Top (A) and side (B) views of the low temperature electron microscopy (cryoEM) volume at 26 Å resolution with one (A) and three (B) copies of the crystallographic α_2-dimer of *A. brasilense* NADPH-GltS (1EA0, [51]) fitted into the pillars of the complex indicating the presence of six αGltS subunits in the core region and six βGltS at the periphery. (C) and (D) Top (C) and side (D) views of the final model of NADPH-GltS obtained by fitting of the 9.5 Å resolution cryoEM volume with the crystallographic α_2-dimers and a homology model of βGltS based on the structure of DPD (1H7W, chain A, residues 55-520, [41]). (E) The αβ-protomer extracted from the $(\alpha\beta)_6$-structure (see panel (C), top right). (F) Details of the interface region showing the linear electron transfer chain formed by (from right to left): FAD, the [4Fe-4S] clusters homologous to those of DPD, the [3Fe-4S] center and FMN. Color coding for αGltS domains is the same as in ►Figs. 13.1 and 13.5; βGltS is in yellow. Phe54, which may participate in electron transfer from one of the [4Fe-4S] clusters on βGltS to the [3Fe-4S] center is marked with a blue star. The figure has been reprinted with modifications from [49] with permission. Copyright 2008. ASBMB; for additional views, see movie1.mov available as supplementary material of [49].

The model of the αβ-protomer extracted from the $(\alpha\beta)_6$-hexamer is consistent with the proposal that it corresponds to the catalytically active unit of GltS (►Fig. 13.9E). Most of the intraprotomeric contacts with αGltS involve the N-terminal region of βGltS, which harbors the [4Fe-4S] clusters of NADPH-GltS. This observation supports the key role of the [4Fe-4S] clusters in structuring the βGltS N-terminal region, which is in turn responsible for the formation of the catalytically active αβ-protomer.

In the model of NADPH-GltS αβ-protomer the cofactors form a linear chain in which FAD and FMN are connected by the two [4Fe-4S] clusters encoded by βGltS and conserved in DPD N-terminal βGltS-like domain, and by the [3Fe-4S] cluster of αGltS (►Fig. 13.9F). Thus, electron transfer from FAD to FMN must take place via two sequential 1-electron transfer events with the two electrons following the same path, as depicted in ►Fig. 13.4. Interestingly, Phe54 of βGltS, conserved as a Phe or Tyr in all βGltS, may participate at the electron transfer process between the [4Fe-4S] cluster on the βGltS closest to the interface and the [3Fe-4S] center on αGltS.

Whether the oligomeric state of NADPH-GltS is affected by its substrates, and to what extent it may affect the GltS biological function, was studied by SAXS, dynamic light scattering (DLS) and steady-state kinetics [49]. The $(\alpha\beta)_6$-hexamers are extremely stable so that reversible dissociation into αβ-monomers is obtained only in the presence of high concentration of NaCl, and, even in the presence of 1 M NaCl, the process is slow taking at least 10 h to reach completion. Reassociation is relatively fast (taking minutes) when NaCl is removed by dilution or centrifugal gel filtration. The $(\alpha\beta)_6$-hexamers are destabilized to a small extent by $NADP^+$, which leads to some dissociation (5–20% at high and low protein concentrations, respectively) into $(\alpha\beta)_2$-dimers. The latter is in agreement with the fact that βGltS is implicated in the interaction of αβ-protomers of adjacent pillars and with the observation that $NADP^+$ altered the sensitivity of sites of αGltS to limited proteolysis [32]. On the contrary, L-Gln, MetS and 2-OG (added in various combinations) had no detectable effect on the GltS oligomerization state, as found also by limited proteolysis. By working with enzyme that had been preincubated under conditions that cause dissociation of the $(\alpha\beta)_6$-hexamer into αβ-protomers (*i.e.* 1 M NaCl for 20 h) and with assays run in 1 M NaCl (to prevent reassociation) steady-state kinetics showed that the protein is catalytically active. When compared to the $(\alpha\beta)_6$-hexamer (with activity assays carried out in 1 M NaCl, and exploiting its slow dissociation), the αβ-protomer exhibits a similar k_{cat}, but approximately 3-fold lower K_m values for both 2-OG and L-Gln. Such a difference does not seem sufficiently high to indicate that changes in the oligomerization state of GltS may have a significant effect on the reaction catalyzed by GltS in the cells. However, it cannot be ruled out that some properties of GltS may be affected by changes in oligomerization state, including sensitivity to as yet unidentified allosteric regulators or interacting proteins.

This and other aspects of this interesting enzyme still await elucidation. In spite of the complexity of the system, the work shown here demonstrates that experiments are possible thanks to the wealth of information gathered since the pioneering work done in the early 1970s when GltS was discovered [1].

13.8 Acknowledgments

Work on GltS has been supported by grants from the Ministero dell' Università e la Ricerca Scientifica (MIUR), PRIN program and from the Università degli Studi di Milano (FIRST and PUR programs). MAV is indebted to students and post-doctoral associates that participated at the work throughout the years, and to collaborators that contributed to the current knowledge on GltS, especially Drs.Giuliana Zanetti, Dale E. Edmondson, Andrea Mattevi, and Dmitri Svergun. This review is dedicated to the late Prof. Bruno Curti whose continuous support throughout the years is gratefully acknowledged.

13.9 References

[1] Tempest DW, Meers JL, Brown CM. Synthesis of glutamate in *Aerobacter aerogenes* by a hitherto unknown route. Biochem J 1970;117:405–7.

[2] Vanoni MA, Curti B, Zanetti G. Glutamate synthase. In: Mueller F, ed. Chemistry and Biochemistry of Flavoenzymes. Boca Raton: CRC Press; 1992:309–17.

[3] Lea PJ, Miflin BJ. Alternative route for nitrogen assimilation in higher plants. Nature 1974;251:614–6.

[4] Zalkin H. The amidotransferases. Adv Enzymol Relat Areas Mol Biol 1993;66:203–309.

[5] Zalkin H, Smith JL. Enzymes utilizing glutamine as an amide donor. Adv Enzymol Relat Areas Mol Biol 1998;72:87–144.

[6] Vanoni MA, Curti B. Structure-function studies of glutamate synthases: a class of self-regulated iron-sulfur flavoenzymes essential for nitrogen assimilation. IUBMB Life 2008;60:287–300.

[7] Vanoni MA, Curti B. Glutamate synthase: a complex iron-sulfur flavoprotein. Cell Mol Life Sci 1999;55:617–38.

[8] Ratti S, Curti B, Zanetti G, Galli E. Purification and characterization of glutamate synthase from *Azospirillum brasilense*. J Bacteriol 1985;163:724–9.

[9] van den Heuvel RH, Curti B, Vanoni MA, Mattevi A. Glutamate synthase: a fascinating pathway from L-glutamine to L-glutamate. Cell Mol Life Sci 2004;61:669–81.

[10] Vanoni MA, Curti B. Structure--function studies on the iron-sulfur flavoenzyme glutamate synthase: an unexpectedly complex self-regulated enzyme. Arch Biochem Biophys 2005;433:193–211.

[11] Vanoni MA, Dossena L, van den Heuvel RH, Curti B. Structure-function studies on the complex iron-sulfur flavoprotein glutamate synthase: the key enzyme of ammonia assimilation. Photosynth Res 2005;83:219–38.

[12] Knaff DB, Hirasawa M. Ferredoxin-dependent chloroplast enzymes. Biochim Biophys Acta 1991;1056:93–125.

[13] Miflin BJ, Wallsgrove RM, Lea PJ. Glutamine metabolism in higher plants. Curr Top Cell Regul 1981;20:1–43.

[14] Suzuki A, Knaff DB. Glutamate synthase: structural, mechanistic and regulatory properties, and role in the amino acid metabolism. Photosynth Res 2005;83:191–217.

[15] Temple SJ, Vance CP, Gantt JS. Glutamate synthase and nitrogen assimilation. Trends in Plant Sciences 1998;3:51–6.

[16] Muro-Pastor MI, Reyes JC, Florencio FJ. Ammonium assimilation in cyanobacteria. Photosynth Res 2005;83:135–50.

[17] Nesbo CL, L'Haridon S, Stetter KO, Doolittle WF. Phylogenetic analyses of two "archaeal" genes in thermotoga maritima reveal multiple transfers between archaea and bacteria. Mol Biol Evol 2001;18:362–75.

[18] Curti B, Vanoni MA, Verzotti E, Zanetti G. Glutamate synthase: a complex iron-sulphur flavoprotein. Biochem Soc Trans 1996;24:95–9.

[19] Stabile H, Curti B, Vanoni MA. Functional properties of recombinant *Azospirillum brasilense* glutamate synthase, a complex iron-sulfur flavoprotein. Eur J Biochem 2000;267: 2720–30.

[20] Vanoni MA, Nuzzi L, Rescigno M, Zanetti G, Curti B. The kinetic mechanism of the reactions catalyzed by the glutamate synthase from *Azospirillum brasilense*. Eur J Biochem 1991;202:181–9.

[21] Vanoni MA, Edmondson DE, Rescigno M, Zanetti G, Curti B. Mechanistic studies on *Azospirillum brasilense* glutamate synthase. Biochemistry 1991;30:11478–84.

[22] Vanoni MA, Edmondson DE, Zanetti G, Curti B. Characterization of the flavins and the iron-sulfur centers of glutamate synthase from *Azospirillum brasilense* by absorption, circular dichroism, and electron paramagnetic resonance spectroscopies. Biochemistry 1992;31: 4613–23.

[23] Vanoni MA, Verzotti E, Zanetti G, Curti B. Properties of the recombinant β subunit of glutamate synthase. Eur J Biochem 1996;236:937–46.
[24] Vanoni MA, Fischer F, Ravasio S, Verzotti E, Edmondson DE, Hagen WR, Zanetti G, Curti B. The recombinant α subunit of glutamate synthase: spectroscopic and catalytic properties. Biochemistry 1998;37:1828–38.
[25] Chaudhuri BN, Lange SC, Myers RS, Chittur SV, Davisson VJ, Smith JL. Crystal structure of imidazole glycerol phosphate synthase: a tunnel through a $(\beta/\alpha)_8$ barrel joins two active sites. Structure 2001;9:987–97.
[26] Mouilleron S, Golinelli-Pimpaneau B. Conformational changes in ammonia-channeling glutamine amidotransferases. Curr Opin Struct Biol 2007;17:653–64.
[27] Raushel FM, Thoden JB, Holden HM. The amidotransferase family of enzymes: molecular machines for the production and delivery of ammonia. Biochemistry 1999;38:7891–9.
[28] Raushel FM, Thoden JB, Holden HM. Enzymes with molecular tunnels. Acc Chem Res 2003;36:539–48.
[29] Ravasio S, Dossena L, Martin-Figueroa E, Florencio FJ, Mattevi A, Morandi P, Curti B, Vanoni MA. Properties of the recombinant ferredoxin-dependent glutamate synthase of *Synechocystis* PCC6803. Comparison with the *Azospirillum brasilense* NADPH-dependent enzyme and its isolated α subunit. Biochemistry 2002;41:8120–33.
[30] Dossena L, Curti B, Vanoni MA. Activation and coupling of the glutaminase and synthase reaction of glutamate synthase is mediated by E1013 of the ferredoxin-dependent enzyme, belonging to loop 4 of the synthase domain. Biochemistry 2007;46:4473–85.
[31] Vanoni MA, Accornero P, Carrera G, Curti B. The pH-dependent behavior of catalytic activities of *Azospirillum brasilense* glutamate synthase and iodoacetamide modification of the enzyme provide evidence for a catalytic Cys-His ion pair. Arch Biochem Biophys 1994;309:222–30.
[32] Vanoni MA, Mazzoni A, Fumagalli P, Negri A, Zanetti G, Curti B. Interdomain loops and conformational changes of glutamate synthase as detected by limited proteolysis. Eur J Biochem 1994;226:505–15.
[33] Massiere F, Badet-Denisot MA. The mechanism of glutamine-dependent amidotransferases. Cell Mol Life Sci 1998;54:205–22.
[34] Knaff DB, Hirasawa M, Ameyibor E, Fu W, Johnson MK. Spectroscopic evidence for a [3Fe-4S] cluster in spinach glutamate synthase. J Biol Chem 1991;266:15080–4.
[35] Navarro F, Martin-Figueroa E, Candau P, Florencio FJ. Ferredoxin-dependent iron-sulfur flavoprotein glutamate synthase (GlsF) from the cyanobacterium *Synechocystis* sp. PCC 6803: expression and assembly in *Escherichia coli.* Arch Biochem Biophys 2000;379:267–76.
[36] Gosset G, Merino E, Recillas F, Oliver G, Becerril B, Bolivar F. Amino acid sequence analysis of the glutamate synthase enzyme from *Escherichia coli* K-12. Protein Seq Data Anal 1989;2:9–16.
[37] Pelanda R, Vanoni MA, Perego M, Piubelli L, Galizzi A, Curti B, Zanetti G. Glutamate synthase genes of the diazotroph *Azospirillum brasilense.* Cloning, sequencing, and analysis of functional domains. J Biol Chem 1993;268:3099–106.
[38] Sakakibara H, Watanabe M, Hase T, Sugiyama T. Molecular cloning and characterization of complementary DNA encoding for ferredoxin-dependent glutamate synthase in maize leaf. J Biol Chem 1991;266:2028–35.
[39] Navarro F, Chavez S, Candau P, Florencio FJ. Existence of two ferredoxin-glutamate synthases in the cyanobacterium *Synechocystis sp.* PCC 6803. Isolation and insertional inactivation of gltB and gltS genes. Plant Mol Biol 1995;27:753–67.
[40] Morandi P, Valzasina B, Colombo C, Curti B, Vanoni MA. Glutamate synthase: identification of the NADPH-binding site by site-directed mutagenesis. Biochemistry 2000;39:727–35.
[41] Dobritzsch D, Schneider G, Schnackerz KD, Lindqvist Y. Crystal structure of dihydropyrimidine dehydrogenase, a major determinant of the pharmacokinetics of the anti-cancer drug 5–fluorouracil. EMBO J 2001;20:650–60.

[42] Rosenbaum K, Jahnke K, Curti B, Hagen WR, Schnackerz KD, Vanoni MA. Porcine recombinant dihydropyrimidine dehydrogenase: comparison of the spectroscopic and catalytic properties of the wild-type and C671A mutant enzymes. Biochemistry 1998;37:17598–609.

[43] Agnelli P, Dossena L, Colombi P, Mulazzi S, Morandi P, Tedeschi G, Negri A, Curti B, Vanoni MA. The unexpected structural role of glutamate synthase $[4Fe\text{-}4S]^{(+1,+2)}$ clusters as demonstrated by site-directed mutagenesis of conserved C residues at the N-terminus of the enzyme beta subunit. Arch Biochem Biophys 2005;436:355–66.

[44] Dincturk HB, Cunin R, Akce H. Expression and functional analysis of glutamate synthase small subunit-like proteins from archaeon Pyrococcus horikoshii. Microbiol Res 2011;166:294–303.

[45] Dahl C, Engels S, Pott-Sperling AS, Schulte A, Sander J, Lubbe Y, Deuster O, Brune DC. Novel genes of the dsr gene cluster and evidence for close interaction of Dsr proteins during sulfur oxidation in the phototrophic sulfur bacterium *Allochromatium vinosum*. J Bacteriol 2005;187:1392–404.

[46] Jeelani G, Husain A, Sato D, Ali V, Suematsu M, Soga T, Nozaki T. Two atypical L-cysteine-regulated NADPH-dependent oxidoreductases involved in redox maintenance, L-cystine and iron reduction, and metronidazole activation in the enteric protozoan *Entamoeba histolytica*. J Biol Chem 2010;285:26889–99.

[47] Hagen WR, Vanoni MA, Rosenbaum K, Schnackerz KD. On the iron-sulfur clusters in the complex redox enzyme dihydropyrimidine dehydrogenase. Eur J Biochem 2000;267:3640–6.

[48] Ravasio S, Curti B, Vanoni MA. Determination of the midpoint potential of the FAD and FMN flavin cofactors and of the 3Fe-4S cluster of glutamate synthase. Biochemistry 2001;40:5533–41.

[49] Cottevieille M, Larquet E, Jonic S, Petoukhov MV, Caprini G, Paravisi S, Svergun DI, Vanoni MA, Boisset N. The subnanometer resolution structure of the glutamate synthase 1.2 MDa hexamer by cryo-electron microscopy and its oligomerization behavior in solution: functional implications. J Biol Chem 2008.

[50] Hirasawa M, Robertson DE, Ameyibor E, Johnson MK, Knaff DB. Oxidation-reduction properties of the ferredoxin-linked glutamate synthase from spinach leaf. Biochim Biophys Acta 1992;1100:105–8.

[51] Binda C, Bossi RT, Wakatsuki S, Arzt S, Coda A, Curti B, Vanoni MA, Mattevi A. Cross-talk and ammonia channeling between active centers in the unexpected domain arrangement of glutamate synthase. Structure Fold Des 2000;8:1299–308.

[52] van den Heuvel RH, Ferrari D, Bossi RT, Ravasio S, Curti B, Vanoni MA, Florencio FJ, Mattevi A. Structural studies on the synchronization of catalytic centers in glutamate synthase. J Biol Chem 2002;277:24579–83.

[53] van den Heuvel RH, Svergun DI, Petoukhov MV, Coda A, Curti B, Ravasio S, Vanoni MA, Mattevi A. The active conformation of glutamate synthase and its binding to ferredoxin. J Mol Biol 2003;330:113–28.

[54] Shinmura K, Muraki N, Yoshida A, Hase T, Kurisu G. Crystallization and preliminary X-ray studies of an electron-transfer complex of ferredoxin and ferredoxin-dependent glutamate synthase from the cyanobacterium *Leptolyngbya boryana*. Acta Crystallogr Sect F Struct Biol Cryst Commun;68:324–7.

[55] Kim JH, Krahn JM, Tomchick DR, Smith JL, Zalkin H. Structure and function of the glutamine phosphoribosylpyrophosphate amidotransferase glutamine site and communication with the phosphoribosylpyrophosphate site. J Biol Chem 1996;271:15549–57.

[56] Isupov MN, Obmolova G, Butterworth S, Badet-Denisot MA, Badet B, Polikarpov I, Littlechild JA, Teplyakov A. Substrate binding is required for assembly of the active conformation of the catalytic site in Ntn amidotransferases: evidence from the 1.8 A crystal structure of the glutaminase domain of glucosamine 6-phosphate synthase. Structure 1996;4:801–10.

[57] Coiro VM, Di Nola A, Vanoni MA, Aschi M, Coda A, Curti B, Roccatano D. Molecular dynamics simulation of the interaction between the complex iron-sulfur flavoprotein glutamate synthase and its substrates. Protein Sci 2004;13:2979–91.
[58] Bera AK, Chen S, Smith JL, Zalkin H. Interdomain signaling in glutamine phosphoribosylpyrophosphate amidotransferase. J Biol Chem 1999;274:36498–504.
[59] Petoukhov MV, Svergun DI, Konarev PV, Ravasio S, van den Heuvel RH, Curti B, Vanoni MA. Quaternary structure of *Azospirillum brasilense* NADPH-dependent glutamate synthase in solution as revealed by synchrotron radiation x-ray scattering. J Biol Chem 2003;278:29933–9.
[60] Krahn JM, Kim JH, Burns MR, Parry RJ, Zalkin H, Smith JL. Coupled formation of an amidotransferase interdomain ammonia channel and a phosphoribosyltransferase active site. Biochemistry 1997;36:11061–8.

14 The dihydroorotate dehydrogenases

Maria Cristina Nonato and Antonio José Costa-Filho

Abstract

The dihydroorotate dehydrogenase is a flavoenzyme that catalyzes the stereoselective oxidation of (*S*)-dihydroorotate (DHO) to orotate (ORO) as part of the de *novo* pyrimidine biosynthetic pathway. In the present work, we present an overview about the structure, function and mechanism of action of DHODHs. We also discuss about the current use of DHODH inhibitors as immunosuppressant and antiproliferative agents and present the latest findings on the potential use of DHODH as target for drug development against viral, bacterial and parasitic diseases.

14.1 Biological function

The flavoenzyme dihydroorotate dehydrogenases (DHODH; E.C. 1.3.1.14, 1.3.1.15, 1.3.5.2, or 1.3.98.1, depending on the type) catalyzes the stereoselective oxidation of (S)-dihydroorotate (DHO) to orotate (ORO) in the fourth of the six conserved enzymatic reactions involved in the *de novo* pyrimidine biosynthetic pathway. This FMN-dependent catalytic cycle of DHODH consists of two half-reactions (▶Fig. 14.1). In the first and reductive half-reaction, the enzyme-bound FMN is reduced by DHO. In the second half-reaction, the reduced FMN is regenerated by an oxidizing substrate (Ox) whose nature may vary among different organisms (▶Fig. 14.1).

On the basis of sequence similarity, subcellular location as well as their preferences for oxidizing substrates, the DHODHs are divided in two major classes, *i.e.*, Class 1 and Class 2 [1]. Class 1 is further divided in Class 1A and Class 1B which are structurally and mechanistically distinct. Enzymes of Class 1A are homodimeric proteins encoded by the *pyrD* gene. They are located in the cytosol and can be found in Gram-positive bacteria such as *Lactococcus lactis* and *Enterococcus faecalis*, archaea and lower eukaryotes, *e.g.*, the trypanosomatids from the genera *Trypanosoma* and *Leishmania* [2–8]. Class 1B enzymes, also cytosolic, are prevalent in Gram-positive bacteria such as *Bacillus subtilis* [9], *Clostridium oroticum* [10], *L. lactis* [11] and *E. faecalis* are reported to have both Classes 1A and 1B [12]. Class 1B DHODHs are heterotetrameric proteins, a dimer of heterodimers, composed of two different proteins encoded by *pyrD* and *pyrK* genes.

Class 2 enzymes are monomeric proteins which are found bound in the inner membrane of mitochondria of eukaryotes [13,14] and in the cytosolic membrane of some prokaryotes such as the Gram-negative bacteria *E. coli* [15–17]. The cytosolic enzymes utilize the soluble substances fumarate (Class 1A) or NAD^+ (Class 1B), to oxidize $FMNH_2$, whereas the membrane-bound enzymes (Class 2) require quinones as their physiological oxidizing

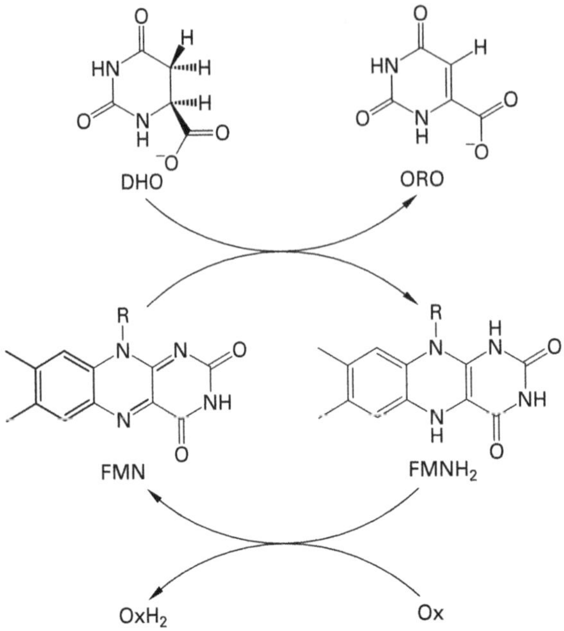

Fig. 14.1: Dihydroorotate dehydrogenase (DHODH) reaction. In the first half-reaction dihydroorotate (DHO) is oxidized to orotate (ORO) with the concomitant reduction of FMN to $FMNH_2$. In the second half-reaction a second substrate (Ox), reoxidizes FMN for a new cycle of catalysis.

agent [18]. Class 1A and Class 1B share about 30% of sequence identity, whereas soluble and membrane-bound DHODHs share less than 20% sequence identity.

A more divergent type of DHODH, named 1S, was described for the thermoacidophilic archaeon *Sulfolobus solfataricus* [19]. Type S DHODH was reported as a new member of Class 1 DHODHs based on its cytosolic cellular localization and was found to be a heteromeric enzyme comprising a catalytic subunit encoded by *pyrD* gene and an electron acceptor subunit, encoded by *orf,* a gene also found in the *pyr* operon [19]. Although type S DHODH is able to use coenzyme Q and molecular oxygen, the physiological electron acceptor, as well as, the quaternary structure of DHODHs are still unknown.

14.2 Protein production, purification and kinetic characterization

14.2.1 Purification

DHODHs have been isolated directly from human liver [20], *Sulfolobus solfataricus* [19] and the epimastigote form of *Trypanosoma cruzi* [7], but the great majority has been overexpressed as recombinant proteins in *E. coli.* Class 1 and Class 2 DHODHs have been purified by a combination of standard chromatographic steps or by using his-tagged fusion systems where the protocol requires a single step by applying immobilized metal affinity chromatography resin utilizing Ni^{2+}, Fe^{2+}, or Co^{2+} [7,13,21–25]. Purification of Class 2 DHODHs requires previous solubilization of membrane bound proteins with detergents such as Triton X-100 for human and *E. coli* DHODH [13, 15] and THESIT for *Plasmodium falciparum* DHODH [23].

14.2.2 Activity test

DHODH activity can be routinely assayed by spectrophotometric methods. Class 1A DHODH activity is assayed by monitoring the orotate formation at 300 nm ($\varepsilon = 2.65\ mM^{-1}\ cm^{-1}$) in a reaction mixture containing 50 mM Tris, pH 8.0; 150 mM KCl with both substrates, dihydroorotate and fumarate [5]. Since Class 2 DHODHs require the presence of detergent for solubilization, the purification protocol developed for this class includes the presence of Triton X-100, which strongly absorbs at 300 nm and impairs the use of orotate absorption at the same wavelength. Thus, an alternative assay, referred to as the DCIP assay, uses the colorimetric reagent 2,6-dichlorophnolindophenol as the final electron acceptor [20]. The reoxidation of OxH_2 into Ox is stoichiometrically equivalent to DCIP reduction which is in turn equivalent to oxidation of dihydroorotate (▶Fig. 14.1). DCIP reduction is monitored by the loss of DCIP absorbance at 610 nm ($\varepsilon = 21{,}500\ M^{-1}cm^{-1}$).

The kinetic parameters, V_{max} and K_m, can be determined by varying the concentrations of both substrates. The reaction can be initiated by the addition of DHODH and the rate of orotate production or DCIP consumption is determined over the time. Data can be analyzed by Lineweaver-Burk plots and the kinetic constants estimated from the fit of the data to the equation that describes the ping-pong mechanism (Eq. 14.1) [26–28]:

$$\frac{1}{v} = \frac{1}{V_{max}} + \frac{K_{DHO}}{V_{max}[DHO]} + \frac{K_{OX}}{V_{max}}\frac{1}{[Ox]}$$

where V_{max} is the maximum velocity and [DHO] and [Ox] and K_{DHO} and K_{Ox} are the concentrations and the K_m values for the variable substrates dihydroorotate and the oxidizing agent, respectively.

14.3 X-ray structures

14.3.1 Crystallization

DHODHs from 10 different organisms have been successfully crystallized. Crystallization of DHODH enzymes have been reported under unrelated conditions; however DHODH seems to best crystallize out of traditional precipitants such as PEG (polyethylene glycol) [6,24,29,30] and sulfate salts [13,22,31–34]. Furthermore, the presence of detergent in the crystallization buffer such as *N,N*-dimethyldecylamine-*N*-oxide (DDAO) used for crystallization of human DHODH [13], pentaethylene glycol monooctyl ether used for crystallization of *P. falciparum* DHODH [35] and n-octyl-β-D-glucoside used for crystallization of both *E. coli* [36] and rat DHODHs [34], appears as obligatory for crystallization of Class 2 DHODH by stabilizing the membrane associate domain located at the N-terminus.

14.3.2 Overall description of the atomic structure

Approximately 60 X-ray structures of dihydroorotate dehydrogenases have been deposited in the Protein Data Bank (PDB) to date. Among these are several mutant protein structures or structures of DHODH in complex with ligands including substrates, reaction products and inhibitors. Class 1A structures have been determined for *L. lactis* (PDB ID 1DOR) [37], *Streptococcus mutans* (PDB ID 3OIX) [30], *Trypanosoma cruzi* (PDB ID 3C3N, 2DJX) [22,38], *Trypanosoma brucei* (PDB ID 2B4G) [39], *Leishmania major* (PDB ID 3GYE) [31], *Leishmania donovani* (PDB ID 3C61) [not published]. Class 1B structure has been solved only for *L. lactis* (PDB ID 3GYE) [40] and Class 2 structures have been solved for *Homo sapiens* (PDB ID 1D3G) [13], *P. falciparum* (PDB ID 1TV5), *Rattus rattus* (PDB ID 1UUO) [34] and *E. coli* (PDB ID 1F76) [41].

In general, DHODH shows the typical α/β-barrel fold of Class 1A DHODH, with a central barrel composed of eight parallel β-strands (β_1-β_8) wrapped by eight α-helices (α_1–α_8) (►Fig. 14.2). Two N-terminal antiparallel β-strands (β_a-β_b) are found at the bottom

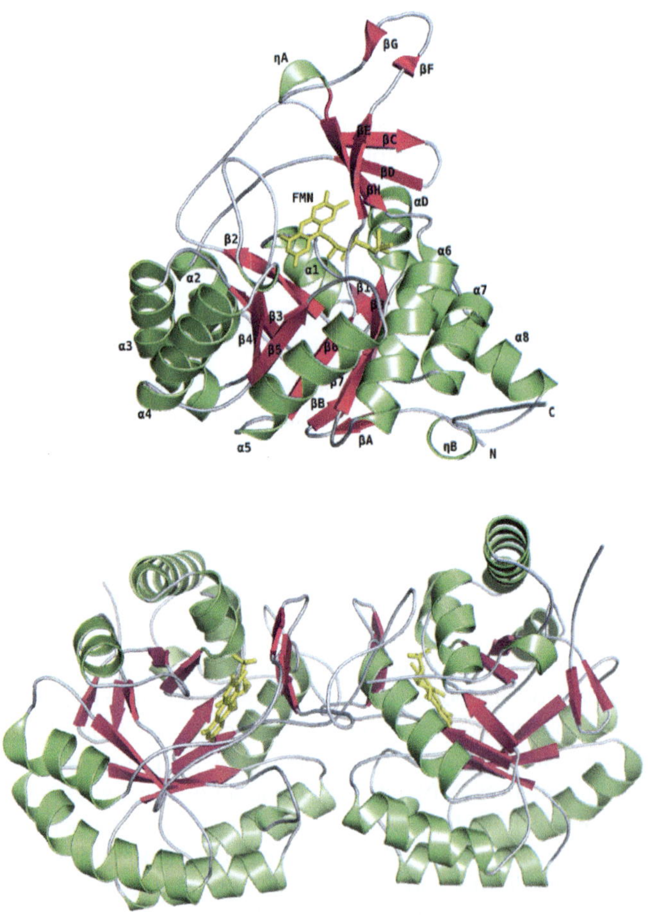

Fig. 14.2: Cartoon representation of Class 1A DHODH (PDB ID: 3TQO was used as a model) Top: Class 1A DHODH monomer. Bottom: Class 1A DHODH functional dimer.

of the barrel, and a conserved four-stranded antiparallel β-sheet (β_c, β_d, β_e and β_f) form a protrusive subdomain present at the top of the barrel. The prosthetic FMN group is located between the top of the barrel and the subdomain formed by these insertions (▶Fig. 14.2).

Class 1A DHODHs fold as homodimers where the two monomers are related by a 2-fold axis with hydrophobic, hydrophilic and, in particular, stacking interactions playing a role in dimer stabilization (▶Fig. 14.2) [31]. The high degree of sequence identity shared by the residues involved in dimer formation within Class 1A DHODHs, which suggests the relevance of the dimeric form for Class 1A DHODHs, corroborates with previous observation that dimer dissociation led to the loss of enzymatic activity in *L. lactis* DHODH [42].

The crystallographic structures of Class 1A DHODH complexes with dihydroorotate, orotate and fumarate reveal that substrates and the product of the first half-reaction bind to the same active site and exploit similar interactions [31,38,43]. In addition, Class 1A-specific inhibitors, currently limited to benzoate pyrimidine analogues and orotate analogues, were characterized to inhibit Class 1A DHODH by interacting at the same active site by a competitive mechanism against dihydroorotate except for 3,5-dihydroxybenzoate described to be a non-competitive inhibitor for *T. brucei* DHODH [6,44–46].

Class 2 DHODH structures are folded in two distinct domains. The N-terminal extension of approximately 40 residues folds into two α-helices connected by a short loop and plays a role in membrane anchoring. This region was described to be the target site for Class 2 inhibitors and supposedly provides the binding site for respiratory quinones that serve as physiological electron acceptors [13,35,41] (▶Fig. 14.3). In agreement with this hypothesis, kinetic and structural studies have demonstrated that the Class 2 DHODH inhibitor teriflunomide, also known as A771726, behaves as a competitive inhibitor of the ubiquinone binding site and is noncompetitive with respect to dihydroorotate [13,47]. In higher eukaryotes, this extended N-terminus also contains the signal peptide that targets DHODHs into mitochondria.

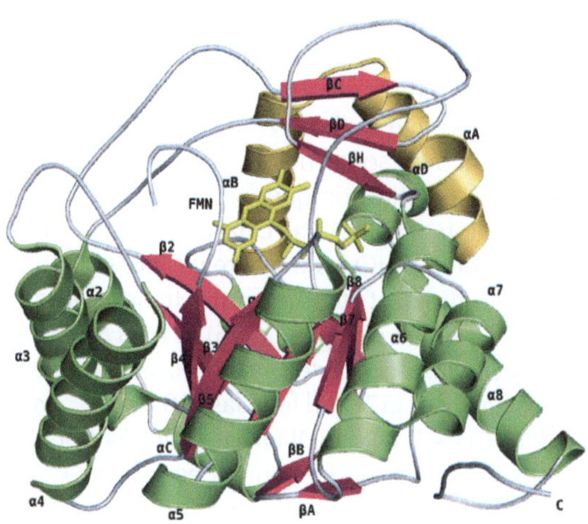

Fig. 14.3: Cartoon representation of the monomeric Class 2 DHODH. (PDB ID 1D3G was used as a model.)

Connected by an extended loop to the N-terminal domain, the C-terminal domain of Class 2 DHODHs resembles the α/β-barrel fold described for Class 1A DHODHs (►Fig. 14.3). In general, despite low sequence identity, most of the secondary structural elements and the residues involved in both FMN and substrate binding are conserved in both classes. However, several punctual regions of C-terminal domain were found to distinguish Class 1A and Class 2 DHODHs: the helix α_C, located in the bottom of the barrel and found replaced by a short loop; and short insertions at the connecting loops between β_3 and α_3, β_4 and α_4, and α_4 and β_5.

In addition, there is a remarkable difference at strands β_E, β_F, β_G, β_H and the η_A short 3_{10} helix found in Class 1A DHODHs and which were not observed in Class 2 enzymes (►Figs. 14.2 and 14.3). In fact, this protuberance is responsible for the dimer formation, a feature only observed in Class 1A DHODHs (►Fig. 14.2).

The only Class 1B DHODH crystal structure is available for the Gram positive bacteria *L. lactis* (PDB ID: 1EP1) [40]. Class 1B DHODH folds as a heterotetramer with two closely dimers of PyrD-PyrK dimers (►Fig. 14.4). The dimer of two PyrD subunits resembles the Class 1A enzymes and forms the central core of DHODB (►Fig. 14.4). PyrK belongs to the ferredoxin-NADP reductase superfamily and can be described as comprised of three distinct domains (►Fig. 14.4): the FAD binding domain (the N-terminal domain), the 2Fe-2S cluster binding domain (the C-terminal domain) and an α/β-structure that resembles the NAD-binding domain of the ferredoxin reductase family and that has been hypothesized to be the site of interaction for the NAD^+ oxidant agent of Class 1B DHODHs. The close proximity of the three redox centers makes it possible to propose electron transfer pathway involving residues conserved among the Class 1B DHODHs.

14.4 Mechanism

Although they all have the α/β-barrel fold as a fingerprint, DHODHs also have significant variations, which correlate to changes in their functional mechanisms. This structural diversity leads to changes in the mechanism of catalysis which have been investigated by a series of methods, such as UV-Vis absorbance, calorimetry, and electron paramagnetic resonance [15,48–56]. Information derived from the crystal structures, along with these kinetics and spectroscopic studies, uncover some details in the catalytic mechanism that are shared by all DHODHs and also details unique for each class.

In the reductive half-reaction (oxidation of DHO), the catalytic base, located in a loop, is brought to an adequate position closer to DHO from where it is possible to abstract the proton from DHO carbon 5 (C5) with the next step being the transfer, as a hydride, of the hydrogen from DHO carbon 6 (C6) to the nitrogen 5 (N5) of the FMN isoalloxazine. The two hydrogen transfers may occur in a concerted reaction, as observed in kinetic studies of Class 2 DHODH from bovine liver [57] and for Class 1B from *C. oroticum* [10], or as sequential steps as reported for Class 1A DHODH from *Crithidia fasciculate* [58].

The active site bases that deprotonate DHO are different for each class: Class 1A and B DHODHs use a cysteine [21,22,24], whereas in Class 1S and Class 2 DHODHs a serine residue is used instead [1,25]. Interestingly, site-direct mutagenesis studies by using *L. lactis* DHODH as a model suggest that the two basis are not interchangeable [8].

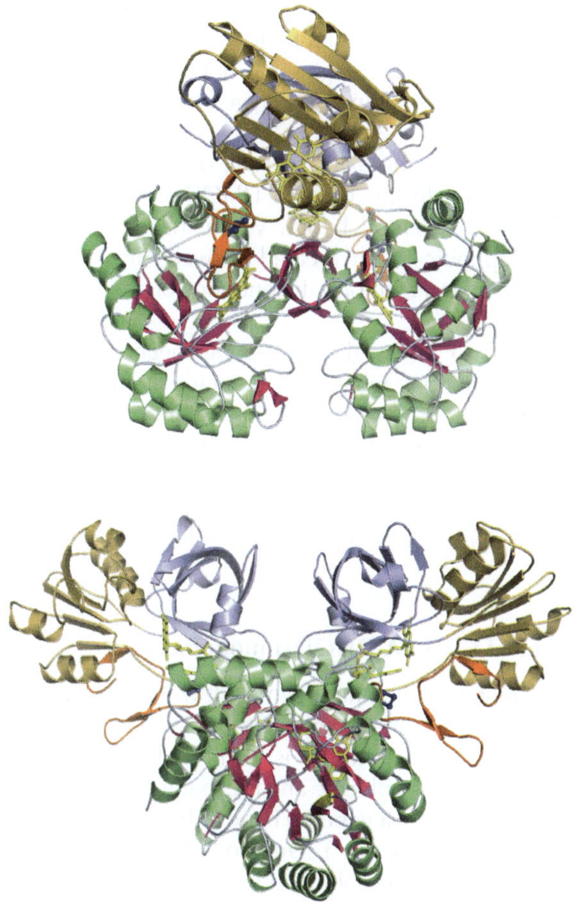

Fig. 14.4: Cartoon representation of Class 1B DHODH in two orthogonal views (PDB ID 1EP1 was used as a model). Class 1B DHODH folds as a heterotetramer with two closely dimers of PyrD-PyrK dimers. PyrD homodimeric protein (green and pink) resembles Class 1A DHODH fold where the FMN (yellow) molecule is located. The PyrK protein consists of three distinct domains, the flavin binding domain (gold) where the FAD molecule is located (yellow), the iron-sulfur cluster binding domain (orange) where it is found a 2Fe-2S cluster (dark blue). In light blue is suggested to be the NAD-binding domain where the enzyme is oxidized by NAD^+.

The oxidative half-reactions performed by DHODHs from different classes are dramatically different. The differences in 3D arrangements assumed by members of Classes 1A, 1B, and 2 are suggested to be intrinsically correlated to this specific step of the enzymatic reaction. In the dimeric structure presented by Class 1A DHODHs, each monomer catalyzes the full reaction in an apparent independent way (see below), with FMN, dihydroorotate and fumarate reacting at the same active site. Class 1B DHODHs contain three different cofactors (FMN, FAD, and an iron-sulfur cluster) distributed in the two heterodimers (PyrD-PyrK subunits) that form the tetrameric structure of members of this subclass. Here a much more sophisticated pathway is used to transfer electrons from the oxidation of DHO to the final electron acceptor. The electrons follow a transfer pathway that starts in the FMN binding site located in the PyrD subunit, passes through the [2Fe-2S] cluster in the PyrK subunit to the FAD molecules and finally reaches the NAD^+ electron acceptor that, although it has not been observed in the only crystal structure reported for a Class 1B enzyme, is believed to bind in the PyrK subunit (►Fig.14.4) [40].

Members of Class 2 DHODH restore the oxidized state of the enzyme by oxidizing $FMNH_2$ when bound to membranes. In this class of DHODHs, there are two redox sites: DHO binds to the top of the central barrel, while the quinone binding site has been proposed as a hydrophobic patch located between the two helices that constitute the N-terminal extension and that form the entrance of a tunnel ending at the FMN site [13].

DHODHs have been reported to follow the ping-pong mechanism [48,57], also called a double-displacement reaction, which is characterized by the change of the enzyme into an intermediate form when the first half-reaction occurs. This non-sequential mechanism of catalysis involves the two substrates: dihydroorotate and the oxidizing agent, with the flavin mononucleotide (FMN) serving as an intermediate in the electron transfer. Hence, Class 2 and Class 1B DHODHs exhibit a two-site ping-pong mechanism [34,35], whereas Class 1A is reported to follow a single-site type of mechanism [21,38,43].

14.4.1 Asymmetric behavior of Class 1A DHODH monomers

One interesting functional characteristic that has been dealt with in a series of papers is related to the behavior of the monomers that compose the overall structure of Class 1A DHODHs during catalysis. Since each monomer is a full DHODH unit in terms of catalytic activity, *i.e.* it contains the FMN molecule and is capable of binding the electron acceptor and performing the full reaction, a genuine issue to be raised is whether or not it would be possible to have each monomer as a standalone DHODH unit.

Class 1A DHODH functional dimers are stabilized by a combination of hydrophilic and hydrophobic interactions between residues from both polypeptide chains. Thus, in order to investigate the relevance of the oligomeric state in Class 1A, a series of Class 1A *L. lactis* mutants were introduced to promote dimer destabilization and monomer dissociation [42]. Mutations in residues E206 and K296, which are involved in salt bridges at the dimer interface, were changed, thus generating mutants with the salt bridges either disrupted or with its polarity inverted (in this case a double-mutant E206K-K296E was produced). These results show that the salt-bridges were not critical for enzyme activity but were likely to play an essential role in the stabilization of the dimer. However, the monomeric forms observed upon dilution of concentrated solutions of both the wild-type enzyme and the double-mutant were shown to be inactive, thus indicating that the presence of the dimer is needed to support catalysis.

This issue was addressed by regarding the conformations of loops relevant for enzymatic function [59]. By applying site directed mutagenesis, kinetic, and structural characterization, it was observed that the *cis*-proline loop (β4-α4-loop), which helps to define the active site, plays a relevant role in catalysis. The modifications of sequential residues were shown to influence kinetics very differently and the crystal structures showed significant differences in terms of mobility of the active site loop as inferred from the variations of B-factors for those regions [59]. The active site loop was found in open and closed conformations, and asymmetric conformations were also seen for the active site loop of the K136E mutant. In this case, one of the active site loops was in a closed conformation, whereas the other was modeled to coexist in both open and closed conformations, in agreement with concentrations of high and low affinity binding sites obtained from the kinetic assays. Those results suggested a model where the binding of substrate to one of the monomers could cause alteration in the conformations and dynamics of the loop regions in the structure, especially the active site loop, making those alterations reach the second active site and leading to apparent negative cooperativity [59]. In this model, although the active sites are apart from each other and also from the dimer interface, an inter-unit communication mediated by the mobility of the loops could play an essential role for Class 1A DHODH function.

The asymmetric behavior of Class 1A DHODH monomers was also demonstrated by single-molecule fluorescence [51]. By monitoring the kinetics of individual enzymes it was observed that Class 1A DHODH dimers can exist in three redox states: fully oxidized (maximum fluorescence with both monomers oxidized), half-oxidized (or half-reduced), and fully reduced (both monomers reduced and thus non-fluorescent). These results were rationalized within a half-sites reactivity scheme in which only one of the monomers is active at a time. Starting with the enzyme in the fully oxidized state ($E_{ox}{*}E_{ox}$), one of the monomers (marked with an asterisk) reacts with DHO and is reduced to ($E_{red}{*}E_{ox}$), which is in turn either oxidized in the active subunit or the enzyme isomerizes to become ($E_{red}E_{ox}{*}$) that can be reduced in the second monomer ($E_{red}E_{red}{*}$). The opposite way coming from the fully reduced state to the fully oxidized would demand the same sort of reactions.

The data discussed above shed light on the details of Class 1A DHODH mechanism of function. The different conformations observed for the loops that define and control access to the active site in one of the Class 1A DHODH mutants [59] give support to the half-sites reactivity scheme [51] and point to orchestrated dynamics of the loops to control catalysis. Furthermore, the coexistence of different conformations of loops and other structural elements in proteins has attracted a lot of attention in the last few years due to the increasing importance of the so-called conformational selection mechanism for enzyme kinetics [60–62]. Unlike the more traditional induced-fit mechanism, in that mechanism the protein structure (or at least the part relevant for activity) exists as a set of conformations in equilibrium including the one that has the right atomic disposition for ligand binding. Upon binding, the conformation is selected and the chemical equilibrium is shifted towards that specific conformation.

Whether or not Class 1B DHODHs also present the same sort of monomer behavior as that observed for Class 1A is an issue that still needs investigation. Nevertheless, understanding the dynamics of loops and/or other structural elements in DHODHs seems of paramount relevance not only to have a full picture of enzyme kinetics but also to open up new possibilities when it comes to the development of new drug design approaches.

14.4.2 Class 2 DHODHs and the interaction with membranes

The N-terminal region of Class 2 DHODHs is the main structural difference between Class 1 and 2 enzymes. Instead of forming dimers or tetramers as observed for Class 1 DHODHs, Class 2 members make use of a longer N-terminal to create the appropriate environment for catalysis. In eukaryotic enzymes, such as in human and rat DHODHs, the N-terminal is composed by sequences that are believed to contain a signal peptide to direct the enzyme to the proper compartment in the cell, a transmembrane domain for membrane insertion, and a shorter domain, which is the only region also present in the bacterial N-terminal, that hosts the electron acceptor binding site. Rawls *et al.* [63] reported a series of studies on the roles of the N-terminal sequence as requisites for cell localization. In several other studies, truncated forms, lacking the signal peptide and transmembrane sequences, of DHODHs were used and showed that the truncated form of Class 2 DHODH maintained catalytic activity, which indicates that not the whole N-terminal sequence is needed for catalysis or membrane association but rather for the correct cell localization [20,52].

Structures of human and rat DHODHs in a truncated form [34], in which the N-terminal contained neither the signal peptide nor the transmembrane sequences, and of the *E. coli* DHODH [41] helped understanding differences between members of Class 2 DHODHs. The crystal structures of truncated DHODHs bound to inhibitors with chemical structures similar to the natural electron acceptors showed that the inhibitor-binding site was always at the N-terminal extension, which suggests that the quinones would occupy that site during catalysis. Furthermore, the available structural data along with the sequence information for the N-terminus show that this region presents very few conserved residues and significant variations in the length and orientations of the helices. For example, in *E. coli* DHODH N-terminal there is an extra kink that forms a 3_{10} helix that is not present in *H. sapiens* DHODH [13]. All together these discrepancies could explain why inhibitors that bind to the human enzyme do not inhibit *E. coli* DHODH or rat DHODH.

In terms of the molecular mechanism underlying the interaction of Class 2 DHODHs with membranes, the amphipathicity and distribution of hydrophobic and charged residues in the N-terminal extension is compatible with its association with membrane (*H. sapiens* DHODH and *E. coli* DHODH). The investigation of the detailed interaction between *E. coli* DHODH and model membranes using a combined approach of spin-labeled phospholipids inserted in the membranes, electron spin resonance experiments and detailed spectral simulations was conducted by Couto *et al.* [15]. The most significant outcome found was that, upon binding of *E. coli* DHODH to mixed model membranes, a structural defect was observed in the hydrophobic part of the model membrane. This was likely the result of the insertion of the N-terminal extension into the core of the model membrane, creating such defect that could make it possible for the quinones to go in and out of their binding site located in the helices that compose the extension. In a subsequent paper, Couto *et al.* [49] selectively placed spin-probes on the side-chains of cysteines introduced in the N-terminal extension via site directed mutagenesis. In all cases, a coexistence of two conformations (open and closed) for the N-terminal helices were observed, thus indicating that the N-terminus inserts into the core of the vesicles and undergoes dynamical changes from a closed to an open state that would allow the electron acceptors (or inhibitors) to access their binding site. Once more, evidence suggests that the dynamics of structural elements of the protein play a fundamental role in catalysis.

14.5 Therapeutic potential

Nucleotides are vital molecules for cell survival and proliferation. Synthesized from small metabolites by the *de novo* pathway and, to a greater or lesser extent, recycled by the salvage pathway, the purine and pyrimidine nucleotides are used to synthesize DNA and RNA molecules and are precursors of activated forms of both carbohydrates and lipids [64]. The metabolic requirements for the nucleotides make both *de novo* biosynthetic and salvage pathways potential targets for drug development. In particular, DHODH has been shown to be the rate limiting enzyme in the *de novo* pyrimidine synthesis.

In addition to the relevance of DHODH in *de novo* pyrimidine biosynthesis, DHODH reactions are coupled to energy metabolism. The reduction of ubiquinone by Class 2 DHODHs links the pyrimidine biosynthetic pathway to the mitochondrial respiratory chain. Also, the coupling of Class 1A DHODH to fumarate reductase activity, important for maintaining cellular redox balance, provides insights about additional mechanisms of regulation to be exploited by selective DHODH inhibition.

The therapeutic potential of depleting the intracellular pyrimidine pool by inhibiting DHODH, the rate limiting step of the *de novo* pyrimidine biosynthesis, has been deeply exploited in the past decade. Selective Class 2 DHODH inhibition has provided the basis for drugs developed to treat cancer, transplant rejection, rheumatoid arthritis, psoriasis and autoimmune diseases. For example, Leflunomide (Arava) was approved by the FDA as a compound for the treatment of mild to moderate rheumatoid arthritis, but broader applications as immunosuppressive [65], antipsoriatic and antiviral properties have already been demonstrated. Leflunomide, a DHODH inhibitor, is a synthetic isoxazole-derivative which behaves as a pro-drug, *i.e.*, upon absorption, it is rapidly and non-enzymatically converted into its active open ring malononitrile metabolite A771726 (teriflunomide). Brequinar, also a DHODH inhibitor, was discovered in the mid-1980s at DuPont as an antimetabolite in cancer and immunosuppression therapies [66]. More recently, Brequinar was also reported to be effective in the treatment of viral-mediated diseases [67].

P. falciparum, *Plasmodium berghei* and *Plasmodium vivax* species, the parasites responsible for human malaria, have also been described as susceptible to class 2 DHODH inhibition [68,69]. The lack of salvage pathway in *Plasmodium* species suggest that the absolute requirement for the *de novo* pyrimidine biosynthesis. In fact, Malarone, used for the treatment and prevention of malaria, is a combined preparation with proguanil hydrochloride and atovaquone (a hydroxy-1,4-naphthoquinone), the latter an analog of ubiquinone, which selectively inhibits Class 2 DHODHs.

DHODH inhibitors have also been suggested as antibiotics against *Escherichia coli* [70] and *Helicobacter pylori* [71,72], antifungal [73], and antiviral agents [74]. Very recently, it was demonstrated that the antiviral compound NITD-982 acts as a human DHODH inhibitor, suggesting the inhibition of Dengue virus can be achieved through suppression of host pyrimidine biosynthesis [74].

The relevance of DHODH was also investigated for the different species among trypanosomatids. RNAi knockdown of DHODH expression in the bloodstream form of *T. brucei*, the causative agent of African Trypanosomiasis (Sleeping sickness) showed that interference of DHODH expression inhibits growth of the parasite when access to pyrimidines in the environment is blocked [6]. In *T. cruzi*, responsible for Chagas disease, the results are even more striking. Knock out technology has been used to show that the gene interruption of the three gene loci encoding cytosolic DHODH in *T. cruzi* led to cell non-viability and addition of pyrimidine nucleosides to the medium did not rescue the DHODH knockdown cells [7].

At present, there are huge efforts in the development of new generations of DHODH inhibitors where a combination of high-throughput screening of chemical libraries and computational approaches are being extensively exploited in the development of highly potent and selective inhibitors [75]. We strongly believe that with the advances in structural biology and medicinal chemistry one can expected major breakthroughs in the scientific field regarding not only a better comprehension of the mechanism of action of the different DHODH classes, their respective roles in the cell metabolism, but also in the use of their therapeutic potential.

14.6 References

[1] Bjornberg O, Rowland P, Larsen S, Jensen KF. Active site of dihydroorotate dehydrogenase A from Lactococcus lactis investigated by chemical modification and mutagenesis. Biochemistry 1997;36:16197–205.

[2] Marcinkeviciene J, Jiang WJ, Locke G, Kopcho LM, Rogers MJ, Copeland RA. A second dihydroorotate dehydrogenase (type A) of the human pathogen Enterococcus faecalis: Expression, purification, and steady-state kinetic mechanism. Archives of Biochemistry and Biophysics 2000;377:178–86.

[3] Zameitat E, Gojkovic Z, Knecht W, Piskur J, Loeffler M. Biochemical characterization of recombinant dihydroorotate dehydrogenase from the opportunistic pathogenic yeast Candida albicans. Febs Journal 2006;273:3183–91.

[4] Zameitat E, Pierik AJ, Zocher K, Loeffler M. Dihydroorotate dehydrogenase from Saccharomyces cerevisiae: spectroscopic investigations with the recombinant enzyme throw light on catalytic properties and metabolism of fumarate analogues. Fems Yeast Research 2007;7:897–904.

[5] Feliciano PR, Cordeiro AT, Costa AJ, Nonato MC, Ol. Cloning, expression, purification, and characterization of Leishmania major dihydroorotate dehydrogenase. Protein Expression and Purification 2006;48:98–103.

[6] Arakaki TL, Buckner FS, Gillespie JR, Malmquist NA, Phillips MA, Kalyuzhniy O, Luft JR, DeTitta GT, Verlinde CLMJ, Van Voorhis WC, Hol WGJ, Merritt EA. Characterization of Trypanosoma brucei dihydroorotate dehydrogenase as a possible drug target; structural, kinetic and RNAi studies. Molecular Microbiology 2008;68:37–50.

[7] Annoura T, Nara T, Makiuchi T, Hashimoto T, Aoki T. The origin of dihydroorotate dehydrogenase genes of kinetoplastids, with special reference to their biological significance and adaptation to anaerobic, parasitic conditions. Journal of Molecular Evolution 2005;60: 113–27.

[8] Fagan RL, Jensen KF, Bjornberg O, Palfey BA. Mechanism of flavin reduction in the class 1A dihydroorotate dehydrogenase from Lactococcus lactis. Biochemistry 2007;46:4028–36.

[9] Kahler AE, Nielsen FS, Switzer RL. Biochemical characterization of the heteromeric Bacillus subtilis dihydroorotate dehydrogenase and its isolated subunits. Archives of Biochemistry and Biophysics 1999;371:191–201.

[10] Argyrou A, Washabaugh MW, Pickart CM. Dihydroorotate dehydrogenase from Clostridium oroticum is a class 1B enzyme and utilizes a concerted mechansim of catalysis. Biochemistry 2000;39:10373–84.

[11] Nielsen FS, Andersen PS, Jensen KF. The B form of dihydroorotate dehydrogenase from Lactococcus lactis consists of two different subunits, encoded by the pyrDb and pyrK genes, and contains FMN, FAD, and FeS redox centers. Journal of Biological Chemistry 1996;271:29359–65.
[12] Marcinkeviciene J, Tinney LM, Wang KH, Rogers MJ, Copeland RA. Dihydroorotate dehydrogenase B of Enterococcus faecalis. Characterization and insights into chemical mechanism. Biochemistry 1999;38:13129–37.
[13] Liu SP, Neidhardt EA, Grossman TH, Ocain T, Clardy J. Structures of human dihydroorotate dehydrogenase in complex with antiproliferative agents. Structure with Folding & Design 2000;8:25–33.
[14] Loffler M, Knecht W, Rawls J, Ullrich A, Dietz C. Drosophila melanogaster dihydroorotate dehydrogenase: the N-terminus is important for biological function in vivo but not for catalytic properties in vitro. Insect Biochemistry and Molecular Biology 2002;32:1159–69.
[15] Couto SG, Nonato MC, Costa-Filho AJ. Defects in vesicle core induced by Escherichia coli dihydroorotate dehydrogenase. Biophysical Journal 2008;94:1746–53.
[16] Couto SG, Nonato MC, Costa-Filho AJ. Site directed spin labeling studies of Escherichia coli dihydroorotate dehydrogenase N-terminal extension. Biochemical and Biophysical Research Communications 2011;414:487–92.
[17] Fagan RL, Palfey BA. Roles in Binding and Chemistry for Conserved Active Site Residues in the Class 2 Dihydroorotate Dehydrogenase from Escherichia coli. Biochemistry 2009;48:7169–78.
[18] Nagy M, Lacroute F, Thomas D. Divergent evolution of pyrimidine biosynthesis between anaerobic and aerobic yeasts. Proceedings of the National Academy of Sciences of the United States of America 1992;89:8966–70.
[19] Sorensen PG, Dandanell G. A new type of dihydroorotate dehydrogenase, type 1S, from the thermoacidophilic archaeon Sulfolobus solfataricus. Extremophiles 2002;6:245–51.
[20] Copeland RA, Davis JP, Dowling RL, Lombardo D, Murphy KB, Patterson TA. Recombinant human dihydroorotate dehydrogenase - expression, purification, and characterization of a catalytically functional truncated enzyme. Archives of Biochemistry and Biophysics 1995;323:79–86.
[21] Feliciano PR, Cordeiro AT, Costa AJ, Nonato MC. Cloning, expression, purification, and characterization of Leishmania major dihydroorotate dehydrogenase. Protein Expression and Purification 2006;48:98–103.
[22] Pinheiro MP, Iulek J, Nonato MC. Crystal structure of Trypanosoma cruzi dihydroorotate dehydrogenase from Y strain. Biochemical and Biophysical Research Communications 2008;369:812–17.
[23] Hurt DE, Sutton AE, Clardy J. Brequinar derivatives and species-specific drug design for dihydroorotate dehydrogenase. Bioorganic & Medicinal Chemistry Letters 2006;16:1610–15.
[24] Nielsen FS, Rowland P, Larsen S, Jensen KF. Purification and characterization of dihydroorotate dehydrogenase A from Lactococcus lactis, crystallization and preliminary X-ray diffraction studies of the enzyme. Protein Science 1996;5:852–56.
[25] Bjornberg O, Gruner AC, Roepstorff P, Jensen KF. The activity of Escherichia coli dihydroorotate dehydrogenase is dependent on a conserved loop identified by sequence homology, mutagenesis, and limited proteolysis. Biochemistry 1999;38:2899–2908.
[26] Cleland WW. Kinetics of enzyme-catalyzed reactions with 2 or more substrates or products. 1. nomenclature and rate equations. Biochimica et Biophysica Acta 1963;67:104.
[27] Cleland WW. Kinetics of enzyme-catalyzed reactions with 2 or more substrates or products. 2. Inhibition - nomenclature and theory. Biochimica et Biophysica Acta 1963;67:173.
[28] Cleland WW. Kinetics of enzyme-catalyzed reactions with 2 or more substrates or products. 3. Prediction of initial velocity and inhibition patterns by inspection. Biochimica et Biophysica Acta 1963;67:188.
[29] Inaoka DK, Takashima E, Osanai A, Shimizu H, Nara T, Aoki T, Harada S, Kita K, Au. Expression, purification and crystallization of Trypanosoma cruzi dihydroorotate

dehydrogenase complexed with orotate. Acta Crystallographica Section F-Structural Biology and Crystallization Communications 2005;61:875–78.

[30] Liu Y, Gao Z-Q, Liu C-P, Xu J-H, Li L-F, Ji C-N, Su X-D, Dong Y-H. Structure of the putative dihydroorotate dehydrogenase from Streptococcus mutans. Acta Crystallographica Section F-Structural Biology and Crystallization Communications 2011;67:182–87.

[31] Cordeiro AT, Feliciano PR, Pinheiro MP, Nonato MC. Crystal structure of dihydroorotate dehydrogenase from Leishmania major. Biochimie.

[32] Cordeiro AT, Feliciano PR, Nonato MC, Wo. Crystallization and preliminary X-ray diffraction analysis of Leishmania major dihydroorotate dehydrogenase. Acta Crystallographica Section F-Structural Biology and Crystallization Communications 2006;62:1049–51.

[33] Rowland P, Nielsen FS, Jensen KF, Larsen S. Crystallization and preliminary X-ray diffraction analysis of the heterotetrameric dihydroorotate dehydrogenase B of Lactococcus lactis, a flavoprotein enzyme system consisting of two PyrDB subunits and two iron-sulfur cluster containing PyrK subunits. Acta Crystallographica Section D-Biological Crystallography 1997;53:802–04.

[34] Hansen M, Le Nours J, Johansson E, Antal T, Ullrich A, Loffler M, Larsen S. Inhibitor binding in a class 2 dihydroorotate dehydrogenase causes variations in the membrane-associated N-terminal domain. Protein Science 2004;13:1031–42.

[35] Hurt DE, Widom J, Clardy J. Structure of Plasmodium falciparum dihydroorotate dehydrogenase with a bound inhibitor. Acta Crystallographica Section D-Biological Crystallography 2006;62:312–23.

[36] Rowland P, Norager S, Jensen KF, Larsen S. Crystallization and preliminary X-ray studies of membrane-associated Escherichia coli dihydroorotate dehydrogenase. Acta Crystallographica Section D-Biological Crystallography 2000;56:659–61.

[37] Rowland P, Nielsen FS, Jensen KF, Larsen S. The crystal structure of the flavin containing enzyme dihydroorotate dehydrogenase A from Lactococcus lactis. Structure 1997;5:239–52.

[38] Inaoka DK, Sakamoto K, Shimizu H, Shiba T, Kurisu G, Nara T, Aoki T, Kita K, Harada S. Structures of Trypanosoma cruzi dihydroorotate dehydrogenase complexed with substrates and products: Atomic resolution insights into mechanisms of dihydroorotate oxidation and fumarate reduction. Biochemistry 2008;47:10881–91.

[39] Arakaki TL, Buckner FS, Gillespie JR, Malmquist NA, Phillips MA, Kalyuzhniy O, Luft JR, DeTitta GT, Verlinde C, Van Voorhis WC, Hol WGJ, Merritt EA. Characterization of Trypanosoma brucei dihydroorotate dehydrogenase as a possible drug target; structural, kinetic and RNAi studies. Molecular Microbiology 2008;68:37–50.

[40] Rowland P, Norager S, Jensen KF, Larsen S. Structure of dihydroorotate dehydrogenase B: Electron transfer between two flavin groups bridged by an iron-sulphur cluster. Structure 2000;8:1227–38.

[41] Norager S, Jensen KF, Bjornberg O, Larsen S. E-coli dihydroorotate dehydrogenase reveals structural and functional distinctions between different classes of dihydroorotate dehydrogenases. Structure 2002;10:1211–23.

[42] Ottosen MB, Bjornberg O, Norager S, Larsen S, Palfey BA, Jensen KF. The dimeric dihydroorotate dehydrogenase A from Lactococcus lactis dissociates reversibly into inactive monomers. Protein Science 2002;11:2575–83.

[43] Rowland P, Bjornberg O, Nielsen FS, Jensen KF, Larsen S. The crystal structure of Lactococcus lactis dihydroorotate dehydrogenase A complexed with the enzyme reaction product throws light on its enzymatic function. Protein Science 1998;7:1269–79.

[44] Wolfe AE, Thymark M, Gattis SG, Fagan RL, Hu Y-c, Johansson E, Arent S, Larsen S, Palfey BA. Interaction of benzoate pyrimidine analogues with class 1A dihydroorotate dehydrogenase from Lactococcus lactis. Biochemistry 2007;46:5741–53.

[45] Cheleski J, Rocha JR, Pinheiro MP, Wiggers HJ, da Silva ABF, Nonato MC, Montanari CA. Novel insights for dihydroorotate dehydrogenase class 1A inhibitors discovery. European Journal of Medicinal Chemistry 2010;45:5899–5909.

[46] Palfey BA, Bjornberg O, Jensen F. Specific inhibition of a family 1A dihydroorotate dehydrogenase by benzoate pyrimidine analogues. Journal of Medicinal Chemistry 2001;44:2861–64.
[47] Davis JP, Cain GA, Pitts WJ, Magolda RL, Copeland RA. The immunosuppressive metabolite of leflunomide is a potent inhibitor of human dihydroorotate dehydrogenase. Biochemistry 1996;35:1270–73.
[48] Bjornberg O, Jordan DB, Palfey BA, Jensen KF. Dihydrooxonate is a substrate of dihydroorotate dehydrogenase (DHOD) providing evidence for involvement of cysteine and serine residues in base catalysis. Archives of Biochemistry and Biophysics 2001;391:286–94.
[49] Couto SG, Cristina Nonato M, Costa-Filho AJ. Site directed spin labeling studies of Escherichia coli dihydroorotate dehydrogenase N-terminal extension. Biochemical and Biophysical Research Communications 2011;414:487–92.
[50] Palfey BA, Bjornberg O, Jensen KF. Insight into the chemistry of flavin reduction and oxidation in Escherichia coli dihydroorotate dehydrogenase obtained by rapid reaction studies. Biochemistry 2001;40:4381–90.
[51] Shi J, Dertouzos J, Gafni A, Steel D, Palfey BA. Single-molecule kinetics reveals signatures of half-sites reactivity in dihydroorotate dehydrogenase A catalysis. Proceedings of the National Academy of Sciences of the United States of America 2006;103:5775–80.
[52] Malmquist NA, Baldwin J, Phillips MA. Detergent-dependent kinetics of truncated Plasmodium falciparum dihydroorotate dehydrogenase. Journal of Biological Chemistry 2007;282:12678–86.
[53] Baldwin J, Farajallah AM, Malmquist NA, Rathod PK, Phillips MA. Malarial dihydroorotate dehydrogenase. Journal of Biological Chemistry 2002;277:41827–34.
[54] Aleman V, Handler P, Palmer G, Beinert H. Studies on dihydroorotate dehydrogenase by electron paramagnetic resonance spectroscopy. 2. Electron paramagnetic resonance and optical spectra and titrations. Journal of Biological Chemistry 1968;243:2560.
[55] Jordan DB, Bisaha JJ, Picollelli MA. Catalytic properties of dihydroorotate dehydrogenase from Saccharomyces cerevisiae: Studies on pH, alternate substrates, and inhibitors. Archives of Biochemistry and Biophysics 2000;378:84–92.
[56] Cheleski J, Wiggers HJ, Citadini AP, da Costa Filho AJ, Nonato MC, Montanari CA. Kinetic mechanism and catalysis of Trypanosoma cruzi dihydroorotate dehydrogenase enzyme evaluated by isothermal titration calorimetry. Analytical Biochemistry 2010;399:13–22.
[57] Hines V, Johnston M. Mechanistic studies on the bovine liver mitochondrial dihydroorotate dehydrogenase using kinetic deuterium-isotope effects. Biochemistry 1989;28:1227–34.
[58] Pascal RA, Walsh CT. Mechanistic studies with deuterated dihydroorotates on the dihydroorotate oxidase from crithidia-fasciculata. Biochemistry 1984;23:2745–52.
[59] Norager S, Arent S, Bjornberg O, Ottosen M, Lo Leggio L, Jensen KF, Larsen S. Lactococcus lactis dihydroorotate dehydrogenase A mutants reveal important facets of the enzymatic function. Journal of Biological Chemistry 2003;278:28812–22.
[60] Devenish SRA, Gerrard JA. The role of quaternary structure in (beta/alpha)(8)-barrel proteins: evolutionary happenstance or a higher level of structure-function relationships? Organic & Biomolecular Chemistry 2009;7:833–39.
[61] Henzler-Wildman KA, Lei M, Thai V, Kerns SJ, Karplus M, Kern D. A hierarchy of timescales in protein dynamics is linked to enzyme catalysis. Nature 2007;450:913–U27.
[62] Henzler-Wildman K, Kern D. Dynamic personalities of proteins. Nature 2007;450:964–72.
[63] Rawls J, Knecht W, Diekert K, Lill R, Loffler M. Requirements for the mitochondrial import and localization of dihydroorotate dehydrogenase. European Journal of Biochemistry 2000;267:2079–87.
[64] Huang M, Graves LM. De novo synthesis of pyrimidine nucleotides; emerging interfaces with signal transduction pathways. Cellular and Molecular Life Sciences 2003;60:321–36.
[65] Claussen MC, Korn T. Immune mechanisms of new therapeutic strategies in MS - Teriflunomide. Clinical Immunology 2012;142:49–56.

[66] Batt DG. Inhibitors of dihydroorotate dehydrogenase. Expert Opinion on Therapeutic Patents 1999;9:41–54.
[67] Qing M, Zou G, Wang Q-Y, Xu HY, Dong H, Yuan Z, Shi P-Y. Characterization of Dengue Virus Resistance to Brequinar in Cell Culture. Antimicrobial Agents and Chemotherapy 2010;54:3686–95.
[68] Booker ML, Bastos CM, Kramer ML, Barker RH, Jr., Skerlj R, Sidhu AB, Deng X, Celatka C, Cortese JF, Bravo JEG, Llado KNC, Serrano AE, Angulo-Barturen I, Belen Jimenez-Diaz M, Viera S, Garuti H, Wittlin S, Papastogiannidis P, Lin J-W, Janse CJ, Khan SM, Duraisingh M, Coleman B, Goldsmith EJ, Phillips MA, Munoz B, Wirth DF, Klinger JD, Wiegand R, Sybertz E. Novel Inhibitors of Plasmodium falciparum Dihydroorotate Dehydrogenase with Anti-malarial Activity in the Mouse Model. Journal of Biological Chemistry 2010;285:33054–64.
[69] Patel V, Booker M, Kramer M, Ross L, Celatka CA, Kennedy LM, Dvorin JD, Duraisingh MT, Sliz P, Wirth DF, Clardy J. Identification and Characterization of Small Molecule Inhibitors of Plasmodium falciparum Dihydroorotate Dehydrogenase. Journal of Biological Chemistry 2008;283:35078–85.
[70] Marcinkeviciene J, Rogers MJ, Kopcho L, Jiang WJ, Wang K, Murphy DJ, Lippy J, Link S, Chung TDY, Hobbs F, Haque T, Trainor GL, Slee A, Stern AM, Copeland RA. Selective inhibition of bacterial dihydroorotate dehydrogenases by thiadiazolidinediones. Biochemical Pharmacology 2000;60:339–42.
[71] Haque TS, Tadesse S, Marcinkeviciene J, Rogers MJ, Sizemore C, Kopcho LM, Amsler K, Ecret LD, Zhan DL, Hobbs F, Slee A, Trainor GL, Stern AM, Copeland RA, Combs AP. Parallel synthesis of potent, pyrazole-based inhibitors of Helicobacter pylori dihydroorotate dehydrogenase. Journal of Medicinal Chemistry 2002;45:4669–78.
[72] Copeland RA, Marcinkeviciene J, Haque TS, Kopcho LM, Jiang WJ, Wang K, Ecret LD, Sizemore C, Amsler KA, Foster L, Tadesse S, Combs AP, Stern AM, Trainor GL, Slee A, Rogers MJ, Hobbs F. Helicobacter pylori-selective antibacterials based on inhibition of pyrimidine biosynthesis. Journal of Biological Chemistry 2000;275:33373–78.
[73] Gustafson G, Davis G, Waldron C, Smith A, Henry M. Identification of a new antifungal target site through a dual biochemical and molecular-genetics approach. Current Genetics 1996;30:159–165.
[74] Wang Q-Y, Bushell S, Qing M, Xu HY, Bonavia A, Nunes S, Zhou J, Poh MK, de Sessions PF, Niyomrattanakit P, Dong H, Hoffmaster K, Goh A, Nilar S, Schul W, Jones S, Kramer L, Compton T, Shi P-Y. Inhibition of Dengue Virus through Suppression of Host Pyrimidine Biosynthesis. Journal of Virology 2011;85:6548–56.
[75] Vyas VK, Ghate M. Recent Developments in the Medicinal Chemistry and Therapeutic Potential of Dihydroorotate Dehydrogenase (DHODH) Inhibitors. Mini-Reviews in Medicinal Chemistry 2011;11:1039–55.
[76] Takashima E, Inaoka DK, Osanai A, Nara T, Odaka M, Aoki T, Inaka K, Harada S, Kita K. Characterization of the dihydroorotate dehydrogenase as a soluble fumarate reductase in Trypanosoma cruzi. Molecular and Biochemical Parasitology 2002;122:189–200.

15 Ferredoxin-NADP+ reductases

Daniela L. Catalano-Dupuy, Daniela V. Rial and Eduardo A. Ceccarelli

Abstract

Ferredoxin-NADP+ reductases (FNRs) comprise a widespread family of hydrophilic proteins that contain firmly non-covalently bound FAD as prosthetic group. These flavoenzymes catalyze a broad range of redox reactions and participate in key metabolic pathways. In chloroplasts, FNRs take part in the last step of the photosynthetic electron transfer process, shuttling electrons from reduced ferredoxin to NADP+ in order to generate the NADPH required for CO_2 assimilation. FNRs also participate in other processes such as isoprenoid biosynthesis, steroid metabolism, nitrogen fixation, xenobiotic detoxification, iron-sulfur cluster biogenesis and oxidative-stress response. This chapter focuses on the extensively-studied subfamily of plant-type FNRs found in plants, some bacteria and apicomplexan parasites. We offer a detailed description of FNR structural features, the differences encountered within the subfamily and a comprehensive outlook of the way in which the natural substrates (NADP+ and ferredoxin) interact with FNR. In this regard, we emphasize the mode in which NADP+ may enter the active site and adopt a productive conformation for catalysis. In this chapter, we compile relevant information for proper expression and purification of FNRs and discuss the spectroscopic techniques most commonly used for the analysis of these flavoproteins. Finally, we highlight significant aspects of plant-type FNRs that position them as appealing targets for biotechnological applications in the future.

15.1 Introduction

Ferredoxin-NADP+ reductases (FNRs, EC 1.18.1.2) are hydrophilic and monomeric enzymes that bind FAD as a prosthetic group in a non-covalent fashion. FNRs participate in electron transfer processes in plastids, mitochondria and bacteria, delivering NADPH or low-potential one-electron carriers (ferredoxin, flavodoxin, adrenodoxin) to redox-based metabolism. In chloroplasts, FNR catalyzes the final step of photosynthetic electron transport [1], a reaction that generates the NADPH required for CO_2 fixation in plants and cyanobacteria. In this reaction, FNR catalyzes the transfer of two electrons from two equivalents of reduced ferredoxin (Fd) to NADP+ (Eq. 15.1) [1]:

$$2\ Fd(Fe^{2+}) + NADP^+ + H^+ \leftrightarrow 2\ Fd\ (Fe^{3+}) + NADPH \qquad \text{(Eq. 15.1)}$$

FNRs also participate in other oxidation-reduction processes, including nitrogen fixation, isoprenoid biosynthesis, steroid metabolism, xenobiotic detoxification, oxidative-stress response and iron-sulfur cluster biogenesis [2–6]. However, in these processes

FNRs mediate the electron exchange in the opposite direction, *i.e.*, the oxidation of NADPH and the reduction of Fd and/or another electron acceptor. For this reason, photosynthetic FNRs are sometimes referred as autotrophic and all the others as heterotrophic FNRs [7,8].

The diaphorase (NAD(P)H oxidase) activity in chloroplasts was first described by Avron and Jagendorf [9]. FNRs show a strong preference for NADPH over NADH, but are more versatile regarding electron acceptors. Several compounds can replace Fd or flavodoxin *in vitro*. Amongst them, potassium ferricyanide, indophenols and tetrazolium salts are widely used as artificial substrates for diaphorase activity. Thus, FNR-catalyzed reactions can be postulated to occur in two steps (Eq. 15.2 and 15.3): first a hydride exchange with the pyridine nucleotide, followed by an electron transfer to the final electron acceptor.

$$\text{NADPH} + \text{H}^+ + \text{FNR}_{\text{ox}} \leftrightarrow \text{NADP}^+ + \text{FNRH}_2 \qquad \text{(Eq. 15.2)}$$

$$\text{FNRH}_2 + n\text{A}_{\text{ox}} \rightarrow \text{FNR}_{\text{ox}} + n\text{A}_{\text{red}};\ \text{A, electron acceptor; } n = 1, 2 \qquad \text{(Eq. 15.3)}$$

►Fig.15.1A shows the structure of FAD. The isoalloxazine is the chemically active moiety of the molecule [10]. FAD can exist in three oxidation states (an oxidized form, a semiquinone radical reduced by one electron and a hydroquinone reduced by two electrons) (►Fig. 15.1B), and it is this feature that accounts for the ability of FNRs to exchange electrons between obligatory one- and two-electron carriers [11].

The protein environment modulates the reduction potentials of the isoalloxazine in such a manner that allows the enzyme to adapt to specific metabolic conditions.

Two totally unrelated families of FNRs exist: the plant-type FNRs and the glutathione reductase-type FNRs. This chapter provides a comprehensive description on plant-type ferredoxin-NADP(H) reductases, their structural and kinetic properties as well as their metabolic relevance. We also offer a vast compilation of successful expression and purification protocols.

15.2 Classification of FNRs

As just noted, FNRs can be grouped into two phylogenetically and structurally unrelated protein families referred as the plant-type and glutathione reductase-type FNRs. The two FNR families can be further subdivided on the basis of their three-dimensional structures and amino acid sequences. Plant-type FNRs comprise the plastidic and bacterial enzymes, whereas the adrenodoxin reductase-like flavoproteins, the oxygenase-coupled NADH-ferredoxin reductase and the recently included thioredoxin reductase-like FNR belong to the glutathione reductase-type FNRs [7,12,13] (►Fig. 15.2). The structures of FNRs of both families possess a two-domain organization with the active site located at the interface between distinct FAD- and NADP-binding domains. Four general folds for FAD binding have been identified in 32 families of FAD-containing proteins [14]. The FAD-binding domain of the glutathione reductase-type and the plant-type FNRs have different general folds. While in plant-type FNRs the FAD-binding domain is formed by the N-terminal portion of the polypeptide chain, in glutathione reductase-type enzymes two discontinuous segments of the polypeptide form the FAD-binding domain. On the

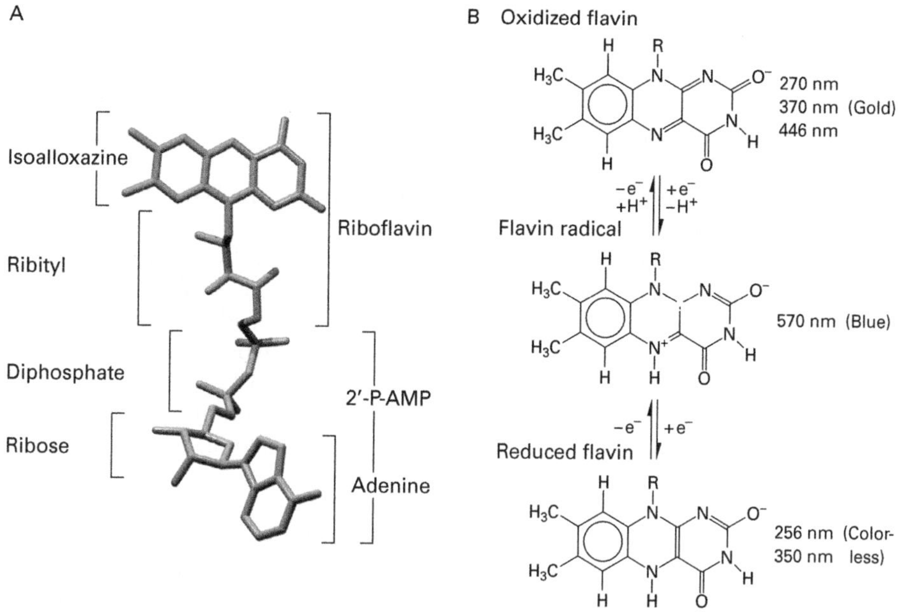

Fig. 15.1: Structure and redox states of FAD. (A) View of FAD and its different moieties. (B) Redox states adopted by the FAD.

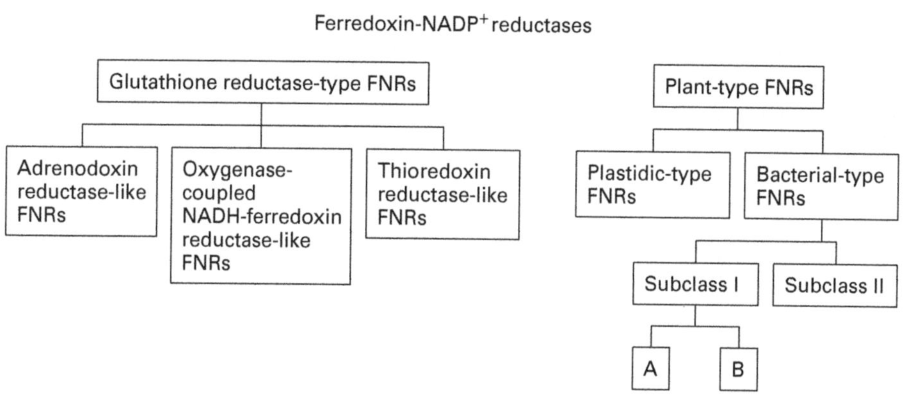

Fig. 15.2: Classification of FNRs.

other hand, the NADP-binding domain of both FNR families has the same topology, although these domains differ in several details in the two families of FNRs, particularly in the precise manner in which NADP(H) is bound.

There are some flavoproteins like the bacterial oxygenase-coupled NADH-ferredoxin reductase that, although NAD-dependent, are structurally and functionally related to the adrenodoxin reductase-like FNRs. Oxygenase-coupled NADH-ferredoxin reductase-like enzymes mainly differ from adrenodoxin reductase-like FNRs in having both FAD- and

NAD-domains sharing the same organization and in possessing a third C-terminal domain that is involved in protein homodimerization. Recently, a homodimeric FNR with significant sequence homology with NADPH-dependent thioredoxin reductase has been identified [12,13]. This FNR is distinct from plant-type FNR and on the basis of amino acid sequence comparisons has been classified as a glutathione reductase-type FNR [12,13].

In the present work we will focus exclusively on the plant-type FNRs. For a detailed description of oxygenase-coupled NADH-ferredoxin reductase or adrenodoxin reductase-like FNRs , the reader is referred to the excellent review by Aliverti *et al.* [7].

By comparing three-dimensional structures and amino acid sequences of several plant-type FNRs, six highly conserved sequence motifs have been identified (▶Fig. 15.3). These motifs have been used to define groups in which plant-type FNRs can be further subdivided, each characterized by specific structural features [6,15].

Members of the plastidic and bacterial classes differ not only in sequence alignment but also in the environment of the active site, FAD conformation and catalytic efficiency [4].

Enzymes of the plastidic class possess a tyrosine residue at the carboxy terminus that is absolutely conserved. In these enzymes, the FAD is bound in an extended conformation, with the adenine moiety interacting with a second tyrosine residue. A common amino acid cluster is always present, corresponding to the protruding sheet-loop-sheet interacting with the 2′-P-AMP region of FAD (▶Fig. 15.4A). A rotated conformation of the adenine moiety of FAD has been observed in the plastidic-type enzyme from *Leptospira interrogans* (▶Fig. 15.4B and E) [16]. In bacterial FNRs the aforementioned cluster is absent. Accordingly, the loop responsible for the extended FAD binding is missing and the prosthetic group adopts a twisted conformation, as first pointed by Ingelman *et al.* [17] (▶Fig. 15.4C, D and E). In all bacterial FNRs, an aromatic residue is found to the C-terminal side of the residue facing the isoalloxazine

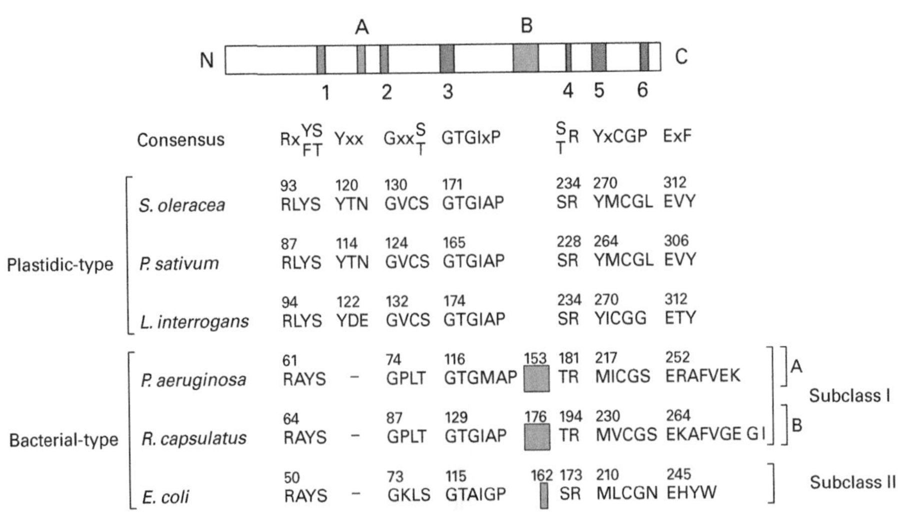

Fig. 15.3: Specific sequence motifs of plant-type FNRs.

ring that interacts with the adenine moiety of FAD. The adenine is in fact involved in aromatic interactions in all FNRs, although the aromatic residue involved in this interaction is provided by the N-terminal domain in the plastidic class enzymes rather than the C-terminal residue seen in the bacterial ones. A strictly conserved tryptophan fulfils this role in subclass II of the bacterial FNRs, and a phenylalanine in subclass I. In the latter reductases, the interacting amino acid facing the isoalloxazine is a conserved alanine, while in the plastidic, cyanobacterial and *Escherichia coli* flavoenzymes it is a tyrosine. Another structural feature of the subclass II bacterial FNRs is the presence of an amino acid gap [6], shown as gray boxes in ▸Fig. 15.3.

Recently, it has been suggested that subclass I bacterial FNRs can be further subdivided into two new groups: subclass IA, which in ▸Fig. 15.4 is represented by the FNR from *Pseudomonas aeruginosa*; and subclass IB, for which the representative member is the FNR from *Rhodobacter capsulatus*. The main structural differences between the two subclasses are located in the carboxy-terminal region. While enzymes from subclass IA have a conserved lysine (Lys258 in the *P. aeruginosa* enzyme), the

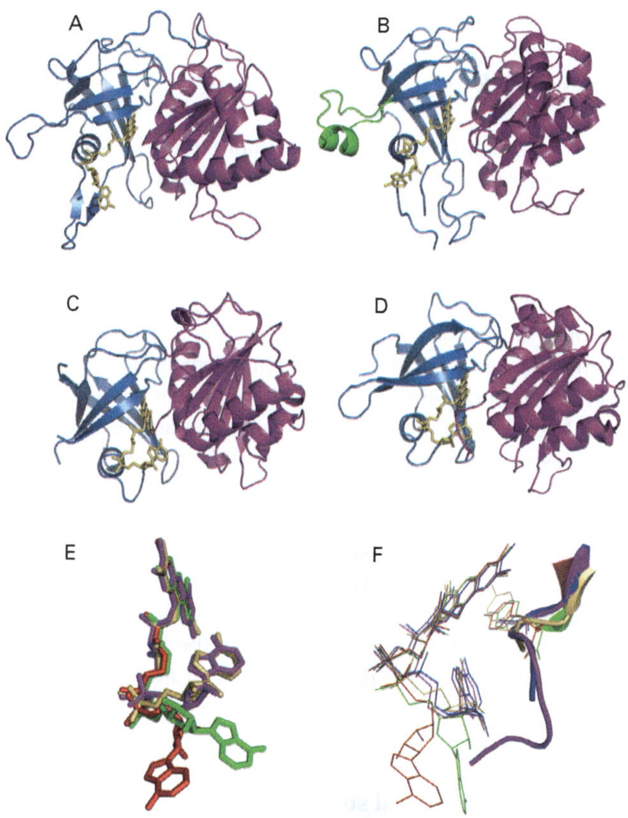

Fig. 15.4: FNR structural models. FNRs from (A) spinach, (B) *L. interrogans,* (C) *E. coli,* (D) *R. capsulatus*. The amino-terminal domain has been colored in blue, carboxy-terminal domain in magenta and the FAD in yellow. Detailed view of (E) FAD conformation and (F) carboxy-terminal region for the FNRs from spinach (green), *L. interrogans* (red), *E. coli* (yellow), *P. aeuruginosa* (blue), and *R. capsulatus* (magenta).

subclass IB FNR proteins have a glutamate or an aspartate at the equivalent position and a longer carboxy-terminal region. Consequently, subclass IA is defined by the carboxy-terminal sequence VEK and the subclass IB by the sequence (V/A)G(E/D) G(I/V) [18] (▶Fig. 15.3).

Finally, bacterial-type FNRs differ from plastidic-type FNRs in having physiological functions related to nitrogen fixation (in *Azotobacter* and *Rhodobacter*) and detoxification of reactive oxygen species by using both Fd and flavodoxin as electron acceptors, and by having lower k_{cat} values than their plastidic-type counterparts [6]. Flavodoxin substitutes for Fd under conditions in which the iron-sulfur cluster of Fd cannot be assembled (*e.g.*, under limiting iron availability) [19], but the use of flavodoxin as a redox partner is not restricted to bacterial-type FNRs as it is also an alternative substrate for the plastidic-type FNRs of cyanobacteria and some algae [6], and in transgenic plants [20]. In plants and cyanobacteria, optimization for FNR activity may be related to the demands of the photosynthetic process that require a very fast electron flow to sustain CO_2 fixation rates. In organisms growing heterotrophically or by anoxygenic photosynthesis, FNR is involved in pathways that proceed at a much lower rates, acting as a shuttle between the abundant NAD(P)H pool and the low potential electron carriers.

15.3 Structural features of FNR

The first crystal structure for a plant-type FNR was obtained for the spinach enzyme in 1991 [21]. Several structures from different sources have subsequently been obtained by X-ray crystallography or NMR, up to a refined resolution of 1.05 Å. This has contributed significantly to our understanding of the catalytic mechanism and function of FNRs.

Three-dimensional models of oxidized and fully reduced forms of different FNRs have been reported [16,21–27], including that from *Anabaena variabilis* [28].

All FNRs are made of two structural domains, each containing approximately 150 amino acids (▶Fig. 15.4). The amino-terminal FAD binding domain, depicted in blue in ▶Fig. 15.4, is structured as a β-barrel. The carboxy-terminal domain (magenta in ▶Fig. 15.4) includes the residues involved in NADP(H) binding and displays a characteristic α-helix/β-strand fold. FAD is found in an extended conformation in plastidic-type FNRs [23,28], due to the interaction of the adenine moiety with a strand-loop-strand β-hairpin motif that is always present in this family of FNRs. The isoalloxazine is shielded from the environment by the phenol group of a tyrosine in plastidic and subclass II bacterial-type FNRs. The equivalent residue is an alanine in subclass I bacterial FNRs. These residues are stacked on the *re*-face of the isoalloxazine and are likely displaced by the nicotinamide ring of NADP+ upon binding [21,24,29,30]. The edge of the dimethylbenzene ring of the isoalloxazine is exposed to solvent and participates in electron transfer to other protein substrates [23] (▶Fig. 15.4).

15.4 Interaction of FNR with its natural substrates

The first attempt to elucidate the binding mode of NADP+ was done by the group of Karplus [23], who determined the crystal structure of spinach FNR with 2′-P-AMP bound. Subsequently, structural information of wild-type and mutant FNR-NADP(H) complexes were obtained for enzymes from *Anabaena* [28,30,31], pea [24], maize [27,32], *R. capsulatus* [33], *L. interrogans* [16], *P. aeruginosa* [34], and *E. coli* [35]. Different $NADP^+$ conformations have been obtained in complex with FNR which may reflect different intermediates during binding and catalysis [28,30,31].

During enzyme catalysis the $NADP^+$ nicotinamide ring must be close to the isoalloxazine moiety so as to allow direct hydride transfer [30,36–39]. As the isoalloxazine is shielded by a tyrosine/alanine in FNRs, these residues and probably the carboxy-terminal region of FNR must reorient to facilitate substrate entrance. $NADP^+$ binds to the enzyme in a bipartite fashion [39]: the nucleotide is initially bound to the enzyme in an unproductive conformation, where the 2′-P-AMP portion first interacts with a preformed binding site, but with the nicotinamide far away from the catalytic site. The nicotinamide moiety then moves into proximity with the isoalloxazine and establishes a productive conformation (▶Fig. 15.5). This conformation has only been observed in crystal structures of pea and *Anabaena* mutant FNRs, where the tyrosine facing the isoalloxazine has been replaced by a serine [24,30]. Moreover, structural rearrangements occurring in the enzyme upon $NADP^+$ binding may be important for enhancing the catalytic efficiency of the reductase [29,31,40,41]. NMR evidence for the movement of the carboxy-terminal region of the enzyme from maize has been obtained [32]. Analysis of hydrogen-deuterium exchange shows that FNR activity is also modulated by pH, and that the $NADP^+$ binding domain has a flexible region that controls the binding affinity of FNR for $NADP^+$ [42]. Molecular dynamics calculations have resulted in a model of bound $NADP^+$ in the catalytic site in a productive conformation [43]. In future, this type of approach may be a powerful tool to elucidate the catalytic mechanism of the reductases.

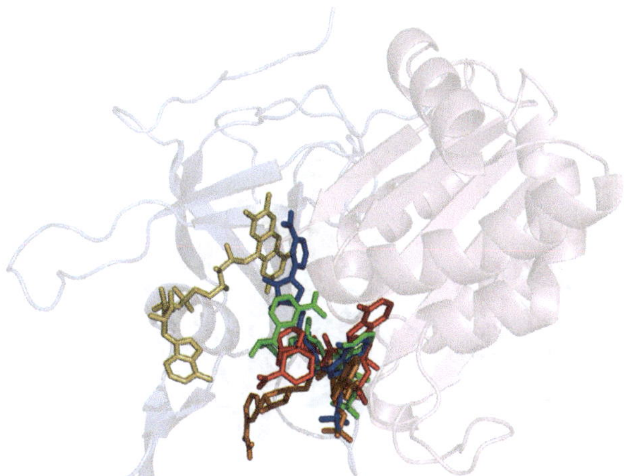

Fig. 15.5: $NADP^+$ conformation in the FNR-$NADP^+$ complex. Superposition of the different conformations observed in crystal structures from wild type *Anabaena* (green and orange) and in *L. interrogans* (red) FNRs, and pea Y308S mutant (blue).

Formation of a complex between FNR and Fd involves specific portions of the two proteins that accommodate their respective FAD and iron-sulfur prosthetic groups at an appropriate distance and orientation for electron transfer (►Fig. 15.6). A good depiction of their interacting molecular surfaces and of the role of specific side-chains in electron transfer has been obtained by studying the complexes of Fd with wild-type and several site-directed mutants of both *Anabaena* and maize FNR [25,44]. The complex interface is formed by a hydrophobic core surrounded by charged residues, where basic side-chains are mostly provided by the FNR and acidic ones by Fd.

The crystal structures of Fd complexes with three different plastidic-type FNR have been determined by X-ray crystallography [25,45,46]. In each case, the iron–sulfur protein binds at a concave surface formed by the two reductase domains. It has been proposed that small induced-fit conformational changes may occur in the complexes that would favor the electron transfer steps of the catalytic cycle [45–47]. In addition, Karplus [15] has proposed that, under physiological conditions, the photosynthetic FNR would always bind either NADP$^+$ or NADPH, while Fd would interact with the reductase in a collisional manner with no discrete protein–protein complex populated during the catalytic cycle [7].

By comparing the crystal structure of the available FNR–Fd complexes it was observed that, although the structures of each protein are highly conserved, the geometry of their interaction is not [45]. Indeed, whereas the interacting surfaces are the same in the complexes, the proteins are oriented somewhat differently in each case. ►Fig. 15.6 shows the models of the complexes from *Anabaena* and maize in which the reductases have been positioned in the same orientation. It is clearly seen that the Fd of each complex is oriented somewhat differently, although in each case the iron-sulfur center is maintained in a proper orientation for electron transfer. Aliverti *et al.* have concluded that the heterogeneity in the protein–protein interaction observed in the different plant-type FNR–Fd complexes probably reflects the absence of strict geometric constrains in the interaction between the partners rather than actual species-specific differences [7].

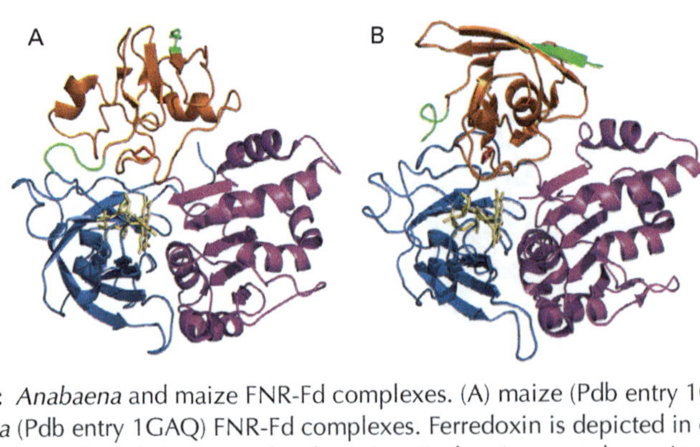

Fig. 15.6: *Anabaena* and maize FNR-Fd complexes. (A) maize (Pdb entry 1GAW) and (B) *Anabaena* (Pdb entry 1GAQ) FNR-Fd complexes. Ferredoxin is depicted in orange and its Fe-S prosthetic group in red. Amino and carboxy-terminal regions are shown in green.

15.5 The metabolic roles of FNR

Plant-type FNRs are present in higher plants and algae, bacteria, cyanobacteria and apicoplasts of intracellular parasites, and participate in a variety of processes, including photosynthesis, steroid hydroxylation, anaerobic pyruvate assimilation, nitrate reduction and fatty acid desaturation. They have been also involved in xenobiotic detoxification, amino acid and deoxyribonucleotide synthesis, iron-sulfur cluster biogenesis and in the regulation of some metabolic pathways.

FNR is a key enzyme of photosynthetic electron transport. This enzyme transfers reducing equivalents between the one-electron carrier Fd and the two-electron carrier NADP(H) at the last step of the photosynthetic electron transport chain. FNR also participates in other relevant processes, such as the electron cyclic flow around Photosystem I and in NADPH/$NADP^+$ homeostasis of stressed chloroplasts [48].

It has been demonstrated that FNR catalyzes one of the rate-limiting steps in photosynthesis, and controls the balance between the demand for reducing equivalents and photosynthetic activity under different environmental conditions [49]. For this reason, FNR may be considered a possible target for crop improvement and weed control. It has been proposed that the transhydrogenase activity of FNR may function *in vivo* as an intra-chloroplastic source of NADH [50,51], although NADH in chloroplasts is most commonly generated via the malate/oxalacetate shuttle [52,53].

It has been shown that FNR is an oxygen-sensitive nitroreductase [54,55]. Shah and Spain have found that nitrite is released from the explosive 2,4,6-trinitrophenylmethylnitramine by the action of FNR [56]. FNR has also been implicated in electron transfer in the last two steps of the non-mevalonate pathway for isoprenoid biosynthesis [2,57–59], a reactivity that has been proposed as an attractive target for development of antibacterial compounds since it is present in clinically important pathogens but absent in humans [2,61,62].

Some pathogenic microorganisms utilize the heme oxygenase enzyme to acquire iron from host's heme. In bacteria, it has been suggested that the NADPH/FNR/Fd system could donate reducing equivalents to heme oxygenases [63–65]. Moreover, in some cases the catalytic activity of these enzymes is efficiently supported by FNR without the need of a Fd, as it was reported for the *P. aeruginosa* heme oxygenase [66]. Different authors have suggested that iron-sulfur cluster assembly in the apicoplast of some organisms is an essential event that could not be replaced by mitochondrial machinery and implies at least five different Fds, the redox protein partners of FNR [67]. In addition, Fds are the source of reducing power for several metabolic processes in bacteria, donating electrons initially provided by FNRs [53].

15.6 Activities of ferredoxin-$NADP^+$ reductase

FNR is able to catalyze different reactions, and Fd as well as several other electron acceptors can participate in the oxidation of NADPH catalyzed by the reductase. FNR transfers two electrons from NADPH to the final acceptors and produces $NADP^+$ [4]. The *in vitro* reaction, which involves electron transfer from NADPH to Fd, can be followed by using cytochrome *c* as final electron acceptor in a coupled assay known as the cytochrome *c* reductase activity. The reaction is most often described as consisting of two half-reactions: the FNR-catalyzed reduction of Fd by NADPH, and the subsequent reoxidation of the iron-sulfur protein by cytochrome *c*. It has been also proposed that a 1:1 Fd:cytochrome *c* complex is the true substrate for the reductase. Significant rates of oxygen uptake have been found to occur along with cytochrome *c* reductase activity under aerobic conditions [53].

Equation 15.3 (see the Introduction section) shows the oxidized and reduced forms of the electron acceptor, A_{ox} and A_{red} respectively, where *n* equals one or two depending whether the oxidant behaves as a one- (ferricyanide, cytochromes) or two-electron (NAD^+, 2,6-dichlorophenol indophenol) carrier. ▶Tab. 15.1 (taken from [68]) summarizes the different reactions catalyzed by FNR, which can be assayed *in vitro* using NADPH. The reduced nucleotide can be alternatively obtained *in situ* by a regenerating system containing $NADP^+$, glucose- 6-phosphate (0.3 mM) and glucose-6-phosphate dehydrogenase (1 unit/ml).

The optimum pH for FNR activity is in the range of 7.5–8.0 [69,70]. The FNR isoelectric point varies in the range of 5.5–6.8 [71], and thus the reductase is expected to be negatively charged at the assay pH, thereby favoring its ability to interact with redox partners.

It has been shown that the diaphorase reaction proceeds via a "ping-pong" mechanism for either ferricyanide, indophenols dyes and tetrazolium salts [72–76]. The same kinetic mechanism holds for the transhydrogenase activity [77,78]. The diaphorase activity is probably not physiologically significant, but it has contributed to the study of FNR function and catalytic mechanism. Recently, however, metabolic relevance has been assigned for these activities in bacterial FNRs [79,80]. Moreover, these artificial reactions might have technological applications for bioremediation and pharmaceutical industry [4,81,82].

Tab. 15.1: Activities catalyzed by ferredoxin-NADP+ reductase

Electron donor	Electron acceptor	Activity	K_{cat} (s^{-1})
Ferredoxin	$NADP^+$	Fd-NADP(H) reductase	500
NADPH	Potassium ferricyanide indophenols viologens tetrazolium salts	Diaphorase	100–500 (depending on the electron acceptor)
NADPH	Fd-cytochrome *c*	cytochrome *c* reductase	50–100
NADPH	NAD^+	Transhydrogenase	5–10
NADPH	O_2, Fd-O_2, Fld-O_2	Oxidase	0.5

The mechanism of electron transfer from reduced Fd to $NADP^+$, as in the photosynthetic reaction, has been characterized *in vitro* and it has been postulated to occur via a ternary complex mechanism [4]. In the first step, the $NADP^+$ binds to FNR, followed by the binding of reduced Fd, which reduces FNR to the semiquinone state. The oxidized Fd dissociates and a second equivalent of reduced Fd binds to the FNR(semiquinone)-$NADP^+$ complex and reduces the enzyme to the hydroquinone form. Finally, the pair of reducing equivalents is transferred from the enzyme to $NADP^+$. The reactivity of reduced FNR toward O_2 is very low [53,83]. This reaction is enhanced several-fold by different electronic acceptors, including one-electron reduced Fd or flavodoxin, viologens, nitroderivates and quinones, that can readily engage in oxygen-dependent redox cycling leading to superoxide formation [56,83]. In addition to these reactions, and as mentioned above, FNR catalyzes transhydrogenation between NADPH and NAD^+ [84].

The stopped-flow technique has been employed to characterize intermediates in the reaction of oxidized FNR and NADPH, reduced FNR and $NADP^+$ and reduced Fd or flavodoxin and oxidized FNR [85–91]. The first reduction of *Anabaena* FNR by Fd produces a semiquinone form that has been characterized by rapid mixing techniques [88,89]; this reaction is too fast to be measured in pea and other FNRs [4]. The electron transfer processes that involve the dissociation of oxidized Fd, the binding of reduced Fd and flavin reduction occur more slowly and have been characterized in wild-type and mutant enzymes [37,92].

15.7 Purification procedures

FNRs are purified from diverse biological sources, including plants, cyanobacteria and transgenic *E. coli*. General procedures for transgenic expression, extract preparation and protein purification are discussed in this section.

15.7.1 Transgenic expression in *E. coli*

FNRs can be obtained by recombinant expression of the corresponding bacterial gene or cDNA in *E. coli*. For this purpose, different vectors have been used depending on the subsequent purification strategy planned. If a His-tagged protein is desired the most commonly used vectors are derived from the pET (Novagen, USA) series, with expression controlled by the T7 promoter [93,94]. In other cases, expression is carried out under the control of *lac* or *tac* promoters, as in the pUC or pGEX vector series, respectively [95,96]. Recombinant expression of wild-type plant FNRs has been performed in Luria Bertani and 2 YT (16 g/l tryptone, 10 g/l yeast extract, 5.0 g/l NaCl) by addition of variable IPTG concentrations (0.2–1 mM) for 2 to 5 h at temperatures between 25 and 37 °C [41,97,98]. However, milder conditions are used in other cases. Expression of unstable mutants of different FNRs can be induced with 0.1 mM IPTG for longer periods and at 15 °C [29,35], while expression of the cDNA from *A. variabilis* FNR has been carried out overnight at 30 °C with no inducer [40].

15.7.2 Preparation of soluble protein extracts

15.7.2.1 Preparation of soluble protein extracts from recombinant *E. coli*

Induced cultures of *E. coli* are centrifuged and the bacterial pellets resuspended in lysis buffer and disrupted using a sonicator or a French Press. The composition of the lysis buffer may vary upon the features of the FNR under study but in general it contains 25–50 mM Tris-HCl pH 7.5–8.0 and 100–150 mM NaCl. Two additives may be beneficial as well: benzamidine hydrochloride (5 mM) and phenylmethylsulfonyl fluoride (1 mM), which help to protect FNR from proteolytic degradation. Addition of nucleases (*i.e.* DNAase) is highly recommended especially for large scale preparations. Finally, cellular debris and unbroken cells should be removed by centrifugation.

15.7.2.2 Preparation of soluble protein extracts from natural sources

Plant leaves and roots as well as cyanobacteria are natural sources of FNRs. FNR was purified from paprika leaves [26], spinach leaves [99], soluble extracts of wheat chloroplasts [71] and plant roots [97,100,101]. In addition, Sancho *et al.* have reported the purification of FNR from cell cultures of cyanobacteria [102].

15.7.2.3 Purification of FNR

FNRs have been purified from soluble protein extracts using different experimental strategies, most consisting of a combination of ammonium sulfate precipitation and chromatographic steps. Affinity and ionic-exchange chromatography are widely used based on the properties of the FNR under study.

Some groups have taken advantage of the affinity of FNRs for the adenosine-like structure of some dyes or for Fd. Dorowski *et al.* have reported the purification of paprika FNR by affinity chromatography using a Reactive Red 120 column eluted in a gradient of NaCl (from 0.05 to 1 M in 25 mM Tris-HCl pH 8.0 and 50 mM NaCl). The FNR containing-fractions are then dialyzed and applied to a DEAE-Sepharose column. After washing, elution is carried out in a gradient of NaCl. FNR containing-fractions are applied to a phenyl-Sepharose column and eluted in a linear gradient of $(NH_4)_2SO_4$ (1.5 to 0 M in 25 mM Tris-HCl pH 8.0). Finally, FNR samples are concentrated and dialyzed against 25 mM Tris-HCl pH 8.0 and 50 mM NaCl [26]. Another dye employed for the purification of FNRs is Cibacron Blue [103]. Elution of FNR from these dyes-based columns can be carried out with $NADP^+$ as well. Moreover, chromatography on a Cibacron Blue matrix efficiently removes the nucleotide from FNR molecules [104]. Besides, Fds were immobilized on cyanogen bromide-activated Sepharose 4B and used to capture FNR from extracts [105,106], to analyze the complex formation of FNR and Fd and to determine the dissociation constant of these complexes [97,105].

Purification of FNRs is also possible using either a carrier protein such glutathione *S*-transferase (GST) [96] or a polyhystidine tag [41]. If required, the recognition site for a protease (as Factor Xa, Thrombin or TEV) can be placed between the FNR and the tag, which allows digestion of the fusion and removal of the tag or carrier. Serra *et al.* have obtained a GST-FNR fusion protein from a pGEX derived vector and bound it to a glutathione-agarose matrix. After washing, the fusion is eluted in buffer containing 10 mM reduced glutathione, dialyzed and digested with Factor Xa. The mixture is incubated again with the glutathione-agarose matrix and the free FNR eluted then subjected to gel filtration [96]. A GST-FNR fusion containing a thrombin recognition site between the carboxyl terminus of GST and the first amino acid of FNR can be purified according to the procedure described above but using thrombin to separate the FNR from the GST. FNR containing samples are then subjected to anion-exchange chromatography in a DEAE Macroprep column and eluted in a linear gradient of NaCl in 50 mM Tris-HCl pH 8. Finally, purified FNR is dialyzed and concentrated [94]. It is worth noting that due to the participation of the carboxy-terminal region of FNRs in substrate binding and catalysis, amino-terminal fusions are preferred [41]. All purification steps are recommended to be carried out at 4 °C to preserve FNR quality. FNR can be stored for several months at –20 °C and for years at –70 °C. Addition of 2-mercaptoethanol or dithiothreitol helps to minimize aggregation of some FNRs [107].

15.7.2.4 Deflavination and reconstitution

Procedures to facilitate removal and replacement of the FAD have been reviewed in [108]. Conventional techniques for the preparation of reconstitutable apoflavoproteins are mainly focused on strategies that lower the pH, increase the salt concentration or temperature and change the solvent. Modern protocols are based on chromatographic procedures to facilitate deflavination and reconstitution. Techniques such as ion exchange chromatography, hydrophobic interaction chromatography, hydroxyapatite chromatography, dye affinity chromatography, covalent chromatography and immobilized metal affinity chromatography were employed.

Preparation of FNR apoprotein from spinach leaves was reported for the first time in 1978 [109]. Zanetti *et al.* described the preparation of apoFNR from spinach leaves by incubation with 2.5 M $CaCl_2$ at pH 7.5 and 2 °C in the presence of 1 mM dithiothreitol, 0.1 mM EDTA, 17% glycerol (v/v) and 0.1 M guanidine/HCl. They proposed that Ca^{2+} ions may induce a conformational change which exposes the FAD and eventually promotes the release of the prosthetic group. The authors also achieved the reactivation of FNR apoprotein in a solution containing an excess of FAD in 0.1 M Hepes pH 7.5 in the presence of 1 mM dithiothreitol, 0.1 mM EDTA, 17% glycerol (v/v), 0.1 M guanidine/HCl. Under these conditions, reactivation was complete in minutes at 0 °C [110]. Maeda *et al.* prepared apoFNR from maize leaves by a similar procedure in the presence of 3 M $CaCl_2$ in 100 mM Tris-HCl pH 8.5, 1 mM dithiothreitol, 0.1 mM EDTA, 17% glycerol (v/v), and 100 mM guanidine/HCl by incubation at 4 °C for 2 h. The solution was then filtrated and the amount of apoFNR estimated to be greater than 99% [111]. The authors compared the spectral and physicochemical properties of this apoFNR with those of a mutant that lacks FAD due to the absence of one of the stacking aromatic rings. They postulated the formation of partially folded intermediates which retained the ability to bind $NADP^+$ in a native-like binding domain but contained a disordered FAD binding domain. To date, no FNRs have been refolded *in vitro* from a completely unfolded polypeptide. A possible explanation is that molecular chaperones are necessary for the proper folding of these enzymes *in vivo* [112].

15.7.2.5 Determination of dissociation constant for ApoFNR-FAD complexes

The dissociation constant for FAD bound to apoFNR can be determined as described in [110]. The authors employed two titration methods to evaluate the binding of FAD to apoFNR. One approach was based on the fact that FAD fluorescence is quenched in

FNR holoenzyme [113]. Consequently, the binding of FAD to apoFNR was measured by following the quenching of flavin fluorescence at 10 °C [114]. Emission was measured at 530 nm by excitation at 450 nm. A 1:1 molar ratio of apoprotein and flavin was found and a dissociation constant of *circa* 4×10^{-9} M was calculated. In the other approach, differential spectroscopy experiments were performed to follow the formation of apoFNR-FAD complex [110].

15.7.2.6 Determination of FAD dissociation rate constants

The dissociation rate constants of FAD (k_{off}) from plastidic and bacterial wild type and mutants FNRs have been determined recently [35,90]. Before the experiments, the enzymes were separated from free FAD by filtration through a Sephadex G-10 column (Sigma, St Louis). The authors employed 5 µM enzyme to perform a continuous measurement of FAD fluorescence emission at 520 nm ($\lambda_{exc.}$ 450 nm) during 5 h at 25 °C in 1 cm path length cells. Then, the samples were denatured with 0.2% (w/v) SDS to release the FAD and the fluorescence was determined. Assuming a 1:1 molar ratio apoFNR-FAD, the dissociation rate constant was calculated as the number of micromoles of FAD released per micromole of enzyme per hour. The initial fluorescence of the samples was discounted in each experiment.

15.7.3 Spectroscopic properties of FNR

Various spectroscopic techniques are useful to detect perturbations in the transition energies of the isoalloxazine moiety of FAD caused by alterations in the environment, especially in close proximity of the isoalloxazine ring. These experiments provide useful information about the structure and the functionality of the flavin and the FNR as we will discuss below.

15.7.3.1 UV-visible spectroscopy of FNR

In solution, free FAD at neutral pH displays a spectrum with maxima at 270, 370 and 446 nm. However, a typical UV-visible spectrum of a plant-type FNR shows peaks around 385 and 456 nm (►Fig. 15.7). This spectral shift is caused by the interaction of the protein scaffold with the isoalloxazine of FAD. Other informative features in the flavoprotein spectra are typical peaks or shoulders, which may indicate the presence of nucleotides bound to the enzyme [39], or chemical variations in the flavin or in its protein environment. Binding of substrates, substrate analogues and inhibitors to FNRs can also be analyzed by UV-visible spectroscopy [39,85,90,115]. Difference spectra are very useful for these studies since slight changes produced by interaction of a substrate with FNR are readily detectable upon subtraction of the spectrum of the pure enzyme from that of the complex with substrate. Dissociation constants for complexes with various substrates and substrate analogs have been determined by titration of the enzyme with the corresponding compound [90]. Moreover, the shape of the spectrum can provide information concerning the nature of the interaction with the substrates. Experimental data indicate that when $NADP^+$ is present at saturating concentrations, the degree of nicotinamide ring occupancy of the binding site in pea FNR is 14–15% [39]. In addition, the UV-visible spectral properties of FNRs also depend on the oxidation state of the flavin and are useful as tools to study these enzymes [37,116].

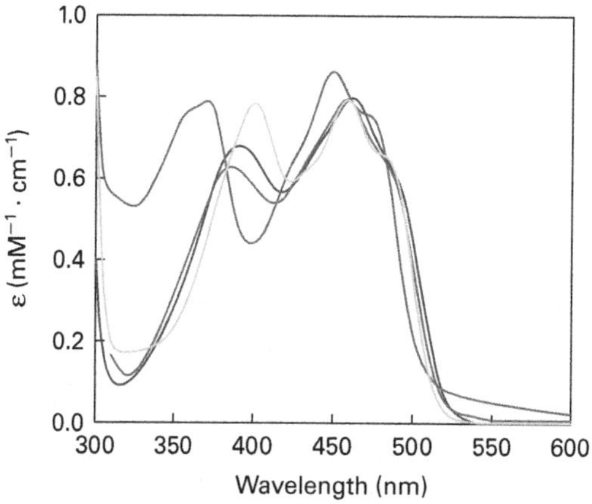

Fig. 15.7: Analysis of FNR by UV-visible spectroscopy. Absorption spectra of pea FNR (green), *E. coli* (yellow), *L. interrogans* (red) and *Xanthomonas citri* pv. citri (magenta).

15.7.3.2 Fluorescence spectroscopy of FNR

Fluorescence spectra of plant FNRs are typically recorded with excitation in the 450–460 nm region and display characteristic emission peaks in the range of 525–535 nm. As mentioned above, the fluorescence emitted by FAD is largely quenched when the prosthetic group binds to FNR. The fluorescence of FAD bound to the native oxidized form of wild-type FNR is found to be only 0.5–0.75% of that of free FAD in solution [72,113]. Analysis of changes in the emission spectrum can be very useful in studying interactions between FNR and substrates such as $NADP^+$, Fd or flavodoxin as explained in [29,94]. Titration curves based on FAD fluorescence can be performed to calculate dissociation constants of FNR-$NADP^+$ complexes [29,94]. Similarly, dissociation constants for FNR-Fd complexes in the presence or absence of $NADP^+$ can be determined by recording the quenching of flavoprotein fluorescence at 340 nm (excitation at 270 nm) after each addition of Fd to pure FNR. The experimental data are typically fitted to a theoretical equation for a 1:1 complex to determine the dissociation constant [29,94,117].

The degree of accessibility of the flavin in FNR to the solvent can be estimated by titrating the enzyme with a dynamic quencher, such as potassium iodide. When the FAD fluorescence is plotted versus the quencher concentration the exposure of the prosthetic group to the solvent can be estimated. A fluorescence change with a steeper slope indicates a higher accessibility of FAD [90]. Although extremely pure protein preparations are mandatory for the experiments described above, the high sensitivity of the fluorescence spectroscopy allow low concentrations of FNR.

15.7.3.3 CD spectroscopy of FNR

The α-helical and β-sheet content of FNRs can be analyzed by circular dichroism spectroscopy. Information about the secondary structure of the protein and the influence of mutations or external factors such as temperature, solvents and binding of small molecules can be obtained from CD spectra. FAD bound to FNR shows peaks in the near-UV and visible region (250–600 nm). Signals in this region reflect perturbations in the environment of the prosthetic group [35].

The denaturation of FNR can also be followed by CD spectroscopy. When FNR is incubated at increasingly high temperatures or denaturant concentrations and the CD signal at 220 nm followed, a sigmoidal behaviour is detected. From this curve it is possible to calculate parameters associated with the folded-to-unfolded transition and the stability of the enzyme. Changes in the far-UV region may indicate the disruption of secondary structure and in the visible region may reflect the integrity of flavin binding site [118]. Both the FAD and $NADP^+$ binding domains contain α-helical structure. Consequently, CD allows the assessment of changes in any of these domains. FNR from *Toxoplasma gondii* [119] and pea leaf [29,35,36], have been analyzed using this technique. The unfolding of pea FNR showed a typical cooperative process with an apparently complete disruption of the structure. Significant differences in stability may exist among FNRs. Indeed, FNR from *L. interrogans* shows persistent secondary structure up to 8 M urea and no changes at low chaotrope concentration (0–2 M urea) [61]. All other FNRs tested up to date are denatured at urea concentrations below 6 M. Using a different methodological approach, it was found that the global unfolding induced by urea in maize leaf FNR is a cooperative process in which unfolding of the secondary structure and release of FAD occur concomitantly [42].

15.8 Conclusions

Ferredoxin-$NADP^+$ reductases have been extensively studied and an enormous amount of relevant information has been accumulated over the time. This knowledge has driven the progress on FNR and other flavoenzymes as well. Plant-type FNRs are found in plant, bacteria and in apicomplexan parasites. They are absent in animals and yeast, where their functions are fulfilled by a different type of FNRs. The differences among these two broad groups of enzymes can be used to develop valuable tools to control pathogens and to generate safe herbicides. Moreover, altering the catalytic efficiency and the substrate preferences of these enzymes could provide tools to improve crops and develop new industrial processes.

The crystal structures of several plant-type FNRs have been solved, their catalytic mechanisms intensively studied and many of their biological roles elucidated. However, there is still a long road ahead and several issues must be addressed. The way in which the substrate enters the active site and reaches a conformation conducive to electron transfer has not been elucidated, and structures of wild-type FNRs with the nucleotide in the productive position have yet to be reported. Substrate specificity is poorly understood and development of enzymes with increased preferences for NADH may open important biotechnological possibilities for xenobiotics conversion, environmental pollutants degradation and other processes. The role of FNR in photosynthesis is well-known, but the participation of the enzyme in other processes is less well understood. In bacteria and parasites, more research is needed to reveal the metabolic function of FNR. This enzyme might be an essential metabolic step in some pathogens and thus an appealing target for antibiotic development.

15.9 Acknowledgments

This work was supported by grants from CONICET, Agencia de Promoción Científica y Tecnológica (ANPCyT), Universidad Nacional de Rosario and Fundación Bunge y Born, Argentina. DLCD, DVR and EAC are members of CONICET, Argentina.

15.10 Abbreviations

FNR	ferredoxin-NADP(H) reductase
Fd	ferredoxin
GST	glutathione *S*-transferase

15.11 References

[1] Shin M, Arnon DI. Enzymatic mechanisms of pyridine nucleotide reduction in chloroplast. J Biol Chem 1965;240:1405–11.

[2] Seeber F, Aliverti A, Zanetti G. The plant-type ferredoxin-NADP$^+$ reductase/ferredoxin redox system as a possible drug target against apicomplexan human parasites. Curr Pharm Des 2005;11:3159–72.

[3] Rohrich RC, Englert N, Troschke K, et al. Reconstitution of an apicoplast-localised electron transfer pathway involved in the isoprenoid biosynthesis of *Plasmodium falciparum*. FEBS Lett 2005;579:6433–38.

[4] Carrillo N, Ceccarelli EA. Open questions in ferredoxin-NADP$^+$ reductase catalytic mechanism. Eur J Biochem 2003;270:1900–15.

[5] Medina M, Gomez-Moreno C. Interaction of Ferredoxin-NADP$^+$ Reductase with Its Substrates: Optimal Interaction for Efficient Electron Transfer. Photosynth Res 2004;79:113–31.

[6] Ceccarelli EA, Arakaki AK, Cortez N, Carrillo N. Functional plasticity and catalytic efficiency in plant and bacterial ferredoxin-NADP(H) reductases. Biochim Biophys Acta 2004;1698: 155–65.

[7] Aliverti A, Pandini V, Pennati A, de Rosa M, Zanetti G. Structural and functional diversity of ferredoxin-NADP(+) reductases. Arch Biochem Biophys 2008;474:283–91.

[8] Arakaki AK, Ceccarelli EA, Carrillo N. Plant-type ferredoxin-NADP$^+$ reductases: a basal structural framework and a multiplicity of functions. FASEB J 1997;11:133–40.

[9] Avron M, Jagendorf AT. A TPNH diaphorase from chloroplast. Arch Biochem Biophys 1956;65:475–90.

[10] Heelis PF. The photophysical and photochemical properties of flavins (isoalloxazines). Chem Soc Rev 1982;11:15–39.

[11] Dudley KH, Ehrenberg A, Hemmerich P, Muller F. Spektren und Strukturen der am Flavin-Redoxsystem beteiligten Partikeln. Studien in der Flavinreihe IX [1]. HCA 1964;47:1354–1383.

[12] Mandai T, Fujiwara S, Imaoka S. A novel electron transport system for thermostable CYP175A1 from *Thermus thermophilus* HB27. FEBS J 2009;276:2416–29.

[13] Muraki N, Seo D, Shiba T, Sakurai T, Kurisu G. Asymmetric dimeric structure of ferredoxin-NAD(P)+ oxidoreductase from the green sulfur bacterium *Chlorobaculum tepidum*: implications for binding ferredoxin and NADP$^+$. J Mol Biol 2010;401:403–14.

[14] Dym O, Eisenberg D. Sequence-structure analysis of FAD-containing proteins. Protein Sci 2001;10:1712–28.

[15] Karplus PA, Faber HR. Structural Aspects of Plant Ferredoxin: NADP(+) Oxidoreductases. Photosynth Res 2004;81:303–15.

[16] Nascimento AS, Catalano-Dupuy DL, Bernardes A, et al. Crystal structures of *Leptospira interrogans* FAD-containing ferredoxin-NADP$^+$ reductase and its complex with NADP$^+$. BMC Struct Biol 2007;7:69.

[17] Ingelman M, Bianchi V, Eklund H. The three-dimensional structure of flavodoxin reductase from *Escherichia coli* at 1.7 A resolution. J Mol Biol 1997;268:147–57.

[18] Tondo ML, Musumeci MA, Delprato ML, Ceccarelli EA, Orellano EG. Structural-functional characterization and physiological significance of ferredoxin-NADP$^+$ reductase from *Xanthomonas axonopodis* pv. citri. PLoS One 2011;6:e27124.

[19] Razquin P, Fillat MF, Schmitz S, et al. Expression of ferredoxin-NADP$^+$ reductase in heterocysts from Anabaena sp. Biochem J 1996;316 (Pt 1):157–60.

[20] Tognetti VB, Zurbriggen MD, Morandi EN, et al. Enhanced plant tolerance to iron starvation by functional substitution of chloroplast ferredoxin with a bacterial flavodoxin. Proc Natl Acad Sci U S A 2007;104:11495–500.
[21] Karplus PA, Daniels MJ, Herriott JR. Atomic structure of ferredoxin-NADP+ reductase: prototype for a structurally novel flavoenzyme family. Science 1991;251:60–66.
[22] Correll CC, Ludwig ML, Bruns CM, Karplus PA. Structural prototypes for an extended family of flavoprotein reductases: comparison of phthalate dioxygenase reductase with ferredoxin reductase and ferredoxin. Protein Sci 1993;2:2112–33.
[23] Bruns CM, Karplus PA. Refined crystal structure of spinach ferredoxin reductase at 1.7 A resolution: oxidized, reduced and 2′-phospho-5′-AMP bound states. J Mol Biol 1995;247:125–45.
[24] Deng Z, Aliverti A, Zanetti G, et al. A productive NADP+ binding mode of ferredoxin-NADP+ reductase revealed by protein engineering and crystallographic studies. Nat Struct Biol 1999;6:847–53.
[25] Kurisu G, Kusunoki M, Katoh E, et al. Structure of the electron transfer complex between ferredoxin and ferredoxin-NADP+ reductase. Nat Struct Biol 2001;8:117–21.
[26] Dorowski A, Hofmann A, Steegborn C, Boicu M, Huber R. Crystal structure of paprika ferredoxin-NADP+ reductase - implications for the electron transfer pathway. J Biol Chem 2000;276:9253–63.
[27] Tronrud DE, Berkholz DS, Karplus PA. Using a conformation-dependent stereochemical library improves crystallographic refinement of proteins. Acta Crystallogr D Biol Crystallogr 2010;66:834–42.
[28] Serre L, Vellieux FM, Medina M, Gomez-Moreno C, Fontecilla-Camps JC, Frey M. X-ray structure of the ferredoxin:NADP+ reductase from the cyanobacterium Anabaena PCC 7119 at 1.8 A resolution, and crystallographic studies of NADP+ binding at 2.25 A resolution. J Mol Biol 1996;263:20–39.
[29] Musumeci MA, Arakaki AK, Rial DV, Catalano-Dupuy DL, Ceccarelli EA. Modulation of the enzymatic efficiency of ferredoxin-NADP(H) reductase by the amino acid volume around the catalytic site. FEBS J 2008;275:1350–66.
[30] Tejero J, Perez-Dorado I, Maya C, et al. C-terminal tyrosine of ferredoxin-NADP+ reductase in hydride transfer processes with NAD(P)+/H. Biochemistry 2005;44:13477–90.
[31] Hermoso JA, Mayoral T, Faro M, Gomez-Moreno C, Sanz-Aparicio J, Medina M. Mechanism of coenzyme recognition and binding revealed by crystal structure analysis of ferredoxin-NADP+ reductase complexed with NADP+. J Mol Biol 2002;319:1133–42.
[32] Maeda M, Lee YH, Ikegami T, et al. Identification of the N- and C-terminal substrate binding segments of ferredoxin-NADP+ reductase by NMR. Biochemistry 2005;44:10644–53.
[33] Nogues I, Perez-Dorado I, Frago S, et al. The ferredoxin-NADP(H) reductase from *Rhodobacter capsulatus*: molecular structure and catalytic mechanism. Biochemistry 2005;44:11730–40.
[34] Wang A, Rodriguez JC, Han H, Schonbrunn E, Rivera M. X-ray crystallographic and solution state nuclear magnetic resonance spectroscopic investigations of NADP+ binding to ferredoxin-NADP+ reductase from *Pseudomonas aeruginosa*. Biochemistry 2008;47:8080–93.
[35] Musumeci MA, Botti H, Buschiazzo A, Ceccarelli EA. Swapping FAD Binding Motifs between Plastidic and Bacterial Ferredoxin-NADP(H) Reductases. Biochemistry 2011;50:2111–22.
[36] Calcaterra NB, Pico GA, Orellano EG, Ottado J, Carrillo N, Ceccarelli EA. Contribution of the FAD binding site residue tyrosine 308 to the stability of pea ferredoxin-NADP+ oxidoreductase. Biochemistry 1995;34:12842–48.
[37] Nogues I, Tejero J, Hurley JK, et al. Role of the C-terminal tyrosine of ferredoxin-nicotinamide adenine dinucleotide phosphate reductase in the electron transfer processes with its protein partners ferredoxin and flavodoxin. Biochemistry 2004;43:6127–37.
[38] Orellano EG, Calcaterra NB, Carrillo N, Ceccarelli EA. Probing the role of the carboxyl-terminal region of ferredoxin-NADP+ reductase by site-directed mutagenesis and deletion analysis. J Biol Chem 1993;268:19267–73.

[39] Piubelli L, Aliverti A, Arakaki AK, et al. Competition between C-terminal tyrosine and nicotinamide modulates pyridine nucleotide affinity and specificity in plant ferredoxin-NADP(+) reductase. J Biol Chem 2000;275:10472–76.
[40] Tejero J, Martinez-Julvez M, Mayoral T, et al. Involvement of the pyrophosphate and the 2′-phosphate binding regions of ferredoxin-NADP$^+$ reductase in coenzyme specificity. J Biol Chem 2003;278:49203–14.
[41] Catalano-Dupuy DL, Orecchia M, Rial DV, Ceccarelli EA. Reduction of the pea ferredoxin-NADP(H) reductase catalytic efficiency by the structuring of a carboxyl-terminal artificial metal binding site. Biochemistry 2006;45:13899–909.
[42] Lee YH, Tamura K, Maeda M, et al. Cores and pH-dependent dynamics of ferredoxin-NADP$^+$ reductase revealed by hydrogen/deuterium exchange. J Biol Chem 2007;282:5959–67.
[43] Peregrina JR, Lans I, Medina M. The transient catalytically competent coenzyme allocation into the active site of Anabaena ferredoxin NADP(+)-reductase. Eur Biophys J 2011.
[44] Hurley JK, Morales R, Martinez-Julvez M, et al. Structure-function relationships in Anabaena ferredoxin/ferredoxin:NADP(+) reductase electron transfer: insights from site-directed mutagenesis, transient absorption spectroscopy and X-ray crystallography. Biochim Biophys Acta 2002;1554:5–21.
[45] Hanke G, Kurisu G, Kusunoki M, Hase T. Fd: FNR electron transfer complexes: evolutionary refinement of structural interactions. Photos Research 2004;81:317–27.
[46] Morales R, Charon MH, Kachalova G, et al. A redox-dependent interaction between two electron-transfer partners involved in photosynthesis. EMBO Rep 2000;1:271–76.
[47] Morales R, Charon MH, Hudry-Clergeon G, et al. Refined X-ray structures of the oxidized, at 1.3 A, and reduced, at 1.17 A, [2Fe-2S]ferredoxin from the cyanobacterium Anabaena PCC7119 show redox-linked conformational changes. Biochemistry 1999;38:15764–73.
[48] Palatnik JF, Valle EM, Carrillo N. Oxidative stress causes ferredoxin-NADP$^+$ reductase solubilization from the thylakoid membranes in methyl viologen-treated plants. Plant Physiol 1997;115:1721–27.
[49] Hajirezaei MR, Peisker M, Tschiersch H, et al. Small changes in the activity of chloroplastic NADP(+)-dependent ferredoxin oxidoreductase lead to impaired plant growth and restrict photosynthetic activity of transgenic tobacco plants. Plant J 2002;29:281–93.
[50] Chopowick R, Israelstam GF. Pyridine nucleotide transhydrogenase from *Chlorella*. Planta 1971;101:171–73.
[51] Krawetz SA, Israelstam GF. Kinetics of pyridine nucleotide transhydrogenase from Chlorella. Plant Science Letters 1978;12:323–26.
[52] Krause GH, Heber U. Energetics in intact chloroplasts. In: Barber H, editors. The intact chloroplast.Elsevier, Amsterdam.1976; pp. 171–214.
[53] Carrillo N, Vallejos RH. Ferredoxin-NADP$^+$ oxidoreductase. In *The Light Reactions*. In: Barber J, editors. *Topics in Photosynthesis*. Amsterdam-New York-Oxford: Elsevier.1987; pp. 527–60.
[54] Orna MV, Mason RP. Correlation of kinetic parameters of nitroreductase enzymes with redox properties of nitroaromatic compounds. J Biol Chem 1989;264:12379–84.
[55] Qu Y, Zhou H, Li A, Ma F, Zhou J. Nitroreductase activity of ferredoxin reductase BphA4 from *Dyella ginsengisoli* LA-4 by catalytic and structural properties analysis. Appl Microbiol Biotechnol 2011;89:655–63.
[56] Shah MM, Spain JC. Elimination of nitrite from the explosive 2,4,6-trinitrophenylmethylnitramine (tetryl) catalyzed by ferredoxin NADP oxidoreductase from spinach. Biochem Biophys Res Commun 1996;220:563–68.
[57] Wolff M, Seemann M, Tse Sum BB, et al. Isoprenoid biosynthesis via the methylerythritol phosphate pathway: the (E)-4-hydroxy-3-methylbut-2-enyl diphosphate reductase (LytB/IspH) from *Escherichia coli* is a [4Fe-4S]protein. FEBS Lett 2003;541:115–20.
[58] Grawert T, Groll M, Rohdich F, Bacher A, Eisenreich W. Biochemistry of the non-mevalonate isoprenoid pathway. Cell Mol Life Sci 2011;68:3797–3814.

[59] Seemann M, Bui BT, Wolff M, et al. Isoprenoid biosynthesis through the methylerythritol phosphate pathway: the (E)-4-hydroxy-3-methylbut-2-enyl diphosphate synthase (GcpE) is a [4Fe-4S]protein. Angew Chem Int Ed Engl 2002;41:4337–39.

[60] Okada K, Hase T. Cyanobacterial non-mevalonate pathway: (E)-4-hydroxy-3-methylbut-2-enyl diphosphate synthase interacts with ferredoxin in *Thermosynechococcus elongatus* BP-1. J Biol Chem 2005;280:20672–79.

[61] Catalano-Dupuy DL, Musumeci MA, Lopez-Rivero A, Ceccarelli EA. A highly stable plastidic-type ferredoxin-NADP(H) reductase in the pathogenic bacterium *Leptospira interrogans*. PLoS One 2011;6:e26736.

[62] Seeber F. Biosynthetic pathways of plastid-derived organelles as potential drug targets against parasitic apicomplexa. Curr Drug Targets Immune Endocr Metabol Disord 2003;3: 99–109.

[63] Muramoto T, Tsurui N, Terry MJ, Yokota A, Kohchi T. Expression and biochemical properties of a ferredoxin-dependent heme oxygenase required for phytochrome chromophore synthesis. Plant Physiol 2002;130:1958–66.

[64] Gohya T, Zhang X, Yoshida T, Migita CT. Spectroscopic characterization of a higher plant heme oxygenase isoform-1 from *Glycine max* (soybean)--coordination structure of the heme complex and catabolism of heme. FEBS J 2006;273:5384–99.

[65] Wegele R, Tasler R, Zeng Y, Rivera M, Frankenberg-Dinkel N. The heme oxygenase(s)-phytochrome system of *Pseudomonas aeruginosa*. J Biol Chem 2004;279:45791–802.

[66] Wang A, Zeng Y, Han H, et al. Biochemical and structural characterization of Pseudomonas aeruginosa Bfd and FPR: ferredoxin NADP+ reductase and not ferredoxin is the redox partner of heme oxygenase under iron-starvation conditions. Biochemistry 2007;46:12198–211.

[67] Seeber F. Malaria and anti-malarials - a focused view. In: P. Selzer, editors. Antiparasitic and Antibacterial Drug Discovery in Infectious Diseases - From Molecular Targets to Drug Candidates. Wiley-VCH, Weinheim.2009; pp. 277–98.

[68] Musumeci MA, Ceccarelli EA, Catalano-Dupuy DL. The Plant-Type Ferredoxin-NADP+ Reductases. In: Dr Mohammad Najafpour (Ed.), editors. Advances in Photosynthesis - Fundamental Aspects. InTech. 2012; pp. 539–62.

[69] Melamed-Harel H, Tel-Or E, Pietro AS. Effect of Ferredoxin on the Diaphorase Activity of Cyanobacterial Ferredoxin-NADP Reductase. Plant Physiology 1985;77:229–31.

[70] Masaki R, Yoshikawa S, Matsubara H. Steady-state kinetics of oxidation of reduced ferredoxin with ferredoxin-nadp+ reductase. Biochim Biophys Acta 1982;700:101–9.

[71] Grzyb J, Malec P, Rumak I, Garstka M, Strzalka K. Two isoforms of ferredoxin:NADP(+) oxidoreductase from wheat leaves: purification and initial biochemical characterization. Photosynth Res 2008;96:99–112.

[72] Zanetti G, Forti G. Studies on the triphosphopyridine nucleotide-cytochrome f reductase of chloroplasts. J Biol Chem 1966;241:279–85.

[73] Forti G, Sturani E. On the structure and function of reduced nicotinamide adenine dinucleotide phosphate-cytochrome f reductase of spinach chloroplasts. Eur J Biochem 1968;3:461–72.

[74] Nakamura S, Kimura T. Studies on spinach ferredoxin-nicotinamide adenine dinucleotide phosphate reductase. Kinetic studies on the interactions of the reductase and ferredoxin and a possible regulation of enzyme activities by ionic strength. J Biol Chem 1971;246:6235–41.

[75] Masaki R, Wada K, Matsubara H. Isolation and characterization of two ferredoxin-NADP+ reductases from Spirulina platensis. J Biochem (Tokyo) 1979;86:951–62.

[76] Zanetti G, Curti B. Interactions between ferredoxin-NADP reductase and ferredoxin at different reduction levels of the two proteins. FEBS Lett 1981;129:201–4.

[77] Shin M, Pietro AS. Complex formation of ferredoxin-NADP reductase with ferredoxin and with NADP. Biochem Biophys Res Commun 1968;33:38–42.

[78] Böger P. Relationship of transhydrogenase and diaphorase activity of ferredoxin-NADP+ reductase with photosynthetic NADP+ reduction. Z Naturforsch B 1971;26:807–15.

[79] Takeda K, Sato J, Goto K, et al. *Escherichia coli* ferredoxin-NADP+ reductase and oxygen-insensitive nitroreductase are capable of functioning as ferric reductase and of driving the Fenton reaction. Biometals 2010;23:727–37.
[80] Yeom J, Jeon CO, Madsen EL, Park W. Ferredoxin-NADP+ reductase from *Pseudomonas putida* functions as a ferric reductase. J Bacteriol 2009;191:1472–79.
[81] Cenas N, Nemeikaite-Ceniene A, Sergediene E, Nivinskas H, Anusevicius Z, Sarlauskas J. Quantitative structure-activity relationships in enzymatic single-electron reduction of nitroaromatic explosives: implications for their cytotoxicity. Biochim Biophys Acta 2001;1528:31–38.
[82] Tognetti VB, Monti MR, Valle EM, Carrillo N, Smania AM. Detoxification of 2,4-dinitrotoluene by Transgenic Tobacco Plants Expressing a Bacterial Flavodoxin. Environmental Science & Technology 2007;41:4071–76.
[83] Gomez-Moreno C, Medina M, Hurley JK, et al. Protein engineering for the elucidation of the mechanism of electron transfer in redox proteins. Biochem Soc Trans 1994;22:796–800.
[84] Böger P. Einfluß von Ferredoxin auf Ferredoxin-NADP-Reduktase. Planta 1971;99:319–38.
[85] Medina M, Luquita A, Tejero J, et al. Probing the determinants of coenzyme specificity in ferredoxin-NADP+ reductase by site-directed mutagenesis. J Biol Chem 2001;276: 11902–12.
[86] Tejero J, Peregrina JR, Martinez-Julvez M, et al. Catalytic mechanism of hydride transfer between NADP+/H and ferredoxin-NADP+ reductase from Anabaena PCC 7119. Arch Biochem Biophys 2007;459:79–90.
[87] Anusevicius Z, Miseviciene L, Medina M, Martinez-Julvez M, Gomez-Moreno C, Cenas N. FAD semiquinone stability regulates single- and two-electron reduction of quinones by Anabaena PCC7119 ferredoxin:NADP+ reductase and its Glu301Ala mutant. Arch Biochem Biophys 2005;437:144–50.
[88] Hurley JK, Fillat M, Gomez-Moreno C, Tollin G. Structure-function relationships in the ferredoxin/ferredoxin: NADP+ reductase system from Anabaena. Biochimie 1995;77:539–48.
[89] Martinez-Julvez M, Hermoso J, Hurley JK, et al. Role of Arg100 and Arg264 from Anabaena PCC 7119 ferredoxin-NADP+ reductase for optimal NADP+ binding and electron transfer. Biochemistry 1998;37:17680–91.
[90] Paladini DH, Musumeci MA, Carrillo N, Ceccarelli EA. Induced fit and equilibrium dynamics for high catalytic efficiency in ferredoxin-NADP(H) reductases. Biochemistry 2009;48: 5760–68.
[91] Serrano A, Medina M. Fast Kinetic Methods with Photodiode Array Detection in the Study of the Interaction and Electron Transfer Between Flavodoxin and Ferredoxin NADP+-Reductase. In: Mohammad Mahdi Najafpour, editors. Advances in Photosynthesis - Fundamental Aspects. InTech. 2012.
[92] Medina M. Structural and mechanistic aspects of flavoproteins: photosynthetic electron transfer from photosystem I to NADP+. FEBS J 2009;276:3942–58.
[93] Rial DV, Ceccarelli EA. Removal of DnaK contamination during fusion protein purifications. Protein Expr Purif 2002;25:503.
[94] Catalano Dupuy DL, Rial DV, Ceccarelli EA. Inhibition of pea ferredoxin-NADP(H) reductase by Zn-ferrocyanide. Eur J Biochem 2004;271:4582–93.
[95] Ceccarelli EA, Viale AM, Krapp AR, Carrillo N. Expression, assembly, and processing of an active plant ferredoxin- NADP+ oxidoreductase and its precursor protein in *Escherichia coli.* J Biol Chem 1991;266:14283–87.
[96] Serra EC, Carrillo N, Krapp AR, Ceccarelli EA. One-step purification of plant ferredoxin-NADP+ oxidoreductase expressed in *Escherichia coli* as fusion with glutathione S-transferase. Protein Expr Purif 1993;4:539–46.
[97] Onda Y, Matsumura T, Kimata-Ariga Y, Sakakibara H, Sugiyama T, Hase T. Differential interaction of maize root ferredoxin:NADP(+) oxidoreductase with photosynthetic and non-photosynthetic ferredoxin isoproteins. Plant Physiol 2000;123:1037–45.

[98] Aliverti A, Jansen T, Zanetti G, Ronchi S, Herrmann RG, Curti B. Expression in *Escherichia coli* of ferredoxin:NADP+ reductase from spinach. Bacterial synthesis of the holoflavoprotein and of an active enzyme form lacking the first 28 amino acid residues of the sequence. Eur J Biochem 1990;191:551–55.

[99] Grzyb J, Waloszek A, Latowski D, Wieckowski S. Effect of cadmium on ferredoxin: NADP+ oxidoreductase activity. J Inorg Biochem 2004;98:1338–46.

[100] Morigasaki S, Takata K, Suzuki T, Wada K. Purification and Characterization of a Ferredoxin-NADP Oxidoreductase-Like Enzyme from Radish Root Tissues. Plant Physiol 1990;93: 896–901.

[101] Green LS, Yee BC, Buchanan BB, Kamide K, Sanada Y, Wada K. Ferredoxin and Ferredoxin-NADP Reductase from Photosynthetic and Nonphotosynthetic Tissues of Tomato. Plant Physiology 1991;96:1207–13.

[102] Sancho J, Peleato ML, Gomez-Moreno C, Edmondson DE. Purification and properties of ferredoxin-NADP+ oxidoreductase from the nitrogen-fixing cyanobacteria *Anabaena variabilis*. Arch Biochem Biophys 1988;260:200–07.

[103] Carrillo N, Vallejos RH. Interaction of ferredoxin-NADP+ oxidoreductase with triazine dyes. A rapid purification method by affinity chromatography. Biochim Biophys Acta 1983;742: 285–294.

[104] Martinez-Julvez M, Tejero J, Peregrina JR, et al. Towards a new interaction enzyme:coenzyme. Biophys Chem 2005;115:219–24.

[105] Sakihama N, Nagai K, Ohmori H, Tomizawa H, Tsujita M, Shin M. Immobilized ferredoxins for affinity chromatography of ferredoxin-dependent enzymes. J Chromatogr 1992;597: 147–53.

[106] Onda Y, Hase T. FAD assembly and thylakoid membrane binding of ferredoxin:NADP+ oxidoreductase in chloroplasts. FEBS Lett 2004;564:116–20.

[107] Nakajima M, Sakamoto T, Wada K. The complete purification and characterization of three forms of ferredoxin-NADP(+) oxidoreductase from a thermophilic cyanobacterium *Synechococcus elongatus*. Plant Cell Physiol 2002;43:484–93.

[108] Hefti MH, Vervoort J, van Berkel WJ. Deflavination and reconstitution of flavoproteins. Eur J Biochem 2003;270:4227–42.

[109] Bookjans G, San Pietro A, Boger P. Resolution and reconstitution of spinach ferredoxin-NADP+ reductase. Biochem Biophys Res Commun 1978;80:759–65.

[110] Zanetti G, Cidaria D, Curti B. Preparation of apoprotein from spinach ferredoxin-NADP+ reductase. Studies on the resolution process and characterization of the FAD reconstituted holoenzyme. Eur J Biochem 1982;126:453–58.

[111] Maeda M, Hamada D, Hoshino M, Onda Y, Hase T, Goto Y. Partially folded structure of flavin adenine dinucleotide-depleted ferredoxin-NADP+ reductase with residual NADP+ binding domain. J Biol Chem 2002;277:17101–07.

[112] Dionisi HM, Checa SK, Krapp AR, et al. Cooperation of the DnaK and GroE chaperone systems in the folding pathway of plant ferredoxin-NADP+ reductase expressed in Escherichia coli. Eur J Biochem 1998;251:724–28.

[113] Shin M. Complex formation by ferredoxin-NADP reductase with ferredoxin or NADP. Biochim Biophys Acta 1973;292:13–19.

[114] Edmondson DE, Tollin G. Circular dichroism studies of the flavin chromophore and of the relation between redox properties and flavin environment in oxidases and dehydrogenases. Biochemistry 1971;10:113–24.

[115] Aliverti A, Bruns CM, Pandini VE, et al. Involvement of serine 96 in the catalytic mechanism of ferredoxin-NADP+ reductase: structure--function relationship as studied by site-directed mutagenesis and X-ray crystallography. Biochemistry 1995;34:8371–79.

[116] Macheroux P. UV-Visible Spectroscopy as a Tool to Study Flavoproteins. In: Stephen K. Chapman, Graeme A. Reid, editors. Flavoprotein Protocols. Springer Science. 1999.

[117] Davis DJ. Tryptophan fluorescence studies of ferredoxin:NADP reductase indicate the presence of tryptophan in or near the ferredoxin binding site. Arch Biochem Biophys 1990;276:1–5.
[118] Munro AW, Kelly S, Price N. Circular Dichroism studies of flavoproteins. In: Chapman S. RG, editors. Flavoprotein protocols. Humana press. 1999; pp. 111–29.
[119] Singh K, Bhakuni V. Guanidine hydrochloride- and urea-induced unfolding of *Toxoplasma gondii* ferredoxin-NADP+ reductase: stabilization of a functionally inactive holo-intermediate. J Biochem 2009;145:721–31.

16 Flavoprotein dehalogenases

Steven E. Rokita

Abstract

Dehalogenation of organic halides is promoted in biology through a range of transformations involving either oxidation, hydrolysis or reduction. Flavoproteins are not the exclusive catalysts of these reactions, but representative flavoproteins have been discovered within each dehalogenation category. Examples in this chapter illustrate the ability of flavin to contribute indirectly to these processes by shuttling electrons between reduced nicotinamides and enzymes that act as either monooxygenases or reductases. Direct involvement of flavin is also presented in examples of dehalogenation based alternatively on monooxygenation and reductive deiodination. Although significant precedence is available for the mechanism of the oxidative reactions, little is yet known about their reductive counterparts. Previous chemical models did not accurately anticipate the reductive chemistry of iodotyrosine deiodinase. Even less information is currently available for the unique flavoprotein haloacrylate hydratase that catalyzes hydrolytic dehalogenation.

16.1 Organic halides and biological dehalogenation

The popular media typically describe organic halides as a scourge produced by the chemical industry, but their origins are much more diverse and their properties are not always so malevolent. To the contrary, numerous synthetic and natural compounds containing halogen atoms have become highly valuable in fighting a range of diseases. Bacteria, fungi, plants and animals all contribute to the formation and distribution of organic halides [1–3]. Even our own bodies generate halogenated compounds including the thyroid hormone thyroxine (3,3′,5,5′-tetraiodothyronine, T4) (▸Fig. 16.1) and its tri- and diiodinated derivatives [4]. Among the simplest examples of organic halides released into the environment are methyl bromide used as a crop fumigant, tetrachloroethylene used as a dry-cleaning solvent, and 1,2-dibromoethane used as a gasoline additive. Interestingly, these same compounds are also produced by marine algae [2]. At another extreme, the structural complexity of halogen-containing natural products can be very high, as illustrated by the glycosidic region of calicheamicin and the aglycone of vancomycin (▸Fig. 16.1).

Strategies for incorporating halogens into natural products may vary but most involve oxidation [5]. In contrast, the complementary process of dehalogenation utilizes a broad range of transformations [6,7]. Oxidative reactions still contributes significantly, but many hydrolytic, thiolytic and reductive dehalogenations have also been described. Some of the degradative pathways likely evolved long ago to enhance or merely protect

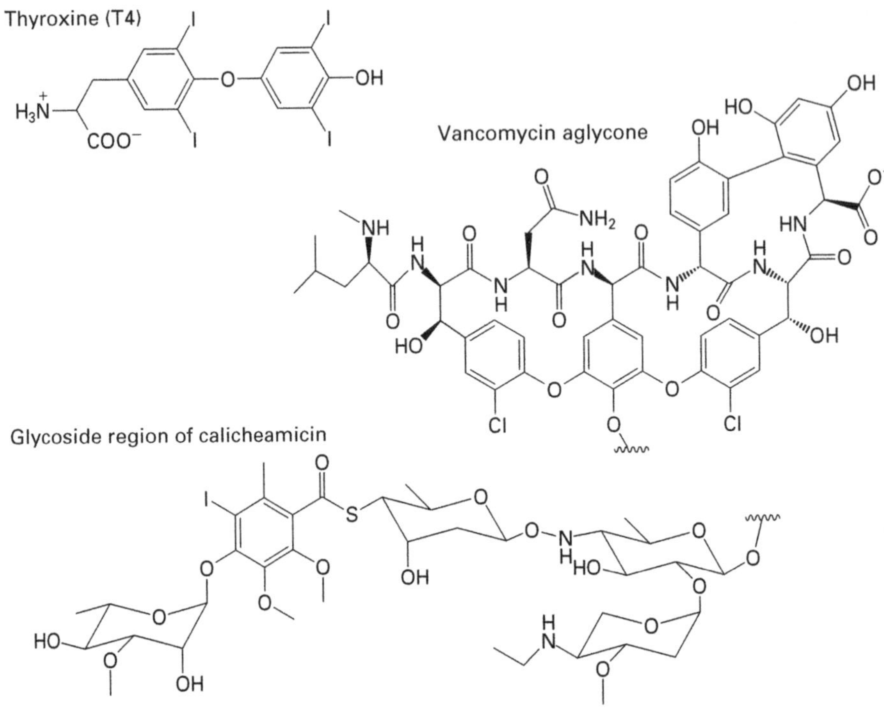

Fig. 16.1: Examples of organic halides produced by nature.

cell viability as illustrated by the dehalogenase responsible for detoxifying fluoroacetate by hydrolysis of the carbon-fluorine bond [8]. Other pathways have likewise arisen quite recently in response to novel industrial compounds such as pentachlorophenol [9]. The adaptability of microorganisms to consume such compounds inspires much of the research on bioremediation [6,10,11]. From a mechanistic perspective, attention should also be directed to flavin-dependent chemistry due to its demonstrated flexibility in promoting dehalogenation alternatively through oxidative, hydrolytic and reductive mechanisms. Examples of each are described below after a brief section describing these general processes and their associated enzymes. This chapter is designed to highlight the diversity of catalysis supporting dehalogenation and does not attempt to review this topic comprehensively. Accordingly, not all of the excellent contributions could be included in this review and for that, the author apologizes.

16.1.1 Strategies for dehalogenation

Oxidative dehalogenation is most common in aerobic organisms and usually involves an O_2-dependent hydroxylation or epoxidation of the carbon bound to the halogen. Subsequent enzymatic or non-enzymatic steps result in halide elimination (►Fig. 16.2). Cytochromes P450 are most often associated with this process in organisms ranging from microbes to humans [12,13]. Related dehalogenation by certain heme-dependent peroxidases has also been discovered (►Fig. 16.2) [14]. Alternative hydrolytic dehalogenation requires significant activation of reactants since most carbon-halogen

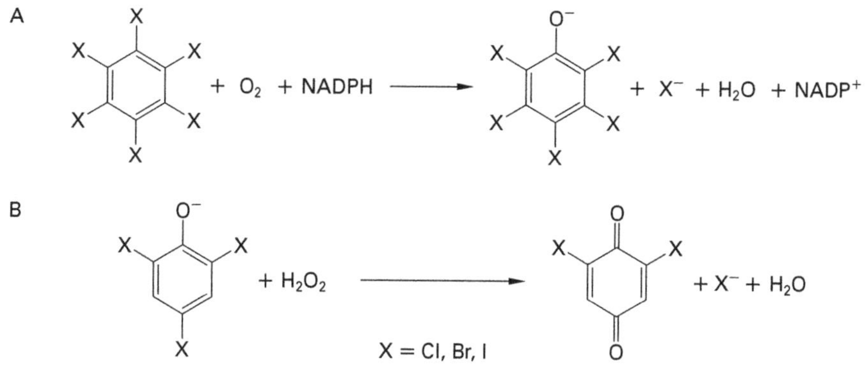

Fig. 16.2: Oxidative dehalogenation catalyzed by (A) P_{450} monooxygenase and (B) dehaloperoxidase.

bonds are relatively stable to substitution by water under physiological conditions. Haloalkane dehalogenases circumvent this constraint by recruiting an active site carboxylate for nucleophilic displacement of chloride, bromide and iodide from alkyl halides (▶Fig. 16.3) [15,16]. Remarkably, this same strategy is even effective for hydrolyzing the relatively inert carbon-fluorine bond during turnover of fluoroacetate dehalogenase [17,18]. Aromatic carbon-chlorine bonds can also be hydrolyzed through a substitution reaction with an active site carboxylate although further activation of the aromatic system is often required as well. For example, *para*-chlorobenzoate is initially converted to its CoA thioester to stabilize the intervening Meisenheimer complex formed by transient addition of the carboxylate group to the aromatic system during dehalogenation (▶Fig. 16.3) [19].

Both aliphatic and aromatic halides, including various chloroethylenes and chlorophenols, can also be dehalogenated by reductive mechanisms under anearobic conditions [20]. These processes often require vitamin B_{12} and/or iron centers and may play a central role in certain strategies of energy production termed halorespiration. When studied *in vitro*, the enzymes responsible for reductive dehalogenation require the presence of very strong electron donors. Benzyl viologen is not usually sufficient to drive reaction, and instead methyl viologen or chelated titanium(III) are necessary [21,22]. Much hope has been placed on organisms that express these activities since they may directly colonize sites of environmental contamination. However, reductive dehalogenation is not limited to anaerobes. Aerobic metabolism of pentachlorophenol

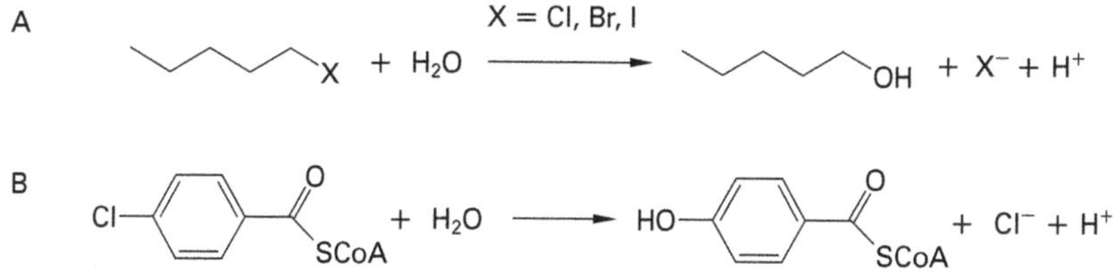

Fig. 16.3: Hydrolytic dehalogenation catalyzed by (A) haloalkane dehalogenase and (B) 4-chlorobenzoyl-CoA dehalogenase.

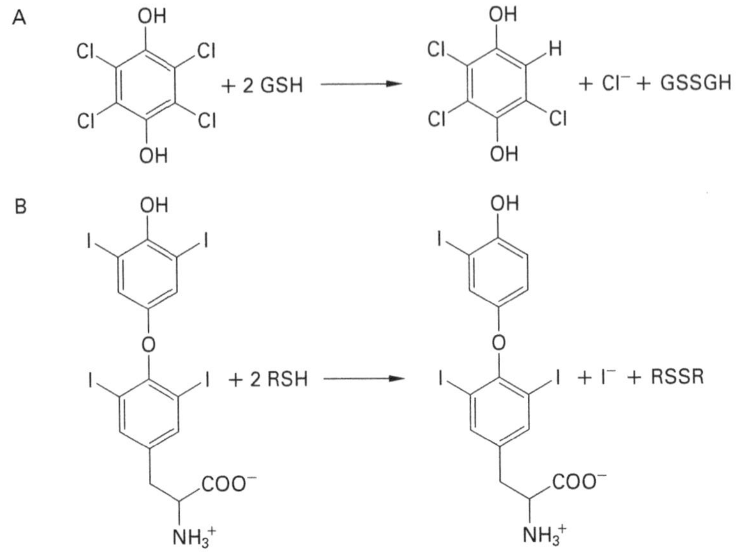

Fig. 16.4: Thiol-dependent reductive dehalogenation under aerobic conditions catalyzed by (A) tetrachlorohydroquinone dehalogenase and (B) iodothyronine deiodinase.

by a soil bacterium also includes a reductive dechlorination of tetrachlorohydroquinone (►Fig. 16.4). The dehalogenase responsible for this activity is a member of the glutathione S-transferase superfamily and relies on the reducing power of glutathione to complete turnover [23]. A related dechlorinase has also now been identified in the eukaryote *Dictyostelium* [24]. In each case, an active site cysteine likely promotes reaction through its nucleophilic and oxidation-reduction activities. One of the two reductive dehalogenases so far discovered in higher organisms may share a similar mechanism. Deiodination of the thyroid hormone thyroxine and its metabolic derivatives appears to be driven by thiols, but the deiodinase contains an active site selenocysteine rather than a cysteine and belongs to the thioredoxin structural superfamily (►Fig. 16.4) [25]. The second reductive dehalogenase promotes an almost equivalent deiodination of iodotyrosine. However, this enzyme requires no thiol for reaction and contains neither cysteine nor selenocysteine in its active site [26]. Instead, FMN is required for activity [27].

16.2 Flavin-dependent dehalogenation

Flavins and flavin-dependent reactions obviously need no introduction at this point in a compilation entitled Handbook of Flavoproteins, but the continual discovery of new catalytic roles for flavins deserves repeated emphasis. The diversity of reactions that this cofactor promotes is already remarkable [28–30], and further expansion of its activities is even more impressive. Recent investigations have linked flavoproteins to certain halogenation reactions such as that catalyzed by tryptophan 7-halogenase [31,32].

An active site flavin has also been implicated in an acid-base, rather than oxidation-reduction, mechanism of catalysis for a type 2 isopentenyl-diphosphate isomerase [33,34]. In other examples, the requirement for flavin may be well established, and yet its exact function in catalysis is only poorly defined, as illustrated by iodotyrosine deiodinase (see below, section 16.2.3). Much of flavin's chemical repertoire is collectively displayed by the flavoproteins that are distributed among the oxidative, hydrolytic and reductive dehalogenases.

16.2.1 Oxidative dehalogenation by flavoproteins

Flavins participate both indirectly and directly in oxidative dehalogenation reactions. Flavin-dependent electron transferases and reductases indirectly support activation of molecular oxygen and subsequent substrate dehalogenation by providing the necessary reducing equivalents to enzymes such as cytochrome P450 [35]. Flavoproteins additionally mediate electron transport from reduced nicotinamides to flavin-dependent monooxygenases that promote hydroxylation and dehalogenation of halophenols (▶Fig. 16.5) [36]. Alternatively, two component systems transfer reduced flavin from one protein acting as a reductase to another protein acting as a monooxygenase [37]. Together, these systems can also catalyze oxidative dehalogenation of chloro- and fluorophenols (▶Fig. 16.5) [38,39]. Other flavoproteins express both the flavin reductase and monooxygenase activities within a single polypeptide [40]. The final steps of molecular oxygen activation and halide elimination are likely similar within the range of these enzyme variants. All are expected to form a common 4a-hydroperoxy derivative of flavin for electrophilic aromatic hydroxylation (▶Fig. 16.6).

16.2.2 Hydrolytic dehalogenation catalyzed by flavoproteins

The most recent addition to the gamut of flavin-dependent dehalogenations is the fascinating enzyme 2-haloacrylate hydratase. As the name implies, this catalyzes

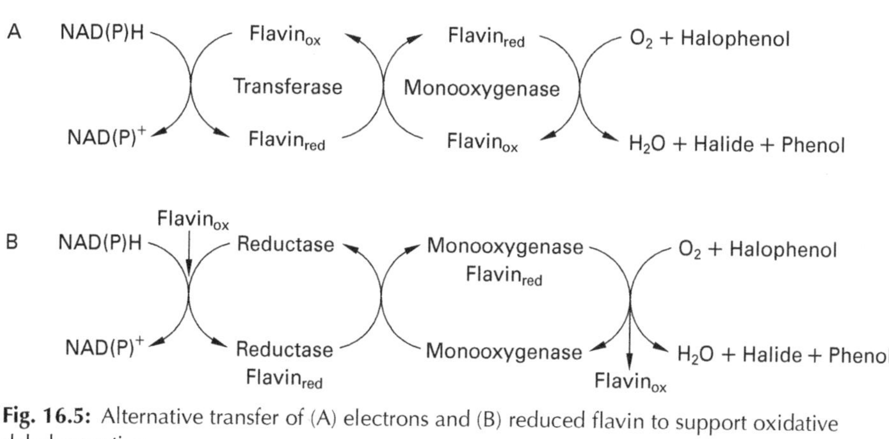

Fig. 16.5: Alternative transfer of (A) electrons and (B) reduced flavin to support oxidative dehalogenation.

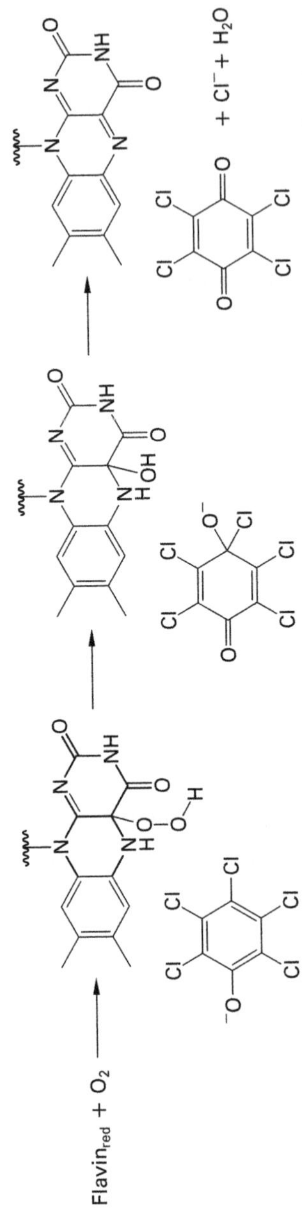

Fig. 16.6: Hydroxylation and dehalogenation of pentachlorophenol by the flavoprotein pentachlorophenol hydroxylase.

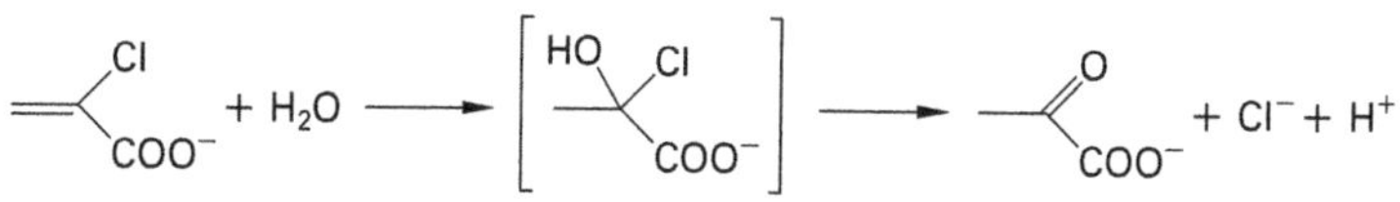

Fig. 16.7: Hydrolytic dehalogenation catalyzed by the $FADH_2$-dependent 2-haloacrylate hydratase.

hydrolytic dehalogenation of 2-chloro and 2-bromoacrylate and requires $FADH_2$ for activity even though no net oxidation-reduction chemistry is apparent (▶Fig. 16.7) [41]. Curiously, analogues lacking a halogen are not substrates. Neither acrylate nor methacrylate are hydrated by this enzyme [41]. The role of the reduced flavin has not yet been determined, and two possible roles are currently under investigation [41]. Flavin may act as an acid-base catalyst similar to that proposed for an isopentenyl-diphosphate isomerase mentioned above [33,34]. Alternatively, flavin may transiently donate an electron to substrate as proposed in the flavin-dependent enzyme chorismate synthase [42,43]. Of course, the hydratase may yet reveal an entirely new aspect of flavin chemistry as well. Further studies on this enzyme are eagerly awaited.

16.2.3 Reductive dehalogenation catalyzed by flavoproteins

Similar to flavin's contributions to oxidative dehalogenation, flavin also has the potential to drive reductive dehalogenation, either indirectly or directly. Indirect contributions derive from the use of flavoproteins in electron transport for energy production during halorespiration. This process utilizes organic halides as the ultimate electron acceptors and results in their reductive dehalogenation [44]. Flavin-dependent reductases also donate electrons to cytochromes P450. As described above, this promotes oxidative reactions under aerobic conditions, but reductive dehalogenation may occur under anaerobic conditions. For example, cytochrome $P450_{CAM}$ reduces chlorinated methane and ethane derivatives and releases chloride in the absence of O_2 [45]. Thus, flavoproteins contribute to P450-dependent reductive as well as oxidative transformations.

Few flavoproteins have yet been ascribed to a direct role in reductive dehalogenation. In one example, the role of flavin remains uncertain although addition of $FADH_2$ to membrane fractions of *Escherichia coli* was observed to stimulated reductive dechlorination of dichlorodiphenyltrichloroethane (DDT) to form dichlorodiphenyldichloroethane (TDE) [46]. Two other reports offer tantalizing hints of a flavin-dependent reductive debromination of α-bromoisovalerylurea by intestinal bacteria [47] and a reductive dechlorination of 3-chloro-1-hydroxypropan-2-one by a strain of *Alcaligenes* (▶Fig. 16.8) [48]. In contrast, the very well-characterized glucose oxidase has been shown to function as a reductive dehalogenase in the presence of an appropriately configured substrate. Since the goal of this effort was to develop a mechanism-based inactivator, successful reduction and halide elimination generated a flavin adduct lacking further catalytic activity (▶Fig. 16.9) [49].

The flavin-dependent reductive dehalogenase investigated most extensively is iodotyrosine deiodinase (▶Fig. 16.10) [50]. This enzyme is responsible in part for the

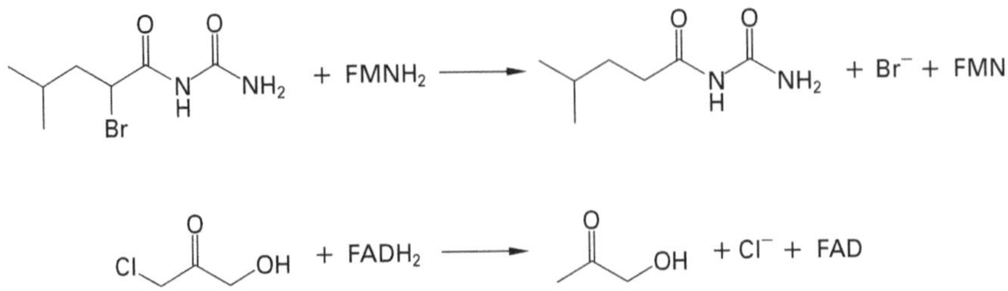

Fig. 16.8: Current examples of reactions catalyzed by reductive dehalogenases that are identified as flavoproteins in bacteria.

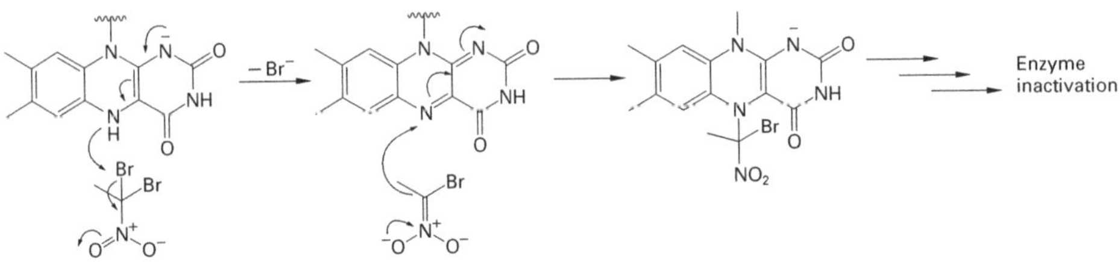

Fig. 16.9: Reductive debromination of dibromonitromethane by glucose oxidase.

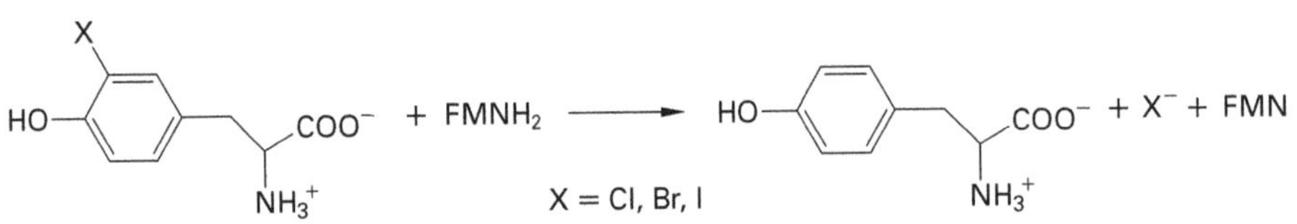

Fig. 16.10: Iodotyrosine deiodinase promotes dehalogenation of halotyrosines.

homeostasis of iodide that is required in higher organisms for production of the iodine-containing hormone thyroxine. Generation of this hormone consumes many equivalents of iodide and releases substantial quantities of mono- and diiodotyrosine [51]. Iodide is recycled from these derivatives by the deiodinase, and loss of its activity in humans leads to hypothyroidism [52–54]. *In vivo,* NADPH provides the reducing equivalents for the deiodinase, but not directly [26,55,56]. Instead, a membrane-bound reductase likely mediates NADPH oxidation and FMN reduction. The ability of the deiodinase to utilize NADPH is lost once the deiodinase is separated from the membrane. *In vitro* studies consequently rely most typically on dithionite to provide the necessary reducing equivalents [50].

Significant quantities of the deiodinase have only recently become available. Originally, its preparation required extraction from thyroid microsomes and multiple types of chromatography [27]. Heterologous expression of a truncated derivative lacking the N-terminal membrane anchor now provides a soluble form of the enzyme in sufficient quantities to begin detailed mechanistic investigations [57]. This advance has already facilitated discovery of the deiodinase's ability to catalyze reductive elimination of bromide and chloride from their corresponding halotyrosines [58]. Only fluorotyrosine remains inert to the deiodinase. Crystallization and X-ray diffraction of the soluble enzyme

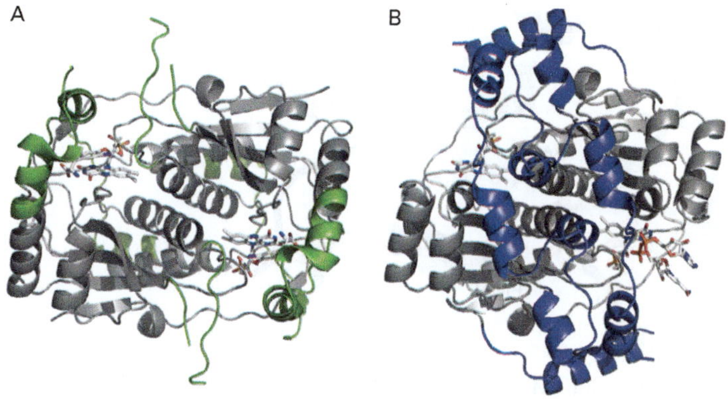

Fig. 16.11: Crystal structures of (A) iodotyrosine deiodinase with monoiodotyrosine in its active site (PDB: 3GFD) and (B) flavin reductase with NAD^+ in its active site (PDB: 2BKJ). Their unique domains are highlighted in green and blue, and their common structural framework is marked in grey.

have confirmed its earlier assignment as a member of the NADH oxidase/flavin reductase superfamily (►Fig. 16.11) [59]. Structural analysis of co-crystals formed by the deiodinase and monoiodotyrosine indicates that an active site flap closes around this substrate and sequesters it from solvent. Such a flap is common to only one other member of the same structural superfamily (BluB introduced below). Within the deiodinase, monoiodotyrosine was found to stack directly above the isoalloxazine portion of FMN with the carbon-iodine bond proximal to the central C4a carbon (►Fig. 16.12). The most unusual aspect of the co-crystal is a direct chelation of the pyrimidine ring of flavin by the zwitterionic region of iodotyrosine (►Fig. 16.13). This suggests that the substrate more than the surrounding protein may control the redox chemistry and reactivity of the flavin.

Many of the active site characteristics of the deiodinase resemble those of BluB, a flavin destructase enzyme that is unique in its ability to degrade $FMNH_2$ to 5,6-dimethylbenzimidazole in the presence of O_2 [60,61]. Unlike the other members of the structural superfamily, BluB is expected to stabilize the one electron chemistry of

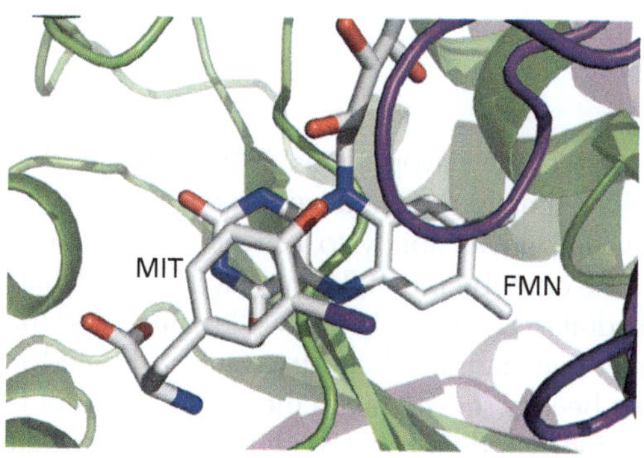

Fig. 16.12: Monoiodotyrosine (MIT) stacks directly above the flavin (FMN) of iodotyrosine deiodinase. Both polypeptides of the homodimer (marked in green and magenta) contribute to the active sites.

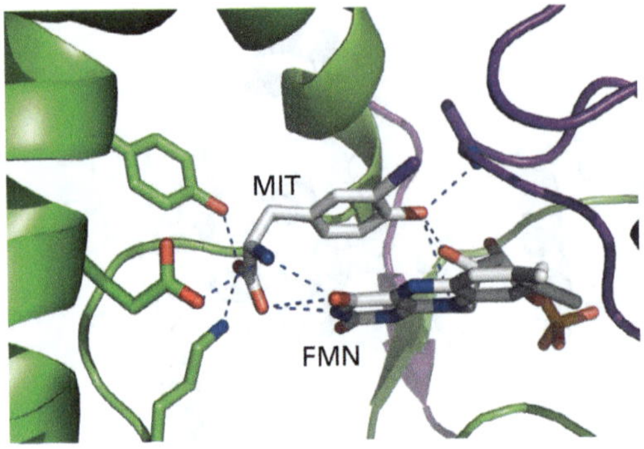

Fig. 16.13: Monoiodotyrosine (MIT) coordinates to the pyrimidine ring of FMN and side chains. of amino acids from the active site flap. The polypeptides of the homodimer are marked in green and magenta.

flavin to induce reaction with molecular oxygen. Most enzymes within this superfamily derive from bacteria and are thought instead to suppress one electron chemistry while they act as either flavin reductases or oxygen insensitive nitroreductases [62,63]. Similarities between BluB and iodotyrosine deiodinase provide one of the first hints that dehalogenation may also be catalyzed through a series of one electron steps.

Initial precedence considered for reductive dehalogenation all suggested catalysis would occur through two electron chemistry. Model studies first demonstrated the ability of cysteine to reductively deiodinate diiodotyrosine non-enzymatically [64], and similarly cysteine residues were later discovered to be crucial in the reductive dehalogenation catalyzed by tetrachlorohydroquinone dehalogenase (▶Fig. 16.4) [23]. Analogous chemistry has also been proposed for the selenocysteine-dependent deiodination of thyroxine [65–67]. Surprisingly, cysteine residues are not required for iodotyrosine deiodinase activity [58], and no other nucleophiles are present in the active site to act in kind [57,59,68]. Thus, one-electron chemistry now appears most likely for the catalysis of iodotyrosine deiodinase, but definitive experiments will be necessary to confirm this.

16.3 Conclusions

The breadth of chemistry supported by flavoproteins is evident in the multiple strategies used to catalyze dehalogenation. Oxidative processes recruit both the electron transfer properties of flavin and its ability to activate molecular oxygen for electrophilic monooxygenation. The contribution of flavin to hydrolytic dehalogenation as catalyzed by 2-haloacrylate hydratase is currently less well defined and may involve either acid/base or reversible electron transfer chemistry. Only a limited number of reductive dehalogenases have yet been associated with the direct action of flavins and flavoproteins, and only iodotyrosine deiodinase has received significant attention to date. Still, much remains to be learned about this process. Discovery of even more flavin-dependent dehalogenases seems inevitable as the field continues to explore the full realm of flavoproteins and their role in nature.

16.4 References

[1] Gribble GW. Naturally Occurring Organohalogen Compounds -- A Comprehensive Update, in Progress in the Chemistry of Organic Natural Products. New York, NY, USA: Springer, 1996
[2] Gribble GW. Naturally Occurring Organohalogen Compounds. Acc Chem Res 1998;31: 141–52.
[3] Valverde-RC, Orozco A, Becerra A, Jeziorski MC, Villalobos P, Solis-S. JC. Halometabolites and cellular dehalogenase systems: an evolutionary perspective. Int Rev of Cytology 2004;234:143–99.
[4] Burrows GN, Oppenheimer JH, Volpé R. Thyroid Function and Disease. Philadelphia, PA, USA: Saunders, 1989.
[5] Vaillancourt FH, Yeh E, Vosburg DA, Ganeau-Tsodikova S, Walsh CT. Nature's inventory of halogenation catalysts: oxidative strategies predominate. Chem Rev 2006;106:3364–78.
[6] Fetzner S, Lingens F. Bacterial dehalogenases: biochemistry, genetics, and biotechnological applications. Microbiol Rev 1994;58:641–85.
[7] van Pée K-H, Unversucht S. Biological dehalogenation and halogenation reactions. Chemosphere 2003;52:299–312.
[8] Kurihara T, Esaki N, Soda K. Bacterial 2-haloacid dehalogenases: structure and reaction mechanisms. J Mol Cat B 2000;10:57–65.
[9] Copley SD. Evolution of efficient pathways for degradation of anthropogenic chemicals. Nat Chem Biol 2009;5:559–66.
[10] Häggblom MM, Bossert ID, eds. Dehalogenation: Microbial Processes and Environmental Applications. Kluwer Academic Publishers: Boston, 2003.
[11] Lal R, Pandey G, Sharma P, Kumari K, Malhotra S, Pandey R, Raina V, Kohler H-PE, Holliger C, Jackson C, Oakeshott JG. Biochemistry of microbial degradation of hexachlorocyclohexane and prospects for bioremediation. Microbiol Mol Biol Rev 2010;74:58–80.
[12] Isin EM, Guengerich P. Complex reactions catalyzed by cytochrome P450 enzymes. Bioch Biophys Acta 2007;1770:314–29.
[13] Sakaki T, Munetsuna E. Enzyme systems for biodegradation of polychlorinated dibenzo-*p*-dioxins. Appl Microbiol Biotechnol 2010;88:23–30.
[14] Feducia J, Dumarieh R, Gilvery LBG, Smirnova T, Franzen S, Ghiladi RA. Characterization of dehaloperoxidase compound ES and its reactivity with trihalophenols. Biochemistry 2009;48:995–1005.
[15] Janssen DB. Evolving haloalkane dehalogenase. Curr Opin in Chem Biol 2004;8:150–59.
[16] Gehret JJ, Gu L, Geders TW, Brown WC, Gerwick L, Gerwick WH, Sherman DH, Smith JL. Structure and activity of DmmA, a marine haloalkane dehalogenase. Protein Sci 2012;21: 239–48.
[17] Chan PWY, Yakunin AF, Edwards EA, Pai EF. Mapping the reaction coordinates of enzymatic defluorination. J Am Chem Soc 2011;133:7461–68.
[18] Nakayama T, Kamachi T, Jitsumori K, Omi R, Hirotsu K, Esaki N, Kurihara T, Yoshizawa K. Substate specificity of fluoroacetate dehalogenase: an insight from cyrstallographic analysis, fluorescence spectroscopy and theoretical computations. Chem Eur J 2012;18:8392–8402.
[19] Wu J, Xu, D, Lu X, Wang,C, Guo H, Dunaway-Mariano D. Contributions of long-range electrostatic interactions to 4-chlorobenzoyl-CoA dehalogenase catalysis: a combined theoretical and experimental study. Biochemistry 2006;45:102–12.
[20] Smidt H, de Vos WM. Anaerobic microbial dehalogenation. Annu Rev Microbiol 2004;58: 43–73.
[21] DeWeerd KA, Suflita JM. Anaerobic aryl reductive dehalogenation of halobenzoates by cell extracts of "*Desulfomonile tiedjei*". Appl Environ Microbiol 1990;56:2999–3005.

[22] Muller JA, Rosner BM, von Abendroth G, Meshulam-Simon G, McCarthy PL, Spormann AM. Molecular identification of the catabolic vinyl chloride reductase from *Dehalococcoides* sp. strain VS and its environ. Appl Environ Microbiol 2004;70:4880–88.

[23] Warner JR, Copley SD. Pre-steady-state kinetic studies of the reductive dehalogenation catalyzed by tetrachlorohydroquinone dehalogenase. Biochemistry 2007;46:13211–22.

[24] Velazquez F, Peak-Chew SY, Fernández IS, Neumann CS, Kay RR. Identification of a eukaryotic reductive dechlorinase and characterization of tis mechanism of action on its natural substrate. Chem Biol 2001;18:1252–60.

[25] Callebaut I, Curcio-Morelli C, Mornon J-P, Gereben B, Buettner C, Huang S, Castro B, Fonseca TL, Harney JW, Larsen PR, Bianco AC. The iodothyronine selenodeiodinases are thioredoxin-fold family proteins containing a glycoside hydrolase clan GH-A-like structure. J Biol Chem 2003;278:36887–96.

[26] Watson JA, Jr., McTamney PM, Adler JM, Rokita SE. The flavoprotein iodotyrosine deiodinase functions without cysteine residues. ChemBioChem 2008;9:504–06.

[27] Rosenberg IN, Goswami A. Purification and characterization of a flavoprotein from bovine thyroid with iodotyrosine deiodinase activity. J Biol Chem 1979;254:12318–25.

[28] De Colibus L, Mattevi A. New frontiers in structural flavoenzymology. Curr Opin Struct Biol 2006;16:722–28.

[29] Mansoorabadi SO, Thibodeaux CJ, Liu H-W. The diverse roles of flavin coenyzmes--nature's most versatile thespians. J Org Chem 2007;72:6329–42.

[30] Macheroux P, Kappes B, Ealick SE. Flavogenomics -- a genomic and structural view of flavin-dependent proteins. FEBS J 2001;278:2625–34.

[31] van Pée K-H, Patallo EP. Flavin-dependent halogenases involved in secondary metabolism in bacteria. Appl Microbiol Biotechnol 2006;70:631–41.

[32] Yeh E, Blasiak LC, Koglin A, Drennan CL, Walsh CT. Chlorination by a long-lived intermediate in the mechanism of flavin-dependent halogenases. Biochemistry 2007;46:1284–92.

[33] Kittleman W, Thibodeaux CJ, Liu Y-n, Zhang, H, Liu H-W. Characterization and mechanistic studies of type II isopentenyl diphosphate:dimethylallyl diphosphate isomerase from *Staphylococcus aureus.* Biochemistry 2007;46:8401–2558.

[34] Unno H, Yamashita S, Ikeda Y, Sekiguchi S-Y, Yoshida N, Yoshimura T, Kusunoki M, Nakayama T, Nishino Y, Hemmi H. New role of flavin as a general acid-base catalyst with no redox function in type 2 isopentenyl-diphosphate isomerase. J Biol Chem 2009;284:9160–67.

[35] Rietjens IMCM, den Beste C, Hanzlik RP, van Bladeren PJ. Cytochrome P450-catalyzed oxidation of halobenzene derivatives. Chem Res Toxicol 1997:629–35.

[36] Xun L, Orser, CS. Purification and properties of pentachlorophenol hydroxylase, a flavoprotein from *Flavobacterium sp.* strain ATCC 39723. J Bacteriol 1991;173:4447–53.

[37] Ellis HR. The FMN-dependent two-component monooxygenase systems. Arch Biochem Biophys 2010;497:1–12.

[38] Xun L, Webster CM. A monooxygenase catalyzes sequential dechlorinations of 2,4,6-trichlorophenol by oxidative and hydrolytic reactions. J Biol Chem 2004;279:6696–6700.

[39] Ferreira MIM, Iida T, Hasan SA, Nakamura K, Fraaije MW, Janssen DB, Kudo T. Analysis of two gene clusters involved in the degradation of 4-fluorophenol by *Arthrobacter* sp. strain IF1. Appl Environ Microbiol 2009;75:7767–73.

[40] Hlouchova K, Rudolph J, Pietari JMH, Behlen LS, Copley SD. Pentachlorophenol hydroxylase, a poorly functioning enzyme required for degradation of pentachlorophenol by *Sphingobium chlorophenolicum.* Biochemistry 2012;51:3848–60.

[41] Mowafy AM, Kurihara T, Kurata A, Uemura T, Esaki N. 2-Haloacrylate hydratase, a new class of flavoenzyme that catalyzes the addition of water to the substrate for dehalogenation. Appl Environ Microbiol 2010;76:6032–37.

[42] Osborne A, Thorneley RNF, Abell C, Bornemann S. Studies with substrate and cofactor analogues provide evidence for a radical mechanism in the chorismate synthase reaction. J Biol Chem 2000;275:35825–30.

[43] Maclean J, Ali S. The structure of chorismate synthase reveals a novel binding site fundamental to a unique chemical reaction. Structure 2003;11:1499–1511.
[44] Pral L, Malliard J, Grimaud R, Holliger C. Physiological adaptation of *Desulfitobacterium hafniense* strain TCE1 to tetrachloroethene respiration. Appl Environ Microbiol 2011;77: 3853–59.
[45] Li S, Wackett LP. Reductive dehalogenation by cytochrome $P450_{CAM}$: substrate binding and catalysis. Biochemistry 1993;32:9355–61.
[46] French AL, Hoopingarner RA. Dechlorination of DDT by membranes isolated from *Escherichia coli.* J Econ Entomol 1970;63:756–59.
[47] Kitamura S, Kuwasako M, Ohta S, Tatsumi K. Reductive debromination of (α-bromoiso-valeryl) urea by intestinal bacteria. J Pharm Pharmacol 1999;51:79–84.
[48] Suzuki T, Kasia N. Generation of optically active glycerol derivatives by microbial resolution or development of useful synthetic units for pharmaceuticals. Trends Glycosci Glycotechnol 2003;15:329–49.
[49] Porter DJT, Voet JG, Bright HJ. Active site generation of a protonically unstable suicide substrate from a stable precursor: glucose oxidase and dibromonitromethane. Biochemistry 2000;39:11808–17.
[50] Rokita SE, Adler JM, McTamney PM, Watson JA, Jr. Efficient use and recycling of the micronutrient iodide in mammals. Biochimie 2010;92:1227–35.
[51] Nunez J, Pommier J. Formation of thyroid hormones. Vitam Horm 1982;39:175–229.
[52] Moreno JC, Klootwijk W, van Toor H, Pinto G, D'Alessandro M, Lèger A, Goudie D, Polak M, Grüters A, Visser TJ. Mutations in the iodotryosine deiodinase gene and hypothyroidism. N Engl J Med 2008;358:1811–18.
[53] Afink G, Kulik W, Overmars H, de Randamie J, Veenboer T, van Cruchten A, Craen M, Ris-Stalpers C. Molecular characterization of iodotyrosine dehalogenase deficiency in patients with hypothyroidism. J Clin Endocinol Metab 2008;93:4894–4901.
[54] Burniat A, Pirson I, Vilain C, Kulik W, Afink G, Moreno-Reyes R, Corvilain B, Abramowicz M. Iodotyrosine deiodinase defect identified via genome-wide approach. J Clin Endocrinol Metab 2012;97:E1276–E1283.
[55] Goswami A, Rosenberg IN. Studies on a soluble thyroid iodotyrosine deiodinase: activation by NADPH and electron carriers. Endocrinol 1977;101:331–41.
[56] Gnidehou S, Caillou B, Talbot M, Ohayon R, Kaniewski J, Noël-Hudson M-S, Morand S, Agnangj, D, Sezan A, Courtin F, Virion A, Dupuy C. Iodotyrosine dehalogenase 1 (DEHAL1) is a transmembrane protein involved in the recycling of iodide close to the thyroglobulin iodination site. FASEB J 2004;18:1574–76.
[57] Buss JM, McTamney PM, Rokita SE. Expression of a soluble form of iodotyrosine deiodinase for active site characterization by engineering the native membrane protein from *Mus musculus.* Protein. Sci 2011;21:351–61.
[58] McTamney PM, Rokita SE. A mammalian reductive deiodinase has broad power to dehalogenate chlorinated and brominated substrates. J Am Chem Soc 2009;131:14212–13.
[59] Thomas S, McTamney PM, Adler JM, LaRonde-LeBlanc N, Rokita SE. Crystal structure of iodotyrosine deiodinase, a novel flavoprotein responsible for iodide salvage in thyroid glands. J Biol Chem 2009;284:19659–67.
[60] Taga ME, Larsen NA, Howard-Jones AR, Walsh CT, Walker GC. BluB cannibalizes flavin to form the lower ligand of vitamin B_{12}. Nature 2007;446:449–53.
[61] Yu T-Y, Mok KC, Kennedy KJ, Valton J, Anderson KS, Walker GC, Taga ME. Active site residues critical for flavin binding and 5,6-dimethylbenzimidazole biosynthesis in the flavin destructase enzyme BluB. Protein Sci 2012;21:839–49.
[62] Tu SC. Reduced flavin: donor and acceptor enzymes and mechanisms of channeling. Antioxid. Redox Signaling 2001;3:881–97.
[63] Koder RL, Haynes CA, Rodgers ME, Rodgers DW, Miller A-F. Flavin thermodynamics explain the oxygen insensitivity of enteric nitroreductase. Biochemistry 2002;41:14197–205.

[64] Hartmann K, Hartmann N. Untersuchungen zum reaktionsablauf der dejodierung von 3,-5-dijod-tyrosin durch cystein. Z Chem 1971;11:344–45.
[65] Berry MJ, Kieffer JD, Harney JW, Larsen PR. Selenocysteine confers the biochemical properties characteristic of the type I iodothyronine deiodinase. J Biol Chem 1991;266:14155–14158.
[66] Bayse CA, Rafferty ER. Is halogen bonding the basis for iodothyronine deiodinase activity? Inorg Chem 2010;49:5365–67.
[67] Manna D, Mugesh G. A chemical model for the inner-ring deiodination of thyroxine by iodothyronine deiodinase. Angew Chem Int Ed 2010;49:9246–49.
[68] Talekar RS, Chen GS, Lai S-Y, Chern J-W. Nonreductive deiodination of *ortho*-iodo-hydroxylated arenes using tertiary amines. J Org Chem 2005;70:8590–93.

Index

Printed and bound by CPI Group (UK) Ltd, Croydon, CR0 4YY
13/09/2026
14960851-0004